STUDIES IN THE HISTORY OF CIVIL ENGINEERING

General Editor: Joyce Brown

with the assistance of M.M. Chrimes, A.W. Skempton, N.A.F. Smith and R.J.M. Sutherland

---

Twelve volumes are to appear from the Autumn of 1997:

1. **The Engineering of Medieval Cathedrals**
   Edited by LYNN T. COURTENAY, University of Wisconsin

2. **Masonry Bridges and Viaducts**
   Edited by TED RUDDOCK, University of Edinburgh

3. **Land Drainage and Irrigation**
   Edited by SALVATORE CIRIACONO, University of Padova

4. **Dams**
   Edited by DONALD C. JACKSON, Lafayette College

5. **Water Supply and Public Health Engineering**
   Edited by DENIS SMITH, formerly University of East London

6. **Port and Harbour Engineering**
   Edited by ADRIAN JARVIS, Merseyside Maritime Museum and Liverpool University

7. **The Civil Engineering of Canals and Railways before 1850**
   Edited by MIKE CHRIMES, Institution of Civil Engineers

8. **The Development of Timber as a Structural Material**
   Edited by DAVID T. YEOMANS, Liverpool University

9. **Structural Iron, 1750–1850**
   Edited by R.J.M. SUTHERLAND, Consultant, Harris and Sutherland, London

10. **Structural Iron and Steel, 1850–1900**
    Edited by ROBERT THORNE, Editor, *Construction History Journal*

11. **Early Reinforced Concrete**
    Edited by FRANK NEWBY, formerly F.J. Samuely and Partners, London

12. **Structural and Civil Engineering Design**
    Edited by WILLIAM ADDIS, University of Reading

Studies in the History of Civil Engineering

Volume 4

Dams

Studies in the History of Civil Engineering
General Editor: Joyce Brown

Volume 4

Dams

edited by
Donald C. Jackson

Routledge
Taylor & Francis Group
LONDON AND NEW YORK

First published 1997 by Ashgate Publishing

Published 2016 by Routledge
2 Park Square, Milton Park, Abingdon, Oxon OX14 4RN
52 Vanderbilt Avenue, New York, NY 10017

Routledge is an imprint of the Taylor & Francis Group, an informa business

British Library CIP data
Dams. – (Studies in the History of Civil
Engineering; 4).
1. Dams–History.
I. Jackson, Donald C.
627.8'09

US Library of Congress CIP data
Dams/edited by Donald C. Jackson.
p. cm. – (Studies in the History of Civil Engineering; v. 4).
Includes bibliographical references.
ISBN 0–86078–753–2 (alk. paper)
1. Dams–Design and construction.
I. Jackson, Donald C. (Donald Conrad), 1953– . II. Series.
TC540. D37 1998 97–34227
627'. 8–dc21 CIP

ISBN 13: 978-0-86078-753-2 (vol 4, hbk)
ISBN 13: 978–0–754–60654–3 (set)

STUDIES IN THE HISTORY OF CIVIL ENGINEERING – VOL. 4

This volume is dedicated to

N.A.F. Smith

for his contributions to
the study of dam history

Contents

Acknowledgements

The chapters in this volume are taken from the sources listed below, for which the editors and publishers wish to thank their authors, original publishers or other copyright holders for permission to use their material as follows:

Chapter 1: G. Garbrecht, 'Sadd-el-Kafara, the world's oldest large dam', *International Water Power and Dam Construction*, 27 (July 1985), pp. 71–76. Copyright © 1985 by Wilmington Publishing, Wilmington, Kent.

Chapter 2: Norman A.F. Smith, 'The Roman dams of Subiaco', *Technology and Cutlure*, 11, no. 1 (1970), pp. 58–68. Copyright © 1970 by The University of Chicago Press.

Chapter 3: J.A. Garcia-Diego, 'The chapter on weirs in the Codex of Juanelo Turriano: a question of authorship', *Technology and Culture*, 17, no. 2 (1976), pp. 217–234. Copyright © 1976 by The University of Chicago Press.

Chapter 4: Nicholas Adams, 'Architecture for fish: the Sienese dam on the Bruna River—Structures and designs,ca. 1468–1530',*Technology and Culture*, 25, no. 4 (1984), pp. 768–797. Copyright © 1984 by The University of Chicago Press.

Chapter 5: N.J. Schnitter, 'The evolution of the arch dam', *International Water Power and Dam Construction*, 28 (October/November 1976), pp. 34–40, 19–21. Copyright © 1976 by Wilmington Publishing, Wilmington, Kent.

Chapter 6: N.J. Schnitter, 'The evolution of buttress dams', *International Water Power and Dam Construction*, 36 (June/July 1984), pp. 38–42, 20–22. Copyright © 1984 by Wilmington Publishing, Wilmington, Kent.

Chapter 7: Geoffrey Binnie, 'The evolution the British dams', *Transactions of the Newcomen Society*, 47 (1974–76), pp. 207–224. Copyright © 1974–76 by The Newcomen Society, London.

Chapter 8: Robert F. Legget, 'The Jones Falls Dam on the Rideau Canal, Ontario,Canada', *Transactions of the Newcomen Society*, 31 (1957–59), pp. 205–218. Copyright © 1957–59 by The Newcomen Society, London.

Chapter 9: Peter M. Molloy, 'Nineteenth-century hydropower: design and construction of the Lawrence Dam, 1845–1848', *Winterthur Portfolio*, 15, no. 4 (1980), pp. 315–343. Published by The University of Chicago Press on behalf of The Henry Francis du Pont Winterthur Museum, Delaware. Copyright © 1980 by The Henry Francis du Pont Winterthur Museum.

Chapter 10: Norman A.F. Smith, 'The failure of the Bouzey Dam in 1895', *Construction History*, 10 (1994), pp. 47–65. Copyright © 1994 by The Chartered Institute of Building, Ascot, Berks.

Chapter 11: I. Davidson, 'George Deacon (1843–1909) and the Vyrnwy works', *Transactions of the Newcomen Society*, 59 (1987–88), pp. 81–95. Copyright © 1987–88 by The Newcomen Society, London.

Chapter 12: Donald C. Jackson, 'Engineering in the progressive era: a new look at Frederick Haynes Newell and the U.S. Reclamation Service', *Technology and Culture*, 34, no. 3 (1993), pp. 539–574. Copyright © 1993 by The University of Chicago Press.

Chapter 13: Terry S. Reynolds, 'A narrow window of opportunity: the rise and fall of the fixed steel dam', *Journal of the Society for Industrial Archaeology*, 5, no. 1 (1989), pp. 1–20. Copyright © 1989 by the Society for Industrial Archaeology, Houghton, Michigan.

Chapter 14: Donald C. Jackson, 'Considering the multiple arch dam: theory, practice and the ethics of safety in a case of innovative hydraulic engineering', *Natural Resources Journal*, 32 (1992), pp. 77–100. Copyright © 1992 *Natural Resources Journal*, c/o The University of New Mexico, Albuquerque.

Chapter 15: A.W. Skempton, 'Historical development of British embankment dams to 1960', *Clay Barriers for Embankment Dams*. Proceedings of the conference organized by the Institution of Civil Engineers and held in London on 18 October 1989 (London, 1990), pp. 15–52. Copyright © 1990 Thomas Telford Publishing, London.

Chapter 16: Richard Guy Wilson, 'Machine-age iconography in the American west: the design of Hoover Dam', *Pacific Historical Review* 54, no. 4 (1985), pp. 463–493. Copyright © 1985 by The University of California Press, Berkeley, California.

Chapter 17: Donald Worster, 'Hoover Dam: a study in domination', *Under Western Skies: Nature and History in the American West* (New York/Oxford,

1992), pp. 64–78, 263–265. Copyright © 1992 by Oxford University Press, New York.

***

Chapters 7, 8 and 11 are reproduced from the *Transactions of the Newcomen Society* by permission of the Society's Council. The Newcomen Society promotes, encourages and co-ordinates the study of the history of all types of engineering and technology from the earliest times to the present day. Details from the Executive Secretary, The Newcomen Society, The Science Museum, London SW7 2DD, Tel/Fax 0171-589-1793.

Every effort has been made to trace all the copyright holders, but if any have been inadvertently overlooked the publishers will be pleased to make the necessary arrangement at the first opportunity.

General Editor's Preface

Joyce Brown

Civil engineering has a long history. Although practitioners did not begin to describe themselves as 'civil engineers' until the eighteenth century, the origins of their work lie in the construction skills of the ancient world and in the works of military engineers. Shortly after British civil engineers had formed themselves into a professional body, the Institution of Civil Engineers in London, their role was described in 1828 by Thomas Tredgold in these words:

> Civil engineering is the art of directing the great sources of power in Nature for the use and convenience of man.[1]

He thus neatly encapsulated what the civil engineer's job is about – the provision of the means to give society what it wants for its survival.

As time has gone by, the civil engineer's skills have had to diffuse over an ever-widening field and at the same time become more specialised. In all periods he has had to be the ingenious 'fixer', but more than a mere artisan – a point observed in the Institution's motto, *scientia et ingenio.*

The history of civil engineering is a fascinating one. Certainly 'the great sources of power in Nature' are not always directed without a certain amount of opposition by them, and the provision of the basic requirements – habitation, water supply, main drainage, harbours bridges, places of worship, transport systems by land and water – has called for inventiveness and sometimes for daring and courage.

Yet interest in the history of civil engineering is not of long standing, and it seems fair to state that most of the important work has been done in the last forty years. With a few notable exceptions, most of the major books on the subject have appeared in that era; many specialist societies with their own transactions have been founded; and many of the journals in which articles might be published have come into existence. While archaeologists and architectural historians have long made their contribution to this subject, their ranks have been swelled by a large number of serious scholars – including engineers themselves – now interested in recording and evaluating the engineering achievements of earlier times. In particular, there is a lively interest in the study of our industrial past, at least in part because some of the evidence of it are already disappearing from the landscape.

Our intention in creating the series of which this volume forms part is to provide, through the reproduction of important contributions to this subject, an

---

[1] *Minutes of Council, Institution of Civil Engineers*, 4 January 1828.

invaluable reference collection for its study. The series encompasses many different branches of engineering from early times to the beginning of this century; its perspective is global; and the chosen articles have an international authorship, to the extent that this can be achieved for an essentially English-language series. The introduction to each volume has allowed the volume editor to set the pages selected in the context of the whole history of the subject and its historiography, while the provision of references and a bibliography will enable the reader to go further into the study of the topic.

This series will have succeeded if it gives civil engineers some unexpected insights into their own craft and other readers a new way of looking at engineering structures.

***

The volume editors were chosen for their particular knowledge and expertise in the field. They have worked with enthusiasm and good will to put together collections of articles intended to be both informative and stimulating, and I am grateful to them for their efforts. Where possible, they have incorporated the work of the best scholars, while at the same time giving preference to articles often obscurely published and not readily available. The learned journals are represented, as also are conference proceedings and essay chapters contributed to books.

In all of this, I have been much assisted by my Advisory Panel. Emeritus Professor A.W. Skempton (Imperial College) has made himself freely available to me for consultation on many aspects of this venture. Mike Chrimes (Librarian of the Institution of Civil Engineers), Dr. Norman Smith (Emeritus Reader, Imperial College) and James Sutherland (formerly a Senior Partner in Harris and Sutherland, London, consulting engineers) have all contributed valuable advice based on their extensive knowledge of particular fields. The idea for the series was developed in discussion with Dr John Smedley of Ashgate, and Publisher of the Variorum Collected Studies Series, who has given me much personal support and the benefit of his experience.

I am grateful also for the help given by the staff of several libraries, notably Kay Crooks and Susannah Parry in the library of the Department of Civil Engineering, Imperial College, and Mike Chrimes and staff of the library of the Institution of Civil Engineers, London. Many friends have given me encouragement and useful advice; in particular, I want to thank Paula Kahn and Mark Baldwin.

Formerly Department of Civil Engineering,
Imperial College, London

Introduction

Donald C. Jackson

The technology of water control appears so simple, so basic, that it can easily fade from a society's collective consciousness. We may marvel at advanced developments in communication/information systems or the latest advances in medical technology, all the while taking for granted how we conserve and allocate water, one of Earth's most ubiquitous – and important – natural resources. If directly confronted, we will probably acknowledge the importance of dams in underpinning the water supply systems of the modern world. But how often – or seldom – do most people think about such things?

For thousands of years, human societies have used technology to alter riverflow and store water. The oldest known dams are a collection of small earthen structures built in the fourth millennium BCE at Jawa in present-day Jordan; they were designed to capture rainfall for agricultural and domestic use before it could evaporate or sink into the desert sands.[1] By 2800 BCE Egyptian builders had expanded upon the basic embankment technology used at Jawa and erected the 20-metre high Sadd-el-Kafara Dam (chapter 1). Since then, cultures throughout the world have practiced the art of dam-building and the technology of water control has evolved in myriad ways. That evolution is the subject of this volume.

In bringing together the articles reprinted in this volume, my goal has been to illustrate the history of dam technology from ancient to modern times. Technical issues that underlie differences in various dam designs represent an important – at times even critical – component of these articles. However, articles have also been selected that help illuminate *why* various human societies have built dams and how 'social' (or seemingly 'non-technical') factors have influenced the process of dam design. Engineers and professionals responsible for modern water control systems can certainly learn much about the historical dimensions of specific technical issues. Nonetheless, general readers interested in water resources and environmental history need not fear that any special technical aptitude is necessary to appreciate this book.

Dam-building is not an activity confined to any specific region or culture. Because this compendium presents articles written in English and because many important developments associated with large-scale reservoirs have occurred within the Mediterranean/European region and in North America, the focus is on activities within these areas. Rather than regard dams in Asia, India, sub-Sahara

---

[1] S.W. Helm, 'Jawa, a fortified town of the fourth millennium BC', *Archaeology*, 27 (1974), 136–7; and Helm, *Jawa: Lost City of the Black Desert* .

Africa, South America, or Australia as less important than those built in Europe and North America, readers should consider this book as a foundation for studying water control projects throughout the globe.

Historiography

Dam history is not widely studied and is represented by a relatively limited historiography. Its bibliography is also circumscribed, as least when compared with fields of interest such as bridges, building construction, steam engines, canals, or railroads that comprise a prominent place in the history of technology and industrial archaeology. Yet, the subject bears great relevance to the form of the ancient and modern worlds, and the seventeen articles reprinted in this volume document the importance of dams and their place in cultural development. These articles also bear witness to the skills and professional orientation of the people who have made (and continue to make) contributions to this evolving field of research.

Dam historians come from a variety of backgrounds. Some of the first were nineteenth-century civil engineers and irrigation specialists who became interested in older systems of water control as part of their work in planning new systems. Although not historians *per se*, these engineers (including Charles Markham, C.C. Scott-Moncrieff and James D. Schuyler) documented the dimensions of important early dams and their efforts served as inspiration for future researchers.[2] By the mid-twentieth century the role of the engineer-as-historian reached an advanced stage as several engineers used their technical skills not simply to document the accomplishments of early builders, but to *analyse* and *assess* how the technology of dams had evolved. Among the most important of these authors are A.W. Skempton, Geoffrey Binnie, and N.A.F. Smith of Britain, Nicholas Schnitter of Switzerland, and J.A. Garcia-Diego of Spain; not surprisingly, their work is well represented in this volume (chapters 2, 3, 5, 6, 7, 10 and 15).

By the 1960s the discipline of the history of technology yielded studies focusing on how technologies interact with – and are influenced by – larger social and cultural forces. In the field of dam history, N.A.F. Smith played a critical role in incorporating the study of water control technology into this more culturally-oriented approach. Smith's books, *A History of Dams* (1971) and *Man and Water* (1975), established scholarly benchmarks in the field and stimulated

[2] For example, see Charles R. Markham, *Report on the Irrigation of Eastern Spain* (London, 1867); C.C. Scott-Moncrieff, *Irrigation in Southern Europe* (London, 1868); and Schuyler, *Reservoirs for Irrigation...* (1909). The importance of engineers in developing the nascent field of dam history is discussed in Smith, *A History of Dams*, 128–9; and Smith, *Man and Water*, 49.

further work by a new generation of technological historians.[3] These new scholars are not necessarily trained as engineers, but they bring an appreciation and understanding of technical detail to their work that fosters insight into the technological decision-making process. Among these I include Terry Reynolds, Peter Molloy, and myself, all of whose work is represented in this volume (chapters 9, 12, 13 and 14).

Other researchers involved in dam history have been archaeologists studying the remains of ancient civilizations. For example, our knowledge of the oldest known dam at Jawa is based upon the research and publications of S.W. Helm.[4] In assembling their broad synthetic works on dam history both N.A.F. Smith and Nicholas Schnitter (*A History of Dams: The Useful Pyramids*) have drawn upon archaeological investigations in documenting the dimensions of many ancient water control systems.[5] Architectural historians have also contributed to the field through their interest in the form of the built environment; in this volume, Nicholas Adams and Richard Guy Wilson document both dam-building in Renaissance Italy (an era and locale where the professions of architect and engineer largely overlapped) and the basic design, construction and surface treatment of the Hoover Dam (chapters 4 and 16).

Dams and Environment

Dams are frequently considered symbols of civilization and advancing cultural development, but they can be viewed as bringing unwanted 'progress' to pastoral landscapes. One of the first controversies involving this issue arose in the 1870s when Manchester's municipal water supply system expanded into the Lake District of northwest England. The Lake Thirlmere Dam evoked opposition from a range of traditional interest groups (including railway and canal companies, industrial factions, and citizens of Manchester who would have to pay for the project). In addition, as N.A.F. Smith notes in *Man and Water*:

> The Thirlmere scheme... was bitterly opposed on aesthetic grounds. [It threatened] the beauty and solitude of the Lake District, more highly prized for its scenery than any other part of Britain [and] incensed feelings throughout the country. There was total conviction on the part of many that Thirlmere's dam and aqueduct would cause ugly and permanent disfigurement of the landscape.[6]

---

[3] Smith, see Bibliography.

[4] Helm, see note 1. Other archaeological studies relevant to dam history include J.M. Balcer, 'The Mycenaean dam at Tiryns', *American Journal of Archaeology*, 78 (April 1974), 141–9; and G. Garbrecht, 'The water supply system of Tuspa (Urartu)', *World Archaeology*, 82 (1980), 306–12.

[5] Smith, *A History of Dams*; and Schnitter, *A History of Dams* (1994).

[6] Smith, *Man and Water...*, 121–2.

Similarly, in the early-twentieth century San Francisco's plans to utilize Hetch Hetchy Valley in America's Yosemite National Park for a storage reservoir provoked the wrath of naturalist John Muir and other early members of the Sierra Club. Muir's crusade ultimately proved futile, but not before gaining support from citizens throughout the United States.[7] And in the 1950s, a defining moment in the history of America's environmental movement focused around plans – eventually abandoned – to build Echo Park Dam within the confines of Dinosaur National Monument in northeastern Utah.[8]

Just as Muir considered the Hetch Hetchy Dam an undesirable intrusion into Yosemite, so, too, many people concerned about 'the environment' have come to perceive dams as agents of biological destruction and undesirable social change.[9] They are sceptical of claims made for the economic benefits of large-scale water storage projects. And they fear insufficient appreciation of the long-term costs associated with their effects: the flooding of riparian bottom land; the disruption of stream flow and its effect on animal habitats (especially those of spawning fish); the intrusion of salt water into fresh water marshes; and the increases in human population made possible by water storage and aqueduct projects.

Of course, no one should be surprised that dams alter riparian environments and create new ecological systems in their wake. Dams have always been built in order to transform the environment and make water resources available at different times, in different quantities, and at different locations than would be offered by the unaltered hydrology of a river basin. This constitutes the primary social impetus underlying all water control projects. In the twentieth century, the proliferation of major dams around the world (scores of which extend several hundred feet in height and inundate hundreds of square miles within their 'take area') has reached the point where the environmental and cultural costs associated with creating large-scale reservoirs warrant everyone's careful attention.[10] Such

[7] The best analysis of the Hetch Hetchy controversy is Kendrick Clements, 'Politics and the park: San Francisco's fight for Hetch Hetchy, 1908–1913', *Pacific Historical Review*, 48 (May 1979), 185–215.

[8] Mark W.T. Harvey, *A Symbol of Wilderness: Echo Park and the American Conservation Movement* (Albuquerque, New Mexico, 1994).

[9] Books highlighting the environmental costs associated with dam-building include: Constance Elizabeth Hunt and Verne Huser, *Down by the River: The Impact of Federal Water Projects and Policies on Biological Diversity* (Washington, D.C., 1988); Goldsmith and Hild Yard, ed., *The Social and Environmental Effects of Large Dams*; Martin Russell, *The Story that Stands Like a Dam: Glen Canyon and the Struggle for the Soul of the American West* (New York, 1989); Marc Reisner, *Cadillac Desert: The American West and Its Disappearing Water* (New York, 1986); and Worster, *Rivers of Empire...* A good treatment of how dam-building is being debated in the developing world is Dhawan, *Big Dams...*

[10] Probably the most controversial project of this type is the Three Gorges Dam on the Changjiang (Yangtze) River in central China. The enormous ecological and cultural impact of this dam (which is under construction as of 1997) is described in the Chinese Academy of Science report: *Atlas of Ecology and Environment...*

views are acknowledged through the inclusion of Donald Worster's article on Hoover Dam (chapter 17) which highlights how dams can be seen as serving the interests of particular groups of people rather than humankind as a whole.

This compendium is intended to be neither a defence of dam-building as an unmitigated economic good nor a diatribe castigating water storage technologies as purveyors of ecological disaster. Rather, within a larger social context, it seeks to illuminate how dam technology has evolved over the past several centuries. Readers may draw upon this information in formulating their own views about the place of dam-building in the modern world.

Diversion versus Storage

In simplest terms, dams are defined as either diversion or storage structures. Diversion dams (also known as 'weirs' or 'barrages') are designed to raise the height of a river so that water can be diverted out of a streambed and into a canal, flume, penstock or other 'artificial' waterway. Diversion dams usually block only a portion of a river's flow; the rest passes over the top of the structure and continues downstream. Storage dams, which impound water for relatively long periods of time, are intended to block riverflow completely and generally release water only through specially built gates or valves; during times of heavy floods, water in rising reservoirs is often directed over spillways especially designed to prevent unintended overtopping of the dam.

Diversion dams are usually smaller – and thus easier to build – than storage dams. In contrast, storage dams can store large floods (created by seasonal snowmelt or infrequent large storms that may occur for only a few weeks per year) and discharge water over a period of several months into a canal or aqueduct that would dry up if it depended upon a simple diversion dam. As a result of their ability to 'capture the floods' and increase long-term water supplies within a particular watershed, storage dams are often considered particularly important within arid environments. Of course, diversion dams can also provide for a modicum of water storage (even if the water they impound is sufficient to fill a canal system for only a few days) and storage dams can also be used to divert or redirect the flow of a river. Thus, the line separating these two types of structures can at times become blurred.

Most articles reprinted in this book describe storage dams, but some focus on diversion structures. For example, J.A. Garcia-Diego examines a sixteenth-century treatise on Spanish dam-building concerned with practical issues involved in constructing diversion dams for small-scale irrigation systems (chapter 3). Peter Molloy analyses a structure used to power mid-nineteenth century textile mills in Lawrence, Massachusetts, that is much larger than the designs discussed by Garcia-Diego (chapter 9). Nonetheless, the Lawrence Dam functions almost exclusively as an overflow diversion dam and shares similarities with diversion dams dating back centuries.

Storage dam technology is amply represented by structures built over a period of more than 4000 years. These include the Sadd-el-Kafara Dam (chapter 1); designs for the sixteenth-century masonry dam erected near Siena, Italy (chapter 4); many British earthen embankment dams described by A.W. Skempton (chapter 15); the Jones Falls Dam in Ontario, Canada (chapter 8); Liverpool's Vyrnwy Dam in northern Wales (chapter 11); the Roosevelt Dam near Phoenix, Arizona (chapter 12), and the Hoover Dam across the Colorado River near Las Vegas, Nevada (chapters 16 and 17). While the dams discussed in the above articles can vary dramatically in terms of design, materials, and purpose, they share a common trait in impounding water for long-term use.

Reasons for Building Dams

Irrigation

Most early dams were built in arid environments with large tracts of arable land, but only limited amounts of water. By diverting water out of streambeds and into man-made canals, crops could be cultivated on otherwise barren land. Large-scale irrigation systems, such as those built in the river valleys of ancient Mesopotamia (present-day Iraq), covered thousands of acres of land and supported 'irrigation cities' populated by thousands of people.

Since the first appearance of 'hydraulic civilization' five thousand years ago, irrigation has flourished in a variety of regions including China, India, Egypt, Meso-America, the greater Mediterranean region and the American Southwest.[11] Usually, the earliest irrigation systems in a region depended upon diversion dams (and sometimes – as in ancient Egypt – upon the control of seasonal river floods). Only later were these systems enhanced by large-scale storage structures. For example, the Indian subcontinent featured an expansive irrigation system based around diversion dams prior to the coming of British colonial rule in the nineteenth century; the first large-scale storage reservoir was not built until almost the beginning of the twentieth century.[12]

[11] See Smith, *Man and Water* for evocative and far-ranging discussion of irrigation systems throughout the world. Schnitter, *A History of Dams* provides detailed descriptions of early irrigation systems in the Near East, India, South-east Asia, China, Mexico, and elsewhere.

[12] Smith, *Man and Water*, 46–50, describes Indian irrigation and the eventual construction of the 65-metre high Periyar Dam near Madras. Also see J.M. Brown, 'Contributions of the British to irrigation engineering in Upper India in the nineteenth century', *Transactions of the Newcomen Society*, 54 (1983–84), 85–112.

In the modern world, irrigation remains one of the most common reasons for building dams and water control systems. Today, the term 'reclamation' is often used as a synonym for 'irrigation'. However, the former term often connotes larger, more bureaucratic projects as opposed to small-scale water distribution systems.

Municipal Supply

In early 'irrigation cities' water for the domestic needs of city dwellers was often supplied by wells or irrigation canals. By the Roman era, aqueducts delivered water expressly for municipal use and these systems often utilized dams for diversion and some storage.[13] After the decline of the Roman Empire, European cities generally did not achieve sufficient size to reinvigorate interest in large-scale municipal water supply systems until the eighteenth century (or even later). In this volume, the importance of municipal demand is highlighted in articles focused on Liverpool's Vyrnwy Dam and Los Angeles' association with the Hoover Dam in the American Southwest (chapters 11, 16 and 17). In addition, the desire of citizens in Siena to increase fish supplies for consumption by its citizens reflects another way in which urban growth fostered dam construction (chapter 4).[14]

Power

Along with irrigation and municipal growth, a primary reason for building dams is to generate power.[15] By building up a height of water, dams can help turn water-wheels attached to mills. By the fifteenth century, water-powered grist mills were common in Europe and hydro-power systems used to generate mechanical energy were easily adapted to textile factories in the eighteenth and early nineteenth centuries. By the beginning of the twentieth century, hydro-power systems were being used for the generation of electricity, and water power assumed great import in the formation of electric power networks. In fact, the generation of *hydroelectric* power comprises a primary function of many modern dams (chapter 14, 16 and 17).[16]

---

[13] Smith, *Man and Water*, 80–93, describes Roman water supply systems.

[14] Several other fish ponds are also described in Schnitter, *A History of Dams*, 120–3.

[15] Smith, *Man and Water*, 137–99, provides a good overview of hydropower technology from ancient through modern times.

[16] Jackson, *Building the Ultimate Dam...*, 8–9, 47–83, 109–33, provides extensive discussion of how electric power systems prompted interest in dam-building in the early-twentieth century.

Transportation

Dams are frequently built to supply water to transportation canals or otherwise improve the navigability of rivers by flooding out rapids and/or ensuring enough water to prevent ships and barges from 'running aground'. Geoffrey Binnie highlights the important relationship between canals and dams in Britain, Robert Legget documents how the Jones Falls Dam functions as part of the Rideau Canal in Ontario, Canada, and N.A.F. Smith describes the Bouzey Dam, erected to supply water for the Canal de l'Est in northern France (chapters 7, 8 and 10).[17]

Mining

Mines and other extractive industries also comprise a focus for dam-building, as they often rely upon water power systems to haul ore or drive drainage devices.[18] In addition, the technique known as 'hydraulic mining' utilizes large quantities of pressurized water to 'hose down' hillsides; this creates a water-laden muck that can be sluiced to processing facilities and valuable minerals removed. Hydraulic mining was widely practiced in nineteenth-century California where mining companies built several large-scale storage dams to conserve run-off and maximize gold production.[19]

Flood Control and Multi-Purpose Dams

Surging floods can wreak enormous havoc and, as cities and agricultural lands occupy ever-larger sectors of low-lying areas, interest has grown in protecting such land by building specially designed flood control dams. Of course, all storage dams control floods to some extent; in the twentieth century many large dams were promoted as 'multi-purpose' structures that could serve more than one function. For example, the Hoover Dam in the American Southwest is often extolled because it provides:

1) water for irrigation in the California's Imperial Valley;
2) water for municipal users in greater Los Angeles and central Arizona;
3) hydroelectric power to consumers throughout the region; and
4) flood control for riparian land in the lower Colorado River Valley (chapter 16).

---

[17] Wegmann and Noetzli, *The Design and Construction of Dams*, includes extended discussion of navigation dams. See also Johnson, *The Davis Island Lock and Dam...*

[18] Schnitter, *A History of Dams*, 109–15, describes what he terms 'postmedieval power dams for mining'.

[19] Many of California's early mining dams are described in Bowie, *Practical Treatise on Hydraulic Mining*. Also see Schuyler, *Reservoirs for Irrigation...*

While the above categories constitute the primary reasons for building dams, they do not represent the only justifications for impounding water. For example, in the first century AD the Roman emperor Nero built a 40-metre high dam at his villa in Subiaco simply to create a lake that would enhance the beauty of the site and provide for pleasure (chapter 2). Later, the reservoir impounded by Nero's dam was connected to the intake for Rome's Anio Novus aqueduct, but this did not comprise the original purpose of the dam. In the light of objections that dams can destroy the beauty of natural canyons, it is worth noting that, at times, they have been built expressly to enhance the visual character of a setting.[20]

Design Traditions

In its most elementary form, a dam consists of a mass of material that by its weight holds back a volume of water. Structures that are too heavy for the impounded water to move, tip over or otherwise dislodge are usually referred to as gravity dams. This name is appropriate because the force of gravity pulling down on the material in the dam is what provides resistance against the water pressure. Over time, a *massive tradition* of gravity dam design developed, fostering construction of structures that resist hydrostatic pressures merely by means of their size.[21]

Designs exemplifying the massive tradition can be based upon sophisticated engineering analysis, but the basic principle underlying the tradition is simple: accumulate as much material as possible, thus increasing the likelihood that the structure will achieve long-term stability. Because of its conceptual simplicity, the massive tradition was employed for the earliest known 'embankment dams' (such as Sadd-el-Kafara) which consisted of piled up earth and/or rock, enough to resist the hydrostatic pressure exerted by a reservoir.

Earth and rock are often available in large quantities and can be excellent materials to use for a permanent water storage structure. However, to impound water successfully, a matrix of rockfill or a mound of soil frequently requires a special impervious barrier (e.g. a layer of dense earthen clay, timber planks, a concrete slab, etc.) on top of the upstream face or within the interior of the structure. In grappling with problems related to percolation and seepage, early dam builders developed techniques for placing dressed masonry blocks that were similar to methods used for building solid walls. By filling the space between

---

20 Schnitter, *A History of Dams*, 134–6, notes how 'at various times peculiar kinds of water supply systems were developed for the pools and fountains of the palace gardens of ostentatious rulers'.

21 The concept of 'traditions' as applied to dam design is discussed in Jackson, *Building the Ultimate Dam...*, 18–21.

stones with mortar, cement or concrete, a waterproof barrier could be formed that constituted a good upstream face for a dam. This type of design is usually termed a masonry gravity dam, a name that refers to solid structures comprised of stone blocks, concrete, or a mixture of these two materials.

The *structural tradition* offers alternatives to gravity designs that do not rely upon bulk for stability. Instead, dams adhering to the structural tradition depend upon their shape – and not simply their mass – for stability in resisting hydrostatic pressure. In essence, the amount of material in a structural dam is not as important as developing a design that utilizes the strength (as opposed to the weight) of the material used to build the dam. For example, in narrow canyons with hard rock foundations, arch dams can resist the water pressure by transferring some of the hydrostatic load to the canyon walls. An arch dam does not extend straight across a valley but is built along an arc that curves upstream into the reservoir; the pressure exerted by the reservoir is carried by 'arch action' into the canyon walls. Because of this 'arch action', the thickness (and hence bulk) of the dam's profile can be much less than that of a gravity dam of the same height.

A note of clarification: gravity dams and arch dams depend upon different structural principles for their basic stability, but confusion can arise when a structure contains enough material to function as a gravity dam but is built along a curved axis. In a *curved gravity* dam the dimensions of the profile are large enough to function as a gravity structure. In contrast, the profile of an arch dam is insufficient to resist a hydrostatic load simply by the force of gravity. In other words, if a curved gravity dam were straightened it would be able to impound water without tipping over. But if an arch dam were straightened, the structure would collapse backwards under the pressure of a full reservoir.

The structural tradition also includes designs known as 'buttress dams'. Early Roman dams of this type featured buttresses built perpendicular to the downstream side of a relatively thin masonry wall. By the nineteenth century, designs that employed buttresses primarily as supplements to bulky profiles were superseded by ones in which buttresses comprised the entire structure behind the upstream face. Buttress structures which utilize a flat surface for the upstream face are called flat-slab dams; those featuring a series of arches are known as multiple arch dams. Flat-slab and multiple arch buttress dams are not solid monoliths that present a constant cross-section along the length of the structure. Often called 'hollow dams' because of the extensive empty space that lies between adjacent buttresses, they require much less material than gravity dams of comparable height.

The design principles underlying the structural tradition are not so obvious (at least compared with gravity designs) and structural dams have not been built in great abundance. Roman engineers built the first known arch dam and several buttress dams but, after the collapse of the Roman Empire, the structural tradition (along with large-scale dam building in general) waned in Western civilization.

Nonetheless, it did not die out completely. During the thirteenth and fourteenth centuries, Ilkhanid Mongols in Persia also built several noteworthy arch dams including the 60-metre high Kurit Dam that stood as the world's tallest dam for 500 years (chapter 5).

Massive gravity dams may be simple to conceptualize, but they make profligate use of construction materials and are often expensive to build. In contrast, structural dams require relatively small amounts of material while demanding more sophisticated design and construction techniques in order to achieve financial savings. Either tradition can foster 'safe' designs and it is not a question of one tradition being 'correct' and the other 'incorrect'. The massive and structural traditions may be based upon contrasting design methodologies but, as outlined below, the history of both reflects engineers' interest in rationalizing the design and construction process.

Rational Design

Achievements in sixteenth-century Spain mark the beginning of modern dam-building in Western civilization. During this time, principles to guide construction began to be written down; portions of the Codex of Juanelo Turriano represent the first known manuscript to describe dam-building (chapter 3). In the late-eighteenth and early-nineteenth centuries published treatises on the theory and construction of gravity dams began to be published in Europe.[22] Although these publications did not immediately transform the way dams were built, they established important precedents for adapting mathematical theory to the practice of dam design. In essence, they began to 'rationalize' the way in which engineers conceived and developed designs.

Embankment Dams

The erection of embankment dams using huge quantities of earth or rockfill has been common from ancient to modern times, largely because these materials are often readily available near dam sites (chapters 1 and 7). The building of early embankment dams did not involve the use of mathematical formulae. But by the nineteenth century, basic theories of engineering statics (such as those used to calculate where the 'resultant force' of water pressure combined with the weight of a dam will act upon the foundation) offered a way to ascertain mathematically the amount of material necessary to build a safe embankment dam. Nonetheless,

[22] See Molloy, 'Design and construction of the Lawrence Dam' (chapter 9) for references to early dam design treatises by Bernard Belidor, Charles Bossut, Charles Augustin Coulomb, John Smeaton and Henry Moseley.

in 1900 Edward Wegmann could report in his encyclopaedic *The Design and Construction of Dams*:

> The profiles of earthen dams are determined entirely by practical considerations. The general dimensions of the profile depend upon the materials used and the height of the bank, additional strength [i.e. more material] being given to high dams.[23]

Almost 50 years later, a prominent textbook on dam design reaffirmed Wegmann's appraisal by counselling:

> Precise and highly mathematical methods of design or analysis of the various features of earth dams are seldom justified because we know in advance that the assumptions which must be made are sufficiently wide so that no precision is justifiable... [The engineer will] do well to confine himself to the simple, less precise methods of design and analysis.[24]

While the overall dimensions of embankment dams did not change dramatically during the nineteenth and twentieth centuries, methods of transporting and placing material prompted significant innovation. For example, during the first part of the twentieth century techniques adopted from the placer mining industry led to widespread construction of hydraulic-fill dams formed by: 1) hosing down hillsides with pressurized water; 2) transporting the resultant muck by pipe or flume to the dam site; 3) placing it in the embankment; and 4) allowing the excess water to drain away.[25] By the mid-twentieth century, the proliferation of large-scale equipment and motorized trucks fostered a boom in non-hydraulic-fill embankment dams that continued into the latter part of the century.

In terms of basic shape and profile, embankment dams may have remained unchanged, but the technology has inspired much sophisticated engineering analysis within the field of soil mechanics.[26] In particular, since the early twentieth century engineers have paid close attention to the quality of materials used (and the means of placement) in order to minimize the possibility of seepage and settlement (chapter 15).

---

23 Wegmann and Noetzli, *Design and Construction of Dams*, 222–3.

24 Joel Justin, William Creager and Julian Hinds, *Engineering for Dams*, Volume III, (New York, 1900), 655.

25 Schuyler, 'Recent practice in hydraulic-fill dam construction', 196–277.

26 Smith, *A History of Dams*, 225–66; and Schnitter, *A History of Dams*, 164–5, both note the importance of Karl Terzaghi in advancing the discipline of soil mechanics in the 1920s and 1930s.

Masonry Gravity Dams

As exemplified by the construction of the Lawrence Dam in the 1840s, safe and secure masonry gravity dams could be built without reliance upon mathematical formulae (chapter 9). However, by the mid-nineteenth century, engineers realized that this type of structure was amenable to a quantifiable method of design. The key theoretical development involving the masonry gravity dams occurred in the early 1850s when the French engineer Augustin De Sazilly postulated his 'profile of equal resistance'.[27]

Using basic principles of statics, De Sazilly developed a cross-section in which the stresses at the upstream face when the reservoir is empty are equal to the stresses at the downstream face when the reservoir is full. By considering these two extreme conditions, he formulated a design that – at least in cross-section – minimized the amount of material required for a stable gravity dam. De Sazilly never built a dam based upon the 'profile of equal resistance', but in 1858 Emile Delocre employed the theory for the 50-metre high Furens Dam near Saint-Etienne in Southern France.[28] Completed in 1866, this massive curved gravity structure contained over 52 000 cubic yards of masonry; within a few years engineering textbooks lauded it as the first masonry dam 'built in accordance with correct scientific principles'.[29]

In the early 1870s, the British mathematician/engineer W.J.M. Rankine confirmed the validity of De Sazilly's and Delocre's work. In addition, Rankine postulated that for long-term stability, gravity dams must feature a cross-sectional profile sufficiently thick that the combined 'resultant force' of the horizontal hydrostatic pressure and the vertical weight of masonry falls within the middle third of the structure at all elevations.[30] If the resultant falls outside the 'middle third', a gravity dam will develop tension (instead of compression) along the downstream edge and become susceptible to dangerous cracking. The ideas of De Sazilly, Delocre and Rankine served as models for many major structures in the late nineteenth and early twentieth centuries. But, as documented by Norman Smith's analysis of the failure of the Bouzey Dam in northern France in the 1890s, the importance of adhering to the 'middle third' rule was not immediately recognized by all designers of masonry gravity dams (chapter 10).

---

[27] J.A.T. DeSazilly, 'Sur un type de profil d'égale résistance proposé pour les murs des reservoirs d'eau', *Annales des Ponts et Chaussées, Mémoires et Documents* (1853), 191–222. Also see Smith, *A History of Dams*, 195–7.

[28] Emile Delocre, 'Mémoire sur la forme du profil à adopter pour les grands barrages en maçonnerie des reservoirs', *Annales des Ponts et Chaussées, Mémoires et Documents* (1866), 212–72. Also see Smith, *A History of Dams*, 197–200.

[29] Wegmann and Noetzli, *Design and Construction of Dams*, 69.

[30] W.J.M. Rankine, 'Report on the design and construction of masonry dams', *The Engineer*, 33 (5 January 1872), 1–2.

In almost any technologically-oriented field, failures represent an important catalyst for technological change – or at least they often prompt a re-thinking of the methods and assumptions that underlie the design/construction process.[31] Dams are no exception to this rule. In the wake of the Bouzey Dam collapse – and in anticipation of building a large storage dam across the river Nile at Aswan – British engineers began to focus on the suitability of existing methods of designing masonry gravity dams.

Shortly after the turn of the century, L.W. Atcherley and Karl Pearson carried out experiments with small-scale wooden models that appeared to reveal the existence of significant tensile stresses in standard gravity designs.[32] In turn, this prompted further testing of models – built out of indiarubber rather than wood – that allayed apprehension over the basic tenets of gravity dam design.[33] As Edward Wegmann later reported, in a 1908 meeting sponsored by the Institution of Civil Engineers:

> All the experiments on models of dams... were fully discussed by prominent engineers, who were all of the opinion that the design of a masonry dam could not safely be based upon such experiments, nor on pure mathematics, but that, until some better way was discovered, dams could be safely designed by keeping the lines of pressure, reservoir full or empty, within the center-third of the profile, and by limiting the maximum pressures in the masonry to pressures which experience had shown to be safe.[34]

Although an issue of grave concern at the time, the 'model controversy' eventually passed without dramatically transforming established design methods. In fact, gravity dam design proved remarkably resilient in the face of safety concerns. For example, the Bouzey Dam failure (along with other masonry dams that experienced 'sliding' across their foundations) also raised questions about the effect of 'uplift' on massive gravity designs.[35] In simple terms, 'uplift' is caused

---

[31] Books and articles documenting some of the most prominent dam failures include Amey, *The Collapse of the Dale Dyke Dam, 1864*; Binnie, 'The collapse of the Dale Dyke Dam in retrospect', 305–25; McCullough, *The Johnstown Flood*; Outland, *Man-Made Disaster...*; Smith, 'The failure of the Bouzey dam in 1895', chapter 10 in this volume; and John R. Freeman, 'Some thoughts suggested by the Austin dam failures regarding text books on hydraulic engineering and dam design in general', *Engineering News*, 66 (19 October 1911), 462–3. Also see Lars Jorgenson, 'The record of 100 dam failures', *Journal of Electricity*, 44 (15 March 1920), 274–6; and (1 April 1920), 320–1.

[32] Atcherley and Pearson, 'On some disregarded points in the stability of masonry dams'.

[33] Wilson and Gore, 'Stresses in dams...'.

[34] Wegmann and Noetzli, *Design and Construction of Dams*, 10. The 'model controversy' is described in Pippard, 'The functions of engineering research...'; and in Smith, *A History of Dams*, 217–20.

[35] P.S. Bulson et al., *Engineering Structures...*, 229–59, discusses late-nineteenth- and early twentieth-century concerns about the theory of gravity dam design and 'uplift'. Also see John R. Freeman, 'Some thoughts suggested by the Austin Dam failures regarding text books on hydraulic

by water from a reservoir that seeps under a dam's foundation and then pushes upward, increasing the tendency of the structure to slide horizontally downstream. And while 'uplift' attracted the concern of twentieth century hydraulic engineers, in the end it did little to affect the basic design of gravity dams except to: 1) encourage the use of thicker, more massive, profiles; and 2) focus attention on methods of either draining or 'grouting' foundations in order to control/reduce subsurface flow.

In 1913, George Holmes Moore astutely noted that all masonry gravity dams are distinguished by a height-to-thickness ratio of at least 3:2 (e.g. a gravity dam 100 feet tall will feature a base thickness of at least 66 feet).[36] In essence, the technology had reached a point where empirical methods of design (based upon a 3:2 height-to-thickness ratio) could produce results analogous to those derived from theoretical calculations; and if engineers wished to 'strengthen' a design, they simply made the profile thicker. Engineers could develop their own particular gravity profiles, but these represented relatively inconsequential variations on the basic 'profile of equal resistance' first proposed by De Sazilly.

Arch Dams

While gravity dams always have been much more common than structural dams, the Elche Dam in Spain and the Ponte Alto Dam in Italy (both built in the seventeenth century) featured profiles that could not provide stability without a pronounced upstream curve. In the early nineteenth century, the strength of the arch was utilized by Lieutenant Colonel John By of the British Army for a dam in a remote part of Canada. Erected in 1828–1831, the Jones Falls Dam functioned as part of the Kingston Canal connecting Lake Ontario with the Ottawa River. The slender profile of By's design (19 metres high, 6 metres thick) was developed without any apparent reference to mathematical theory; instead, the design was predicated upon a desire to reduce the amount of stone that needed to be quarried and transported to the site, thus accelerating the construction process (chapter 8).

Unaware of By's design, in the 1840s François Zola developed a plan to increase the municipal water supply of Aix-en-Province centered around a 45-metre high arch dam.[37] The basic dimensions for Zola's design were determined using a mathematical relation known as the 'cylinder formula'. Conceived by the French mathematician Louis Navier in 1826, this formula states that for cylindrical

engineering and dam design in general', *Engineering News*, 66 (19 October, 1911), 462–3; and Harrison, 'Provision for uplift and ice pressure…'.

[36] Moore, 'Neglected first principles…'.

[37] Smith, *History of Dams*, 181–83; and Schnitter, 'Evolution of the arch dam', chapter 5 in this volume.

shapes placed under hydrostatic pressure, the arch thickness necessary to achieve a given level of stress can be calculated in terms of:

1) the water pressure; and
2) the radius of the arch.

Zola perceived his dam as a stack of horizontal arches. At the top, where water pressure is at a minimum, the structure is thinnest; in lower parts of the dam where water pressure is greatest, the structure is thickest.[38]

Zola Dam was completed in 1854, but arch dam construction languished for many years thereafter. Then, in 1884, the American engineer Frank E. Brown conceived a daring design in Southern California in which the 'resultant force' fell far beyond the downstream edge of the foundation.[39] Brown's 21-metre high Bear Valley Dam featured a maximum thickness of 7 metres and demonstrated that thin arch designs could radically reduce the amount of material required for a stable masonry dam. In 1886, Brown started another arch structure across the Sweetwater River near San Diego, California; shortly after construction began James D. Schuyler took charge of completing the 27-metre high, 14-metre thick design. Completed in 1888, Sweetwater Dam further confirmed that the cylinder formula could foster storage dam construction requiring limited amounts of masonry.[40]

Insight garnered from the cylinder formula soon led to a significant design variation that, although not implemented until 1913, represented the final important innovation derived from the theory. In 1879, Albert Pelletreau published a theoretical study proposing that the radius of the arch at different elevations be varied in order to minimize the necessary thickness of the structure.[41] Most canyons are narrower at the bottom than at the top; thus, it is easy to conceptualize arch dams consisting of a stack of arches with progressively smaller radii. Referred to as either constant angle arch dams (i.e. the angle remains constant while the radius gets smaller) or variable-radius dams (because the radius varies in length), structures of this type resemble inverted cones. And because the cylinder formula correlates thinner arches with smaller radii, constant angle/variable radius arch dams can dramatically reduce the amount of concrete required

---

[38] Schnitter, 'Evolution of the arch dam', chapter 5 in this volume provides a brief analysis of the structural nature of the Zola Dam. In examining its profile it becomes evident that the cylinder formula was not used as a rigid means of proportioning the structure. Rather, Zola used it as a guide to assist him in increasing the thickness of the arch as it descended into the canyon.

[39] Brown, 'The Bear Valley Dam'. Also see Schuyler, *Reservoirs for Irrigation Water Power...*, 246–56, for a good description and photograph of this daring arch dam.

[40] James D. Schuyler, 'The construction of the Sweetwater Dam', *Transactions of the American Society of Civil Engineers*, 19 (1888), 201–32.

[41] Pelletreau, 'Barrages cintrés en forme de voûte'. A constant angle design is also described in John S. Eastwood, 'An arch dam design for the site of the Shoshone Dam', *Engineering News*, 63 (9 June 1910), 678–80.

for construction. The first dam of this type was built near Juneau, Alaska in 1913, when Lars Jorgensen designed the Salmon Creek Dam as part of a mining company's hydroelectric power system.[42] Since that time, constant angle/variable radius designs have been built in locations around the globe.[43]

The cylinder formula can accurately predict stresses in an infinitely thin arch encompassing an arc of 180 degrees supported on rigid foundations. However, as a structure varies from this ideal, the formula becomes less reliable because it fails to consider that external forces can cause deformations that might alter the distribution of stresses. In 1854, Jacques Bresse formulated an *elastic theory of arches* that considered the structural mechanics within a broad mathematical framework and analysed the effect of deformations upon arch stresses.[44] For many years thereafter, elastic theory played no role in dam design. But after several early-twentieth-century concrete arch dams in Australia experienced extensive cracking, the adequacy of the cylinder formula came into serious question.[45] And engineers soon focused attention on how elastic theory could be integrated into the dam design process.

Although the elastic theory of arches is based upon its own set of idealized assumptions, it nonetheless appears to offer an accurate means of analysing stresses.[46] By the 1920s, many dam engineers perceived it as an improvement over the cylinder formula, especially when utilized as part of the 'trial load' design method. First hypothesized in 1889, the 'trial load' method recognizes that an arch dam resists some of the hydrostatic pressure by 'gravity action', especially in the lower parts of the structure where the arch is relatively thick and in close proximity to the foundation.[47]

The 'trial load' method conceives of an arch dam as an interrelated set of horizontal arches and vertical cantilever sections coexisting within the structure. Using the elastic theory, the structure is designed to insure that the deflections (or deformations) of the two systems under the same hydrostatic load are compatible in various parts of the dam. By postulating a way to distribute stresses to both the arch and cantilever systems, the 'trial load' method appears to offer a way to calculate stresses with great mathematical precision.[48]

---

[42] Jorgensen, 'The constant-angle arch dam'. Also see Mannikko, 'Seattle City Light's...'.

[43] Schnitter, *A History of Dams*, 200–1.

[44] Ibid., 195.

[45] Wade, 'Concrete and masonry dam construction...'.

[46] For more on the elastic theory as it applies to arch dams see William Cain, 'The circular arch under normal loads', *Transactions of the American Society of Civil Engineers*, 85 (1922), 233–83.

[47] Vischer and Wagoner, 'On the strains...'. Another early version of the 'trial load' method is described in George Wisner and Edgar T. Wheeler, 'Investigations of stresses in high masonry dams of short spans', *Engineering News*, 54 (10 August 1905), 141–4.

[48] Arch deflections are determined according to elastic arch theory while the cantilever deflections are calculated according to a similar method that engineers term 'beam theory'.

The method's name relates to the use of a trial and error approach in determining the relationship between the dam's deformation and the distribution of the hydrostatic load. After three or four iterations (or trials), a design could be formulated in which stresses could be forecast with reasonable accuracy. However, these designs were not necessarily efficient in their use of materials and could feature profiles ample enough for a gravity dam. For example, the Bureau of Reclamation's Hoover Dam is a good example of a massive curved gravity structure designed using the trial-load method (chapter 16).[49]

Buttress Dams

Masonry dams supplemented by downstream buttresses were built by both the Romans and post-Renaissance Spanish engineers (chapter 6). The upstream faces for these early buttress dams were not thick enough to stand alone as gravity dams, but they were often quite ample.[50]

In the nineteenth century, many wooden buttress dams with inclined upstream faces were built in timber-rich America, but they did not foster widely discussed theories of 'rational' design comparable to those developed for massive gravity dams.[51] In fact, at the close of the nineteenth century, the empirical nature of small-scale buttress dam design bore considerable similarity to the design of embankment dams.

The first major non-wooden buttress dams of the modern era were built using steel; the 13-metre high Ash Fork Dam in northern Arizona (completed in 1898) and the 22-metre high Redridge Dam in Michigan (completed in 1902) appeared to affirm the efficacy of the technology.[52] However, the 24-metre high Hauser Lake Dam in Montana (completed in 1907) collapsed soon after construction because of problems with its foundation. Concern generated by this failure – along

---

[49] The 'trial load' method used by the Bureau of Reclamation is described by Howell and Jacquith, 'Analysis of arch dams...'. Schnitter, *A History of Dams*, 204, notes that 'the somewhat paradoxical caution on the part of the Bureau of Reclamation [in designing Hoover Dam as a massive curved gravity structure] initiated the decline in the construction of true arch dams in the USA. This trend is exemplified by the developments in California, where the number of new arch dams, as well as their proportion in relation to other types, decreased sharply after World War II'.

[50] Schnitter, *A History of Dams*, 57–8; on pp. 147–8 Schnitter describes several eighteenth-century masonry buttress dams in Mexico with vertical upstream faces.

[51] Leffel, *Construction of Mill Dams*, 16–19.

[52] F.H. Bainbridge patented a steel dam in the United States on 16 April 1895 (No 537, 520) that served as the basis for the Ash Fork Dam. See A.T. and S.F. Ry, 'Steel dam at Ash Fork, Arizona', *Engineering News*, 39 (12 May 1898), 299–200; 'The Redridge Dam', *Engineering News*, 46 (15 August 1901), 101–2; and F.H. Bainbridge, 'Structural steel dams', *Journal of the Western Society of Engineers*, 10 (September 1905), 615–37.

with rising steel prices – effectively closed what historian Terry Reynolds terms 'the brief window of opportunity' for steel dam construction (chapter 13).[53]

In place of steel, buttress dam builders turned to reinforced concrete. In 1903 Nils F. Ambursen built the first reinforced concrete flat-slab buttress dam at Theresa Falls in upstate New York.[54] Although small (3 metres high), this structure established an important precedent; in 1909, completion of the 40-metre high La Prele Dam in eastern Wyoming demonstrated that reinforced concrete buttress dams need not be limited in size.[55] While Ambursen dams required more material than comparable multiple arch designs (the upstream face is usually much thicker and the buttresses spaced closer together), flat-slab structures proved important in confirming the viability of reinforced concrete dams.[56]

The sixteenth-century Codex of Juanelo Turriano described multiple arch dams with inclined upstream faces, and several small structures of this type were built in eighteenth-century Spain (chapters 3 and 6).[57] In 1803–04 the British engineer Henry Russle supervised construction of the 2000-foot long multiple arch Meer Allum (Mir Alum) Dam near Hyderabad, India. Although the masonry arches at Meer Allum are 13 metres high, they feature vertical upstream faces and do not utilize a vertical water load to resist horizontal water pressure. Despite its great size, the Meer Allum Dam did not foster further multiple arch dam construction and the concept lay dormant for several decades.[58]

In 1897 Henry Goldmark proposed a large (34-metre high) concrete multiple arch dam for a site near Ogden, Utah.[59] Although never built, this design featured seven inclined upstream arches proportioned in accord with the cylinder formula. The rest of Goldmark's design relied upon conventional standards of gravity dam design, where the resultant force of the combined water pressure and masonry load would fall within the buttresses' 'middle third'. In 1908–09, John S. Eastwood implemented Goldmark's theoretical concept by building the world's first reinforced concrete multiple arch dam at Hume Lake in California.[60] The 20-metre high Hume Lake Dam features arches (designed in accord with the

53 'The Hauser Lake steel dam in the Missouri River near Helena, Montana', *Engineering News*, 58 (14 November 1907), 507–9; and F.L. Sizer, 'The break in the Hauser Lake Dam, Montana', *Engineering News*, 59 (30 April 1908), 491–2. Also see Otis Ellis Hovey, *Steel Dams* (New York, 1935).

54 Ambursen and Sayles, 'A hollow concrete-steel dam...'.

55 'The La Prele Dam; a 130-ft hollow reinforced concrete dam near Douglas, Wyoming', *Engineering News*, (10 November 1910), 499–501.

56 Jackson, *Building the Ultimate Dam...*, 146–8, 297–8, discusses why flat-slab designs required more concrete (and more money) to build than multiple arch designs.

57 See Schnitter, *A History of Dams*, 118–20.

58 Ibid., p.51 describes the Meer Allum Dam.

59 Goldmark, 'The power plant...'.

60 Eastwood, 'Hume Lake Dam'.

cylinder formula) that vary in thickness from 0.3 metres at the top of the structure to 0.75 metres at the deepest parts of the dam (chapter 14).

After his success at Hume Lake, Eastwood built more than a dozen additional multiple arch dams in the American West over the next fifteen years.[61] In fact, he came to devote all of his professional energies to the task of promoting the multiple arch dam as 'The Ultimate Dam' based upon its economic and structural advantages.[62] Specifically, these related to the strength of the arch in forming an impervious barrier and the ability of buttress dams to obviate problems associated with 'uplift'. Other engineers soon developed their own approaches to multiple arch dam design, including methods that incorporated the elastic theory of arches.[63] And, although Eastwood's vision of a hydraulic infrastructure abounding with multiple arch dams never came to pass, the final edition of Wegmann's *The Design and Construction of Dams* (1927) features a lengthy new section on the technology, attesting to its importance in the world of hydraulic engineering.[64]

Of course, mathematically-based formulae do not guarantee safe dams. Nor can formulae reflect with absolute accuracy the stresses that exist within structures that operate within the real world. As engineer/historian David Billington of Princeton University has counselled:

> As the 1920s wore on, science-based formulas gained prestige. The major results of these formulas were to create a scientific discipline of analysis isolated from design... this was not a question of intuition versus rigor, or of approximate and uncertain estimates versus precise and scientific predictions. The primary uncertainties in reinforced concrete behavior have never been removed by mathematical analysis however rigorously consistent.[65]

Since the 1920s, mathematically-based approaches to dam design have remained a key part of the discipline. In addition, the use of small-scale models to aid the design process has proliferated, especially in regard to thin arch structures.[66] The use of models and formulae is – and will be for the foreseeable future – an integral part of the dam design process. But this does not negate the wisdom of an observation made by Eastwood in 1916:

---

61 Jackson, *Building the Ultimate Dam...*, Jackson offers in-depth documentation of Eastwood's work as a hydraulic engineer.

62 Eastwood, 'The Ultimate Dam'.

63 For example, see Noetzli, 'Improved type of multiple arch dam'.

64 Wegmann and Noetzli, *The Design and Construction of Dams*, 439–526. [Note: the new section on multiple arch dams was written by Fred Noetzli; it is for this reason that the 8th edition is referenced as having co-authors.]

65 David P. Billington, *The Tower and the Bridge: The New Art of Structural Engineering* (New York, 1983), 214–15. This quotation specifically refers to thin-shell structures, but it bears relevance to all areas of structural design.

66 Smith, *A History of Dams*, 237–8; and Schnitter, *A History of Dams*, 202–4.

> It is undoubtedly the proper procedure to fit theory to practice but the factors of the theory must be based on the actual physical conditions as they are in practice or else theory will lead us astray.[67]

Influences on Design

When developing a dam design, a builder must confront a range of factors related to a particular site. These include:

1) the topography and geology of the site;
2) the dam height necessary to impound various quantities of water;
3) the amount of excavation necessary to reach bedrock;
4) the cost of transporting materials (e.g. earth, rock, cement) to the site;
5) the cost of building formwork and/or placing material.

Beyond seemingly objective 'technical' criteria such as those listed above, other factors influence what type of design is selected for a particular project. For example, the most significant drawback of masonry gravity dams involved the enormous amount of material and capital outlay they required. The 'profile of equal resistance' may have offered a mathematically rational basis of design, but this did not mean that structures adhering to this precept would be inexpensive to build. This becomes clearly evident in the description of Liverpool's Vyrnwy Dam, a structure completed in 1890 at a cost of more than £2 000 000 (chapter 11).

Because the economic benefits of increased urban water supplies can compensate for the expense of massive gravity dams, well-financed government entities did not necessarily find such designs particularly burdensome. In consequence, early-twentieth-century governmental officials (and the hydraulic engineers who advised them) often considered costly gravity structures to represent the most conservative, and hence most appropriate, method of dam building. For example, Frederick H. Newell (the first director of the U.S. Reclamation Service), explicitly championed gravity dam designs because their thick profiles supposedly provided a greater sense of security for the citizens whose taxes funded construction (chapter 12).

But some engineers – most notably, John S. Eastwood – felt constrained by the great costs of building gravity dams and began promoting multiple arch dams as a better and less expensive alternative. In response, prominent advocates

[67] John S. Eastwood to Paul M. Downing, 27 September 1916, Folder no. 4, John S. Eastwood Papers, Water Resources Center Archives, Berkeley, California.

of gravity dam – in particular, John R. Freeman (who served as President of both the American Society of Civil Engineers and the American Society of Mechanical Engineers) – vigorously opposed multiple arch designs not simply on technical grounds, but because their visual appearance was deemed inappropriate for a large storage dam (chapter 14). Referring to a 'psychology' of dam design, Freeman did not rely upon technical arguments to denigrate structural dams. Instead, he drew upon characterisations of visual appearance and 'public opinion' to justify the use of massive gravity designs in the place of less expensive multiple arch dams.[68]

In a broader context, the conflict that arose between massive dams and early multiple arch designs highlights the importance that 'monumentality' can play in implementing dam projects. It also reveals how 'influences' affecting the dam-building process are more socially complex than might appear on the surface. Whether rightfully or not, large dams represent major public structures that can become signposts of a society and its achievements. In particular, they can be promoted as symbols of strength and beneficence displaying the skills and resourcefulness of those people responsible for their construction (chapter 16). Or, alternatively, they can be portrayed as symbols of cultural domination (chapter 17).

As we enter the twenty-first century, we should be aware of the great heritage of dam-building that extends back 5 000 years. And we should acknowledge the value – in both real and symbolic terms – that dams continue to hold for societies around the globe. But we should not allow a fascination with 'monumentality' to blind us to the costs exacted by large-scale water storage in transforming riparian environments. Informed citizenry will need to make important choices in the years ahead regarding the place of dams in the world's socio-economic culture. The articles reprinted in this volume may not make those decisions any easier, but they will provide a solid base for understanding the place of dams in human history.

Bibliography

Addison, Herbert, *Sun and Shadow at Aswan* (London, 1959).

Ambursen, Nils F., and A.W. Sayles, 'A hollow concrete-steel dam at Theresa, N.Y.', *Engineering News*, 50 (5 November 1903), 403.

Amey, G., *The Collapse of the Dale Dyke Dam, 1864* (London, 1873).

[68] Jackson, *Building the Ultimate Dam...*, 109–33, provides detailed analysis of how arguments based upon a 'psychology' of dam design were used by John R. Freeman to halt construction of Eastwood's Big Meadows Dam in 1913.

Atcherley, L.W., and Karl Pearson, 'On some disregarded points in the stability of masonry dams', University of London, *Drapers' Company Research Memoirs*, Technical Series II, 1904.

Balcer, J.M., 'The Mycenaean dam at Tiryns', *American Journal of Archaeology*, 78 (April 1974), 141–9.

Battison, Edwin, 'The Ascutney gravity-arch mill dam, Windsor, Vermont, 1834', *Journal for the Society of Industrial Archeology*, 1 (1975), 53–8.

'The Belubula Dam, New South Wales', *Engineering News*, 40 (8 September 1898), 148.

Binnie, Geoffrey, M., 'The collapse of the Dale Dyke Dam in retrospect', *Quarterly Journal of Engineering Geology*, 11 (1978), 305–25.

—, *Early Victorian Water Engineers* (London, 1981).

—, *Early Dam Builders of Britain* (London, 1987).

Bowie, Aug. J., *Practical Treatise on Hydraulic Mining*, 11th edn. (New York, 1910).

Brown, F.E., 'The Bear Valley Dam', *Engineering News*, 19 (23 June 1888), 513–14.

Brown, J.M., 'Contributions of the British to irrigation engineering in Upper India in the nineteenth century', *Transactions of the Newcomen Society*, 54 (1983–84), 85–112.

[BUREAU OF RECLAMATION], *Dams and Control Works* (Washington D.C., 1929).

[BUREAU OF RECLAMATION], *Water and Power Resources Service: Project Data* (Washington, DC., 1981).

Charles J.A. and J.B. Boden, 'The failure of embankment dams in the United Kingdom', in *Symposium of Failures in Earth-works* (London, 1985), 181–202.

[CHINESE ACADEMY OF SCIENCE], Chen Yu, editor-in-chief, *Atlas of Ecology and Environment in the Three Gorges Area of the Changjiang River* (Beijing, 1990).

Coyne, André, 'Arch dams: their philosophy', *Journal of the Power Division, Proceedings of the American Society of Civil Engineers*, 82 (April 1956), Paper no. 959, 1–32.

Danilevski, V.V., *History of Hydroengineering in Russia Before the Nineteenth Century* (Jerusalem, 1968).

Davis, Arthur Powell, *Irrigation Works Constructed by the United States Government* (New York, 1917).

Delocre, E., 'Mémoire sur la forme du profil à adopter pour les grande barrages en maçonnerie des reservoirs', *Annales des Ponts et Chaussées, Mémoires et Documents* (1866), 212–72.

De Sazilly, J.A.T., 'Sur un type de profil d'égale résistance proposé pour les murs des reservoirs d'eau', *Annales des Ponts et Chaussées*, (1853), 191–222.

Dhawan, B.D., ed., *Big Dams: Claims, Counter Claims* (New Delhi, 1990).
Dillman, George L., 'The great hydraulic principle: or how to make engineering structures hold water', *Engineering News*, 69 (June 12 1913), 1225.
Eastwood, John S., 'Hume Lake Dam', *Journal of Electricity Power and Gas*, 23 (October 30 1909), 398–404.
—, 'The ultimate dam', *Western Engineering*, 3 (September 1913), 175–9.
Garbrecht, G., 'The water supply system at Tuspa (Urartu)', *World Archaeology*, 82 (1980), 306–12.
Garcia-Diego J.A., 'Old dams in Extremadura', *History of Technology*, 2 (1977), 95–124.
Goldmark, Henry, 'The power plant, pipe line and dam of the Pioneer Electric Power Company at Ogden, Utah', *Transactions of the American Society of Civil Engineers*, 38 (1897), 302–5.
Goldsmith, Edward, and Nicholas Hild Yard, ed., *The Social and Environmental Effects of Large Dams* (San Francisco, 1984).
Harrison, Charles L., 'Provision for uplift and ice pressure in designing masonry dams', *Transactions of the American Society of Civil Engineers*, 65 (1912), 142–225.
Helm, S.W., *Jawa: Lost City of the Black Desert* (Ithaca, New York, 1981).
Howell, C.H., and A.C. Jaquith, 'Analysis of arch dams by the trial load method', *Transactions of the American Society of Civil Engineers*, 93 (1929), 1191–1316.
Hundley, Norris, *The Great Thirst: Californians and Water, 1770–1990* (Berkeley, California, 1992).
Hunter, Louis, *A History of Industrial Power in the United States, 1780–1930*, Vol. 1, *Water Power* (Charlottesville, Virginia, 1979).
Introcaso, David, 'The politics of technology: the unpleasant truth about Pleasant Dam', *Western Historical Quarterly, 25 (Autumn 1994), 333–55.*
Jackson, Donald C., *Great American Bridges and Dams* (Washington, D.C., 1988).
—, *Building the Ultimate Dam: John S. Eastwood and the Control of Water in the West* (Lawrence, Kansas, 1995).
Johnson, Leland R., *The Davis Island Lock and Dam, 1870–1922* (Pittsburgh, 1985).
Jorgenson, Lars, 'The constant-angle arch dam', *Transactions of the American Society of Civil Engineers*, 78 (1915), 685–733.
Kollgaard, Eric B., and Wallace L. Chadwick, ed., *Development of Dam Engineering in the United States* (New York, 1988).
Leffel, James, *Leffel's Construction of Mill Dams* (Springfield, Ohio, 1881).
Mannikko, Nancy Farm, 'Seattle City Light's constant angle arch dam at Diablo Canyon', *Journal of the Society for Industrial Archeology*, 17 (1991), 17–31.
McCullough, David G., *The Johnstown Flood* (New York, 1968).

Moore, George Holmes, 'Neglected first principles of masonry dam design', *Engineering News*, 70 (September 4 1913), 442–5.

Noetzli, Fred, 'Improved type of multiple arch dam', *Transactions of the American Society of Civil Engineers*, 87 (1924), 342–413.

Outland, Charles F., *Man-Made Disaster: The Story of the St. Francis Dam* (Glendale, California, 1963).

Ozis, U., 'The ancient dams of Istanbul', *International Water Power and Dam Construction*, 20 (July, 1977), 49–51; and (August 1977), 44–7.

Pelletreau, A., 'Barrages cintrés en forme de voûte', *Annales des Ponts et Chaussées*, 198–218.

Pippard, A.J.S., 'The functions of engineering research in the university', *Journal of the Institution of Civil Engineers*, 33 (February 1950), 265–86.

Pitzer, Paul C., *Grand Coulee: Harnessing a Dream* (Pullman, Washington, 1994).

Rankine, W.J.M., 'Report on the design and construction of masonry dams', *The Engineer*, 33 (5 January 1872), 1–2.

Robinson, Michael C., *Water for the West: The Bureau of Reclamation, 1902–1977* (Chicago, 1979).

Russell, Peter, 'John Frederic La Trobe-Bateman (1810–1889), water engineer', *Transactions of the Newcomen Society*, 52 (1980–81), 199–238.

Schnitter, Nicholas, 'Roman Dams', *Water Supply and Management*, 3 (1979), 29–39.

—, *A History of Dams: The Useful Pyramids* (Rotterdam, 1994).

Schuyler, James D., 'Recent practice in hydraulic-fill dam construction', *Transactions of the American Society of Civil Engineers*, 57 (1907), 196–277.

—, *Reservoirs for Irrigation, Water Power and Domestic Water Supply*, 2nd edn (New York, 1909).

Semenza, Carlo, 'Arch dams: developments in Italy', *Journal of the Power Division, Proceedings of the American Society of Civil Engineers*, 82 (June 1956), Paper no. 1017, 1–41.

Shallot, Todd, *Structures in the Stream: Water, Science, and the Rise of the U.S. Army Corps of Engineers* (Austin, Texas, 1994).

Skempton, A.W., 'Alexandre Collin (1808–1890), pioneer in Soil Mechanics', *Transactions of the Newcomen Society*, 25 (1945–47), 91–104.

Smith, N.A.F., *The Heritage of Spanish Dams* (Madrid, 1970).

—, *A History of Dams* (London, 1971).

—, *Man and Water: A History of Hydro-Technology* (New York, 1975).

Stevens, Joseph, *Hoover Dam: An American Adventure* (Norman, Oklahoma, 1988).

Vaughan-Lee, A.G., 'The Mohammed Aly Barrages, Egypt' *Journal of the Institution of Civil Engineers*, 15 (1940–41), 237–79.

Vischer, Herbert, and Luther Wagoner, 'On the strains in curved masonry dams', *Transactions of the Technical Society of the Pacific Coast*, 6 (December 1889), 75–86.

Vita-Finzi, C., 'Roman dams in antiquity', *Antiquity*, 35 (1961), 14–20.

Wade, L.A.B., 'Concrete and masonry dam-construction in New South Wales', *Minutes of the Proceedings of the Institution of Civil Engineers*, 178 (1909), 1–110.

Watts, W., 'Concrete and puddle for reservoir embankments', *Transactions of the British Association of Waterworks Engineers*, 9 (1904), 57–70.

Wegmann, Edward, and Fred Noetzli, *The Design and Construction of Dams*, 8th edn (New York, 1927).

Willingham, William F., *Waterpower in the Wilderness* (Washington D.C., 1985).

Wilson, J.S., and W. Gore, 'Stresses in dams an experimental investigation by means of India-rubber models', *Proceedings of the Institution of Civil Engineers*, 172 (1907–08), 107–33.

Worster, Donald, *Rivers of Empire: Water, Aridity, and the Growth of the American West* (New York, 1985).

1

Sadd-el-Kafara, the world's oldest large dam

G. Garbrecht

In the Spring of 1982, a team of German and Egyptian engineers and scientists began investigating the remains of the Sadd-el-Kafara dam – discovered exactly 100 years ago by the explorer Georg Schweinfurth. The investigations reveal that a natural disaster overtook the construction but, nevertheless, a great deal of respect must be paid to the original builders for the depth of vision in both design and construction.

IN 1885 the explorer and geographer Georg Schweinfurth discovered in the Wadi Garawi on the eastern Nile bank near Heluan, some 30 km south of Cairo, the remains of an ancient dam. He wrote: ". . . the ancient Egyptians had constructed in the valley . . . a water-retaining, masonry structure for the purpose of catching the water masses which in wet winters often descended with great turbulence from the plateaux of the eastern mountain desert, and of transforming the valley closely surrounded by steep rock faces on a large section into a permanent storage reservoir. . .".

This dam consists of two rockfill sections with the space between them being filled up with rock waste and rubble. The surface of the rockfill is covered with limestone ashlars arranged in the form of stairs. The strong erosion shown by the stones led Schweinfurth to believe the dam was very old, as such disintegration processes usually take a long time.

Regular stone heaps found near the dam were interpreted to be the remains of cottages — workers' accommodation. Ceramics found in and near the stone walls suggested the time of construction as being during the Old Empire. This dam might thus be the oldest structure for a large-scale water retention, or storage, in the history of hydraulic engineering.

After the rediscovery by Schweinfurth, the dam was described in a number of publications by: Mackay (1915)[1], Prelini (1921), Murray (1947, 1955)[2], Hellstrom (1951, 1952)[3,4] Hathaway (1958)[5], Klemm (1979, 1982)[6], Garbrecht (1981)[7] and Troitsch, Weber (1982). All these publications and data given in other papers are, however, scarcely more than general descriptions and interpretations.

During a visit in 1981 it became obvious that the remains of the dam had suffered greatly during the last few decades from the processes of natural disintegration and also to a large extent from human interference. To document the parts still existing of this unusual monument to the history of engineering, the structure was surveyed and investigated in the Spring of 1982 by an interdisciplinary working group of Egyptian and German engineers and scientists.

The Wadi Garawi with the remains of Sadd-el-Kafara.

Geography and geology of Wadi Garawi

The ancient Sadd-el-Kafara dam is situated in the Wadi Garawi, some 30 km south of Cairo and 11 km south-east of Heluan. The Wadi Garawi is one of the numerous wadis in the eastern mountains bordering the Nile valley, which belong to the Arabian desert. The valley bottom in the immediate vicinity of the dam is at an elevation of approximately 110 m and its gradient is about 1 per cent.

Geomorphologically, this part of the eastern desert is a terraced region linked to the occurrence of a number of differing structural plateaux which superposes an old foundation lying much deeper. By an upward folding motion, or by a lowering of the water level, the series of sedimentary rocks rose above sea level and thus into the zone of subeolian erosion.

Geologically, the immediate vicinity of the dam can be identified as lying in the transition period of the Middle to the Upper Eocene series. The series of strata in the Wadi Garawi is composed of limestone, marl and schistous sandstone with the index fossils carolia placuneides, ostrea reili, and nummulites beaumonti.

The strata from which the rocks for the construction of the dam were broken are structurally different. These differences in structure and chemical composition chiefly influence the strength and homogeneity of the stone.

Basically, the material used for the construction of the dam can be subdivided into two categories:

- soft, chalky limestone; and,
- hard, greyish-brown limestone.

The soft, white limestone (chalky and marly stone or arenaceous marl) is composed of amorphous lime, some fine-structured calcite and some quartz with a relatively high salt content.

The hard, greyish-brown limestone consists of amorphous and crystalline carbonate material with a fine-grained structure. Stones from this stratum are very weather-resistant and difficult to tool.

Hydrology and water management

As precipitation is rare and there is no rain at all some years, the Arabian desert is uninhabitable. At the Heluan meteorological station, rainfall of only 25 to 30 mm/year has been measured on a long-year average basis. However, it is quite possible for more than the mean annual precipitation to occur in a single day.

The maximum precipitation intensities for Cairo are reported to be more than 2 mm/min. Comparative values for the Sahara are of the order of 0.1 to 0.2 mm/min and for the Negev 0.4 to 0.6 mm/min. The cloud size and thus the size of the region affected in the case of heavy rainfalls is unknown.

The mean annual evaporation rate at the Heluan meteorological station is 10.5 mm/day. Temperatures are subject to strong daily and seasonal fluctuations. Observations made between 1904 and 1960 show extreme values of 0°C and +47.5°C and the mean annual temperature is given as 21°C.

Basically, large-scale reservoirs are needed either to cover a heavy demand for water such as for irrigation and drinking water, or to protect a region against an excessive amount of flooding or inundation. It is unlikely that the Sadd-el-Kafara was built for drinking-water supplies or irrigation purposes. The dam lies too far from the alabaster quarries situated upstream to have supplied the labour force with drinking water, and vast stretches of fertile arable land together with an abundant supply of water for agricultural needs are available in the nearby Nile valley.

Owing to the geographical and geological conditions prevailing in the catchment area of the Sadd-el-Kafara, heavy rainfalls can give flash floods with disastrous effects in valleys like the Wadi Garawi. Inhabitants report the occurrence of flood waves several metres in height which have recently destroyed villages and claimed lives. It can, therefore, be assumed that the Sadd-el-Kafara was built as a flood protection measure for the lower Wadi Garawi (possibly to safeguard buildings situated at the Ain Fisha spring) and also to protect the Nile valley situated at the mouth of the wadi where there was in all probability, agricultural and other settlements.

Dating of the dam

Since the rediscovery of the Sadd-el-Kafara in 1885, there has been no doubt that the dam is a very old structure. Discoveries in the remains of buildings northwest of the dam indicated that it was constructed in the era of the Old Empire. When the dam was surveyed in 1982, an attempt was made to verify this date and, above all, to establish a direct relationship between the buildings and the dam. The archaeological investigations led to the following conclusions:

- The buildings northwest of the dam form an interrelated group.
- The location, size, type and construction of the buildings suggest a close relationship with the dam. They can only be interpreted as accommodation for the workers employed for the construction work. A distinction can be made between small buildings and dwelling units built for stone-cutters, master builders and superintendents, and larger units which probably served as mass living quarters for the ordinary workers.
- Fragments of bread moulds indicate that the bread was baked locally. However, the question as to whether the beer pitchers found served for transport or for storage after brewing in the camp must be left unanswered. The similarity of the ceramic material from the various buildings confirms the fact that they were interrelated.
- Finds of similar beer pitcher fragments in the dam fill prove that the camp buildings and the dam belong to the same time period.
- The classification of the pottery dating back mainly to the 3rd to 4th dynasty (in particular dishes with bent rims, polished bowls, highly polished beer pitchers, bread moulds) allows their origin to be pinpointed to about 2550 BC.
- The dendrochronologically corrected results of a carbon-14 analysis of charcoal pieces taken from a fire-place in the workers' camp point to an origin between 2690 and 2950 BC

This dating to about 2600 BC to 2700 BC puts Sadd-el-Kafara among the world's oldest dams. The Jawa drinking water reservoir in Jordan is probably older and also diversion dams in the Kasakh river in the southern Soviet Union are possibly of an earlier date, according to Zonn in 1981. As these structures are, however, much smaller, the Sadd-el-Kafara may indeed with good reason, be called the world's oldest large-scale dam.

Dimensions of the Sadd-el-Kafara dam

From the track which leads from Heluan on the northern terrace of the valley into the wadi, the remains of the Sadd-el-Kafara can, at first, only be seen from a distance. The northern wing of the dam still extends about 24 m into the wadi and some 27 m of the southern wing is still preserved. The breach between the two wings is obviously wide enough to permit the passage of floods from the occasional heavy rainfalls.

In the course of recent road construction work, part of the southern dam wing has been converted into a ramp to permit motor vehicles to climb the 15 m from the wadi bottom to the valley terrace. Here, indications as to the design, the construction and failure of the dam have been lost forever. Despite anthropogenic encroachments of this kind and the effects of periodic floods and other natural processes of decay over the millennia, the original shape and structural details of the dam can still be clearly seen.

In cross-section, the Sadd-el-Kafara consists of three elements which differ in their composition and function:

- a core of rubble, gravel and weathered material;
- two rockfill sections on both sides of the core; and,
- the protection of the slopes, consisting of cut stones placed in steps.

The cross-section dimensions are shown in Fig. 1. From the state of the dam surface it can be concluded that the crest was intended to reach el. 125 m, but that this height was probably never achieved.

The inclination of the upstream and downstream slopes differ from each other. While the downstream side has a 30°slope angle, the northern remains of the dam on the upstream side clearly show different slope angles of 43° to 45° in the lower section and 35° in the middle section.

The shallower slope of the topmost steps, about 25°, was probably not intentional and is a result of the strong erosion.

Core and rockfill

According to the present-day definitions, the central part of the Sadd-el-Kafara can be termed an impervious core. It consists of rubble, gravel and weathered material. In terms of soil mechanics, the core material can substantially be classified as calcareous silty sand and gravel. The permeability was determined to be:

$$10^{-4}\ m/s > k_f > 5 \times 10^{-6}\ m/s.$$

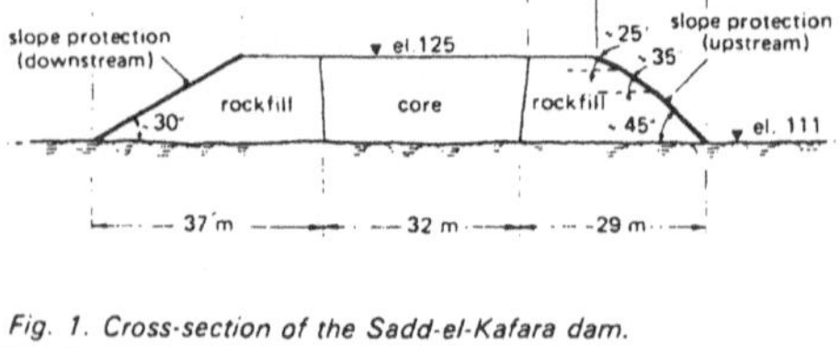

Fig. 1. Cross-section of the Sadd-el-Kafara dam.

As the fill material was chiefly brought from the valley terraces, it can be assumed that the filling progressed from the terrace edges towards the middle of the wadi.

On the upstream and downstream sides, the core is bordered by rockfill sections supporting and protecting the core. The structure of both rockfill sections is generally the same with a stone size of between 0.1 and 0.6 m diameter, mainly 0.3 m having been used. The colour and the mineralogical composition of the stones show that they were broken from the wadi banks upstream and downstream of the dam. As the stone rockfill was dropped haphazardly, the supporting shells give a picture of a very irregular structure. The cavities between the blocks were not filled up with gravel or rock debris.

The face between the irregularly filled core and the drop-fill rock supporting shells consist of carefully placed stones—dry masonry. As revealed by surface and local exploration works, the joints run in a straight line from bank to bank.

One remarkable detail at Sadd-el-Kafara is a distinct offset in the dry-stone wall of the southern dam wing. When the level of el. 118.7 was reached, there seems to have been a turning point in the construction. Not only were the placing and direction of the interface corrected, but also there was clearly an increase in quality and stability.

Slope protection

The facing of the rockfill sections is without doubt one of the most remarkable construction features of the Sadd-el-Kafara dam. On the upstream side, parts of the facing are still quite well preserved. However on the downstream side of the dam, only isolated stone blocks indicate that slope protection corresponding to that on the upstream side was planned and, at least partially, constructed.

On the upstream slope of the southern dam wing, only thirteen rows of stones near the crest are still partially preserved. In a considerably better state is the facing of the upstream slope of the northern dam portion, of which thirty-one steps still exist. The covering blocks are cut stones of different sizes (0.3 m × 0.45 m × 0.8 m on an average) which were broken from the wadi slopes directly upstream and downstream of Sadd-el-Kafara. The coarsely-hewn blocks are placed flat with their smallest face to the outside, forming terraced steps 0.3 m in height. According to profiles obtained by photogrammetry, the slope angle of the facing is 44° for the lower twelve rows of stones and 35° for the upper part.

The individual blocks of the slope protection were not placed directly upon the rockfill of the supporting shells or on the valley bottom. It appears that, possibly to facilitate the placing of the blocks, a layer of loose, weathered material was intercalated between the individual rows. To fix the blocks to one another and to maintain a level elevation, smaller stone chips were placed between and underneath the blocks.

Slope protection; upstream face of the northern wing.

The dating of the dam to the Old Empire is also confirmed by comparing of the covering stones with the limestone blocks of some stepped pyramids in Upper and Middle Egypt built during the middle 3rd to the early 4th dynasty[8]. For the construction of most of these pyramids, blocks with a maximum base size of 0.6 × 0.8 m to 0.6 × 1.20 m were used. The step height of 0.25 m to 0.4 m, too, is in accordance with the dimensions of the blocks facing the slopes of Sadd-el-Kafara.

The phenomenon of hollow block formation, which is typical of Egyptian structures built of coarsely-hewn stone blocks, is very pronounced in the case of the Sadd-el-Kafara. This decay process is caused by the combination of the local rock and climatic parameters which contribute to erosion. The different solubility of various minerals, small irregularities in the original block surface or small hollows in the place of chipped-off pieces of the shell are the points most vulnerable to attacks by chemical weathering. The block faces lying mostly in shadow are most prone to additional physical weathering in the form of spalling.

It is the areas not exposed to the sun which are most affected by weathering and produce the hollowing on the facing blocks. Its effects increase according to the principle of self-intensification: the deeper the weathering of the holes into the rock, the larger the shadowed areas and the longer the unevaporated water is retained and thereby able to chemically dissolve the rock. The weathering finally results in only hollow blocks remaining, according to Wilhelmy in 1958.

Progress of construction work

The visible shape of the layers in the fill material of the core provides an indication as to the way the dam was constructed. The angles of inclination of the strata are of the order of 28° to 34° and correspond to the natural angle of repose of the material used.

An analysis of the local findings leads to the following conclusions:

- The outer supporting shells, of rockfill, were built first.
- Following the construction of the supporting shells, the fill material in the core zone was placed together with the rising rockfill.
- Therefore along the supporting shells, there were layers inclined both towards the middle of the valley and perpendicular to the joint face.
- The subsequent filling of the central core was apparently not carried out according to a strict geometrical plan, but haphazardly.
- The construction of the dam as a whole and also the filling of the core was undertaken, generally, by starting at the wadi edge and working towards the middle of the valley.
- While temporary construction levels (in relation to the horizontal band of rubble limestone at el. 118 ±0.5 m) are observed in the upstream part of the dam, at least the downstream part of the southern dam wing was apparently raised continuously to full height by dropping the rockfill from the edge of the valley terrace towards the middle of the wadi.

The core and supporting shell material were placed directly on the stripped wadi bottom and the cleared wadi flanks. At all four lines of intersection of the dam facing with the wadi flanks, ramps of local fill material can be seen, which may have served during construction work as an access from the valley terraces to the wadi. Their removal would, therefore, have been the last step before

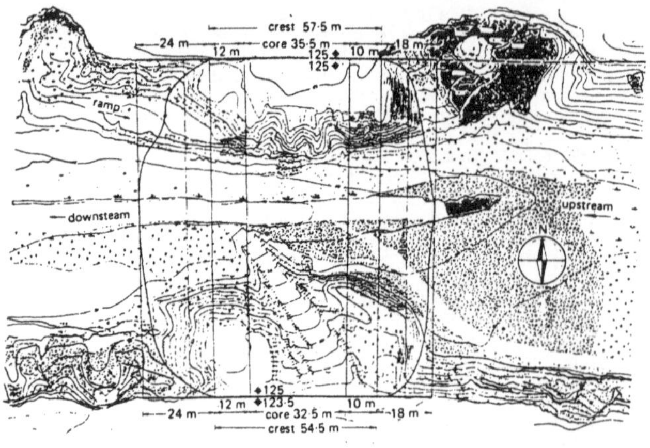

Fig. 2. Plan view to show reconstruction of the dam.

the completion of the dam.

In addition to these four small ramps, there is a ramp-shaped incline made with artificial fill on the downstream side of the northern dam wing. It begins in the region of two large quarries situated below the dam site in the right-hand wadi flank and ends in the core of the dam. From the flat slope (about 5°) and composition of the structure, it may be concluded that this ramp was used to bring up large stones to the required construction height, possibly by the use of transport sleds. The existence of the ramp and its use for the haulage of the slope protection blocks from the quarries to the dam also explains the fact that the completion of the rockfill at the downstream face of the northern wing was behind time as compared with the construction of the other parts of the dam.

Considering the work procedures and working capacity during the Old Empire, and taking into account the volume of fill material that had to be transported from the valley terrace to the dam core and also the amount of rockfill transported from the wadi flanks to the supporting shells, the construction period can be estimated to be of the order of 10 to 12 years.

Reconstruction of the dam

The results of the local investigations, the general survey, and the photogrammetric survey make it possible to reconstruct the envisaged final shape of Sadd-el-Kafara (Fig. 2). The dam was obviously intended to have a crest height of 125 m at the right bank so that it coincided with the northern valley terrace level. Further upstream of the dam axis on the north valley terrace, the ground rises almost immediately to about el. 126 m.

On the left bank, the southern valley terrace near the dam reaches an elevation of only 123.5. Therefore in the case of full impounding, the water would have passed around the dam on this side.

It is not possible to ascertain whether this effect (creating a spillway) was intended, or whether it was planned to fill the gap some 100 m in length up to the 125 m contour on the left bank by a small dam.

The total volume of the reservoir when fully impounded to el. 125 is of the order of 620 000 m^3. Below el. 123.5 m about 465 000 m^3 can be stored. Evaporation, assumed to be 11 mm/day, and percolation of the filled core would result in the emptying of the reservoir within 12 to 14 months. Under the prevailing climatic conditions and in view of the existing dam structure, long-term storage of water was thus not possible.

Assessments of the dam's stability by modern methods yielded the following results:

- in the most unfavourable case of a sudden lowering of the water level, the safety factor for the overall system is $\eta = 2.7$. It exceeds by far the minimum values $\eta = 1.40$ and $\eta = 1.30$ specified for today's standards.
- for the local stability of the upstream slope, a computational safety factor of only $\eta = 1$ is obtained. Here, however, the three-dimensional supporting effect and the stair-like stone facing, both of which imply additional safety features, have not been allowed for. The safety factor of $\eta = 7$ concerning sliding exceeds by far the required minimum value of $\eta = 1.5$.

It is established, therefore, that according to present-day standards, the Sadd-el-Kafara was oversized. The design of the first dam project in history was very conservative. This probably also indicates that no experience was available with structures of this kind during the time of the Old Empire.

Destruction of the dam

From local investigations and knowledge of the geographical conditions of the Wadi Garawi, it can be concluded that the Sadd-el-Kafara was constructed to absorb temporarily flood waves. Emptying of the reservoir would have taken a long time and would be by the processes of evaporation and percolation through the permeable dam body.

As to the filling of the reservoir, two questions are of interest:

- was the rainfall sufficient to fill the reservoir up to el. 125? and,
- what was the maximum discharge that could be expected at the dam site under the condition of a full reservoir?

According to the limited amount of meteorological data available and on the basis of measurements under comparable conditions (North Africa, Negev), peak discharges of 200 m^3/s to 250 m^3/s can occur when rainfall covers 35 per cent of the 170 km^2 catchment area. The probability of such an occurrence is 1 in 30 to 100 years. Discharges of between 50 and 130 m^3/s can be expected if a 1.5 to 5 year probability is assumed.

Owing to the relatively small storage volume of 465 000 m^3 at a water level of 123.5 m, precipitations of some 12 mm over the entire catchment area could result

Northern wing of the Sadd-el-Kafara.

in the reservoir being filled and the water flowing around the dam by the left-bank southern valley terrace "spillway". When the rainfalls cover only 35 to 50 per cent of the catchment area, precipitations of 15 to 18 mm (probability of occurrence between 2.5 to 4 years) will fill the reservoir up to this level.

Hydraulic calculations and model tests have shown the maximum discharge that can be handled by the left bank spillway is 85 m^3/s with the impounded water at the dam crest level of 125 m. If the flood discharge exceeds 85 m^3/s, water will not only by-pass the dam but also overtop the crest. In the case of overtopping, the critical discharge for the stability of the stepped downstream facing is of the order of 120 to 140 m^3/s — equivalent to a water level of 126 m.

Under these conditions, some 175 m^3/s flow around the dam. It can therefore be assumed that the completed dam would have had a safe "spillway capacity" of about 300 m^3/s and withstood all floods that could reasonably be expected.

Formation of hollow blocks on the face of the northern wing.

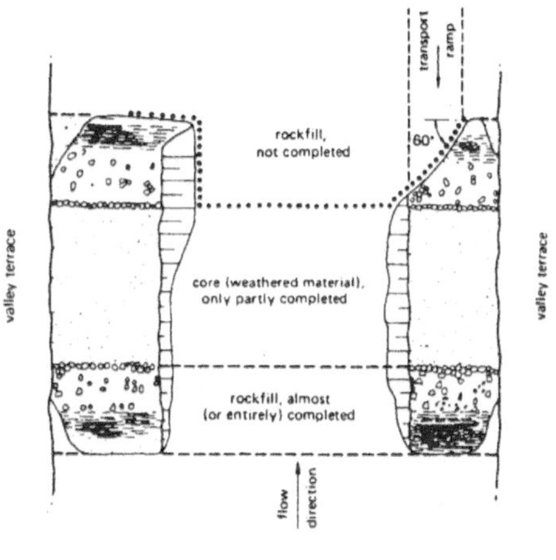

Fig. 3. Stage of construction at the time of failure.

The surveys, together with the state and the shape of the two dam wings, show that the dam was never completed. According to the local findings, it can be assumed that the upstream rockfill section of the dam was almost completed when the structure was abandoned or destroyed. It is also evident there was a large unbuilt gap in the downstream rockfill and perhaps in the core (Fig. 3).

The face of the downstream rockfill of the southern dam section towards the middle of the wadi evidently does not represent an eroded edge, or surface, as is the case on both sides upstream. It shows the state of filling to have been at the natural angle of repose when the construction work was stopped. The downstream side of the rockfill of the northern dam wing does not run transversely to the wadi but deviates from this direction by 60°. It is also inclined at the natural angle of repose of coarse rubble between 44° and 48°.

As the construction period of the dam was about 10 to 12 years and floods occurring every 2.5 to 4 years would result in the reservoir being filled, the dam's destruction during the construction period is very likely. The process of the destruction probably took place as follows:

- The upstream rockfill and its stair-like stone facing were almost completed up to the final height of 125 m.
- The left-hand southern downstream rockfill was almost up to its full height over a length of some 27 m (Fig. 3).
- So that the ramp on the northern bank of the wadi could be used for as long as possible for material transport, the construction of the northern downstream rockfill on the right bank had been delayed and lagged behind the general progress of work.
- The final state of the dam core in the middle of the wadi cannot be reconstructed, as it is completely washed away. It may have occupied any situation between the two rockfill zones and any height between the valley bottom and the final height.
- When this stage of construction work was reached, a medium or heavy flood occurred with a peak discharge of more than 100 m^3/s. The water spilled over the upstream rockfill and that part of the core already completed.
- The unprotected middle portion of the vulnerable core was quickly eroded and the upstream rockfill collapsed with the consequence of a sudden release of 600 000 m^3 of water.

A flood wave, with a severity unknown up until then, resulted in extreme devastation in the lower Wadi Garawi and in the Nile valley downstream from the Wadi outlet. The impression left by this catastrophe must have been so terrible and long lasting that not only the structure was abandoned but, for more than a millennium, Egyptian engineers did not risk constructing a major dam in rivers with unknown flood properties. The second attempt, this time successful, to create a water reservoir was probably the construction of the Quatinah dam on the Orontes, which is assumed to have been built under Sethi I between 1305 and 1290 BC.

Conclusion

The Sadd-el-Kafara, as the structure was called by the inhabitants, which can be translated as "dam of the unbelievers", because its origin dates back to pre-Islamic times, was investigated in detail for the first time in 1982.

The dam was constructed in the era of the Old Egyptian Empire, about 2600 BC, and thus represents, in the history of hydraulic engineering, the first attempt at storing water on a large scale. The dam's purpose was apparently to protect installations in the lower wadi and in the Nile valley from the frequent and sudden floods following heavy rainfalls.

By modern standards, the dam is stable and would have withstood the floods reasonably to be expected if it had been completed. However, a flood during the construction period led to destruction, followed by a catastrophic flood wave. As a consequence, the construction work was abandoned.

The design was basically correct and the dam was adequately protected from damage caused by water bypassing the dam on the left bank or overtopping. It was therefore a tragedy that an unexpected flood during the period of construction should have created such a disaster. Even after 4500 years and on the basis of all our present-day knowledge and experience, the greatest respect must be paid to the Egyptian engineers of the Old Empire for having had the courage to construct this dam with the limited technological means and the scarce scientific knowledge at their disposal.

They have all our sympathy as the failure was not caused by faulty design, but by natural phenomena which could not be foreseen at that time. □

References

1. Mackay, E., "Old Kingdom dam in Wadi Gerrawi", British School of Archaeology in Egypt and Egyptian Research Account, 18th year; 1912. (Also Heliopolis, Kafr Ammar and Shurufa by E. Mackay and W. M. Flinders Petrie, pp38-40.)
2. Murray, G. W., "A note on the Sadd-el-Kafara: The ancient dam in the Wadi Garawi". Bulletin de l'Institut d'Egypte, Vol. 28, Session 1945-1946, pp33-43, Cairo, Egypt; 1947.
3. Hellström, B., "The oldest dam in the world". Bulletin No. 28 of the Institution of Hydraulics at the Royal Institute of Technology, Stockholm, (Reprint); 1951.
4. Hellström, B., "The oldest dam in the world". *La Houille Blanche*, No. 3, May/June; 1952.
5. Hathaway, G. A., Dams, their effect on some ancient civilisations. *Civil Engineering*, S. 58-63., January 1958.
6. Klemm, D. and Klemm R., "Herkunftsbestimmung altägyptischen Steinmaterials". Report of Geländekampagne, September/October 1977. Also: Studies of Ancient Egyptian Civilisation, Vol. 7; 1979.
7. Garbrecht, G., "Wasserspeicher aus der Antike im vorderasiatischen Raum". *Proceedings* Wasser Berlin '81, pp797-822; 1981.
8. Dreyer, G. and Kaiser, W., "Zu den kleinen Stufenpyramiden Oberund Mittelägyptens". *Bulletin of the German Archaeological Institute*, (Cairo Section) Vol. 36, Published by Philipp von Zabern, Mainz, Germany; 1980.

Bibliography

Bulletin No. 81 of the Leichtweiss Institute for Water Research, Braunschweig, Federal Republic of Germany, 2 Vols.
Helms, S. W., "Jawa Excavations 1975. Third preliminary report, Levant; 1977.

2

The Roman dams of Subiaco

Norman A.F. Smith

Although the Romans are famous for their aqueducts, bridges, roads, harbors, and public buildings, their dam-building achievements have received much less attention, despite the fact that they were as competent in this branch of civil engineering as in the others. The purpose of this paper is to shed some light on the highest dam they built—the dam at Subiaco, depicted on the cover.

In 1855 J. S. Renier published in Rome a folio of drawings entitled *Imagerie du Sacro Speco*. The work was dedicated to Pope Pius IX, and is a collection of line drawings representing various frescoes and paintings in the possession of the Monastery of Sacro Speco[1] near Subiaco in central Italy, about 50 miles east of Rome. By all accounts[2] Renier's attempts to render the monastery's art collection as line drawings and to provide an appropriate commentary and analysis was a failure. As a study in the history of art the book is of no value, but for the historian of technology it offers something of great interest; this is shown on the cover, a detail from plate 37 of Renier's work (Fig. 1). The figure on the left is Saint Benedict, depicted fishing from the crest of a big dam.

Figure 2 shows the original painting from which Renier's line drawing is derived.[3] As far as can be ascertained it is the oldest surviving illustration of a dam. This painting, recently restored, hangs in the sacristy of the monastery; it illustrates incidents in the life of Saint

[1] Also known as the Monastery of Saint Benedict. The monastery is built around the cave in which the saint lived for many years; hence the alternative name "Sacro Speco."

[2] See, for instance, F. Hermann, *I Monasteri di Subiaco* (1, pt. 3:411). The present monks at Sacro Speco do not rate Renier's book very highly either.

[3] The author would like to thank the Monastery of Saint Benedict for permission to copy the painting and for much other help.

Benedict, including the years he spent meditating in a cave in the side of the gorge high above the River Aniene. Although we do not know who the painter was, the picture is generally believed to have been commissioned by Ludovico d'Aragona, Bishop of Majorca, in the year 1428.[4] Renier is typically incorrect here in ascribing the work to Magister Conxolus, a 13th-century fresco painter. Also, Renier's drawing is not particularly precise; the inclination of the saint's fishing rod has changed, the proportions of the arched openings in the dam are different, the masonry of the structure is incompletely drawn, and so on. But these discrepancies are not too serious, for our purposes,

[4] This is the view of O. Bonamore, *Il Sacro Speco e S. Scolastica* (Venice, 1884), p. 159, and was confirmed when the painting was restored.

FIG. 1.—This is pl. 37 of the book *Imagerie du Sacro Speco*, prepared by J. S. Renier and published in 1855. It shows Saint Benedict fishing from the big Roman dam of Subiaco while on the right-hand side stands Nero's villa. Saint Benedict appears three times in the drawing: as the fisherman; to the left as a boy with (perhaps) Saint Ignatius; and above that as a hermit in the *sacro speco*. Courtesy of the Victoria and Albert Museum.

even though they are surprising and (one would have thought) easily avoided. The real question is, what can the painting and the drawing tell us about the dam which once spanned the Aniene above Subiaco, very near the monastery of Saint Benedict?

In the middle of the 1st century A.D., Nero established a villa for himself on the banks of the River Anio (the Latin rendering of Aniene) in the gorge above Subiaco. Remains of this villa and

FIG. 2.—The painting which hangs in the Monastery of Saint Benedict (or Sacro Speco). The painting was commissioned in 1428 by Ludovico d'Aragona, Bishop of Majorca. The painter is unknown. Courtesy of the Monastery of Saint Benedict, Subiaco, Italy.

other Roman buildings still survive and have been written about from time to time. In order to add to the attractions of his villa, Nero constructed[5] three dams across the river to create three artificial lakes, and for three reasons these dams are of great interest. First, they are rare among dams, since they were built for recreational purposes rather than the usual utilitarian ones. Second, they were

[5] They are mentioned contemporaneously by Pliny the Elder in the *Natural History*, bk. 3, chap. 12 (Loeb Classical Library, Pliny, *Natural History* [London, 1942], 2:81).

the only dams (as far as we know) which the Romans built in their own country. And last, one of them was the highest dam the Romans built anywhere in their vast empire.

Unfortunately, all three structures have long since disappeared, and not much reliable information about them is available. But they are important enough, especially the biggest one, to merit some discussion. The lakes which the dams created were at different levels, and it seems likely that the upper reaches of the middle and lowest lakes reached up to the bases of the dams above. In other words, the lakes were built in a series of steps, rather like a set of canal locks. So far, it has been generally agreed that the first and third dams were small, forming reservoirs of modest capacity, but the middle dam was large and formed an equivalently large reservoir. This is the dam of greatest interest and is the one depicted in Ludovico d'Aragona's picture.

The diagram (fig. 3) shows the Aniene valley above Subiaco, the area of the dams. If we assume that the ideas of the 19th-century architectural historian Parker were correct, the positions of the two small dams were as shown in the sketch.[6] But what of the intermediate big dam? Its position has never been conclusively identified, but two possibilities have been suggested. The first possible site is directly under the Ponte di San Mauro which carries the road to Guarcino across the river. The other site is some 200 yards upstream from the bridge, spanning the gorge between the shell of Nero's villa on one bank and the remains marked "Roman Arches" on the other bank.

Of these two possible locations, the latter was the choice of Giovannoni.[7] When the road to Jenna was being cut in 1883–84, the remains of what Giovannoni believed to be the left abutment of the dam were unearthed. Giovannoni was able to reach a number of conclusions about the dam. Its crest, he said, was 13.5 meters thick and paved with tiles. So that all normal overflow would be constrained to discharge only at the center of the dam, the crest sloped toward the middle. A bridge providing access across the gorge to Nero's villa was carried along the crest of the dam on arches. Some of the approach spans of this bridge are still visible; Giovannoni has recorded their

[6] J. H. Parker, *The Archaeology of Rome*, pt. 8, *The Aqueducts* (London, 1876), pp. 57–60. Parker (1806–84) was the first keeper of the Ashmolean Museum, Oxford, and wrote widely on architecture. *The Archaeology of Rome*, in eight parts is one of his best-known works.

[7] G. Giovannoni, *L'Architettura nei Monasteri Sublacensi* (Rome, 1904), pp. 275–78. Gustavo Giovannoni was for many years the professor of general architecture, School of Applied Engineering, University of Rome.

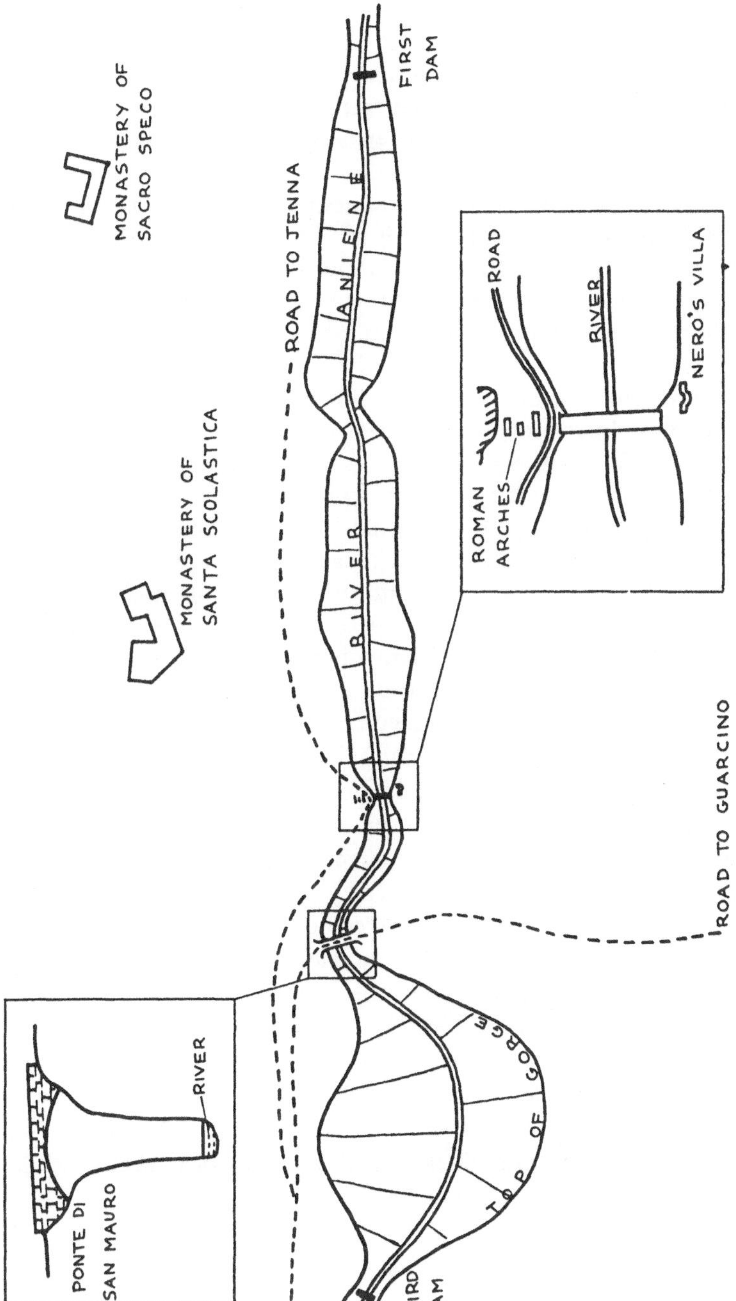

FIG. 3.—A plan of the River Aniene above Subiaco, the area of the dams. Drawn by the author and based on his own survey and other published maps of the area.

dimensions and details of construction.[8] They are made of concrete with travertine voussoirs interspaced with large tiles. The piers which supported these arches are built of massive blocks of travertine and brick-faced concrete.

Was Giovannoni correct in his conclusions? Sir Thomas Ashby at least was fully satisfied,[9] and there is no doubting the existence of the approach spans; they are still there exactly as described. Moreover, as the diagram shows, at the point where the Jenna road swings around the "Roman Arches," there is a distinct promontory jutting out into the valley. Givannoni claimed that it was here that the dam abutted against the side of the gorge, and Ashby agreed. The promontory is a natural formation, making the gorge particularly narrow at this point; this may well have suggested to Nero's engineers that here was a suitable place for a dam—a very big dam.

Unfortunately, the passage of time and the completion of the Jenna road make it impossible to repeat Giovannoni's examination. For instance, the tiled crest cannot now be seen, and nothing on the opposite bank by the villa survives to provide evidence of the dam's existence. However, it is worth remembering that if there ever was a dam on this spot with a bridge on its crest, as Giovannoni describes, then this bridge must have reached right up to the doorstep of Nero's villa. At first sight, Giovannoni's argument is a good one, yet it is not universally accepted. If he was wrong, however, what is to be made of the arches and the tiled pavement? Perhaps the arches were merely the vaults of a second and later villa whose remnants can still be seen. The tiled pavement may have been a promenade from which to view the river below and not the crest of a dam at all. But instead of speculating further on this point, let us consider the proposed alternative site for the dam.

A location under the Ponte di San Mauro was the choice of Parker;[10] this is also favored by the monks of Sacro Speco. The site is unquestionably a good one for a dam. Here the gorge is nearly as deep as at the Giovannoni locality, but is very much narrower. A dam as high as 160 feet could have been built here with a length of perhaps 100 feet at the crest and much less at lower levels.

The crux of Parker's theory is the existence of large lumps of the

[8] G. Giovannoni, p. 275.

[9] T. Ashby, *The Aqueducts of Rome* (Oxford, 1935), p. 253. Ashby (1874–1931) was the director of the British School at Rome from 1906 to 1925. His book on the Roman aqueducts is the standard work.

[10] Parker, p. 58.

dam[11] in the riverbed just below the bridge, which must have fallen there when the structure collapsed. This is a good point. When the dam failed (this will be discussed in a moment) what happened to the considerable volume of material of which it was built? A good deal presumably has been carried away over the centuries by builders. Some has been washed away by the river, which has considerable force when in flood. A little material remains in the riverbed to puzzle historians. While these surviving fragments may have once been part of a dam under the Ponte di San Mauro, they might just as easily have stood in a dam at Giovannoni's site. It is simply a matter of deciding how far they have been pushed by the torrent of the Aniene over a period of several centuries. The evidence of the riverbed fragments then is difficult to interpret.

However, it is interesting to note, that there is a block of concrete among these fragments. Presumably, the Romans used concrete for the core of the dam, the finished structure probably being faced with masonry which was their usual practice. A concrete core would have been a very sound proposition in such a big dam; this appears to be the first time that a Roman dam featured concrete in its construction. Later on, and especially in the 2nd century A.D., concrete was used more often, not only in dams but also in bridges and buildings, although its use was more common outside Italy.

The strength of the Ponte di San Mauro location, in my view, is simply that it is an eminently good site for a dam. The Romans were, when all is said and done, good civil engineers. Would they have chosen to dam a wide emplacement, as Giovannoni believed, when a narrow one, under the San Mauro bridge, was available?

Before appealing to the Bishop of Majorca's painting for evidence, some information on the history of the dam needs to be set out. In A.D. 38, Caligula began the construction of two new aqueducts to Rome, which were completed by Claudius in A.D. 50 and named "Claudia" and "Anio Novus." Originally the Anio Novus drew its water directly from the River Anio; then, because this procedure produced a muddy and discolored supply, a settling reservoir was built, but it was not entirely successful. We have Frontinus's evidence that, during the reign of Trajan, modifications to the Anio Novus's intake were undertaken; Frontinus writes:[12] "He [Trajan] also recognised the possibility of remedying the defects of New Anio,

[11] Plate 4 in Parker's book shows these remains; they are still visible today.

[12] Sextus Julius Frontinus, *The Aqueducts of Rome*, bk. 2, sec. 93. (Loeb Classical Library, Frontinus, *The Strategems and the Aqueducts of Rome* [London, 1961], p. 421.)

for he gave orders to stop drawing directly from the river and to take from the lake lying above the Sublacensian Villa of Nero, at the point where the Anio is clearest." Nero's dam now began serving a useful purpose, supplying water to Rome; presumably it continued to fulfill this role so long as the Anio Novus was in commission.

What next happened to the dams is largely a mystery. Medieval documents mention[13] two lakes at first and then one, implying the ultimate disappearance of the first and third dams, as already mentioned. Documents of the 12th and 13th centuries constantly use the name *pons marmoreus* when referring to the remaining dam, the big middle one. This is extremely interesting, since it lends support to the ideas of Giovannoni, who found traces of a marble parapet during his excavations of the dam and bridge.

In 1305 the dam collapsed. Contemporary records[14] place the blame fairly and squarely on the shoulders of two monks who took it upon themselves to remove stones from the dam, apparently in an attempt to lower the level of the lake which was flooding their fields. Probably the dam was not wiped out at one stroke; some decades were presumably needed for the river to carry away the bulk of the structure, aided no doubt by the scavenging of local builders over a long period. Eventually nothing was left but the fragments which lie in the riverbed and, according to Giovannoni, the remains which were uncovered by the Jenna roadbuilders.

So when Ludovico d'Aragona commissioned the picture in 1428, the dam's failure had occurred 123 years before. Nevertheless, in my view the picture is valuable in throwing light on the nature of the dam and its position. The painting shows a straight masonry wall with two openings through which pours the overflow from the reservoir. These openings presumably represent the arches of the bridge which the Romans built along the top of the dam. At the right-hand end of the structure and immediately adjacent to it stands Nero's villa (see fig. 1). Moreover the painting indicates that the top of the bridge was at a level just below that of the villa.

Thus we find that the painter agrees in principle, if not in detail, with the views of Giovannoni; both men agree on the layout of the structure, its position on the river, and its height relative to Nero's villa. Almost certainly the painter's dam is short of arches; while he could not have known how many there had been on those parts of the dam which had collapsed, it is disturbing that he does not even indicate arches near the left-hand end, the ones which Giovannoni

[13] See Ashby, p. 255.

[14] For details of these, see Ashby, p. 255, n. 2.

mentions. Two other points must be considered. First, even though the dam had been in ruins for 123 years, surely the painter must have known where it once stood. Enough must have been visible in 1428 to leave him in no doubt as to the dam's location if not the details of its construction. Second, if the dam was built at the position claimed by Parker, where the Ponte di San Mauro now stands, Nero's villa should not appear in the painting right at the end of the dam. Not only is the villa a long way from the Ponte di San Mauro, it is not even visible. In conclusion, while it cannot be positively proved that the Aniene gorge was dammed in the way Giovannoni described, the evidence is in his favor. On balance, the writer is inclined to believe that Giovannoni was correct. Of course the problem of explaining Parker's theory of a dam under the Ponte di San Mauro remains.

One opinion has been withheld from the story so far. Giovannoni believed[15] that there could have been a dam under the Ponte di San Mauro as well, but not the big dam that formed the central lake. He believed it to be the third and last dam in the series, and that it was not very large. In figure 3 it will be seen that the third lake, according to Parker, was roughly circular in plan; he believed that this basin had been deliberately excavated by Nero's engineers. Even today, despite thick vegetation, it is evident that, at some time, the valley has been opened out in the area between the bridge and Parker's supposed site for the third dam. It is my belief that this excavation was not made merely to create a pool, but to provide a source of building material, that is, a quarry. The three dams must have required a considerable volume of stone; the Romans would not have carried all this masonry very far if a local supply had been available. Hence, a stone quarry close to the dams is an eminently reasonable notion. A huge amount of excavation would probably not have been undertaken just to make a lake, but it would have been unavoidable in order to open a quarry.

A further point substantiates the view that the dam under the Ponte di San Mauro was the third in the series. As mentioned above, the gorge at this point is very narrow and the required dam could have been constructed with comparative ease. Nor would it have needed to be very high because the required lake would only have had to reach a short distance upstream to the foot of Giovannoni's dam. If it is assumed that the big, middle dam was located where Giovannoni suggests, the site under the bridge is the most logical one for the lowest dam; below it was the quarry.

[15] Giovannoni, p. 278.

From the available evidence, the layout which has now been described seems the most likely one. The location of the first dam, well upstream, has never been disputed. The evidence of Ludovico d'Aragona's painting, the medieval references to a *pons marmoreus*, the excavations of Giovannoni, and the nature of the site itself, all support the likelihood that there was a big central dam close to Nero's villa. Below it and under the Ponte di San Mauro stood the third dam, not very high and set deep down in the gorge. Below that was the circular excavation used as a quarry; beyond this I suggest there was no dam, despite Parker's claim. In any case, the evidence for a third dam below the excavated area is not very substantial. It depends on the interpretation of local nomenclature, while Parker's own opinion is confused, and relies to some extent on the notion[16] that the third lake was used to supply the Aqua Claudia. In fact, the Aqua Claudia was springfed and not connected to the river at all.

Perhaps in the future more explicit documents or more conclusive illustrations will be discovered. In the meantime, it would be instructive if some excavations were carried out at all the suggested dam sites with a view to discovering where the dams once stood, how big they were, and of what they were made.

However, one final question must be considered. Of the three dams at Subiaco, the middle one was the highest the Romans ever built. But exactly how high was it? The answer is that we do not know and perhaps never will for certain. Giovannoni quotes a figure of "perhaps 40 meters" for the overall height of dam and bridge combined. This figure, however, does not accord with the rest of his theory. If, as Giovannoni believed and the painting appears to confirm, the top of the bridge was level with the present Jenna road at one end and the base of Nero's villa at the other, then the total height of the structure must have been greater. To judge from my examination of the gorge, the figure should be nearer 50 meters. But this is still no help in fixing the height of the dam, because one does not know how high the bridge was.

Parker, in putting forward his idea of a big dam under the Ponte di San Mauro, raises an interesting possibility by referring to the level of the water channel in the Anio Novus. From this it should be possible to calculate the height of the dam, because we know that near the end of the 1st century A.D. the aqueduct's intake was moved to the lake. Unfortunately, however, the intake has long since disappeared together with an unknown length of the channel. The first signs of the course of the aqueduct are in fact near the Ponte di

[16] Parker, p. 60.

San Mauro.[17] Precise leveling from this point, then, could be used to determine a minimum figure for the height of the dam but nothing more,[18] because it is impossible to work out from the existing remains of the water channel how much higher the intake was. Even so, we can be sure of one thing; the remains of the Anio Novus are high above the Aniene's present bed, and the crest of the dam could not have been any lower than this point on the aqueduct. In my view, it seems unlikely that the dam was less than 130 feet high. At the same time, we can confidently rule out Lanciani's claim[19] that all three dams at Subiaco were 200 feet high. Such heights would have been not only unnecessary, but topographically impossible.

Had the big dam of Subiaco survived and if we take the estimate that it was about 130 feet high, it would have been easily the highest dam in the world until 1594, when the Alicante dam, 134 feet high, was built in Spain. This structure held the record for height for almost 300 years, and well into the 19th century. That is some measure of the Roman's achievement.

17 Ibid., p. 59 and pl. 6.

18 Neither can the position of the dam be fixed by the remains of the aqueduct. The first signs of the aqueduct are consistent with a dam either at the Ponte di San Mauro site or at Giovannoni's site. But remember also what Frontinus said: "to take [water] from the lake lying *above* the Sublacensian Villa of Nero" (see n. 12). If anything, Frontinus is supporting Giovannoni at this point.

19 R. Lanciani, *Ancient Rome in the Light of Recent Discoveries* (Boston, 1889), p. 274.

3

The chapter on weirs in the Codex of Juanelo Turriano: a question of authorship

J.A. Garcia-Diego

In paying posthumous homage to Ladislao Reti, Bern Dibner made reference to the articles Reti published in this journal concerning the remarkable manuscript of Juanelo Turriano preserved in the Madrid National Library.[1] This manuscript, the Turriano Codex, owes its vital importance to the fact that it is the earliest known treatise on hydraulic engineering. It is, furthermore, of a high technical standard, to the extent that its author may be fairly described as a "rationalist." He bases his arguments, wherever possible, on his own experimental deductions, instead of on tradition; he thus applies to technology that philosophy of science which was born during the Renaissance.

This codex so far has been attributed without question to Juanelo Turriano, a figure who, although not properly studied, is nevertheless famous both as a clockmaker and manufacturer of automatons (in which capacity he served the emperor Charles V) and for the waterworks which he designed and built at Toledo. However, the section of Benito Bails's report to the royal librarian, Señor Santander, which deals with the purchase of the codex raises certain doubts in my mind about the authorship of the manuscript.[2] Interestingly enough, Reti also raised this same question; the main reason for his skepticism was the title page, sketched several decades after Turriano's death.[3]

When I examined the codex carefully—with special attention to the chapter on weirs—I concluded that the attribution to Juanelo clearly

[1]Bern Dibner, "Ladislao Reti (1901–1973)," *Technology and Culture* 15 (1974): 440–42.

[2]Eugenio Llaguno y Amirola, *Noticias de los arquitectos y arquitectura de España desde su restauración,* 4 vols. (Madrid, 1829), 2:250–58.

[3]"The Codex of Juanelo Turriano (1500–1585)," *Technology and Culture* 8 (1967): 53–66.

must be mistaken. I mentioned this to Reti in 1972—when he had set aside his studies on Turriano in order to devote his full energies to editing the Madrid Codices of Leonardo da Vinci; he told me that he had realized this for some time. Then in 1973, a month or two before his untimely death, while we were working together on a possible joint publication on the waterworks of Toledo, he expressed the wish that we might soon reveal our common conclusion that the codex was not really the work of Turriano.

In publishing this article, I trust that I shall have duly fulfilled Reti's wishes. Although his dating of the manuscript to the years 1564–69 may call from some revision, I feel that this will not need to be very drastic.

* * *

When I decided to write a critical study of one of the *libros* of which the codex is composed, I finally decided on the one dealing with weirs, even though there are others that are more impressive, both in style and in graphic illustrations. I opted for weirs partly because I have already published other works on the history of this type of structure, but mainly because the chapter in question appeared several centuries before any other similar work.[4] As regards the design and building of weirs, one actually has to jump from Turriano's codex into the 19th century and the publications of the U.S. Bureau of Reclamation (*Low Dams* [1938] and, later, *Design of Small Dams* [1960]), which either in the original language or in various translations have become by now the daily tools of most hydraulic engineers.

What is more, only very recently has the history of dams and weirs begun to arouse the interest that it deserves. Norman Smith's *History of Dams,* published in 1971,[5] is the first book entirely devoted to this subject. This is an amazing fact when one considers the vast number of books that exist on far less interesting and less important subjects. Smith's contribution is of such great significance that it deserves to be pursued and expanded, especially because it is mainly based on experience gleaned when visiting ancient structures or their remains.

However, Smith's is not a completely appropriate approach, at least for weirs. What frequently happens is that a weir structure is gradually adapted in the course of its life so as to keep up with technological developments. It is probably first used for irrigation, then to provide

[4]Francesco de Giorgio Martini has a section on dams and weirs, but much briefer and poorer than this. References in Leonardo da Vinci manuscripts are scarce.

[5]This is the London edition. The book was published in the United States in 1972 (Secaucus, N.J.) and reviewed by Carl W. Condit in *Technology and Culture* 14 (1973): 621–22.

water for driving a flour or fulling mill or possibly an ironworks, and finally for a small power station. As a rule, this evolution means that the structure loses its original form. A storage dam, on the other hand, may serve the same purpose for many centuries, as is the case with several in Spain.

Consider the example of one of the weirs on the Tagus meander, where it winds around the city of Toledo, reference to which dates back to the 12th century and which, in all probability, stood on that spot even earlier.[6] All that remains today is a structure which can be dated only to the 1920s; and even that structure is broken down and serves no purpose at all, since the power station on one of the abutments has been superseded long since, as the city receives its power from a single supply source. This weir undoubtedly overlies the remains of a whole set of weirs, each of which has broken at some stage in time. Their historical background could only be studied as an exercise in archaeology, invoking their scientific interest; or their complete demolition—at present being contemplated—could be made good use of in order to obtain and preserve as complete a graphic record as possible.

This explains why one is bound, when studying weirs, to resort to the "Theatra machinarum" and other similar published or unpublished works, as well as to the rare original outlines or descriptions that have survived. In this respect, then, "Turriano's" Madrid manuscript is of vital importance, since it can probably provide more information than all else that has been discovered so far.

* * *

The part of the codex dealing with weirs is entitled "Libro 9 que trata de diversos modos de azutes o presas de rios" ("Book 9, which speaks of various types of weirs or river dams"), the reference to river dams having been inserted by another hand than that which wrote what goes before (fols. 153v–180r). *Azuta* (*azud* in modern Spanish, translated here by "weir" or "small dam") means a diversion dam, the main task of which is to raise the water, rather than to store it. Neither in this chapter nor anywhere else in the manuscript is the slightest reference made to storage dams, even though Spanish technology in this field was the most advanced in the world at that time, and Turriano is held to have been engaged then in the building of one of the more important ones.

The chapter in question runs to approximately 11,000 words and

[6]Saelices Mill, mentioned in 1156 in deeds concerning the sale of a vineyard (Julio Porres, *Historia de las calles de Toledo,* Diputación Provincial, Toledo [1970], 2:131).

contains thirty-seven figures. As elsewhere in the codex, these illustrations combine simplicity with accuracy, although, in view of the subject covered by this part of the manuscript, they tend to be somewhat less attractive than those showing machines and the workers operating them.

Figure 1—my own; all the other illustrations are from the codex—shows the different types of weirs. The order, from left to right, is more or less the same as that which would be used in a modern treatise, not the one given in the manuscript. The text offers relatively few difficulties, as compared with some of the other chapters.

The two simplest types of weir raise the water enough for a small amount to be fed off into one or more irrigation ditches. This can be achieved either by simply using stones or by means of wooden piles and stones (fig. 2); in both cases, the basic materials are bound together with branches, tufts, or grass and weeds: this is the obvious explanation why the latter—and I wonder why not the former, too—is called an *azud de selva* (literally, "forest weir") since it can only be constructed where there is abundant vegetation. These structures are not intended to last but have to be rebuilt whenever they are carried away by floodwaters.[7]

Even in these extremely rudimentary types, it is recommended that the dam be at an oblique rather than a right angle to the river's course. This is a constant feature of Spanish straight weirs, based on the argument that reducing the height of the overflowing water has the same effect on material stress. Another practical rule for these weirs is established in the codex: in comparing channels *E* and *G* in figure 2, channel *G* is shown as the more suitable, since it will suffer less sedimentation and be less prone to erosion effects.

The author then goes on to describe weirs that are mainly designed for first raising the water and then diverting it. He examines wooden weirs and those built of wood and stone. In both instances, he recommends that the height be between 1/9 and 1/6 of the base length, as the author is of the view that a proportion of 1/4 would prove "algun tanto agrio y muy enhiesto ("somewhat distasteful [literally, "sour"] and overly tall"; all translations from the codex are my own). The small dam is anchored to the ground by means of a series of sharp pointed or metal-tipped stakes. Since the main beams—those at the

[7]This simple technique has survived. The set of articles published under the editorship of Elwood Mead bearing the all-embracing title *Report of Irrigation Investigations in California* (Washington, D.C., 1910) contains a description, with a photograph, of a weir like the first one described here, on Kings River, California (p. 314 and fig. 27). The name "cobble and brush dam" and the remark that such dams should be "repaired after each freshet" tally precisely with what the codex has to say on this subject.

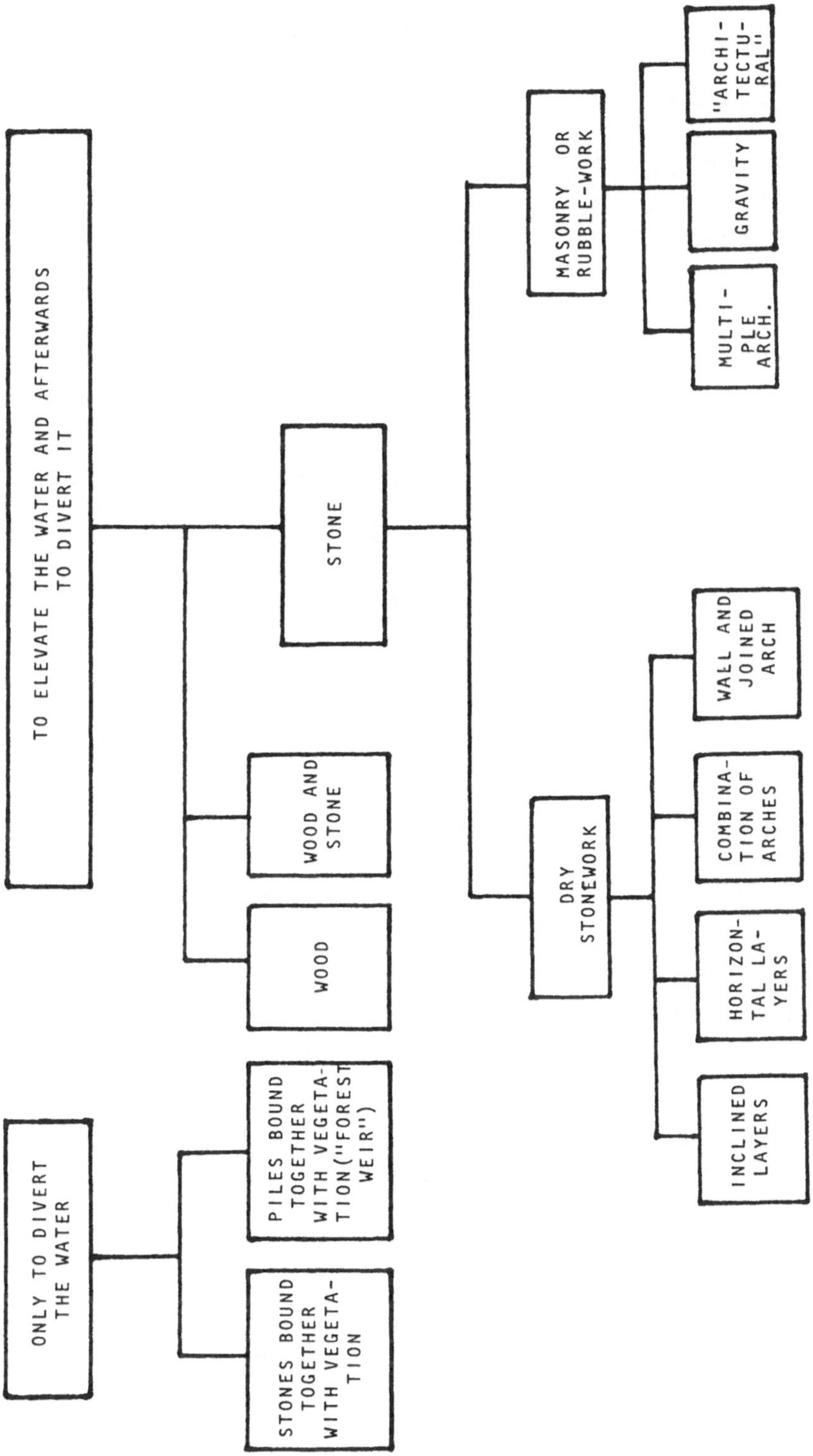

FIG. 1.—Different types of weirs

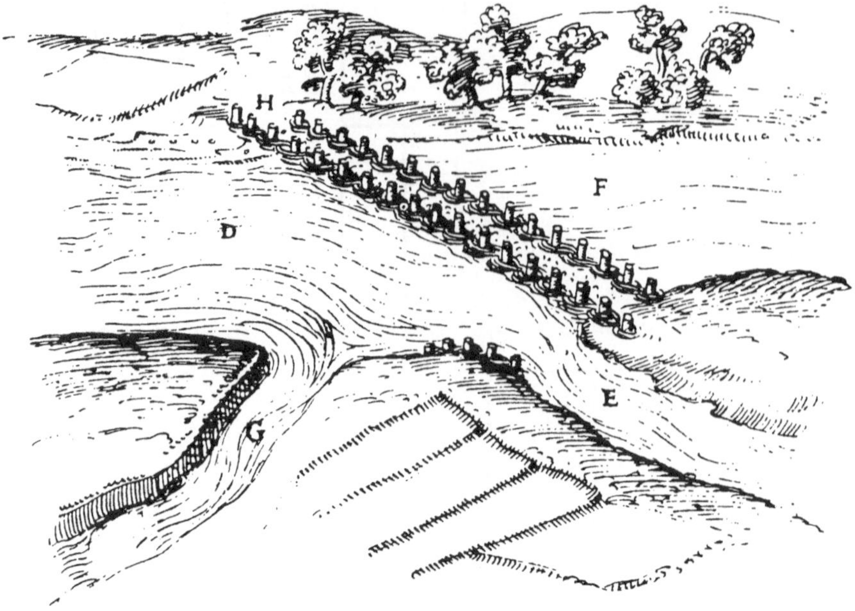

FIG. 2.—"Forest weir"

base and those running from the dam top to the ground—are extremely long, they must be made of several pieces, so the author describes several ways of joining them together. Horizontally, the weir can be straight, arched, or have two straight parts set at an angle to one another; the latter two designs are indicated as preferable.

The most complex example of wooden weirs given by the author is shown in figure 3, although it is pointed out that many parts have not been included in the drawing, for the sake of simplicity. (I have added some schematic illustrations of other varieties of cross-section [I to IV].) The weir is made watertight by covering both faces with wooden paneling. This arrangement offers the advantage of not only sometimes allowing the river to be crossed through the interior of the weir but also enabling leaks to be traced and the damaged part repaired immediately. In the case of wood and stone weirs, on the other hand, some water is always bound to seep through the stones (because, as we shall see below, they are merely lodged against one another). Such reasoning is of great interest because of its essentially modern concept. What it actually calls for is what we should today describe as a highly efficient maintenance service: if a defect appears in one of the faces under floodwater conditions, the crack or hole has to be sealed up immediately from the inside, lest the weir be flooded and very

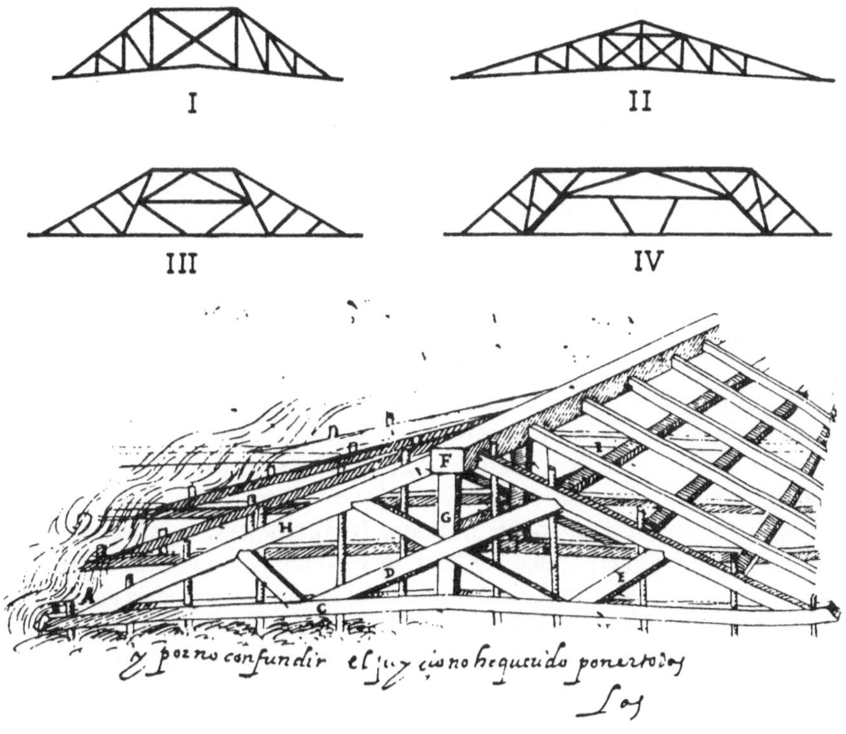

FIG. 3.—Different frames for wood weirs

probably destroyed. The system also means that carefully selected types of wood have to be used. Further on in the book, the author himself admits, "Es necesario que les estén remendando de día en día" ("They need to be patched up from day to day").

The foregoing would not seem, in my view, to tally with common industrial practice at the time of the Renaissance and even less in Spain, where wood—despite the fact that forests were much more abundant then than now—has been employed only on very rare occasions as the basic element in construction work. Even today one finds few wooden houses in the country, and even workers' barracks on rural construction sites (where timber would provide a most convenient solution, being easily transportable elsewhere when the job is finished) are built out of bricks or blocks or concrete. Stone, on the other hand, is prolific in almost every region of Spain, and the dressing—for which there is an excellent tradition of long standing—would not have proved too costly in days gone by. It is therefore my view (considering that the number and details of the solutions prevent one's presuming this to be a case of pure theoretical speculation) that the author drew his inspiration from other countries, perhaps in central Europe.

Similar designs are to be found (and this is surely not the only instance) in a most practical American work which appeared several centuries later: James Leffel's *Construction of Mill Dams* (Springfield, Ohio, 1874).[8] This author's wooden dams observe the same basic principles—using piles (when rock cannot be found near the surface), inner trusswork, and boarded faces; a more extensive comparative study would probably bring many further cases of coincidence to light.

Leffel recommended this system "in many localities where stone is not readily obtained—which is the case in a large portion of the Western States." This prompts us to broach a somewhat daring hypothesis: that the predecessors of Leffel's structures may well have been built by Spanish colonizers. Smith, however, claims that the earliest American weirs are those of New England and date from the 17th century; he considers that they reflect the techniques imported by immigrants hailing from other European nations. Nevertheless, in such a vast country, there is surely space for both possibilities to exist side by side.

Only one type of compound wood-and-stone dam is described in the codex (figure 4). It can be seen how the ashlars are placed in alternating directions inside the wooden framework, thereby forming

FIG. 4.—Wood-and-stone weir

[8]A facsimile of the second edition of 1881 has been published (Noyes Series on History of Technology, no. 1[Park Ridge, N.J., 1972]).

courses in a direction transversal to that of the river. He recommends placing further stones on top of the framework and making the dam watertight by laying esparto grass beneath them. Ancient weirs of this type are still be be found in Spain, although, as I already mentioned, only part of the outer wooden framework is generally visible, due to the fact that coverings of a later date hide the core. In Leffel's book, these weirs are described as "pile and boulder dams."

There follows a description of stone weirs, again beginning with the simplest types: dry-stonework weirs, where no lime is used to bind the stones together. This is a very ancient technique, of which the Roman aqueduct at Segovia is a well-known example. In dams, however, such an arrangement affects impermeability adversely, so the system is only recommended in the case of rivers with little flow. Such weirs are built in sections and involve waiting for the sludge to fill in the joints between the blocks of stone before going on to the next section. Once the weir has been completed, the whole reservoir is filled with sediment. The codex provides four examples of this type of small dam —somewhat impractical nowadays, but in the 16th century people were not in the habit of relating time taken in building to interest on capital.

Although the foregoing are actually all small-gravity dams (i.e., the dam's own weight resists the forces exerted upon it) I have classified them separately, leaving this specific name for the final solution, which is nearest to the present-day type. The first consists of laying courses of stone blocks which rise in a direction opposite to that of the flow. This is followed by a type (fig. 5) which has horizontal courses and stepped faces both upstream and down. The third type is more complex and is described, moreover, in rather imprecise terms. The

FIG. 5.—Dry-stonework weir with horizontal layers

author claims that it offers far greater advantages than the previous two: it is a hollow structure with a series of diminishing arches.

But the fourth is really the most important (fig. 6). Here we have a description of a small-gravity dam embodying all those features which would be required for irrigation or power production, all clad in clearly modern garb. In figure 6, *A* marks the entrance of the irrigation channel, which is duly equipped with its spillway and wooden gate. To the right lies the main spillway, *D*. Under exceptional high-water conditions, however, the author envisages the water flowing over the dam, for which purpose he has included a wall (*G*) about 1.7 meters behind the arch, to ensure that the foundations of the whole structure will not be destroyed. The dam is so designed that the author claims—quite rightly—that a man could stand beneath the arch without getting wet from the free-falling water, although his feet in fact would be standing in the water collecting at the bottom.

The writer then goes on to examine those weirs in which the basic elements form a monolithic whole. This is in fact the type used at present, with the exception of earth or rock-fill dams. He calls these structures *azudes de piedra o presas* ("stone weirs or dams"). This is the first time that thé uses the latter word, and it should be pointed out that, nowadays in Spain, the choice of the word *azud* or *presa* does not depend on the building material being used. In fact, the word *azud* ("weir") is used for diversion dams (usually of reduced height) and "*presa*" ("dam") for storage dams, although in almost every instance the latter word tends to replace the former.

FIG. 6.—Dry-stonework weir with wall and joined arch

In the gravity dam shown in figure 7, the plan is composed of sections in different directions, an arrangement which, to my mind, does not seem justified. The author holds that it is particularly suited to irrigation purposes and goes on to say that it has the appearance of a sort of fortification (*baluart* = "bastion"). One may, however, consider that its most interesting feature is its thoroughly modern-style hydrodynamic profile, at least until one notices that it is quite illogically on the upstream face. However odd this may appear, it nonetheless forms part and parcel of a tradition which survived for many centuries, as can be seen in the Basque dams, which are described as "ancient" by Villareal,[9] writing in the 18th century. These also have

FIG. 7.—Masonry gravity weir

[9]Pedro Bernardo Villareal de Berriz, *Máquinas hidraulicas de molinos y herrerias y govierno de los árboles y montes de Vizcaya* (Madrid, 1736), facsimile ed. (San Sebastian, 1973), with an introduction by myself. Reviewed by Thomas F. Glick in *Technology and Culture* 16 (1975): 293–94.

the upstream face sloping and the downstream one vertical, so as to allow the free fall of water; the codex clearly adopts the rational solution only in the case of buttress dams.[10]

The author recommends straight gravity dams for small streams. These involve building the various elements one after the other while the river is dry, as shown in figure 8, until the whole structure is completed. The gaps (*V*) are sealed with logs slid down grooves into position. Then the hollow core of the dam is first filled in with earth and finally with stones, once the earth has settled. This keeps the dam watertight even when, with time, defects appear in the timberwork.

As I see it, figure 9 is the most interesting example of all—and in my opinion artistically the best in this chapter—for it provides a very complete picture of a multipurpose hydraulic structure. The author states that it may be used for big rivers. The dam is straight, with an opening (called a *puerto,* literally, "harbor") for boats and rafts to sail through, guided by two piers (called *mesas,* or "tables") which have to be quite long, in order to reduce the dangers involved in getting through. Part *F* is allocated for the unloading of timber borne down by the river and also leads to an area reserved for fishing. There is also a diversion canal to be used for irrigation or driving mill wheels,

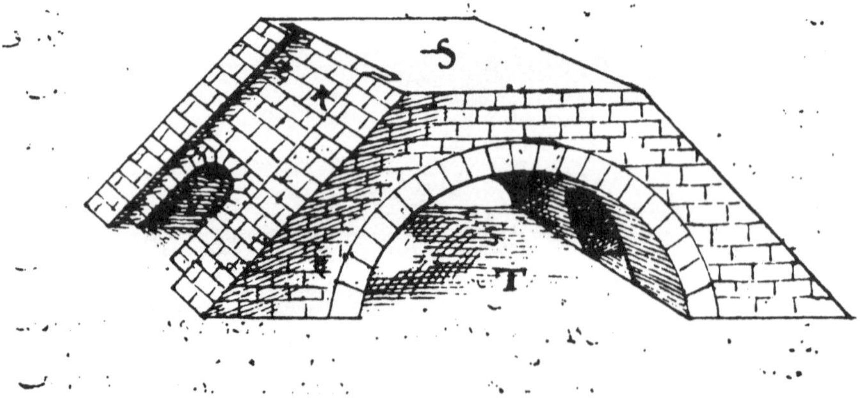

FIG. 8.—Element to be filled with earth and stones

[10]In an even more elaborate drawing (fol. 174v), the hydrodynamic profile is missing. But the central part is curved, instead; and there is also a certain amount of anamorphosis, which in the case of such an excellent draftsman would seem to be intentional, although I cannot understand why. Of course, considering its reduced height in comparison with length, neither this nor the other weirs described make a structural use of the arch effect. The latter merely serves to lengthen the crest, as in oblique dams. It may also be that the arrangement creates a sense of safety in the mind of the designer—or his customer—based on the prestige that arches enjoy as resistant ele-. ments in architecture.

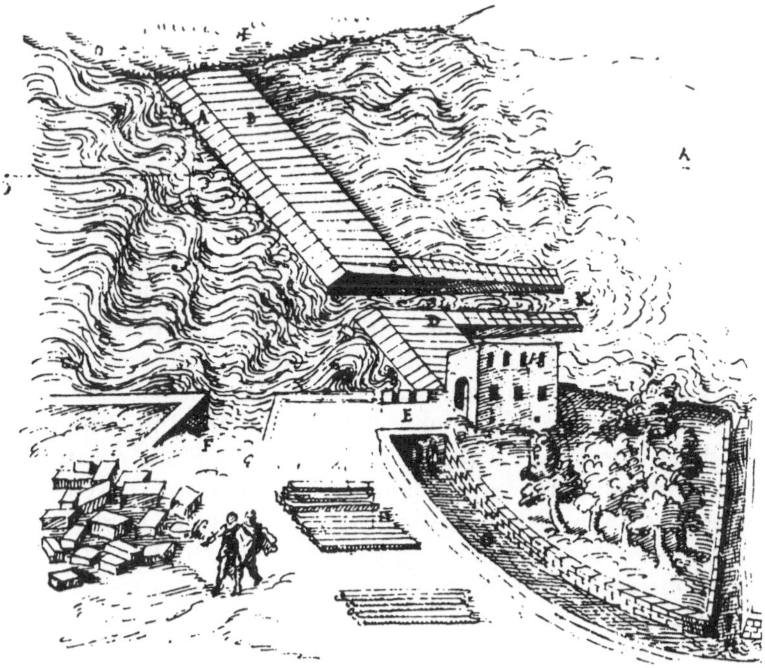

Fig. 9.—Multiple-purpose hydraulic system

provided with a branch channel (*HI*), which the author indicates could be used for draining the canal so as to leave it dry; but surely it would serve also as a spillway during peak flow periods.

The last example from this section is a weir provided with an inner passageway for crossing the river (which could also be filled in with earth). Curved in plan, with the foot of the dam protected against erosion, this is a simple and highly functional structure particularly recommended by the author of the codex.

* * *

So far, the author has made no claim to having invented himself any of these structural types; but now he does, quite justifiably, as he provides us with the first description of a multiple-arch dam. With understandable pride, he declares, ". . . Yo he pensado un modo de açute que me parece que haziendose en esta manera que jamás podrá tener fin por causa de la forma y invención que en ello hay . . ." ("I have thought out a fashion of weir which I believe would last for eternity if built in this manner, considering the shape and invention behind it").

This type of dam is recommended by the author for use on large rivers; as regards the foundations, the reader is referred to that part of the codex devoted to bridge piers,[11] in which dry working is recommended, even if this means diverting the river, rather than trying to build in the actual river flow, as tremendous difficulties then arise: ". . . Es necesario ir mudando los trabajadores que sacan el agua por que ninguno puede durar el trabajo de todo el día, quando más el de 1/2 noche solo el considerar esto pone espanto y terror . . ." ("One has to keep replacing the workers engaged in removing the water, because nobody can withstand such conditions of labor for a whole day, still less half the night; the mere contemplation of it fills one with fear and horror"), as well as other considerations which I shall not quote here. However, if conditions really call for such an arrangement, the weir has to be designed to be built with ashlar masonry and not with rubblework; it needs, then, to be a more carefully constructed and costly structure. What is more, the stones have to be held together with *garfas de hierro o de bronce* ("iron or brass hooks"). On dry soil, the author recommends using concrete for building the weir.

The main advantage of this type of structure, he says, is the fact that its sloping upstream face helps ensure stability, because of the favorable influence of the weight of the water. I do not think that this rational concept appears again until the 18th century, and even then not so manifestly. Indeed, nowadays one still comes across designers of buttress dams who have failed to make use of this obvious advantage.

The arch spans are wider as the height of the dam increases, and the foundations are as deep as the dam is high. The overall design is illustrated by two perspectives (one of which is shown in fig. 10) and a

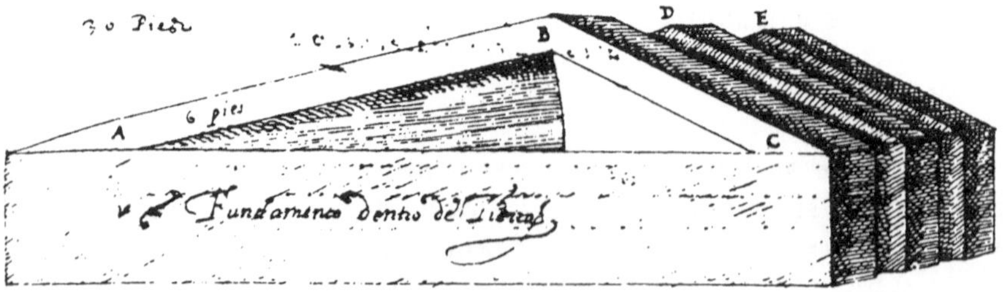

FIG. 10.—Multiple-arch weir

[11] Bk. 18, "De como se han de hazer las pilas de los puentes de piedra en distintas maneras," folio 360r–393v.

plan. Some dimensions have been omitted, and the projection is by no means orthodox.

While reconstructing a series of ancient works in Toledo, I myself designed a small dam of this type, following the norms provided in the codex, proposing that it should replace the present Romaile weir, of which only the ruins of a series of superimposed structures remain.

We are given variations on the theme of this particularly original type of weir, with its downstream face closed, one of which has a truly modern appearance. The interbuttress gaps are closed in, just as one finds in present-day dams of this type,[12] thus arriving at an overfall dam (fig. 11). The river bed is protected by a system which not only deflects the water as far away as possible but also, I believe, constitutes the earliest hint at the possible uses that may be made of hydraulic jump; the author points out that the basin is "un lugar donde ha de cargar el agua . . . [que] queda muerta y con esto el agua no rompe ninguna cosa y es muy mejor" ("a place for the accumulation of the water . . . [which] is now lifeless, so that it does not break anything and is much better").

* * *

So far, I trust that I have shown that the codex reflects experimentally based technical skills that are far superior to those not only of the author's contemporaries but of many who were to succeed him, such as Ramelli,[13] with his intelligently thought out but often impracticable devices. Our author, of course, is a man of the Renaissance; and I

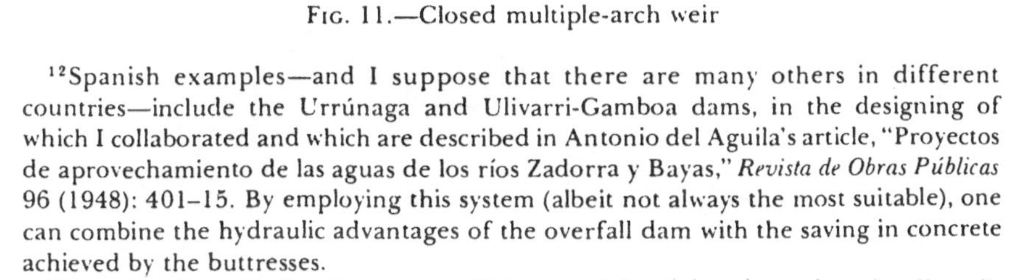

FIG. 11.—Closed multiple-arch weir

[12]Spanish examples—and I suppose that there are many others in different countries—include the Urrúnaga and Ulivarri-Gamboa dams, in the designing of which I collaborated and which are described in Antonio del Aguila's article, "Proyectos de aprovechamiento de las aguas de los ríos Zadorra y Bayas," *Revista de Obras Públicas* 96 (1948): 401–15. By employing this system (albeit not always the most suitable), one can combine the hydraulic advantages of the overfall dam with the saving in concrete achieved by the buttresses.

[13]Agostino Ramelli, "Le diverse et artificiose machine del capitano Agostino Ramelli dal Ponte della Tresia ingegniero del christianissimo re di Francia e di Polonnia" (1558; facsimile ed. Westmead, Farnborough, Hants., 1970).

have the impression that what he is aiming at, as he concludes this chapter, is to show that he can design what we might call "architectural" structures, in which imagination and skilled draftsmanship cause him momentarily to forget the practical aspects, although in so doing he by no means lacks famous predecessors. It should be pointed out also that, even in this instance, he does not fail to mention what he would have us believe are the rational advantages of his design.

Only one such type of weir is reproduced here (fig. 12), although the author illustrates several others, including one which has a plan showing three arches, the direction of one being contrariwise. Finally, we are given descriptions of ways and means of "beautifying" coastal cities, by diverting rivers and turning them into harbors. He also refers to the question of fortifying the shores of Spain. All this, of course, represents a digression from the real subject matter of the present article; but it does tie up with other parts of the codex which undoubtedly contain very important information for the history of maritime works.

* * *

Now let us turn to a basic problem, one, however, which will require a much more exhaustive study if it is to be solved. I refer to the question of finding out what earlier constructions provided the basis of the weirs described in the codex but which were not invented by the author.

He had first hand acquaintance with many Roman hydraulic works such as aqueducts, baths, etc. in both Spain and Italy; but he makes no mention of their dams, not even any of the three main ones that have survived to the present: Proserpina, Cornalvo, and Alcantarilla, the former two being in Mérida and the latter in Toledo. No evidence

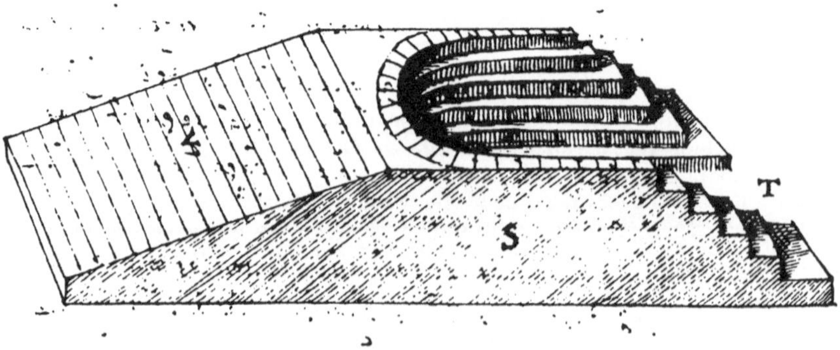

FIG. 12.—"Architectural" weir

exists to suggest that the author visited either of these cites, although the existence of the Mérida aqueduct is referred to in another of the codex books.

One might then consider the possible influence of Arab or Moorish tradition; but here again, supposing it existed, its bearing was so subtle as to be practically indefinable. One only has to compare what has been quoted here with the descriptions and figures contained in Smith's book on Arab-Spanish dams.[14]

At least as a working hypothesis, we must be satisfied with assuming that, where the author does not invent, he is mainly guided by contemporary Spanish and Italian structures and possibly also by the weirs of medieval Christian Spain, of which we still know so little.

* * *

What can be gleaned regarding the identity of the author of this book of the codex, and what was his attitude to the social environment in which he was living? There is actually very little evidence to be found, since our study covers only a small section of the whole codex; but it is interesting to speculate.

Obviously the author was a builder, although one who also allowed himself the intellectual's occasional flight of fancy. Yet in almost every instance one can detect that he lived and worked on actual building sites and himself had to solve practical problems on the job. This background has left its indelible imprint: he makes intelligent remarks about whirlpool effects destroying a structure, but not about the hydraulic theory of whirlpools. I have already quoted part of what he says about constructing foundations in running water and the serious difficulties faced by the laborers as they struggle day and night to remove this water; I am convinced that such a description can only come from the pen of a man who has personally experienced situations of this kind.

The chapter is unique in another way: the author does not quote classical writers in the way that is done elsewhere in the codex (usually Greek and Roman, adding little that is really useful to the context), even if on a noticeably lesser scale than in the works of many famous contemporaries, such as Montaigne. This again would seem to indicate that the author was a practical, not a bookish, man.

Only two city names are mentioned in this *libro,* one being Seville, which, according to his remarks in another section, he never actually visited; and the other, Tortosa. Both appear in the final part, where the author is more concerned with maritime works than with weirs.

[14]Smith, pp. 90–101.

Yet one infers from reading the entire manuscript that he must have acquired his experience mainly in various Italian states and in the Spanish regions of Aragon and Catalonia, although perhaps other regions should be included too.

The author appears to be concerned mainly with the safety of the dams he describes; and, although he does also take the cost of the project into account, it is the former criterion which prevails. Hence his views on this matter provide a suitable conclusion to this study, which covers only a small part of this great codex: "Los açutes deben ser hechos con ánimo sincerissimo y que los que tales cargos tienen de hazerlos conviene que ellos no sean gente que tenga cuenta con el interesse que si esto hay pocas vezes haran lo que deven en lo de la obra, porque estas obras conviene hazerlas a ellas muy riquissimas de materia y de artificio . . ." ("Weirs must needs be built in all sincerity of mind; nor should those who have the charge of building them be burdened with a concern for gain, lest they all too rarely do what they should; for such works demand great lavishness in materials and in skillful devices . . .").

4

Architecture for fish: the Sienese dam on the Bruna River—structures and designs, 1468–ca. 1530

Nicholas Adams

Italy is not generally reckoned to have a climate or a topography conducive to the building of large-scale water-storage dams. Heavy rains in spring and fall, along with winter snowmelt from the mountains, make for two heavy flood seasons. In between, there is a long dry spell during which many rivers and streams shrink to almost nothing. Hence dams must cope with an extreme range of river regime and with extensive silting. Unlike other parts of the Mediterranean, for example Spain and North Africa, where dams were not uncommon in the early

DR. ADAMS, chairman of the Department of Art and Architecture at Lehigh University, received his doctorate from the Institute of Fine Arts at New York University in 1977. He has published articles on 16th-century fortifications, on techniques of minting coins, and on the work of Renaissance estimators, and he is currently completing a study of cultural change in Pienza. He writes: "This article, though partially excerpted from my dissertation, 'Baldassarre Peruzzi: Architect to the Republic of Siena 1527–1535,' owes much to subsequent discussions. In particular I was aided by the opportunity to speak before the Metropolitan Seminar in the History of Technology in New York City and the chance to think about the Bruna River Dam in company with its members: Gustina Scaglia (Queens College) led me to the Munich Taccola, and Thomas Settle (Polytechnic Institute of New York) introduced me to the Florentine tradition of dams and weirs. Comments by Michael Mahoney (Princeton University), Bruce Chandler (City College, New York), George Saliba (Columbia University), and Pamela Long (Washington, D.C.) were also helpful. I am especially grateful for the friendship and generous help of my graduate advisers, Kathleen Weil-Garris Brandt and the late Wolfgang Lotz. Thanks are also due Sonia Fineschi (Archivio di Stato, Siena), Norman Smith (University of London), Douglas Lewis (National Gallery of Art, Washington, D.C.), Maria A. Phillips (Redondo Beach, Calif.), Victoria Steele (Elmer Belt Library, University of California, Los Angeles), Paul Myers (Lehigh University), Carlo Prezzolini (Abbadia San Salvatore), and Daniela Lamberini (University of Florence). Donald Martino, chief of the Project Review Section, and Leonard Baskin, hydraulic engineering consultant of the Division of Dam Safety, Commonwealth of Pennsylvania, Bureau of Dams and Waterway Management (Harrisburg, Pa.), helped me set Peruzzi's proposals into a modern context. During my research I was financially supported by the National Gallery of Art, Washington, D.C., through a Chester Dale Fellowship, and by the Samuel H. Kress Foundation. The College of Arts and Science at Lehigh University and the dean, John Hunt, provided funds for the preparation of the manuscript."

modern period, we can count the number of large-scale Italian storage dams that date before 1700 on the fingers of two hands.[1]

A dam definitely worth including in the count is the massive structure built by the Republic of Siena on the Bruna River, beginning in 1468. This is an interesting work, not just because it coincides with the broader Renaissance effort to domesticate the landscape through law, agriculture, and trade,[2] but also because it involved many noted architects, because it derived from some unusual sources, and because, overall, it sheds much light on the relation between Renaissance building techniques and architectural practice in the gray area of the traditional arts.

In striving to provide sufficient food for citizens, governments in the central Italian hill town of Siena had to make special efforts to obtain supplies of fish.[3] As an inland capital with a population of some 52,000 in the early 14th century—but having no lakes nearby and facing a hazardous and lengthy journey to the sea—Siena had encouraged the construction of fishponds or weirs.[4] In the 15th century, even though the population had shrunk somewhat, these ponds proved insufficient, and Siena turned to the lake of Trasimeno in Perugian territory to supplement local supplies.[5] Perugian fish were costly, however, and the journey only slightly less troublesome than the one from the coast.

In 1468, therefore, two plans were put forward: reclamation of some of the swamps near Castiglione della Pescaia, on the sea near Grosseto, and construction of a dam to create an artificial lake.[6] The former seems to have been judged unsuitable, but on February 21, 1469,

[1]My list derives from Norman Smith, *A History of Dams* (London, 1971), and A. Ghetti and D. Tonini, *Dighe italiane: Guida bibliografica* (Venice, 1961).

[2]See E. Sereni, *Storia del Paesaggio agrario italiano* (Bari, 1961).

[3]On the importance of fish in the medieval and Renaissance diet, see Fernand Braudel, *Capitalism and the Material Life 1400–1800* (New York, 1975), pp. 146–48. The provision of sufficient food was a communal responsibility, and a government often paid dearly for its inability to feed its population.

[4]Provisions were made in favor of the construction of fishponds as early as 1310, as is recorded in the Sienese constitution; see *Il Constituto del comune di Siena*, ed. Alessandro Lisini, 2 vols. (Siena, 1903), 2:310.

[5]See William Heywood, *A History of Perugia* (London, 1910), pp. 123, 289.

[6]All documentary references are to the Archivio di Stato (abbreviated *ASS*). These references are followed by the deliberative body of the Sienese government (Concistoro, Balia, Consiglio Generale), the volume number, and the folio number. *ASS* Concistoro 609, fol. 25 (March 27, 1468); also, Giovanni Cecchini, "Maestri Luganesi e Comaschi a Siena nel secolo XV," *Arte e artisti dei Laghi Lombardi*, ed. E. Arslan, 2 vols. (Como, 1964), 1:140. Castiglione della Pescaia had been the site of a failed dam built by the Florentines. A letter from the architect Michelozzo, dated before 1447, discusses the reasons for the failure. I have not been able to locate this structure precisely. See Harriet McNeal Caplow, *Michelozzo*, 2 vols. (New York, 1977), 1:45–47.

preliminary specifications for the dam were published, and the next year it was begun.[7] Under construction sporadically for twenty-odd years, the dam cracked in 1492 and finally broke. Various plans were submitted for rebuilding, but none was carried through; and today only three short, though imposing, sections of the original wall remain visible. (See fig. 1.)

In terms of design, however, the story does not end in 1492, for we have, in effect, a number of Bruna River dams, built and unbuilt: There is an original proposal, similar to drawings by Mariano Taccola (1381–82 to ca. 1453); a major modification of 1471 by Pietro dell'Abbaco (active 1454–86); and a later series of six proposals, ca. 1530, for a new dam by the artist-architect Baldassarre Peruzzi (1481–1536). The Sienese artist-architect-engineer Francesco di Giorgio Martini (1439–1509) also played a role.[8] In order to understand the dam actually built, let us first review the documents pertinent to construction. With these in mind, we may then move on to consider its likely form on the basis of traditions of Sienese dam and weir building, thereby revealing the insufficiency of those traditions for the Bruna River Dam. We shall then examine the six Peruzzi drawings for the reconstruction of the dam. In conclusion, we shall consider the dams built and proposed by the Sienese in the context of Mediterranean traditions of dam building.

* * *

After the decision had been made to construct a dam for impounding a lake to be stocked with fish, masons were sent to study potential locations during the summer of 1469.[9] Along with them went the mathematician Pietro Cheechi, called Pietro dell'Abbaco. The site chosen for the dam was remote, at the end of the Bruna Valley, at the point before it opens into the Maremma outwash plain between the

[7]*ASS* Concistoro 614, fol. 36; Cecchini, p. 140.

[8]The dam and its history only barely penetrate the scholarly literature of architectural history: Roberto Papini, *Francesco di Giorgio Martini architetto*, 2 vols. (Florence, 1946), briefly discusses Francesco di Giorgio's participation; Mircea Toca, "Un progetto peruzziano per una diga di sbarramento nella Maremma," *Annuale della Scuola Normale Superiore di Pisa* 39 (1970): 107–17, discusses Peruzzi's drawings for the dam but seems not to know the history of the Bruna River Dam; Salvatore di Pasquale, "Riflessioni e note sui progetti di Baldassarre Peruzzi per la formazione di un lago artificiale nel territorio di Siena," *Rilievi di fabbriche attribuite a Baldassarre Peruzzi* (Siena, 1982), pp. 71–88, unfortunately repeats much of Toca's misinformation; Cecchini (n. 6 above), 1:131–150, does not refer to Peruzzi's drawings. In short, while parts of the dam's history have been dealt with, there is no comprehensive discussion. In order to keep the footnotes to a minimum, I have not reproduced lengthy extracts of documents that appear in a generally available source.

[9]*ASS* Concistoro 614, fol. 36; Cecchini, p. 140.

village of Giuncarico and the Castel di Pietra to the northwest of Grosseto (fig. 2). On September 15, the group published specifications. The dam, they estimated, would require 6,000 *canne* of building material (17,508 cubic meters), which ought to be "good stones and lime and good sand." The estimated cost would be between 7,000 and 9,000 *fiorini.*[10] On February 27, 1470, the task of construction was given to Maestro Adamo del Maestro Domenico di San Vico and Maestro Matteo di Maestro Jacopo da Mariano di Val di Lugano.[11] We know little about these masters, but they must have been among the many who emigrated from northern to central Italy in the mid-15th century.[12]

Work began almost immediately on clearing the site and digging the foundations. On May 22, 1471, however, Maestro Adamo appeared before the *Concistoro* in Siena and declared that the dam as it was being built would not stand. Whatever the reasons (Maestro Adamo does not tell us), it was agreed that alterations were necessary, and Pietro dell'Abbaco traveled to Fucecchio, north of Siena near Empoli, to study a Florentine dam there. Subsequently he recommended building "seven buttresses which should stand twelve *braccia* [6.9 meters] below the dam and should follow the slope of the wall." The middle one would be 10 *braccia* (2.9 meters) wide.[13] The addition of these seven buttresses, shown in the reconstruction drawing (fig. 3), provided

FIG. 1.—Drawing of the visible remains of the Bruna River Dam, looking from the northwest, the upstream location of the lake. (Drawing by F. Sancar.)

[10]*ASS* Consiglio Generale 233; fols. 44v–45v. The measurements for the dam will be given in local units—*braccia* (0.583 meters) and *canna* (2.918 meters)—with metric equivalents in parentheses. The measurements given in the documents are probably derived by multiplying the width of the valley by the height of the dam by the average thickness. The value of the *fiorino* is difficult to estimate. Suggestive might be the price of 1,150 *fiorini* paid by the Sienese for a large bronze tabernacle by Vecchietta (1467) to be placed in the church of Santa Maria della Scala. Antonio Federighi received a yearly salary of 80 *fiorini* (1480) as supervisor of the underground aqueduct system. Both examples are cited by Judith Hook, *Siena: A City and Its History* (London, 1979), pp. 104, 113.

[11]*ASS* Concistoro 620, fol. 34v; Cecchini, p. 141.

[12]Cecchini, p. 141.

[13]". . . Audito magistro Adamo conductore murorem dicti lacus dicente et asserente muros dicti lacus secundum locationem sibi factam non habere perfectionem suam. Audito etiam magistro Pietro dell'Albaco transmisso ad videndum compositionem lacus Fuciechi et eius relatione deliberaverunt ut dictus habeat perfectionem suam venustam

support for the barrier, and work recommenced. On May 23, 1473, a governmental deliberation noted that 5,000 *fiorini* had already been spent but nevertheless revised the estimate of building material needed to 12,000 *canne* (35,016 cubic meters).[14] Maestro Adamo died that fall, and new head masons were appointed in the spring of 1474.[15]

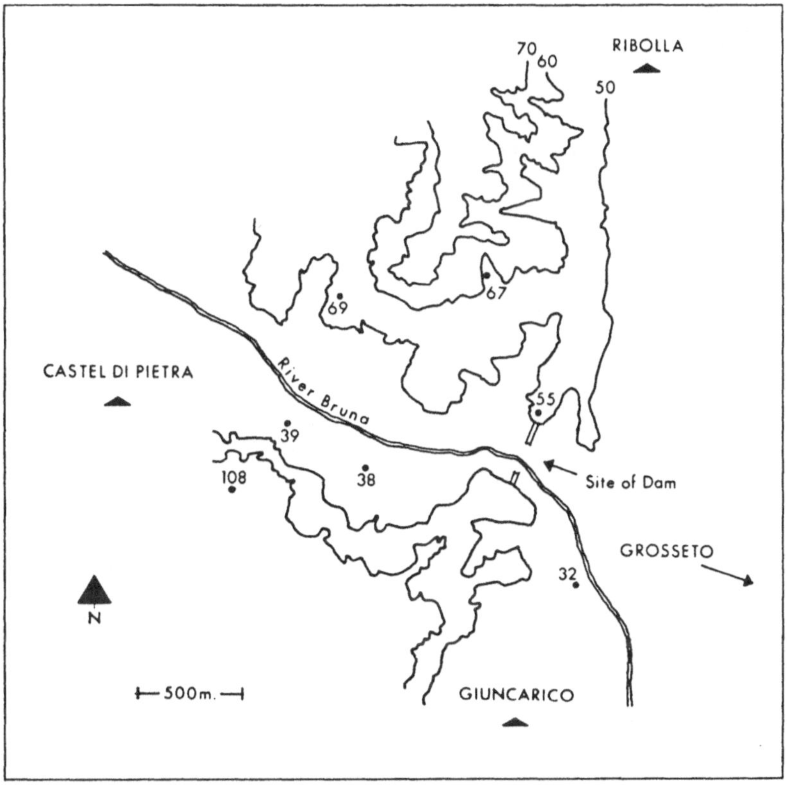

FIG. 2.—Topographic drawing showing site of the Bruna River Dam. (Drawing by Barry Price.)

et remisserunt in dictos operarios qui faciant fieri sette speroni che sieno da pie in fuore dal lato di sotto del serrata braccia dodici et sieno in fuore dal terreno, in su secondo il pendente del muro et sia quello del mezo grosso braccia dieci et gli altri sei sieno grosso braccia cinque" (*ASS* Concistoro 628, fol. 17). It is interesting that the scribe drops from Latin into Italian for the description of the dam. Renaissance Latin was clearly insufficiently rich for this vernacular structure; nothing more clearly shows how out of touch were the Sienese dam makers from the Roman traditions. This document is also noted in Cecchini, p. 141.

[14]*ASS* Consiglio Generale 235, fols. 65–67v. It is clear that the masons were feeling their way through foreign territory. How else could they have been so far off in their estimates of building material?

[15]The new *capomaestri* were Pace Pacini and Antonio di Matteo. *ASS* Concistoro 646, fol. 12v; Cecchini, p. 142.

By September 1475 it appears that the line of the dam crossed the valley, but that there were serious problems—a report to the government stating that "the water of the Bruna fell at the foot [of the dam] and has made a hole eight braccia high [4.6 meters]." The report continued: "a wooden weir or buttressing must be provided; great may be the damage to the wall of said lake."[16] An 8-*braccia* hole caused either by overflow at the air face or from uplift pressure on the foundations was no trifling matter. We do not know the Sienese response, although the situation seems to have stabilized; during 1475–78 we have records of a number of apparently routine visits to the site by Pietro dell'Abbaco to measure the work completed.[17] We will look later at how the builders might have dealt with the water of the Bruna during construction.

On November 30, 1481, the state of the dam was assessed in a petition to the government for money. Pietro dell'Abbaco, new head masons Cristoforo Guidocci and Tomasso Luti, and "other masons and

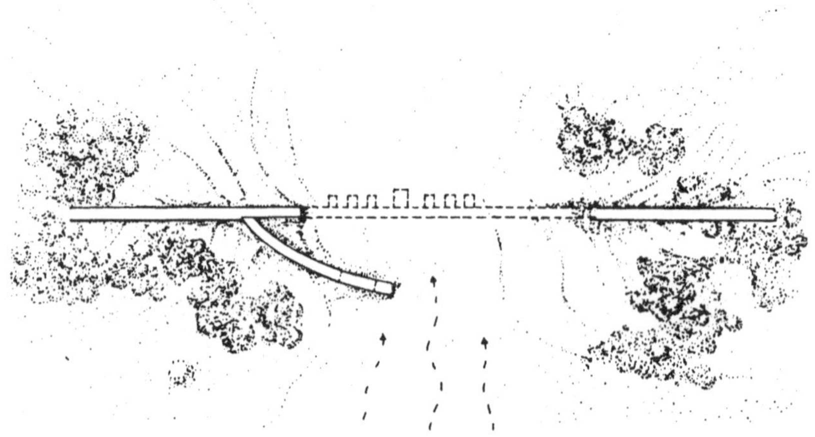

Fig. 3.—Plan of the Bruna River Dam, with dashed lines representing hypothetical reconstruction of the central section with buttresses as specified by Pietro dell'Abbaco. (Drawing by F. Sancar and Barry Price.)

[16]The deliberation in the Concistoro reports "quod aqua Bruna que cadit in pede murit face quandum ghalazam altitudinis octa brachiorum ut circa cui nisi providetur de facto cum una stecharia aut rostatoria magnum posset inferre damnum muro dicti lacus." *ASS* Concistoro 654, fol. 9; see also Cecchini, p. 142.

[17]*ASS* Concistoro 654, fols. 6v–7; Concistoro 656, fol. 30v; Concistoro 666, fol. 7; Concistoro 668, fols. 43v–44. Problems with foundations may have been chronic for Italian dam making during this period. Michelozzo's report on the Florentine dam (n. 6 above) refers to similar problems at Castiglione della Pescaia. Dr. Caplow regards this letter as a parody of officialese, but it reads quite clearly to me. Michelozzo's problems with foundations and river banks sound familiar to the student of the Bruna River Dam.

fishermen," concluded that 5,500 *canne* (16,049 cubic meters) of the dam still remained to be built. Part of this was for portions "which we find the water has cut away [*cavato*]."[18] Despite the government's acceptance of this petition for further appropriations, construction seems to have proceeded more slowly after 1481. The repair of the damaged parts of the dam was probably difficult, and construction work was, as always, slow during the winter. In 1483 the method of keeping accounts was changed, and from then on the documentary record is hard to follow, although the number of *canne* is recorded, and there is also a contradictory notice concerning efforts to "sgombrare" (empty, clear, clean) the dam, efforts that involved Francesco di Giorgio Martini, architect to the Republic of Siena.[19] By 1492 the dam was close to completion.

Early the year before, the Duke of Calabria had written to the Sienese and asked whether Francesco di Giorgio might come south to help with some projects, notably the defenses against the Turks. Pietro dell'Abbaco had retired from the dam project during the 1480s (by early 1487 he was dead), and the Sienese seem to have sent Francesco di Giorgio south with the understanding that he would return when needed. By January 1492, they had decided that they wanted him back, writing ". . . we are about to close [*serrare*] our lake."[20] In January, they wrote to the Duke of Calabria: ". . . we are about to close our lake and without his presence we would not be able to do it." Could Francesco di Giorgio return before March?[21]

Francesco di Giorgio did not return. On July 7 the government wrote directly to him and asked him to return.[22] Again, September 11, they wrote to the Duke of Calabria. Matters were serious at the dam site, "for winter is approaching and several things are needed for its completion."[23] The Duke of Calabria still equivocated; projects were under way and, afraid of the Turks, he felt he needed the expert advice of Francesco di Giorgio for his military affairs. The government's final letter, on December 4, 1492, is the last we possess on this topic and may well have produced the long-desired return of the architect. Now it was

[18]*ASS* Concistoro 238, fols. 254–255.

[19]From September 1483 onward there are few records (see *ASS* Concistoro 696, fols. 8v, 9; Concistoro 699, fol. 15v; Concistoro 702, fols. 8v–9; Concistoro 727, fol. 33). The total number of *canne* noted, 1483–87, was 1,121. The document regarding Francesco di Giorgio, untraceable, was published by Gaetano Milanesi, *Documenti per la storia dell'arte Senese*, 3 vols. (Siena, 1854), 2:424, as "1489?"

[20]*ASS* Balia 411, fols. 18v–19. Also published by Milanesi, 2:444–45.

[21]*ASS* Balia 411, fols. 28v–29 (unpublished).

[22]*ASS* Balia 411, fols. 134–135v. See also Milanesi, 2:446.

[23]*ASS* Balia 411, fols. 153–153v. See also Milanesi, 2:447.

no longer just a question of completing the dam (or lake). It was a call for help: "many things are most urgent and particularly the completion of the lake and which if not done and with great speed, may be ruined. . . ."[24] Whether Francesco di Giorgio arrived before the collapse of the dam we do not know. It was, in any case, too late. In the last days of December 1492 the dam broke.

Riccardo Pini, a Sienese official in nearby Giuncarico, wrote to the government "to offer condolences with Your Lordships for the great damage which you have received for the opening of the wall of the lake. . . . The damage and the misery now and for the future involved in this accident has been seen by me as your official. I have seen the hole created by this break."[25] In his *Sienese Diary* the chronicler Allegretto Allegretti writes: "On the first of January [1493] we had news concerning our lake in the Maremma which had not yet been fished. The wall fell down and flooded much of the countryside. Men and beasts have died and this was the fault of those who completed it, since they bungled it for gain."[26]

* * *

The dam built by the Sienese on the Bruna River seems to have been relatively simple. From the specifications of May 1471 and from site observations, we may conclude that it would have formed little more than a wedge, 12 meters thick at the base and 6 meters at the top. It would have been about 18 meters high at its highest point and stretched like a curtain wall some 300 meters across the valley.[27] In fact, it was probably a little wider at the top than the specifications called for. The interior, as can be seen in the exposed sections, is a mixture of

[24] *ASS* Balia 411, fol. 181v. See also Milanesi, 2:449.

[25] Pini: ". . . condolersi con le Signorie Vostre dello grande danno 'anno ricievuto per le apertura del muro del lago. . . . Lo danno et la miseria nella quale sono et saranno per questo danno involti perche io che so a essi hoffitiale', o visto la oscurita che per decta rottura del lago et sighuita . . ." (*ASS* Balia 549, no. 81).

[26] Allegretti: "Adì primo di gennaro habbemo novelle come il nostro lago di Maremma, il quale non s'era anco cominciato a pescare. Haveva cacciata in terra il muro et allagato molto paese. E morto huomini e bestiame e questo e stato per difetto di chi li ha fatto a perfettione, ma bensì acciabattalo per guadagnare piu assai." Allegretto Allegretti, *Diario Senese delle cose seguite a suo tempo cioe dall'anno 1450 al 1496* (manuscript), Biblioteca Comunale, Siena (C. V. 27): fol. 230.

[27] I am most grateful for the work of Architetto Carlo Prezzolini, who corrected the earlier site surveys. Today the eastern end of the dam measures 104 meters, the open breach between the remains measures 165.2 meters, and the western stump measures 40.8 meters for a total of 310 meters. The top of the eastern end measures 6.5 meters across the crest and stands 15 meters at its highest point. Architetto Prezzolini's survey was undertaken in February 1984.

mortar, alluvial rubble, and small boulders compacted as much as possible. Around the exterior have been set partially squared, rough-cut stones which would have acted as a protective bumper and were intended to contain the inner core (fig. 4). These are laid in courses over one thickness and would not have been impervious to water. Nothing survives to give an indication of the shape of the buttresses, nor is the nature of the washed-out central section known.

Siena was especially famous for her skills at water management. Her aqueducts, the *bottini*, which run for many miles before coming into the city, were matters of great pride, and their path under the city walls was a military and state secret.[28] The Sienese also had much experience with weirs and small storage dams for mills and fishing; although most of these have either been washed away or built over, we are fortunate to have some documentary records of dam-making theory and practice. The earliest is a sketchbook dated 1427 in the Bayerisches Staatsbibliothek, Munich, CLM 197, Part 2, in the hand of Mariano Taccola.[29]

The Taccola manuscript is rich in matters relating to river control and fishponds. It shows various kinds of palisade buttressing for river-

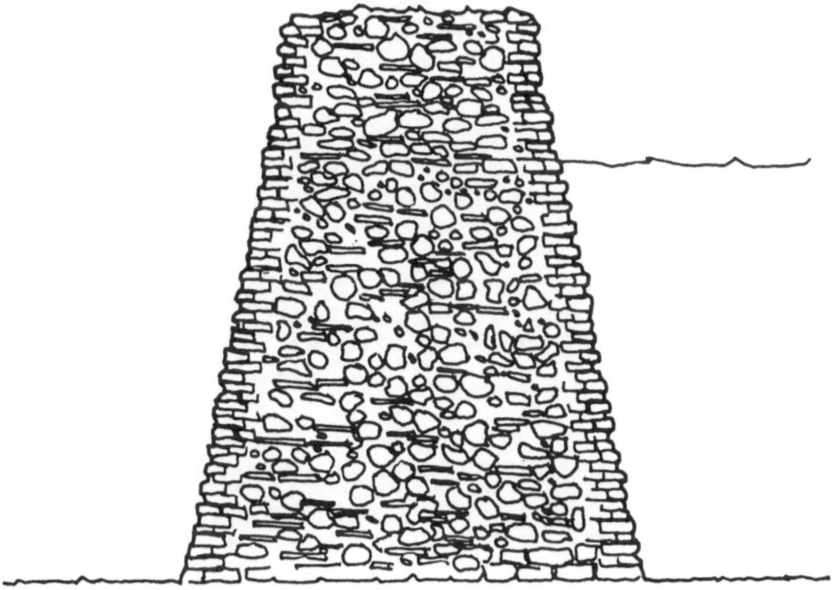

FIG. 4.—Section through Bruna River Dam showing unsorted alluvial boulders within partially squared fieldstones. (Drawing by Barry Price.)

[28]F. Bargagli-Petrucci, *Le Fonte di Siena ed i loro aquedotti*, 2 vols. (Siena, 1906).
[29]This manuscript was kindly shown to me by Gustina Scaglia.

banks and also describes diversion techniques. One of the more amusing drawings, folio 99r, shows a ditch dug in toward a swamp (*stagnus*) along which fish are swimming from the sea (see cover illustration). There is also much material related to dam and weir construction. Folio 99v, a drawing for a tidal mill (fig. 5), shows the kind of scheme that the Sienese may have had in mind during 1468 for the area near the swamps of Castiglione della Pescaia before they began working on the Bruna River site; the area at the top of the sheet is marked *Stagnus*. A guillotine sluice gate controlled the water entry at the right (*Introitus*) and the outlet (*Exitus*) is at the left. In the foreground is the sea (*Mare*). Folio 100v shows a weir of posts and planks (fig. 6). Stakes with fascines woven between them hold the banks in place. The double thickness of the planks on the weir probably was planned with an alluvial infill of some sort. (A similar structure was proposed by Francesco di Giorgio as a cofferdam.) Two outflow channels are provided, one under the dam and the other from a smaller drainage pipe that may come from the lake.

The most interesting of the Taccola drawings from the standpoint of the Bruna River Dam shows what seems to be a large masonry water-storage dam (fig. 7). Its size is suggested by the schematic mountains and carved "cliffs" in the foreground and background. If we figure the sluice gate at a minimum of 1.5 meters across, the dam may approach

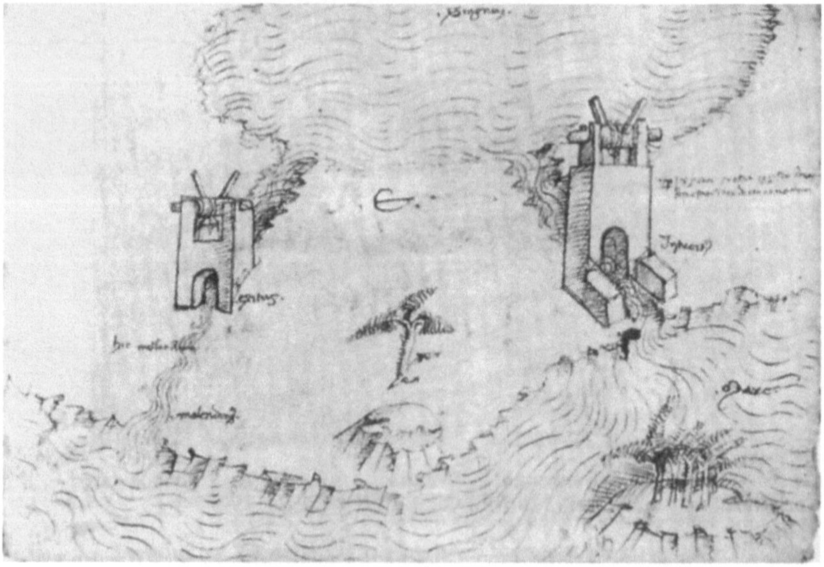

FIG. 5.—Tidal mill drawing by Mariano Taccola, from sketchbook dated 1427 in the Bayerisches Staatsbibliothek, Munich, CLM 197, pt. 2, fol. 99v.

10 meters or so along the crest. All in all it recalls the form of the first Bruna River Dam project of 1468–70 prior to the addition of the buttresses.

Records on dam and weir construction were also left by Francesco di Giorgio Martini in his *Trattati di architettura, ingegneria e arte militare*. Francesco di Giorgio was the heir to many of Taccola's ideas, as Gustina Scaglia has shown, but unfortunately we cannot link him with the earliest phases of the construction of the dam.[50] Moreover, if much of the treatise was written before 1489, it is hard to know whether Francesco di Giorgio was influenced by the course of construction, whether his insights influenced work on the dam, or, as is most likely, whether

FIG. 6.—Drawing of palisade dam by Mariano Taccola, from sketchbook dated 1427 in Bayerisches Staatsbibliothek, Munich, CLM 197, pt. 2, fol. 100v.

[50]Gustina Scaglia and Frank Prager discuss the relation in *Mariano Taccola and His Treatise De Ingeneis* (Cambridge, 1972), pp. 193–201. See also Gustina Scaglia, "Francesco di Giorgio's Chapters on Fortresses and War Machines," *Fort* 10 (1982): 39–69.

both the Bruna River Dam and Francesco di Giorgio's designs derive from common vernacular sources. Whatever the case, some of his designs recall the form of the Bruna River Dam. For example, he recommended the placement of squared stones to cover the exterior, and this technique was likewise followed by the builders of the Bruna River Dam (fig. 8).

Francesco di Giorgio also discussed two techniques for making a dam or weir. The first involves the use of a cofferdam made of posts and cross planks and filled with clay, to divert the stream so that the weir itself could be built in the dry bed (fig. 9). He then noted the various problems that can arise along the banks of such a weir and how they might be avoided. With the cofferdam in place it would be possible to

FIG. 7.—Drawing of masonry dam with central guillotine sluice gate by Mariano Taccola, from sketchbook dated 1427 in Bayerisches Staatsbibliothek, Munich, CLM 197, pt. 2, fol. 114v.

work on the weir itself. This, he suggested, should be constructed on a frame of palisades with crossing planks to create a formwork in which gravel and lime could be placed to create a gravity dam (fig. 10); squared stones cover the surface, and a wooden protective cover allows the structure to set. Water flows over the dam or escapes through a guillotine sluice gate built above it. An alternative technique involved building the dam across the river from the banks toward the middle where, as a final step, a guillotine sluice gate would be installed (fig. 11). Francesco di Giorgio insisted that, though this dam might be built of masonry, it too had to have some form of wooden frame for, if poorly founded at any point, it would shift weight and collapse.

This second type of dam generally recalls the Bruna River Dam. Since there was no feasible outlet upstream because of the nature of the valley, it seems likely that the builders followed the construction principles outlined by Francesco di Giorgio—that is, they built from the banks in toward the center. When the dam was nearing completion, the authorities called on Francesco di Giorgio to *serrare* (close) the dam, which suggests it was time for the placement or manipulation of some sort of sluice gate. Examination of the Bruna site uncovered no sluice gate, but after the collapse it would surely have been carted away for safekeeping or sale as scrap. Moreover, I have not been able to detect any wooden membering within the wall of the dam. Either the builders

FIG. 8.—View of the remains of the Bruna River Dam from the northwest showing the stone cover on the water face. (Photo by author.)

ignored the advice of Francesco di Giorgio (the dam failed, according to the chronicler Allegretto Allegretti, because of the incompetence and greed of the builders), or Francesco di Giorgio's remarks were formulated in light of the problems encountered in 1475.

Comparison with these collateral sources, and an examination of the documents for the Bruna River Dam itself, suggest the following points: (1) The dam probably had a central opening. (2) It is improbable that a cofferdam was used during construction; rather, an opening of some sort allowed water to flow through the dam. (3) Heavy rains or floods may have contributed to the problems in 1475, while the final collapse in 1492 seems to have been provoked by a failure to control the sluice gate. We can hypothesize a number of potential circumstances, any one of which alone or in combination with others would have been sufficient to bring ruin: water flowed over the dam and undermined the air face; the weight of the water became too great and shifted the dam; water seeped through the dam and opened it out; uplift pressure caused by a raised water table and improper foundations forced open the air face. Whatever the case, it is absolutely clear that the dam as

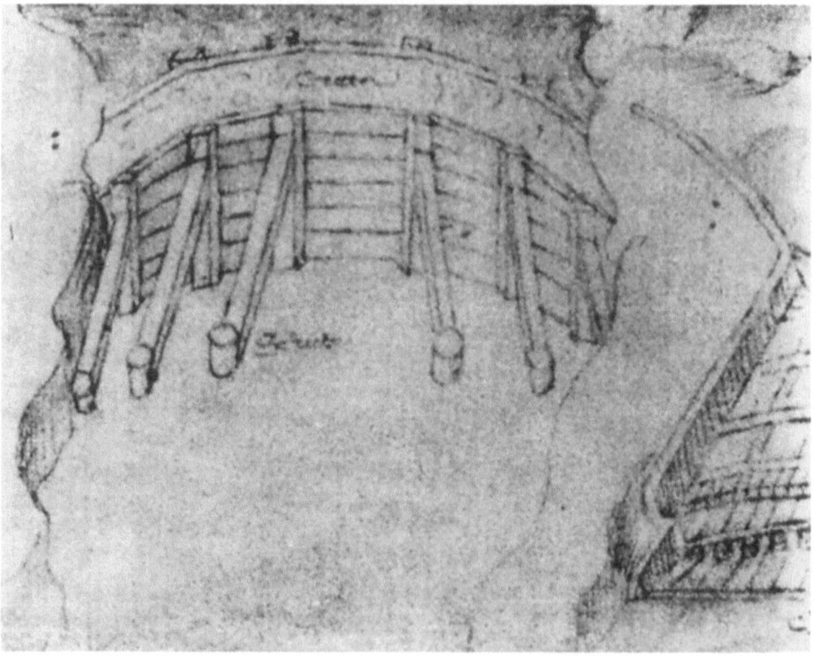

FIG. 9.—Drawing of cofferdam by Francesco di Giorgio Martini, from *Trattati di architettura ingegneria e arte militare*, ed. Corrado Maltese (Milan, 1967), plate 11 (detail). (Reprinted by permission.)

built from the late 1460s onward was inherently unsound largely because the Sienese had no experience with *large-scale* water-storage structures.

Here, indeed—in its scale—was where the Bruna River Dam seems to have departed radically from the traditions of Sienese building, and why its architect-engineer-masons could not deduce answers to the problems of the new form. Nothing that I have been able to find within the Sienese tradition matches the Bruna River Dam in size. There was no experience in founding such a structure and no comprehension of what happened when so massive a weight of water was piled up. Suggestive of the poverty of the Sienese traditions of dam construction was Pietro dell'Abbaco's journey to Fucecchio to examine the Florentine dam. If the Taccola–Francesco di Giorgio Martini drawings represent an adequate survey of local material, as it seems reasonable

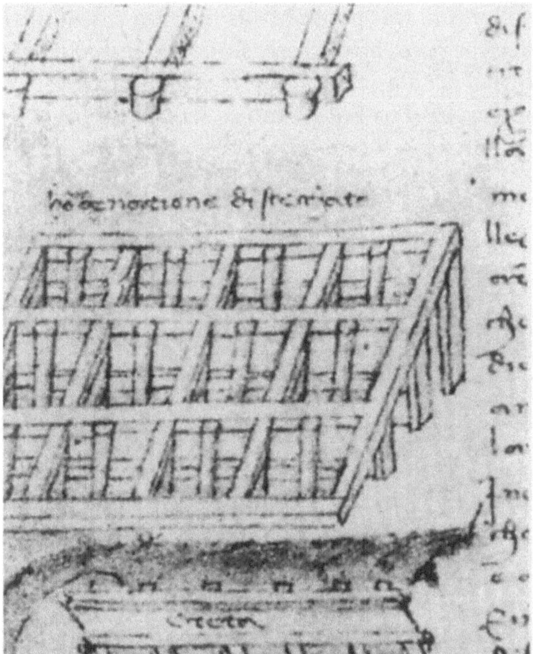

FIG. 10.—Drawing of palisade dams by Francesco di Giorgio Martini, from *Trattati di architettura ingegneria e arte militare*, ed. Corrado Maltese (Milan, 1967), plate 12 (detail). (Reprinted by permission.)

to assume, we can conclude that masonry buttresses were simply not part of Sienese experience. It is possible, in fact, that the Sienese conceived of the dam, at least initially, as not unlike a great wall—the word they often used in Italian to describe it—modified to reflect what little they knew about dams. Theirs was not a very sophisticated effort, but it clearly reflects an adaptive approach to structure of a kind typical of the Middle Ages and the early Renaissance.

The presence of two Lombards as head masons might raise the question whether they brought special skills to Siena from northern Italy, where dam building was, generally speaking, more advanced. That may be, although the mere presence of Lombard masons is no proof that they had any role in the design, or, indeed, that they were even admitted to be of superior skill.[31] Lombard masons were not especially welcome in Siena for they were suspected of taking jobs away from local workers. Their employment outside Siena in the savage wilds of the Maremma was probably as much a result of this attitude as of their special expertise. Moreover, since the initial dam design recalls the drawings of Taccola, and since it was Pietro dell'Abbaco, not a Lombard, who traveled to Fucecchio to study the form and size of the

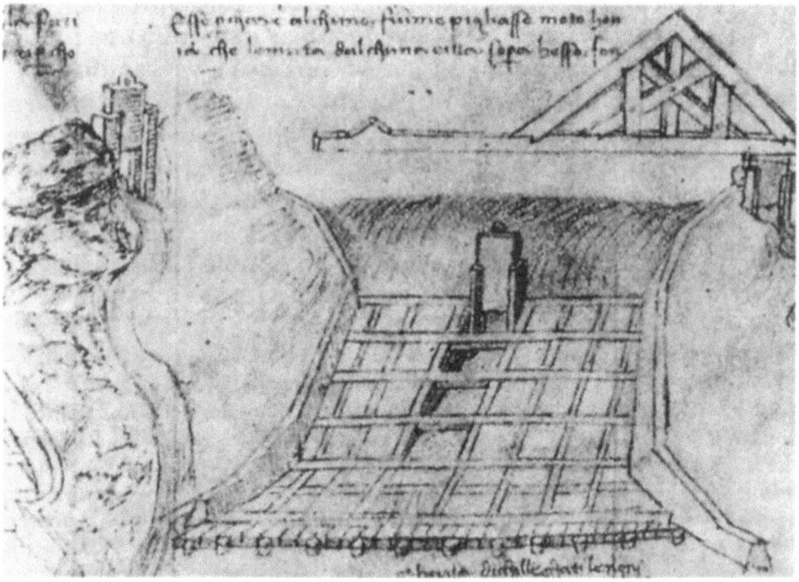

FIG. 11.—Drawing of dam with central guillotine outlet by Francesco di Giorgio Martini, from *Trattati architettura ingegneria e arte militare*, ed. Corrado Maltese (Milan, 1967), plate 12 (detail). (Reprinted by permission.)

[31]There are a number of studies dealing with this matter in *Arte e artisti dei Laghi Lombardi* (n. 6 above).

Florentine dam, we must look elsewhere for the name of an architect. Given that what he had to say about dams is fundamentally inconsistent with parts of the Bruna Dam and given too the absence of his name from the early records, we must exclude Francesco di Giorgio from consideration. In my opinion, the probable designer was the mathematician Pietro dell'Abbaco.

Although neither an architect nor a mason in the traditional sense, Pietro dell'Abbaco was a key figure in Siena. He taught practical mathematics and he was active largely as a quantity surveyor through the third quarter of the 15th century. He seems to have understood fortifications, hydraulic engineering, and construction techniques of all kinds.[32] Further, his trip to Fucecchio marks him as one who could observe, evaluate, and adapt a source to a new formal situation. In that moment, it seems to me, he effectively broke from the Sienese tradition and revealed himself as more than a craftsman copyist. Although he probably did not know Taccola personally, he was not an improbable choice for the overseer and designer of this important civic project.[33]

* * *

Between 1498 and 1535, there were a series of attempts to repair the Bruna River Dam, and the section of wall which is of a different facture and slopes upstream from the water face may date from one of these efforts. The most interesting repair proposal that we can document dates from 1498, when a certain Maestro Giovanni Maria da Parma responded to a call to bid on the work.[34] His letter to the Sienese seems to confirm that one reason the Bruna Dam failed was because of problems with the guillotine sluice gate, for Maestro Giovanni was at pains to point out that his specialty was "locks [*chiave d'acqua*] that close themselves and that will retain whatever amount of water that can be

[32]Pietro dell'Abbaco was born in Montalcino ca. 1435 and came to Siena to practice practical or commercial mathematics and to teach mathematics. As *mensuratore* or quantity surveyor for the Republic of Siena he traveled to all important building sites and oversaw the work undertaken by the masons. This involved not just a knowledge of building techniques but also an understanding of the principles and purposes of most building types. He embodied, in short, a remarkable experience of architectural lore and practice. I have written a biographical sketch of Pietro dell'Abbaco which was delivered as a lecture at the conference, "Les Chantiers de la Renaissance," held at the Centre des Etudes Supérieures de la Renaissance, Tours (June 1984). The acts of the conference are scheduled for publication in 1986.

[33]If the two men met, it was at the end of Taccola's life. Pietro, a young and inexperienced estimator, would probably have had some difficulty in meeting the elderly engineer. I argue in my sketch of Pietro dell'Abbaco's life that he and masters of the traditional arts like him are the true bearers of the "architectural tradition."

[34]Milanesi (n. 19 above), 3:70.

contained through ingeniousness [*ingegnio*] and true art." He noted that he could offer the government drawings made on the spot and thus adapted to the site (could this have been a criticism of the ever-absent Francesco di Giorgio?); that he would study the errors of previous masters; and that his solutions might be compared with those of "any other master and noble fellow [*valentuomo*] who may be an expert in a similar art, not to undervalue anyone." We do not know what became of this proposal, nor of other efforts made around 1515. In 1532 a certain Cesare Marsili was hired to rebuild the dam over a term of four years.[35] Given the poor financial state of the Sienese Republic in these years, however, it is unlikely that much work was done.[36]

While this marks the last explicit documentation pertinent to the dam, there is further graphic evidence which may be associated with it: a group of six drawings in the Gabinetto dei Disegni of the Uffizi in Florence (UA 584–589) in the hand of Baldassarre Peruzzi which show proposals for a large-scale water-storage dam (figs. 12–17). Their authorship by Peruzzi, the leading Sienese architect of the 16th century, has never been in doubt.[37] Moreover, the date of these drawings has been confirmed by Heinrich Wurm as the late 1520s or early 1530s.[38] During the years 1527–35 Peruzzi was active in the Republic of Siena as its communal architect.[39] He traveled frequently for the government and visited the Maremma area on more than one occasion, to inspect fortifications (at Port'Ercole and Orbetello) and to inspect flood damage near Grosseto.

Nothing written on the drawings in the Uffizi explicitly connects them with the Bruna River project,[40] and no documentary material

[35]*ASS* Balia 66, fol. 30; Balia 67.

[36]During the mid-16th century the Sienese were forced to choose (on financial grounds) whether they would decorate the city for the arrival of Charles V or whether they would fortify it with bastions. See Judith Hook, "Fortifications and the End of the Sienese State," *History* 62 (1977): 372–387.

[37]They were first published by Toca (n. 8 above) along with a full transcription of the notations on the drawings.

[38]Heinrich Wurm has used the watermarks on the drawings to help the process of dating, and in his forthcoming *Die Architekturzeichnungen Baldassarre Peruzzi* will date them to ca. 1530. Wurm's dating contradicts Toca's wild hypothesis, seconded by Di Pasquale, that the drawings date to 1509–11, early in Peruzzi's career. Toca's argument is based on the lengthy notations; contra, see below, n. 43.

[39]This phase of Peruzzi's life is documented in Nicholas Adams, "Baldassarre Peruzzi as Architect to the Republic of Siena 1527–1535: Archival Notes," *Bulletino Senese di storia patria* 88 (1981): 256–67.

[40]One piece of evidence that might help associate the drawings with the dam proves to be less than one might have hoped. Only one of the sheets, *Decimo* (fig. 16), has a measurement that may be across the crest from side to side: 65 *canne*, 189.67 meters. In fact, this may only refer to a section of the dam. The exposed face of dam is called 15 *canne* long; the perspectival exaggeration is not credible if the total width is only 65 *canne*.

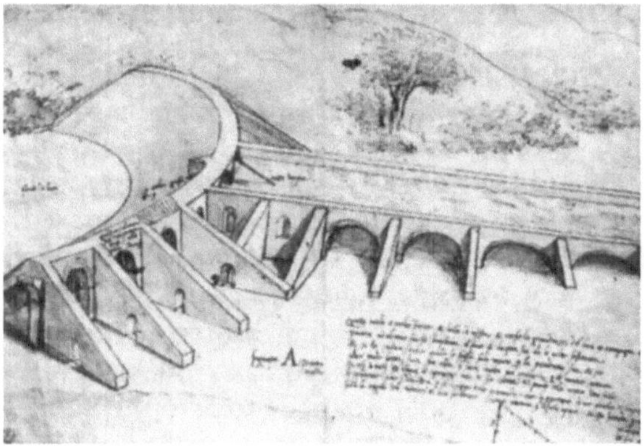

FIGS. 12–17.—Baldassarre Peruzzi, six reconstruction proposals for the Bruna River Dam, Gabinetto Disegni e Stampe degli Uffizi, Florence, Architettura 584, 585, 586, 587,

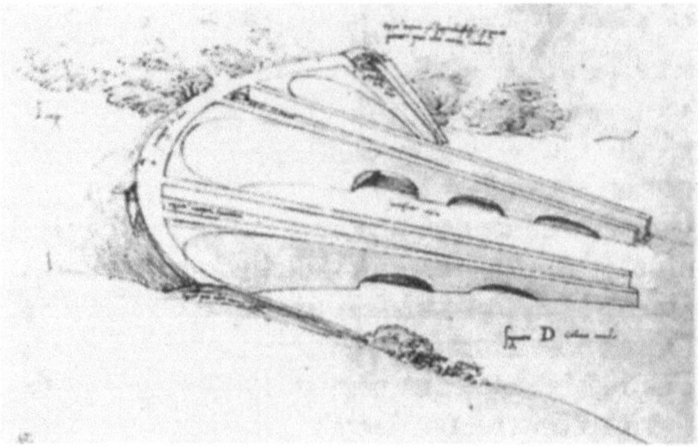

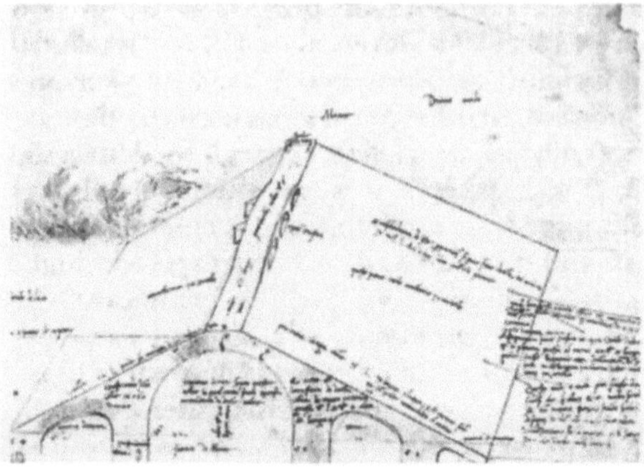

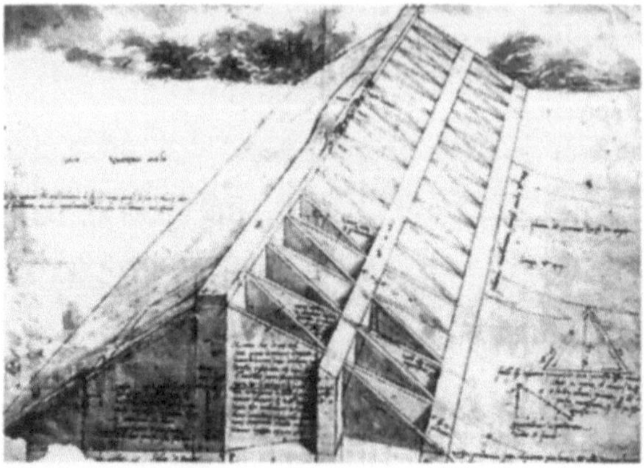

588, 589. These drawings are identified in the text as *A*/1 (top left), *B*/2 (middle left), *C*/5 (bottom left), *D*/8 (top right), *Decimo modo* (middle right), and *Undecimo modo* (bottom right). (Photographs from Gabinetto Fotografico, Soprintendenze Gallerie, Firenze.)

associates Peruzzi with the dam. But in order to avoid the conclusion that these sheets are, in fact, proposed reconstructions of the Bruna River Dam, we should have to posit an entirely undocumented large-scale water-storage dam elsewhere and some way to document Peruzzi's presence during these years at that dam. In favor of the association with the Bruna is the vegetation sketched in by Peruzzi, strongly reminiscent of the Maremma area, and a note on *B*/2 (fig. 13) that the water escape is proposed in a manner to avoid "the great fall [of water] which causes ruin." In short, the coincidences are too great to dismiss; the drawings are reconstructions to the Bruna River Dam and datable to ca. 1530. What do they show us about the development of Sienese understanding of dam building?

It is not possible to say how many projects were prepared by Peruzzi. Evidently there were at least eleven, since UA 589 is called the *Undecimo modo* (eleventh mode). All the sheets have a number and all but the *Decimo* and *Undecimo* (figs. 16, 17) are given a letter designation as well. There appears to be some relation between the letters and numbers: *A*/1, *B*/2, *C*/5, *D*/8. Possibly letters and numbers belong to different stages in the design process, the numbers coming first (one through eleven at least) and the designs then being lettered, singling them out for further consideration. From Peruzzi's commentary on the drawings we can gain an insight into his major concerns with the design of the dam: stability and the water outlet. Figures 12–15 (*A*/1, *B*/2, *C*/5, and *D*/8) have ramp spillways to carry the water away from the toe of the dam. *D*/8 is, perhaps, the most elaborate, with four spillways; the two at either end raised above the central pair by half a *canna* (1.45 meters) in order to take off excess water in case of "great rain" (*gran piove*). Piers are used in some cases to reduce turbulence. The *Decimo* and *Undecimo* resolve the problem of the outlet differently, each being designed to let the water flow over the entire air face as a spillway. In order to protect the surface of the dams, air and water faces are both planned to be laid with bricks *a cortello* (i.e., *a coltello* or on knife-edge with every third or fourth brick laid on edge) to protect the surface.

The baffling variety of ramp and outflow channels provided by Peruzzi is further complicated by the number of different ways that he conceives of the wall itself. There are exposed buttresses (figs. 12, 14 [*A*/1, *C*/5]) and cellular buttresses (fig. 13 [*B*/2]); there is a dam that is curved into the river (fig. 15 [*D*/8]), one that is curved away from the course of the river (fig. 12 [*A*/1]), and dams that run straight across the course of the river (*B*/2, *C*/5, *Decimo modo*, *Undecimo modo*). It is clear that, since the state of Renaissance mathematics did not allow Peruzzi to compute the weight of the water, the strength of the materials, or the mechanics of the soil, he was, roughly stated, letting his architectural fantasy run across a variety of practical solutions to the problems of water storage.

As diverse as the drawings seem at first, Peruzzi followed principles based on the traditions of Sienese dam building, principles of classical architecture, and the traditions of Florentine dam construction. Indeed, variations in plan were common enough among the designs of Peruzzi's Sienese predecessors; curved faces are seen in Taccola and Francesco di Giorgio (figs. 6, 9), and straight faces are also seen (figs. 7, 10). The use of coffers filled with lime and gravel is seen in Francesco di Giorgio, too, much like the elaborate systems for the *Decimo* and *Undecimo*. The *Undecimo* also requires the use of a cellular frame much like Francesco's suggestions.[41] Although Peruzzi probably did not know Francesco di Giorgio personally, he was certainly deeply interested in the older man's work. A series of colored renderings from Francesco di Giorgio's *Trattati* by one of Peruzzi's most able students, Pietro Cataneo, shows that, in the years of his activity for the Republic of Siena, Peruzzi was almost certainly aware of (if not directly responsible for ordering) copies of Francesco di Giorgio's *Trattati*. Dated to 1534, these drawings in the Gabinetto dei Disegni degli Uffizi, Florence (UA 3292), reproduce Francesco di Giorgio's dams. Yet Peruzzi's architectural vision was much wider than that of his predecessor.

The first obvious difference from the Bruna River Dam is the section adopted by Peruzzi for all his designs. While the dam was wedge-shaped, all Peruzzi's proposals are based largely on a wide-based triangle. Since none of the drawings shows extensive attention to foundations, and it is unlikely that Peruzzi thought of excavating down to bedrock, the simple weight of the wall was expected to hold the water. Hence, Peruzzi's apparent understanding of the failure of the Bruna River Dam was that it was too light; his drawings are gravity dams, pure and simple, with no lack of weight. In the lower right-hand corner of the *Undecimo modo* (fig. 17) Peruzzi has drawn an "equilateral pyramid," as he calls it, and a small isosceles triangle with sides 3, 4, 5. On sheet *A*/1 is a similar sketch (fig. 12) with the notation: "So on all these drawings my opinion is that against the weight of the water the slopes are made as is shown with the drawing of the pyramid; that is to say, the proportion of three, four and five, following the rule of the theaters and their slope."

The design principles are derived from the ancient wisdom of Pythagoras and validated by Roman theaters and amphitheaters that had proved their solidity by surviving the ages. (Did Peruzzi also know, one

[41]Francesco di Giorgio Martini, *Trattati di architettura ingegneria e arte militare*, ed. C. Maltese (Milan, 1967), plate 12.

wonders, that an amphitheater was sometimes filled with water, like an artificial lake, for mock naval battles?) Peruzzi's proposals also reflect classical antiquity in other ways. The spillways on a number of sheets (figs. 12, 13, 14 [*A*/1, *B*/2, *C*/5]) recall Roman aqueducts or bridges of the type he undoubtedly had studied for his own projected book on antiquities. Classical models were not only a formal precedent, they also provided structural principles that he could adapt to the problems of the dam.

It is possible too that Peruzzi had seen the great Florentine weir at Remole.[42] Approximately 200 meters long, but only 4 meters across the top, the weir raised the head of the Arno to provide a constant flow of water into a fulling mill. Although the Remole weir was much lower than the Bruna River proposals, its equilateral triangular cross section (or that of weirs like it) could have provided for Peruzzi an alternative to the Bruna River Dam and the Taccola and Francesco di Giorgio proposals. In two of Peruzzi's proposals (figs. 16, 17 [*Decimo* and *Undecimo*]), there is a similar cross section.

Questions of expense were naturally on Peruzzi's mind. He called the *Decimo modo* "not too expensive compared to the other drawings," and the *Undecimo* was recommended as "the least expense compared to all the others I am sending."[43] The sheet Peruzzi singled out for special praise was *B*/2: "this mode is the strongest of all," he wrote. Although we might criticize the open cellular buttresses as a water trap in case of overtopping, *B*/2 seems to incorporate the strongest features of all the proposals: (*a*) a cellular buttressed air face, (*b*) a Pythagorean proportion system called "anterides" and noted by L. B. Alberti in his *Ten Books on Architecture*,[44] and (*c*) resonances of classical antiquity. In short, Peruzzi combined an understanding of the Sienese and Florentine traditions of dam building (from Taccola to Francesco di Giorgio to Remole) with a heightened awareness of ancient architecture—not

[42]Enzo Salvini, "Un flash di archeologia industriale: le gualchiere trecentesche di Remole (FI)," *L'Universo* 62 (1982): 121–46. See also the so-called cavalciotto near Prato; Renato Piattoli, *Lo Statuto dell'arte dei padroni dei mulini sulla destra del fiume Bisenzio* (Prato, 1936). It is typical of the place of the so-called traditional arts within the history of technology that the subject of dams and weirs should have been so infrequently raised. In the otherwise fascinating volumes published by the Touring Club Italiano, *Campagna e industria: I segni del Lavoro* and *Campagna e industria* (Milan, 1981), dams and weirs are mentioned only briefly.

[43]This note suggests to me that Peruzzi was not in Siena for the presentation of the proposals. This would explain the lengthy notations. It might be that a documented 1532 Grosseto trip, known to have been followed by a lengthy stay in Rome, provides the appropriate sequence of events to match the drawings.

[44]See Hans-Karl Lücke, *Alberti index, Leone Battista Alberti de re aedificatoria* (Munich, 1975).

ancient dams, of which Peruzzi seems to have been ignorant, but the traditions of Roman civil architecture.[45]

A number of factors combine to make these sheets remarkable as drawings. For one thing, there is their extraordinary number. I know of no other example of six alternatives for the same project by one hand that has survived from the Renaissance. Furthermore, these drawings are of the highest quality and are probably in a state to be presented to the Sienese building committee, an accident of preservation unusual in itself.[46] Moreover, they maintain a balance between the purely artistic and the architectural for the purpose of making a clear demonstration to people not familiar with architecture or the techniques of architectural drawing.[47] The extensive notes are one aid, but more interesting is the way that Peruzzi controlled the perspective diminution in order to make a clear demonstration of each dam's potential. This gentle violation of the perspective can be seen in all the drawings. For example, on *B*/2 the slope of the water face is accentuated in order to expose the water escape system (fig. 13). On the *Decimo modo* (fig. 16) Peruzzi commented on one technique he adopted. There, the flank of the dam is shown as if it were parallel to the picture plane. But the view is distorted to show us the air and water faces as well.

[45]If one strips off the classical references (aqueducts, bridges, etc.) from Peruzzi's drawings, the dams would seem reasonable today. Mr. Martino and Mr. Baskin of the Office of Dam Safety in Harrisburg registered little surprise at Peruzzi's proposals. Dams such as the Rock Run Dam (1907) near Coatesville, Pa., were shown as equivalent buttress gravity dams, not unlike *C*/5 (fig. 14). Only the spillway is markedly different on Peruzzi's drawing. Peruzzi seems also to have only partially grasped the value of piers at the crest, employed today to control turbulence (see figs. 14, 15 [*C*/5 *D*/8]). The placement of auxiliary spillways, higher than the main spillways (see fig. 15 [*D*/8]) was also noted as a modern feature, but the undirected spillways on the *Decimo* and *Undecimo* (figs. 16, 17) were considered poor design. All in all, Martino and Baskin confirmed the remarkably practical nature of the Peruzzi proposals, judging them to be better proposals than some that crossed their desks today (interview, February 16, 1984). Even chute spillways such as those shown in some of Peruzzi's drawings can be found in a modified form in dams such as Kochess Dam; see William P. Creager, Joel D. Justin, and Julian Hinds, *Engineering for Dams* (New York, 1945), 1:210. Although Peruzzi did not understand the kind of grading of the weights of earthfill practiced today, the profiles he adopted can be seen in H. G. Arthur, "Earthfill Dams," *Design of Small Dams*, United States Department of the Interior, Bureau of Reclamation (Washington, D.C., 1960), pp. 157–223. Buttress dams recalling those of Peruzzi can also be seen on the Big Thompson River in Colorado (Olympus Dam), *Design of Small Dams*, p. 67.

[46]The only comparable survival that I know of is associated with the projects for the facade of San Petronio, Bologna. See G. Zucchini, *Disegni antichi e moderni per la facciata di San Petronio di Bologna* (Bologna, 1935).

[47]On this matter, see Wolfgang Lotz, "The Rendering of the Interior in Architectural Drawings of the Renaissance," *Studies in Italian Renaissance Architecture* (Cambridge, 1977), pp. 1–65.

Peruzzi noted: "It did not seem necessary to make another illustration of the plan because one can recognize it from the flanks." Peruzzi made a presentation that was both comprehensible and visually appealing. Perhaps this explanation was intended to reassure the more literal minded among his audience that the dam was not splayed! One further way Peruzzi tried to maintain his audience's interest in the structure was by setting the dams in their natural landscape—the trees, the scrubby bushes, and the hills of the Maremma.

One might imagine that, with Peruzzi working on the dam, the Sienese were worse off than they had been with Pietro dell'Abbaco. Peruzzi, so far as we know, had no experience with hydraulic engineering earlier in his career; but that probably troubled neither him nor the Sienese. We have noted that the elements put together by Peruzzi came from many sources: Florentine weirs, bridges, aqueducts, amphitheaters. By picking Peruzzi the government expected to get something new, something that would not repeat the old mistakes. Moreover, like Peruzzi's other great project under way at this time, the fortifications for the city of Siena, the designs would display the intelligence, wealth, and civic-mindedness of the Sienese and their government.[48] With his evocation of the structures of the ancient world in particular, he provided something that only a specialist in architecture could provide. Peruzzi was not just another expert in "chiave d'acque" like the *parmigiano* Giovanni Maria, he was an expert in the art of design.

Peruzzi was Sienese too. But in dam construction, did that matter? I think so. Siena's traditions in architecture are not just displayed in certain materials (brick) or characteristic forms (paneled arcades) or favored colors (black and white) but also in answers to practical problems. The experts who built the elaborate fortifications for Siena, for example, faced the same topography century after century and learned from it; so, too, with the famous aqueducts and fountains that watered the city. Architects and builders rubbed shoulders for generations over these matters. In working over the same ground, traditional wisdom was by necessity passed on as new generations tried to improve the techniques and methods of past generations. In that sense, the Bruna River Dam grew out of the collective experience with water in the aqueducts and streams of the *contado*. Pietro dell'Abbaco brought a special mathematical skill and knowledge of building art to the task; Taccola had been his spiritual teacher at least. Francesco di Giorgio must have learned much from Pietro dell'Abbaco. Peruzzi, as Pietro

[48]See Nicholas Adams, "Le fortificazioni di Baldassarre Peruzzi a Siena: 'Cosa veramente fortissima et bella da celebrarsi sempre,' " in *Rilievi di fabbriche attribuite a Baldassarre Peruzzi* (Siena, 1982), pp. 21–52.

Cataneo reminds us, learned from the work of Francesco di Giorgio. Here the Sienese tradition is alive in a way that is sometimes only partially apparent in the civil and ecclesiastical architectural forms of Siena.[49]

In the end we have not one dam but several dams—built, collapsed, and unbuilt. There is an initial structure, its modification by Pietro dell'Abbaco, and the six variations by Peruzzi. While I have, up until now, stressed the Sienese nature of these projects, as a conclusion let us place the Sienese dams in the context of other Mediterranean structures of a similar nature.

* * *

There has been no specialized study of the Italian dam. A bibliographical list published in 1969 noted only a handful of Renaissance works. Although Norman Smith in his *History of Dams* noted a number of others, it is clear that much material remains buried in local historical records.[50] In central Italy, the dam at Fucecchio, a low water-storage structure, was built in the 15th century, but when the marshy area was drained in 1788 by Pietro Leopoldo, the dam was left to ruin.[51] Another Tuscan dam was built near Lucca to create the Lago di Sesto, but in 1561 that dam was lowered to provide more land for cultivation.[52] Insufficient evidence of those structures survives for detailed comparison.

Part of our problem in the study of these matters comes from the general lack of records dealing with dams and weirs. Dam building was one of the traditional arts, unworthy of architects and not generally part of architectural education.[53] There seems to be no systematic way to assess the size of the corpus. Yet even the incomplete state of our

[49]This argument will form the thesis of my discussion of architectural practice in Siena in a forthcoming article.

[50]The problem is not so much finding references to dams but finding information about them. A useful 19th-century study of the northern Italian water supply is R. Baird Smith, *Italian Irrigation*, 2 vols. (London, 1855), which was written as a report for the East India Company.

[51]See Emanuelle Repetti, *Dizionario geografico fisico-storico della Toscana*, 6 vols. (Florence, 1833), 4:13–19.

[52]See Girolamo Tommasi, *Sommario della storia di Lucca* (Florence, 1847), p. 435.

[53]Dams do not appear in the architectural treatises of L. B. Alberti or Vitruvius and seem never to have passed into general architectural or engineering education, not even in the United States. In the academy today there is extensive education in soil mechanics, hydrology, and structures, but the would-be dam maker has to learn his art in the field with the Army Corps of Engineers or the Bureau of Reclamation or one of the state departments of environmental resources. Dams are still partly a traditional art, and engineers are fond of telling stories of local craftsmen and masons who can "eyeball" a

present knowledge suggests that great civic dams were relatively rare. When governments in the Middle Ages or Renaissance chose to make a dam, it seems, more often, to have been associated with a military attack. Filippo Brunelleschi attempted in 1429 to divert the Serchio River in order to flood the city of Lucca.[54] Leonardo da Vinci attempted to divert the Arno during Florence's Pisan Wars.[55] In 1393 a certain Domenico da Firenze tried to dam the river flowing through Mantua so that the sieging army, running along the dry riverbed, could take the city.[56] A similar attempt was made at Pavia in 1521.[57] It is a measure of the limited nature of Italian experience in dam and weir construction that none of these efforts was successful.

The Italian artist most frequently associated with the study of water was Leonardo da Vinci; his work for the Sforza seems to have encouraged these interests. In Milan he was active in work on the Martesana Canal (1494–98), and a drawing in the Codex Atlanticus (fol. 33v) shows locks and weirs along a relatively placid river.[58] Other drawings for the Florence-Pisa Canal (1503) and the Canal of San Cristofano (1509) show similar kinds of work. Leonardo owned a copy of Francesco di Giorgio's *Trattato*, although it is unlikely that the structure of weirs built under his supervision depended in any way on the Sienese master for their form.

In northern Italy there are a number of 15th-century dams and weirs because of the many irrigation and canalization projects along the Po. One in particular, the Cento Dam near Cesena of 1450, has been investigated and gives us some clues that would suggest that Francesco di Giorgio's designs for dams were not so unusual. The structure is made from brick placed within mortar over a wooden framework and held together by wooden bars and oak poles. Although no illustration of its internal structure is known to me, Norman Smith's

---

dam—that is to say, they can estimate the height and thickness of a dam needed for a certain location, estimates which engineers later confirm with more sophisticated calculations.

[54]Cornelius von Fabriczy, *Filippo Brunelleschi: La vita e le opere*, 2 vols. (Florence, 1979), 2:375–79.

[55]L. H. Heydenreich, "The Military Architect," in *The Unknown Leonardo*, ed. L. Reti (New York, 1972), pp. 136–65.

[56]Michael Mallet, *Mercenaries and Their Masters: Warfare in Renaissance Italy* (London, 1974), pp. 167–169.

[57]Charles Oman, *The Art of War in the Sixteenth Century* (New York, 1937), p. 137.

[58]Leonardo owned a copy of Francesco di Giorgio's treatise. See also Carlo Zammattio, "The Mechanics of Water and Stone," in *The Unknown Leonardo*, pp. 190–215; the Codex Atlanticus is richest in such matters. These events are discussed in most general histories of Leonardo; see L. H. Heydenreich, *Leonardo da Vinci*, 2 vols. (Munich, 1954).

description recalls Francesco di Giorgio's weirs[59] (See fig. 10.) The practice of covering the surface of the work with brick, although not endorsed by Francesco di Giorgio, does appear in Peruzzi's *Decimo* and *Undecimo* (figs. 16, 17). The first large-scale water-storage dams that we know about in any detail come from later in the 17th century;[60] this seems to suggest that, at least by its scale alone, the Bruna River Dam was relatively unusual for its time on the Italian peninsula.

Perhaps because of its Muslim and Roman past and a climate that favors irrigation, Spain is the Mediterranean country with the strongest traditions of dam building.[61] Moreover, since the study of the dam as a constructional or architectural type is still relatively new, we are effectively at the mercy of those few earlier studies of dams, and one area where much archaeological work has been done is in Spain. José A. García-Diego's article, "Old Dams in Extremadura," provides many examples of early dams.[62] At Castellar (1500) and Trujillo (1572) are dams that may be compared with that of the Sienese. In section the Castellar Dam seems not unlike the Sienese dam as completed by Pietro dell'Abbaco; rectangular in form (6.80 meters across the top and slightly wider at the bottom), it has buttresses at the air face. Architecturally, the profile of the Castellar Dam is slightly more complex than the Bruna River Dam—its buttresses are an integral part of the dam face in a way they never could have been in the Sienese work—and the addition of a mill into the structure was more elaborate than the simple pond created on the Bruna. At Trujillo, one of the architects, Francesco Becerra, was an architect-engineer, like Francesco di Giorgio and Peruzzi; Spanish practice was thus similar to the Italian.[63] Trujillo is lower (7.10 meters across the top by 10.30 meters) and it is stocky and weighty, not unlike the Bruna River Dam.

More interesting than these early Spanish dams, which seem to grow from local tradition like the Bruna River Dam, are the drawings in the Codex in the Madrid National Library traditionally attributed to Juanelo Turriano.[64] García-Diego has rightly pointed out that these

[59]Smith (n. 1 above), pp. 151–52.

[60]See ibid., pp. 152–153, for a discussion of works such as the Lago di Ternavasso.

[61]Ibid., pp. 102–29; Thomas F. Glick, *Irrigation and Society in Medieval Valencia* (Cambridge, 1970).

[62]José A. García-Diego, "Old Dams in Extremadura," *History of Technology* 2 (1977): 95–124.

[63]It is interesting that we find builders with similar training in architecture as a fine art working on dams during the 16th century in Italy and Spain.

[64]Ladislao Reti, "The Codex of Juanelo Turriano (1550–1585)," *Technology and Culture* 8 (1967): 55–66; J. A. García-Diego, "The Chapter on Weirs in the Codex of Juanelo Turriano," *Technology and Culture* 17 (1976): 217–234.

drawings, dated between 1564 and 1569, prove the advanced state of Spanish knowledge of hydraulics. Yet from an Italian perspective there seem to be many elements comparable to Italian dams. The lacing together of fascines in the forest weir recalls techniques discussed by Francesco di Giorgio.[65] The truss formwork for the wooden weir likewise appears in Francesco's treatise, although in much simplified design; his frame uses mortar (lime and gravel) rather than the cut stone used by Juanelo Turriano.[66] Turriano also illustrates a module for a dam that may be filled with earth and stone.[67] It is not unlike the Peruzzi *Undecimo modo* (fig. 17), although the Spanish dam has a squared masonry cover rather than the bricks described by Peruzzi, and its linking posts are rather more complex than anything on the surface of Peruzzi's work.

García-Diego looked in vain for sources for these dams in the Turriano Codex dams: "At least as a working hypothesis we must be satisfied with assuming that, where the author does not invent, he is mainly guided by contemporary Spanish and Italian structures. . . ."[68] We must still learn much more about Italian dams, but drawings like those of Peruzzi suggest the existence of an "architectural" link between the vernacular dam tradition which emerges into the more academic atmosphere of Taccola, Francesco di Giorgio Martini, and Baldassarre Peruzzi, and an architect-engineer and scientific rationalist like Juanelo Turriano.

The question of the attribution of the Madrid Codex to Juanelo Turriano is one of immense interest in this discussion. If the codex were definitely identified with Turriano (who was from Bologna and whose real name was Giovanni della Torre), we would have the basis for suggesting contact with Italy in a general way and the possibility of direct contact with Peruzzi, who worked at various times for Charles V and visited Bologna on at least one occasion.[69] In place of this traditional attribution, questioned both by García-Diego and by Ladislao Reti, no new name that I am aware of has yet been offered.

* * *

Since I am a historian of architecture, my interest in the dam on the Bruna River is, perforce, different from that of the historian of tech-

[65]García-Diego, "The Chapter on Weirs," fig. 2.

[66]Ibid., fig. 3.

[67]Ibid., fig. 8.

[68]Ibid., p. 233.

[69]The dates of Peruzzi's Bolognese sojourn are not certain, though he may well have been there ca. 1523–25. He was also active at the siege of Florence, 1529–30. The subject of Peruzzi in Bologna is under study by Richard Tuttle, Newcomb College.

nology. I see the dam as an indicator of progressive human skill at owning and controlling the landscape. The changes in architectural form, from the mixture of local tradition and native skill to one of high architectural and artistic complexity, is not just the result of the practical problems posed by the fishpond; the practical matters must be done well but the gritty difficulties need an appropriate architectural and structural language. For architects of the early Renaissance the question could only be answered in terms of tradition and the vernacular, a language that failed when called on to develop new inflections. In the High Renaissance the answer took on new wide-ranging meanings.

Giorgio Vasari, the mid-16th-century biographer and artist, noted that architecture would never have become so great had not necessity forced men to solve problems of comfort and practical need.[70] In the light of Vasari's observation (and one could note many others like it), it is remarkable that historians of art and architecture generally avoid concepts of style in technology and the traditional arts. By the same token, historians of technology may be interested to see that change is not always the product of rational decision; change is also the product of a newfound freedom to manipulate formal architectural language in ways that are not only artistic (that goes without saying) but almost scientific in the degree of their experimentation.

[70]Giorgio Vasari, *Le Opere di Giorgio Vasari*, ed. G. Milanesi, 8 vols. (Florence, 1906), 3:195.

5
The evolution of the arch dam

N.J. Schnitter

PART ONE

The development of arch dams, like that of so many other structures and techniques, was mainly determined by experience, which more often than not preceded the development of the pertinent theories. This account emphasizes the structures marking the principal technical innovations and the concepts leading to the methods most widely used today. Furthermore the presumable or proven transfers of the technology between different regions and periods are stressed, as well as the internationality of its extraordinary development and propagation in the present century.

The forerunners

THE FIRST DAMS go back to the earliest civilizations, the very formation of which was often conditioned by the need for large hydraulic works[1]. The early structures used the more obvious force of gravity to resist the water pressure and were mostly of the earth or rockfill type. These were also preferred by the Romans who appear, however, to have introduced the arch into dam engineering, which they used with such mastery in their buildings and bridges. The corresponding evidence is rather sparse and mostly circumstantial. This is especially true for a small dam in the Vallon de Baume south of Saint-Rémy de Provence in France. Its foundations were discovered and recorded in the 18th century[2], but are now almost completely covered by a curved gravity dam built on the same site in 1891[3]. An old drawing shows that the ancient dam was a relatively strongly curved, sturdy cylinder and consisted probably of two concentrical masonry walls with an impervious earth core in between, as may still be observed at the ruins of the Roman gravity dams near Orükaya and Cavdarhisar in Turkey[4]. Its maximum height must have been about the same as that of the present structure, ie, some 12m†. Roughly at mid-height was the intake for a lead pipe, which carried the water from the small reservoir to the colony of Glanum a few hundred metres to the north.

Several centuries later we find evidence that the basic principle of the arch dam was well understood by the Roman engineers. Procopius of Caesarea relates in his *Buildings* (book II, chapter 3), that when Chryses of Alexandria was commissioned by the Byzantine emperor Justinian I (527-565) to erect a flood control dam for Dara near Mardin on the Turkish-Syrian border, "he did not build this barrier in a straight line, but in the form of a crescent, in order that its arch, which was turned against the stream of the water, might be better able to resist its violence". Nothing else is known about this dam but a picture (Fig. 1) published by A. Poidebard may represent its remains[5].

With the breaking up of the Roman empire and its cultural and economic ties, dam building came to a standstill in the western world, while it continued to flourish in Japan, southern India, Sri Lanka, Arabia and Iran. Especially noteworthy in the present context are the developments in the last mentioned country, since they established a firm tradition in the construction of masonry dams, highlights of which were the Sassanian (3rd century) weir-bridge combinations in the Khuzestan province, the Bujid (10th century) gravity dams for the operation of water-mills in Fars and the Ghasnawid (12th century) gravity dams in Khurassan. The theme was taken up again under the Mongol Ilkhans towards the end of the 13th century, when they reconstructed the country ravaged half a century earlier by the onslaught of their conquering armies.

Fig. 1. The probable remains of a flood control dam on the Turkish-Syrian border built by Chryses of Alexandria (527-565).

Among the Iranian dams from the Ilkhanid period are also some arch dams, the oldest still existing structures of this type. Best known is the Kebar dam, 160km south of Tehran[6] (cover). It has a maximum height of 26m and its sturdy central part shows a modest curvature. While its upstream face is a vertical cylinder, the downstream face is inclined so that the thickness of the masonry increases from 6m at the crest to some 9m at the base of the dam. The masonry consists of limestone blocks set in a lime mortar containing ash from thorn bushes, which like pozzolanas, gives the lime hydraulic properties. Next to the right abutment a spiral staircase descends some 10m into the dam, which has several openings to the upstream face at different levels and a downstream outlet at the bottom. Thus, the withdrawal of water was assured even as the relatively small reservoir silted up progressively.

Less well researched but even more remarkable is the Kurit arch dam in southern Khurassan, 620km southeast of Tehran, which is more than twice as high as Kebar and held the world height record of all dam types until the beginning of the 20th century[7] (Fig. 2). It is located in a narrow gorge and is also completely silted up. Part of the vertical downstream face is badly damaged, thus exposing a shaft and other openings as well as the same

† For additional structural data see Table I.

The 26m-high Kebar arch dam in Iran dating from the end of the 13th Century.

type of masonry as is found at Kebar. Some 40km northwest of the Kurit dam there is a smaller similar structure named after Shah Abbas (1587-1629), but dating from the Mongol period[7]. The dam is likewise located in a narrow gorge, the bottom of which was, however, not cleared completely before construction of the dam. Upon filling of the reservoir, the alluvium left in place washed out together with the lowest part of the dam, the remainder of which was underpinned with a brick vault under the Safavids in the 16/17th century. Under their rule, Iranian dam engineering attained a new peak, although the construction of arch dams was discontinued.

The reasons for the short-lived revival of the arch dam concept, forgotten since the disruption of the Roman world, in Iran under the Ilkhans are not known at present. That it was imported by the Mongols from China is unlikely, since dams as well as the arch in general have never been much used by the Chinese. Quite to the contrary, in 1274 the Mongols sent the Iranian engineer Sa'id Ajall Shams-al-Din (1210-1279) as governor to Kunming in southern China, where he initiated the construction of a dozen dams[8]. That the arch dam concept came from the west, with which the Ilkhans maintained diplomatic as well as cultural relations (eg, the Polo voyages), is equally improbable, as the first of several curved dams built in southeastern Spain is about one century younger. If a transfer of technology ever occurred, it must therefore have been from east to west by the intermediary of the Moslem civilization, in which Iran played an important role at that time.

Southeastern Spain had been settled during the Moslem invasion in the 8th century by Syrians and Yemenites, who brought from their arid homelands not only many irrigation techniques[9] but already a considerable experience in dam engineering[10]. The Moslem period saw the construction of several low diversion weirs, while the earlier mentioned curved dams were built after the area in question had been reconquered by the Christians[11]. Until about the end of the 15th century this amounted, however, mainly to a political-military takeover, leaving unchanged much of the Moslem infrastructure, especially with respect to irrigation procedures and techniques[9].

Fig. 2. The 64m-high Kurit arch dam in Iran built at the end of the 13th century.

The earliest curved dams in southeastern Spain are very sturdy structures, relying for stability mainly on gravity and very little on arch action. The Almansa dam, 80km northwest of Alicante, consists of a 15m high curved portion completed in 1384, to which a 6m high polygonal gravity section was added in 1586. The Tibi dam erected from 1580 to 1594, under the reign of Philip II (1556-1598) and after approval by the famous hydraulician Juanelo Turriano (1500-1585), some 20km north of Alicante is 42m high and exceedingly thick. Both dams were built of rubble masonry set in lime mortar and faced with carefully placed cut stones. Both also show fairly refined hydraulic features. One is the scouring gallery provided through their bases for purging the reservoirs of silt accumulations. In view of this phenomenon the outlet for the irrigation water at Tibi dam was additionally equipped with a vertical intake shaft connected to the reservoir by a multitude of small inlets at various levels, similar to the arrangement used at Kebar and most other Iranian dams since the end of the 13th century.

A true arch dam is the Elche dam (Fig. 3), the construction of which was begun in 1632 some 20km west of Alicante. It consists of a 23m high, cylindrical main arch, which is curved only modestly in relation to its relatively high length-to-height ratio. The upper part of the main arch rests on the right side against an abutment block, an artifice used quite frequently in arch dams, sometimes even on both valley sides. From the abutment block a smaller secondary arch takes off, while on the left side the main arch ends abruptly in front of a narrow gully, a short wing wall closing off the reservoir upstream. The clever utilization of the peculiar topographical features of the site, as well as the satisfactory distribution of the stresses in the main arch (Fig. 4), testifies to the fine structural feeling of its designer(s).

The last mentioned quality is remarkably lacking in the 17th century Relleu dam 30km northeast of Alicante. Located in a narrow gorge, the lowest 7m of which were

Fig. 3. The 23m-high Elche arch dam in south-eastern Spain, the construction of which was begun in 1632.

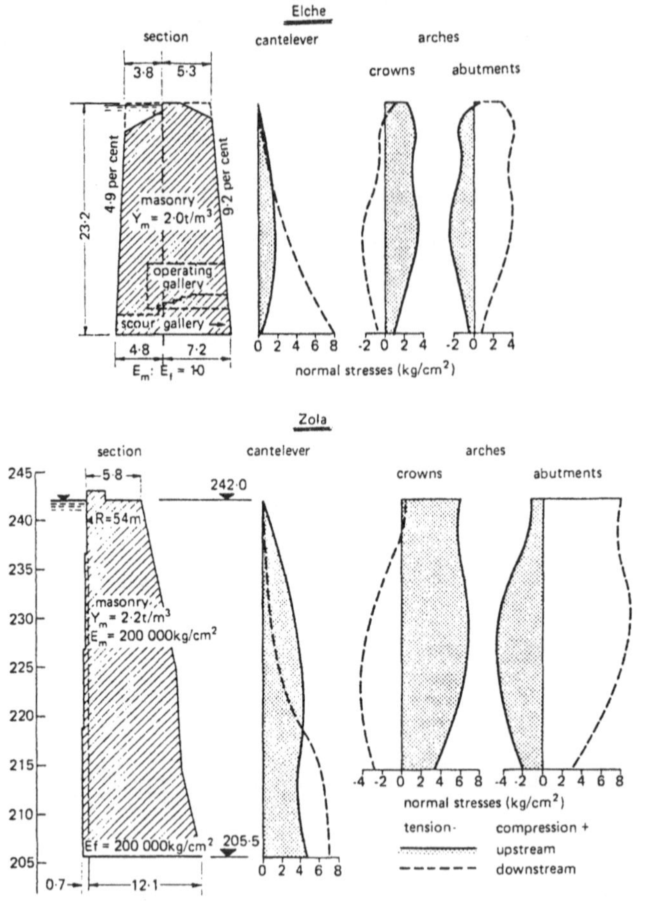

Fig. 4. Results of stress analyses for the Elche and Zola dams derived by the modern "arch and crown-cantilever adjustment" method (contours and dimensions in metres).

barely wide enough to accommodate the scouring gallery, the dam has an almost plane upstream face, while the likewise vertical downstream face is curved only very slightly. Its stability relies therefore not in arch action, but rather in that of a wedged-in triangular plate, the thickness of which is constant in any vertical section. The height of the dam was increased by 5m in 1879, to compensate for the complete silting up of the irrigation reservoir.

After its "degeneration" at Relleu the arch dam concept was abandoned in later Spanish dam construction in favour of buttress, gravity and earth fill structures. The two last mentioned types continued to be used in the old dam building countries like Japan, southern India and Iran, and were also almost exclusively adopted for the dams which began to be built in Turkey, Italy, Austria, Czechoslovakia, Germany, France, Great Britain and the Americas. A notable exception, and again a true arch dam, was the Ponte Alto flood control structure, the building of which was started in 1611 in the extremely narrow canyon of the Fersina near Trento in northern Italy[12]. Originally only 5m high, the dam was raised seven times and in 1887 reached its final height of 39m, over all of which the thickness of the masonry is practically constant. Shortly before the final height increase, the similar Madruzza dam was built downstream to protect the older lower parts of the Ponte Alto dam by creating a backstorage, which on the one hand counteracts the water pressure on them, and on the other, mitigates the shock of the spilled floods on their foundations.

The pioneers

The Ponte Alto dam was probably known at first hand by François Zola (1795-1847)—the father of the famous French writer Emile Zola—for he was born and educated in nearby Venice and in the early 1820s worked on the construction of a railway near Linz in Austria[13]. At that time Austria dominated both Venice and Trento, and the Ponte Alto dam was being raised for the third and fourth times. In 1837, soon after he had established himself as a consulting engineer in Marseilles, France, Zola took part in a competition for a new water-supply scheme for Aix-en-Provence. He proposed an aqueduct from the Infernet creek east of the city, whereby the large seasonal variations of the relatively modest run-off were to be regulated by three consecutive reservoirs. These were all to be formed by arch dams. After a long fight for his ideas, Zola was able to initiate the work on the first of his dams in 1847, but then died just two months later. Deprived of its leader, the work progressed only haltingly and was completed seven years later in 1854.

Named after its designer, the Zola dam is 43m high, modestly curved and fairly sturdy. Also, besides being a very fine structure, still in excellent shape, it has the additional distinction of being the first arch dam designed on the basis of an analysis of the stresses in it. The calculation supposed that the dam consisted of a stack of independent arches, each of which had to resist the total pressure corresponding to its depth below the reservoir water level. The average stress in every arch was determined by the "cylinder formula" as the product of pressure and slenderness*.

Structural statics

This simple relationship had already been found experimentally by the French physicist Edmé Mariotte (1620-1684) and was formulated in 1826 by Louis C. M. Navier (1785-1836), the creator of structural statics, who also laid the foundations for the theory of elastic

* Ratio of radius of curvature of an arch to its thickness.

Table I—Data on arch dams built up to World war I[1]

Name	Country	Period of construction	Height (m)	Length (m)	Structural characteristics[2]			
					L'/H	B/H	R/C	Ø
Baume	France	Roman	12	18	1·5	0·32	4	73
Dara	Turkey	ca. 550	?	?	?	?	7	?
Kebar	Iran	ca. 1300	26	55	0·8	0·35	6	40
Kurit	Iran	ca. 1300	60[3]	27	0·4	?	?	?
Abbas	Iran	ca. 1300	20	?	?	?	?	?
Elche	Spain	1632–?	23	120	3·0	0·50	7	67
Relleu	Spain	17th century	27[4]	23	0·9	0·37	6	22
Ponte Alto	Italy	1611–1887	39	12	0·3	0·10	4	46
Jones Falls	Canada	1828–31	19	107	5·6	0·44	8	115
Zola	France	1847–54	43	66	1·5	0·34	9	77
Parramatta	Australia	1856/98[5]	16	69	4·3	0·29	33	80
Madruzza	Italy	1883	41	15	0·4	0·14	5	52
Bear Valley	USA (California)	1883–84	20	91	4·7	0·31	105	46
Sweetwater	USA (California)	1886–88	30	116	3·9	0·50	18	100
Tamworth	Australia	1898	20	134	?	0·33	83	100
Mudgee	Australia	1899	15	152	10·0	0·22	84	117
Wellington	Australia	1899	15	107	7·0	0·20	50	133
Crosis	Italy	1898–1900	38	40	1·0	0·46	?	?
Upper Otay	USA (California)	1901	23	86	3·7	0·16	90	60
Barossa	Australia	1899–1903	36	144	4·0	0·29	45	136
Lithgow 2	Australia	1906/15[5]	27	68	2·5	0·27	29	126
Medlow	Australia	1907	20	30	1·5	0·14	17	78
Crowley	USA (Oregon)	1908/15[5]	28	49	1·8	0·10	22	181
Gullspang	Sweden	1908	21	90	2·6	0·14	13	80
Pathfinder	USA (Wyoming)	1905–09	65	132	1·4	0·45	14	72
Ponte Serra	Italy	1907–09	44	46	1·0	0·33	11	80
Klinte	Sweden	1909	25	58	2·3	0·20	27	85
Buffalo Bill	USA (Wyoming)	1905–10	99	61	0·6	0·33	15	80
Halligan	USA (Colorado)	1908–10	29/22[6]	107	4·9	0·38	162	62
Gallinas	USA (N.M.)	1910	15	64	4·2	0·31	62	49
North Crow	USA (Wyoming)	1910–11	19	46	2·4	0·19	19	111
Huacal	Mexico	1911–12	30	56	1·4	0·18	20	106
Cebu	Philippines	1912	30	70	2·3	0·13	35	116
Goodwin	USA (California)	1912	24	142	2·9	0·21	16	99
Big Creek 4	USA (California)	1912–13	24	67	2·8	0·35	?	?
Triunfo	USA (California)	1913	15	45	3·0	0·38	?	60
Salmon Creek	USA (Alaska)	1913–14	51	195	3·8	0·25	55	114
Corfino	Italy	1313–14	38	81	1·7	0·18	15	158
Spaulding	USA (California)	1913/16/19[5]	84/64[6]	244	2·8	0·46	44	76
Clear Creek	USA (Washington)	1914/18[5]	26	123	4·7	0·12	42	?

1. After 1880 only dams of over 15m height
2. L' = crest length of arch, excluding wing walls or artificial abutments, H = maximum height, B = base width, R = radius of crest arch, C = crest width, R/C = slenderness and Ø = central angle of crest arch (degrees)
3. Before heightening by 4m about 1850
4. Before heightening by 5m in 1879
5. Subsequent heightening, to which structural characteristics are referred
6. Part above gravity base, to which structural characteristics are referred

arches. The latter was formalized only in 1854 by Jacques A. C. Bresse (1822-1883), like Navier a French teacher and researcher in civil engineering sciences[14]. Notwithstanding the simple calculation Zola had to employ, the stresses in his dam are reasonable even when analysed according to present-day methods (Fig. 6).

Despite this pioneering example, it was almost 80 years before the next French arch dam of similar size was built on the Bromme river in the Massif Central. In part this may have resulted from the fact that the earliest description of the Zola dam appeared only in 1872 and in another context[15], while the deserved wide publicity was given to it in the first standard work on *The Design and Construction of Dams*, published in 1888 by the Swiss-American hydraulic engineer Edward Wegmann (1850-1935).

In the meantime the design of gravity dams had been placed on a rational basis, which resulted in a preference for this type of structure during the following half century. Many of them were given a slight curvature in plan as recommended by the French engineers Auguste M. I. Graeff (1812-1884) and Emile F. X. P. Delocre (1828-1908) in the fundamental papers on the design of masonry dams, which they published in 1866 and in which they also described all the old Spanish structures but not the Zola dam[16]. In his paper, Delocre even outlined a method of analysis for arched dams very similar to that used by Zola, whereby he admitted a maximum permissible stress of $0{\cdot}6MN/m^2$, ie, less than 10 per cent of what is generally allowed today[17]. In an analogous study in 1879 Albert G. Pelletreau (1843-1900)[18] found that to minimize the volume of an arch dam its radius of curvature should decrease from crest to base, which according to the "cylinder formula" also yields smaller arch thicknesses for a given allowable stress.

No publicity at all was received by the small but remarkable Jones Falls dam (Fig. 5), built from 1828 to 1831 in the virgin forests of Canada for a barge canal from Ottawa to Kingston on Lake Ontario[19]. The structure is 19m high, strongly curved and shows almost the same, considerable thickness from crest to base. Especially noteworthy are its unprecedentedly high length-to-height ratio and the fact that the masonry blocks were set in vertical, instead of the usual horizontal, courses. Also unusual is the clay fill placed against the upstream face as an additional water barrier. The dam, as well as the whole canal, was designed and its execution directed by Lieutenant-Colonel John By (1779-1836) of the British Army engineering corps, who was recalled from retirement

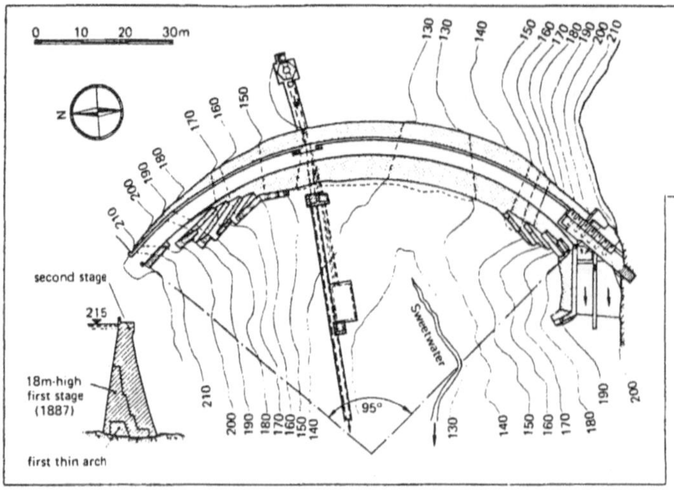

Fig. 6. Plan and median section of the 30m-high Sweetwater arch dam completed in 1888 south-east of San Diego, California (contours in feet).

and sent to Canada for the task. Besides the Jones Falls dam, he built also several low curved weirs like the one constructed by the great English engineer John Smeaton (1724-1792) on the Coquet river north of Newcastle-upon-Tyne in England. Although the curvature of these weirs served to increase the discharge of the overflowing water, they might have inspired By to use the arching effect to strengthen the Jones Falls dam. A similar structure which he had built on a straight alignment at the Hog's Back in Ottawa had failed early in 1829.

At the other end of the British empire a small arch dam was built in 1855/56 near Parramatta east of Sydney Australia[20]. It was designed by Captain Percy Simpson (1787-1877), who had come to Australia in 1822 as commandant of a convict settlement and later became surveyor and land commissioner. The dam shows a considerable length-to-height ratio and is remarkably slender. In 1898 some 3m of concrete was added to the top of the masonry structure under the direction of Cecil W. Darley (1842-1928), who emigrated from Great Britain in 1867 and at the time was chief engineer of the Public Works Department of New South Wales. Concurrently he initiated the construction of a whole series of very thin arch dams in the Blue Mountains behind Sydney to provide small water-supply reservoirs in the newly settled arid areas[21]. These dams all had one vertical cylindrical face, were of moderate height (the highest was the Lithgow No. 2) and were the first arch dams to be built entirely of concrete*, which was batched by volume, hand mixed, transported in wheel barrows, placed in thin layers and consolidated by tamping. It contained roughly 280kg/m^3 of cement, but reached only about one quarter of the strength attainable today with a similar mix. This was sufficient to cover the average stresses of up to 2·6MN/m^2 calculated still with the "cylinder formula" for independent arches, but left a narrow safety margin with respect to some local stress concentrations resulting from the boldness of the dams' design. The most daring is the Mudgee dam which is very thin notwithstanding its extreme length-to-height ratio[22].

Fig. 5. The 19m-high Jones Falls arch dam built in 1831 in eastern Ontario, Canada.

When Darley started the construction of his arch dams he could rely not only on the example set by the Parramatta dam, but also on that of an even bolder structure in California publicized by Wegmann's earlier-mentioned book. This was the (Big) Bear Valley dam built in 1883/84 at over 2000m above sea level in the San Bernardino mountains some 130km east of Los Angeles to the design of the young engineer and land developer Frank E. Brown (1856-1914). Although only moderately curved, it set a record of slenderness. It performed quite satisfactorily but was replaced in 1911, because of fears raised by the considerable volume of its irrigation reservoir of almost 50m^3 × 10^6, by a more conservatively designed buttress dam of the multiple arch type.

Southeast of San Diego the quite similar Upper Otay dam is, however, still in operation. Slightly north of it, the Sweetwater dam had also been started as a thin cylindrical arch, but was changed to a far sturdier structure when the later renowned consulting engineer James D. Schuyler (1848-1912) took over the direction of the works[23]. In contrast to its forerunners the upstream face of the Sweetwater dam is not vertical, having instead a slope similar to the downstream side, so that in place of the upstream face radii of curvature, those of the arch axes are constant from crest to base (Fig. 6). The overall construction costs of the dam amounted to some $15 a cubic metre. When an inflation factor of about six is taken into account, this corresponds almost to present day costs for a similar structure. A short time after its completion the Sweetwater dam successfully withstood almost two days of overtopping by a flood, thus testifying to the inherent strength of a well-built arch dam.

The breakthrough

The Sweetwater dam served as a model for, among others,

* Although concrete had already been used in some Roman dams, its modern form based on Portland cement was introduced into dam engineering in 1866 at the Boyds Corner gravity dam 80km north of New York, USA.

the Pathfinder and Buffalo Bill (formerly Shoshone) dams in central and northwestern Wyoming, USA. Built by the newly formed Bureau of Reclamation of the United States Department of the Interior from 1905 to 1910 these dams reached the unprecedented heights of 65 and 99m respectively. Accordingly they also required considerable quantities of construction materials, ie 50 000m³ of masonry for the Pathfinder and 63 000m³ of concrete for the Buffalo Bill dam. Even more remarkable than their dimensions, however, is the fact that they were the first arch dams, which were not designed merely as a stack of individual arches[24]. Their interdependence was taken into account by considering also the median vertical section of the dam which is statically a cantilever fixed at the base. The water and other external loads were then divided in vertically and horizontally carried parts by adjusting the horizontal deformations of the median cantilever to the radial deflections of the arch crowns. The latter were determined according to the theory of elastic arches. This theory finally became generally accepted, although it still contained imperfections and considered the arches as rigidly fixed at the abutments[25].

The distribution of the loads resulted in a considerable relief for the lower arches, while the upper ones had to support partially the cantilever. The new method of analysis had been developed in the late 1880s by Hubert Vischer and Luther Wagoner (1847-1922) for checking the stresses in the Bear Valley and Sweetwater dams, but was published in a local journal and thus escaped general attention[26]. The latter was achieved only after the publication of the method in 1904 by Silas H. Woodard (1870-1961)[27], who had formulated it anew while designing the Cheesman curved gravity dam 60km south of Denver, Colorado, which incidentally is also the headquarters of the US Bureau of Reclamation.

In the discussion of Woodard's paper, Gardner S. Williams (1866-1931), then professor of hydraulic engineering at the Cornell University in Ithaca, New York, renewed the suggestion of the French engineer Pelletreau to decrease the radius of curvature of an arch dam from crest to base. Besides the advantages already evident from the "cylinder formula", the theory of elastic arches now also taught that the new design would improve the stress distribution within the lower arches, since the shorter radii entailed larger central angles. In contrast to the cylindrical or constant-radius layout used until then, the new design was therefore named "constant-angle" arch dam by its patentee and first applier Lars R. Jorgensen (1876-1938)[28]. A native of Denmark holding German degrees in mechanical and electrical engineering, Jorgensen had emigrated to America in 1901, where he worked as a hydroelectric engineer in California and founded his own consulting firm in San Francisco in 1914.

In the same year, Jorgensen's first constant-angle or variable-radius arch dam was completed for a hydro powerplant on the Salmon Creek near Juneau in southeastern Alaska. Although its median vertical section resembles that of the Sweetwater, Pathfinder and Buffalo Bill dams, the shape of the Salmon Creek dam is strictly the consequence of the new layout and has become standard for the variable-radius type of arch dam. The upstream bulging serves to off-set the undercutting by the stronger curved lower arches near the abutments while the pronounced downstream toe reduces the tensile stresses at the upstream heel. The full realization of the new design was made possible by building the dam completely with concrete, of which a total of 40 000m³ were required and which contained 280kg/m³ of cement. From the aggregate processing and the mixing plants near the upstream heel of the dam, the concrete was lifted to the top of a temporary steel tower, then chuted into the forms. See Fig. 7.

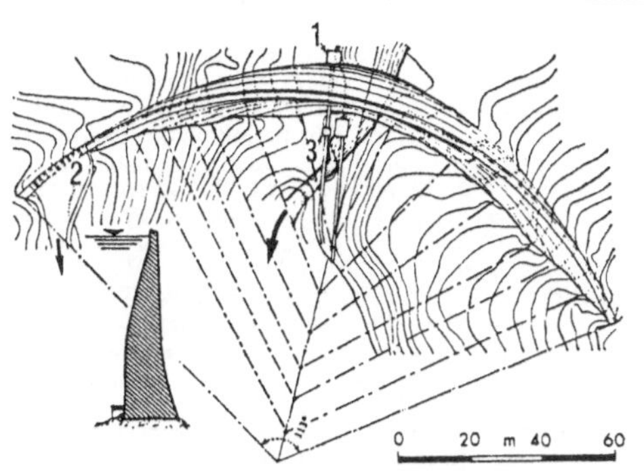

Fig. 7. Plan and section of the Salmon Creek variable-radius arch dam near Juneau, Alaska. Built by Jorgensen in 1914 it became the standard design for this kind of dam. The intake is indicated by the figure 1, the spillway by figure 2 and the valve by figure 3.

To allow for the contraction of the concrete when the heat produced during its set cools off, the Salmon Creek dam was subdivided into three construction blocks by two vertical joints—a further innovation in arch dam engineering. At the Lake Spaulding dam, 110km east of Sacramento, California, which had been begun as a curved gravity dam but whose upper three quarters were continued from 1913 on as a variable-radius arch dam, the contraction joints were provided with keys and filled with grout injected through drill-holes[28]. In later dams special grouting pipes and outlets were embedded in or next to the joints, the spacing of which was moreover reduced to about one quarter of the 50 to 70m used at the Salmon Creek dam.

The technical, as well as economic, advantages of Jorgensen's design were immediately recognized, and it became rapidly very popular in the west of the United States and elsewhere. In this it received additional impetus from the quickening pace in the development of hydroelectric power resources. The variable-radius arch dam was to prove especially suitable for high and large structures, although it never replaced the cylindrical type completely. Scandinavian designers, the Bureau of Reclamation, as well as the French engineers persisted for a long time in the use of the cylindrical design even for large structures.

The Jorgensen type of arch dam was introduced in Europe a few years after its first use by Heinrich E. Gruner (1873-1947). He had inherited a renowned consulting firm in Basle, Switzerland, and had closely followed the developments in America ever since his stay there in 1900/01. In 1917 he was commissioned to design the Montsalvens power dam 20km south of Fribourg, Switzerland. Completed in 1920 the structure is 55m high, strongly curved and relatively sturdy (Fig. 8). Its horizontal sections are not circular but follow the funicular curves for the part of the water load supported by them. Moreover, the arch thickness increases from the crown towards the abutments to take care of the stress concentrations near the latter.

The two innovations resulted from the use of a considerably improved method of analysis published in 1913 by Hugo F. L. Ritter (1883-1956), a son of Wilhelm Ritter (1847-1906), the well-known professor for statics at the institute of technology in Riga, Latvia, and Zurich, Switzerland[29]. Ritter (Jnr) had undertaken a study tour through the United States in 1909/10 and was familiar with Woodard's work. His method was refined by Gruner's collaborators Alfred Stucky (1892-1969) and Henri Gicot (b. 1897), who later became the two most prominent Swiss arch dam designers. In the improved analysis, the radial deflections of the arches were adjusted to those of the median vertical section as well as some additional cantilevers on both sides. This permitted the determination of the variations in the distribution of the loads acting on the dam both in the vertical and in the horizontal sense[30].

Ritter's method of analysis was further expanded between 1923 and 1935 at the Bureau of Reclamation under the direction of John L. Savage (1879-1967), its chief design engineer for over 20 years. The principal additions were the adjustment of the tangential deflections and the rotations in horizontal planes, as well as taking into account of the deformations in the foundation rock. Instead of solving the many linear equations tying the load distribution on arches and cantilevers to the adjustment of their deformations, it was at that time quicker to seek the proper load distribution by trial and error.

The procedure, which was supplemented by a large body of auxiliary tables, was therefore aptly named "Trial Load Method"[31]. After adaptation to electronic data processing and many additional refinements the method remains the one most frequently used. However, it still represents only an approximation of the theory of thick shells, which arch dams actually are. The many attempts since the early 1920s to directly use the latter theory stumbled more often than not over the practical difficulties of its application. These are being overcome only now with the introduction of the analysis by small finite elements and the availability of computers having sufficient storage capacity and speed.

Parallel to these theoretical developments, great efforts were made to collect pertinent data from structures under construction or already in operation. The first measurements of the deflections of an arch dam under load appear to have been those at the 12m high thin arch dam on the Barren Jack Creek in New South Wales, Australia, in 1908/09[22]. They were carried out from scaffolding on the downstream face, while at the Salmon Creek dam surveying methods were introduced for the purpose, which were perfected at the Montsalvens dam and later structures. Montsalvens was also the first arch dam to be equipped with permanent instruments, in this case electric resistance thermometers embedded in the concrete and connected by cables to the measuring stations*. On the initiative of Fred A. Noetzli (1887-1933), a Swiss engineer who emigrated to America in 1915 and became one of the foremost dam designers in the western United States, an 18m-high cylindrical arch dam was built in 1926 on the Stevenson Creek 65km northeast of Fresno, California, for the sole purpose of studying its behaviour under closely controlled loading conditions[32].

The measurements on the Stevenson Creek experimental dam were checked by testing 1/40 and 1/12 scale models of it at the Universities of Princeton, New Jersey, and Boulder, Colorado, respectively[32]. The first model consisted of celluloid, while the second was built of a concrete similar to the one used for the prototype. In both tests the water load was simulated with a thin layer of mercury to obtain exaggerated and hence measurable deformations and strains, from which the stresses could be calculated.

Although two-dimensional models of gravity dams had been tested already at the beginning of the century, the use of this technique in arch dam engineering constituted an innovation which proved highly beneficial. Today the method has evolved along two distinct lines, represented by the two laboratories which are the most prominent in the field. One is the Laboratorio Nacional de Engenharia Civil set up in 1946 in Lisbon, Portugal, which uses small celite-plaster models loaded with mercury. By contrast the other, the Instituto Sperimentale Modelli e Strutture founded in 1951 in Bergamo, Italy, prefers relatively large pumice-concrete models, to which the loads are applied by hydraulic jacks.

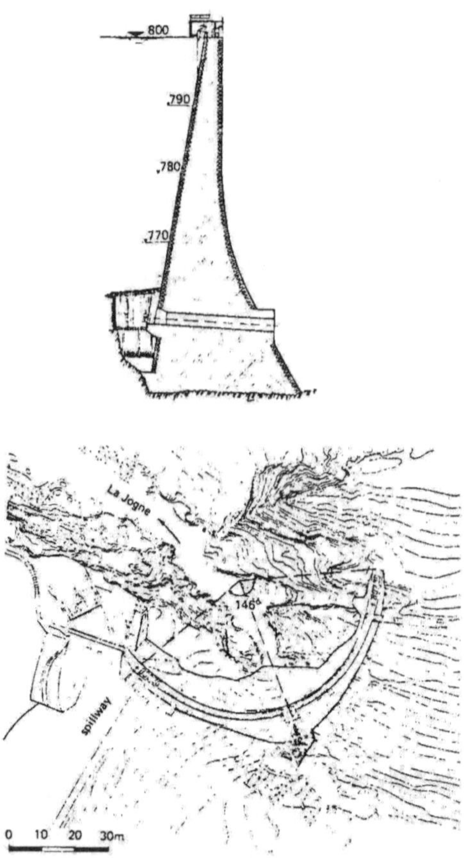

Fig. 8. Plan and median section of the 55m-high Montsalvens dam completed in 1920 south of Fribourg, Switzerland.

Involuntary "tests" of sorts were finally provided late in 1925 and early in 1926 by the 16m-high Moyie arch dam in northern Idaho and by the 19m-high Lake Lanier arch dam in western North Carolina, both of which lost their right abutment over more than two thirds of their height without collapsing[33]. These first failures of arch dams thus attested once more to the great strength reserves inherent in this type of structure.

(To be continued)

* The first deformation measurements on any type of dam were undertaken on the Eschbach curved gravity dam completed in 1891 near Remscheid in the Ruhr, Germany, while the earliest temperature measurements occurred at the Boonton gravity dam built in 1903 in northeastern New Jersey.

PART TWO

The growing use of scale models in the 1920s and the refined analyses of arch dam behaviour have been supplemented by "involuntary" tests, in which existing dams were subject to unanticipated stresses. The widening understanding of their behaviour has been reflected in the increasing number of arch dams being built.

ALTHOUGH two-dimensional models of gravity dams had been tested at the beginning of the century, the use of this technique for arch dams didn't follow until some time later. For instance, the Stevenson Creek dam was built in California in 1926 solely to study its behaviour under controlled loading conditions, and measurements taken were compared with two scale models to validate and refine the process.

In addition to the improved understanding gained by modelling techniques, the behaviour of the Moyie and Lake Lanier arch dams (USA) both of which lost their right abutments without collapsing in 1925 and 1926 respectively, gave added confidence to the designers.

Maturity

Much of the aforementioned great effort to clarify the structural behaviour of arch dams was undertaken in connection with the proposed construction of the Hoover (formerly Boulder) dam on the Colorado river southeast of Las Vegas, Nevada, USA, by the Bureau of Reclamation. Nevertheless, and in understandable deference to its maximum height of 221m (roughly twice as high as any dam then in existence) and to the unprecedented size of its reservoir of $37\,m^3 \times 10^9$, Hoover dam was built (1931-1936) as a massive curved gravity structure. This somewhat paradoxical caution on the part of the leading dam design office initiated the decline in the construction of true arch dams in the United States. This was exemplified by the developments in California (Fig. 9) where the number of new arch dams as well as their proportion in relation to the other types decreased sharply in the 1930s.

At the same time a rise in arch dam construction occurred in Europe and especially in Italy, where it had been almost as popular as in California[34]. Until the 1920s this concerned, however, smaller cylindrical structures located in narrow gorges. These also had relatively conservative dimensions, except for the Corfino dam completed in 1914 some 50km north of Pisa. It was designed, like several other arch dams built in the following decade, by Angelo Omodeo (1876-1941), in whose renowned consulting office Claudio Marcello (1901-1969) worked for some time. Marcello was later chief engineer for hydro powerplants in the Edison Company of Milan and designed many dams, including some major arch dams.

The variable-radius principle was used in Italy for the first time by Gaetano Ganassini (1875-1932) at the 50m-high Gurzia dam, completed in 1925, 40km north of Turin, with an important variation: the upstream bulging of the central part of the dam was carried so far as to result in a pronounced downstream overhang of the crest. Thus, the structure depended on arch action for its stability under the forces of gravity alone and the upstream heel was liable to crack and leak under the water load. This drawback was corrected by Fabio Niccolai (1892-1944) at the Osiglietta dam built 60km west of Genoa just before World War II. The correction consisted of undercutting the upstream heel, thus compensating to some extent for the downstream overhang of the crest and resulting in a double-curved dome-structure (Fig. 10). Williams had proposed this arrangement in 1904 in his previously mentioned discussion of Woodard's paper[27]. On the basis of a model test—one of the earliest to be carried out in Europe—the Osiglietta dam was furthermore provided with a hingelike perimetric joint, which

Fig. 9. Development of arch dam construction from 1900 to 1970 in California, Italy, France and Switzerland.

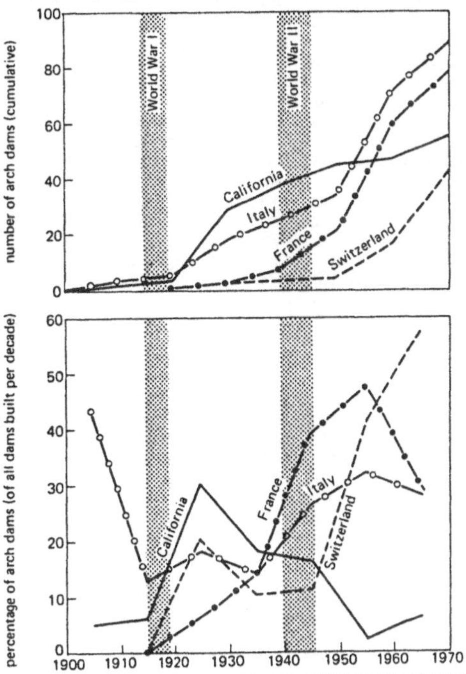

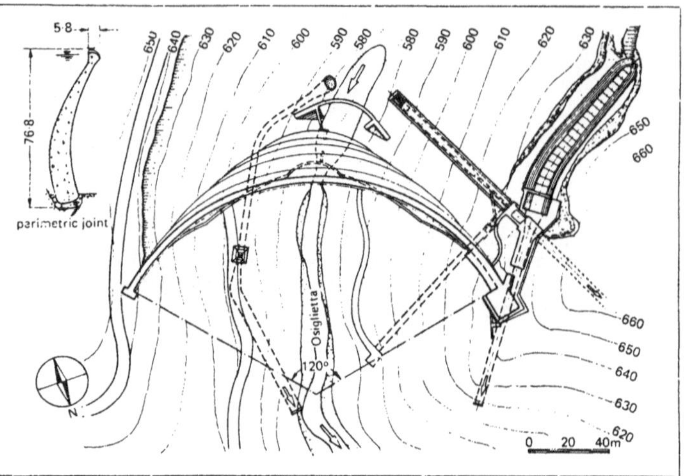

Fig. 10. Plan and median section of the 77m-high Osiglietta dome dam built in 1939 west of Genoa, Italy (contours and dimensions in metres).

separates the dome from a foundation pad and reduces the upstream tensile stresses along the abutments.

The last mentioned provision, which was rarely adopted outside Italy, became characteristic for the many important arch dams designed by Carlo Semenza (1893-1961), for a long time chief engineer for hydro powerplants of the Società Adriatica di Elettricità in Venice[35]. His most ambitious work, and still the highest completed arch dam in the world, is the 262 m-high Vajont dam built from 1956 to 1960 in a narrow gorge 90 km north of Venice. Its reservoir of $169m^3 \times 10^6$ content was filled up to a great extent on October 9, 1963, by the largest recorded rock-slide, which was triggered by the water impoundment. The resulting wave overtopped the dam and killed more than 2000 people downstream, but did little damage to the structure itself.

A similarly tragic end marred the career of the most prominent French arch dam designer André Coyne (1891-1960), who first rose in the ranks of the Ponts et Chaussées public works department and after World War II founded his own consulting firm specializing in dam engineering[36]. In his first large structure, the 90m high Marèges arch dam completed in 1935 in the Massif Central, Coyne used the variable-radius principle with an undercut upstream heel. But thereafter he reverted, especially in his high dams, to a rather sturdy cylindrical type, the downstream face of which he defined by horizontal spirals and which he analyzed on the basis of independent arches cut from the dam by inclined instead of horizontal planes. Characteristic examples of this design are the Aigle, Bort and Chastang dams built soon after World War II on the Dordogne river up- and downstream of the Marèges dam. The highest of them, and the highest French dam, is the 180m high Tignes dam completed in 1952 100km northeast of Grenoble.

Simultaneously with these large structures Coyne designed many smaller arch dams, on which he experimented with new ideas and sometimes adopted extremely bold dimensions. The eventual result was several failures, the most spectacular of which was the collapse on December 2, 1959, of the 66m high Malpasset dam 45km southwest of Nice, when the poor foundation rock gave way on the left flank upon the first complete filling of the reservoir. The 47m high Gage dam, 65km west of Valence, had to be replaced after some years of operation by a more conventional arch dam farther upstream, while the 90m high, sheet-thin Tolla dam 10km east of Ajaccio, Corsica, was transformed into an arch-gravity structure soon after its completion[37].

Like most failures these had, however, their beneficial consequences, as they taught lessons which obviously had still to be learned. This is especially true of the Malpasset accident, which made engineers drastically aware of the importance of a sound foundation for even the most refined dam design and gave a great boost to the still relatively young science of rock mechanics

As is evident from Fig. 9, the proportion of new arch dams constructed in France decreased sharply after the Malpasset failure and the death of Coyne at the end of the 1950s, although some important structures were built afterwards. By contrast the great development which had begun in Switzerland after World War II continued through the 1960s. The great majority of the Swiss arch dams are of the variable-radius type with a moderate vertical curvature. The most important structure is the 237m-high Mauvoisin dam, 26km south of Sion, completed in 1957 to the design of Stucky. Among the smaller dams, Gicot's 51m-high Vieux Emosson dam, constructed in 1954/55, 40km southwest of Sion, is noteworthy for the first use of parabolic arches as an approximation of the funicular curves for the water load. At the 86m-high Les Toules dam completed in 1963 some 30km south of Sion, Gicot introduced elliptical arches, while his 57m high Schiffenen dam finished the same year, 10km north of Fribourg, is remarkable for its very high length-to-height ratio and its foundation on soft sandstone[38].

The 1960s also brought a revival of arch dam construction in the United States, especially after the adoption of the double-curvature design by the Bureau of Reclamation for some of its most important structures like the 153m-high Flaming Gorge dam completed in 1964 in northeastern Utah, the 160m-high Yellowtail dam built in 1966 in southern Montana and the 143-m-high Morrow Point dam completed in 1968 in western Colorado. Furthermore, arch dams became established in such centres of great dam building activity as the Iberian peninsula, Japan and the Soviet Union and in general spread around the globe.

In this period both the designs and the construction techniques reached a considerable uniformity. The latter now include the complete removal of all soil and rock unsuitable for the dam foundation, which is then tightened and consolidated by injecting grout into the underlaying rock. Additional drill-holes are provided to drain any remaining small water leakages. During this preparatory work the concreting plant is erected and a pit or quarry opened up for the procurement of the concrete aggregates. These are carefully washed, screened, batched and mixed with the cement, sometimes with a proportion as little as $160kg/m^3$. From the mostly automatic mixing plant, the concrete is transported to the dam in buckets suspended, in most cases, from cableways spanning the dam from one end to the other. The concrete is then spread by bulldozers and consolidated with powerful vibrators.

Table II—Percentage and number of high arch dams				
Period of completion	Maximum height (m)			Number of arch dams
	100–149	150–199	Over 200	
	per cent*	per cent*	per cent*	
Before 1940	28	–	–	5
1940–1949	50	33	–	5
1950–1959	43	75	100	28
1960–1969	32	66	56	51
1970–1979 (estimated)	24	39	44	29

*of all dams of the same height built in the decade.

Several thousand cubic metres of concrete are thus usually placed per day by a workforce of a few hundred men.

In spite of the high perfection and productivity attained by the construction procedure described, the arch dam has been losing ground since the early 1960s in relation to the other dam types, as is evident from the percentage of arch dams among the structures of more than 100m height built throughout the world in the last decades, Table 11.

Summary

Arch dams belong to those demanding civil works, the engineering of which depends to a considerable extent on intuition and the exchange of ideas and experiences. Such exchanges probably occurred within the Roman empire, where the earliest traces of arch dams can be found. After the disruption of the Roman world the arch dam concept was lost for almost a thousand years until it reappeared at the end of the Middle Ages, first in Iran and then in southeastern Spain.

The technique disappeared again in Iran as well as Spain and thereafter was used only sporadically and with little apparent connection in such disparate places as Italy, Canada, France and Australia. A definite relationship existed between the more or less systematic construction of arch dams in the last mentioned country towards the end of the 19th century and the simultaneous propagation of the concept in California. Both laid the basis for the subsequent breakthrough in the technique. Other important foundations were the development of methods for the structural analysis and the liberty of shaping afforded by the first use of concrete.

The breakthrough took place between about 1905 and 1925, mainly in the fast developing west of the United States. Important contributions also came from Switzerland, Italy and France, which took the lead in the field upon the decline of arch dam construction in the United States in the 1930s. By the end of that decade all important elements had been assembled for the spectacular qualitative as well as quantitative development of arch dams after World War II and their proliferation around the globe. At present the age of the arch dam seems, however, to be drawing to an end, except for very high structures, as shown in Fig. 11.

Competitors are the earth and rockfill dams, which are very flexible in their design and more amenable to highly mechanized construction procedures.

Fig. 11. Worldwide development of high dam construction since World War II, showing the successive replacement of the gravity type by the arch and of the latter by the earth and rock fill dams.

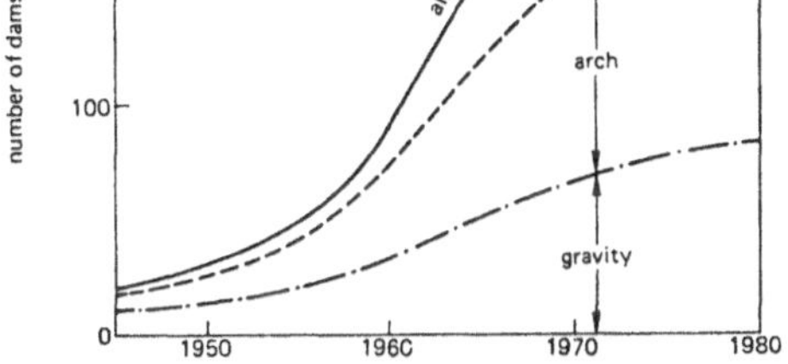

References

1. SMITH, N. "A History of Dams", P. Davies, London 1971 and SCHNITTER, N. J. "A Short History of Dam Engineering", *Water Power* 1967.
2. BENOIT, F. "Le barrage et l'aqueduc romains de Saint-Rémy de Provence", *Revue des études anciennes* 1935.
3. GOBLOT, H. "Sur quelques barrages anciens et la genèse des barrage-voûtes", *Revue d'histoire des sciences* 1967, and SCHNITTER, N. "Wasserbaugeschlichtliche Exkursion in die Provence", *Wasser- und Energiewirtschaft* 1972.
4. STARK, H. "Geologische und technische Beobachtungen an alten anatolischen Talsperren", *Die Wasserwirtschaft* 1957/58.
5. POIDEBARD, A. "La trace de Rome dans de désert de Syrie", P. Geuthner, Paris 1934.
6. GOBLOT, H. "Kébar en Iran, sans doute le plus ancien des barrages-voûtes; l'an 1300 environ", *Art et manufactures* 1965, No. 154.
7. GOBLOT, H. "Du nouveau sur les barrages iraniens de l'époque mongole" *Art et manufactures* 1973, No. 239.
8. NEEDHAM, J., LING, W. AND GWEI-DJEN, L. "Science and Civilisation in China", University Press, Cambridge 1971, Volume 4, part III.
9. GLICK, T. F. "Irrigation and Society in Medieval Valencia", Belknap Press, Cambridge, Massachusetts, USA, 1970.
10. VINCK, F. "Alte Talsperren im Yemen", *Wasser und Boden* 1962.
11. SMITH, N. "The Heritage of Spanish Dams", *Spanish National Committee on Large Dams*, Madrid 1970; and SCHNITTER, N. "Alte Talsperren in Spanien", *Wasser- und Energiewirtschaft* 1973.
12. NOETZLI, F. A. "Pontalto and Madruzza Arch Dams", *Western Construction News*, 1932.
13. RIGAUD, J. "L'ingénieur François Zola, sa famille, sa vie et son oeuvre", Académie, Aix-en-Provence, France, 1957.
14. TIMOSHENKO, S. P. "History of Strength of Materials", McGraw-Hill Co. Inc., New York, USA. 1953.
15. DE TOURNADRE. "Note A annexée à la description d'un barrage exécuté sur la rivière Verdon . . .", *Annales des ponts et chaussées* 1872, 1er semestre.
16. GRAEFF, A. "Sur la forme et le mode de construction du barrage du Gouffre d'Enfer sur le Furens et des grands barrages en général". *Annales des ponts et chaussées* 1866, 2me semestre.
17. DELOCRE, E. "Sur la forme du profil à adopter pour les grands barrages en maçonnerie des réservoirs", *Annales des ponts et chaussées* 1866, 2me semestre.
18. PELLETREAU, A. "Barrages cintres en forme de voûte", *Annales des ponts et chaussées* 1879, 1er semestre.
19. LEGGET, R. F. "The Jones Falls Dam on the Rideau Canal, Ontario/Canada", *Transactions Newcomen Society* 1957/59, and "Rideau Waterway", University of Toronto Press, Toronto, Canada, 1955 and 1972.
20. AIRD, W. V. "The Water Supply, Sewerage and Drainage of Sydney", Metropolitan Water, Sewerage and Drainage Board, Sydney, Australia, 1961.
21. NIMMO, W. H. R. "Historical Review of Dams in Australia", *Bulletin Australian National Commission on Large Dams*, January 1966.
22. WADE, L. A. B. "Concrete and Masonry Dam Construction in New South Wales", *Proceedings Institution of Civil Engineers*, 1908/9, part 4.
23. SCHUYLER, J. D. "The Construction of the Sweetwater Dam", *Transactions American Society of Civil Engineers* 1888.
24. WISNER, G. Y. AND WHEELER, E. T. "Investigation of Stresses in High Masonry Dams of Short Spans", *Engineering News* 1905, part II.
25. BAZANT, Z. "Entwicklung der Berechnung von Bogenstaumauern", 2nd Congress *International Association for Bridge and Structural Engineering* 1936, Preliminary Publication.
26. VISCHER, H. AND WAGONER, L. "On the Strains in Curved Masonry Dams", *Transactions of the Technical Society of the Pacific Coast*, December 1889.
27. HARRISON, C. L. AND WOODARD, S. H. "Lake Cheesman Dam and Reservoir", *Transactions American Society of Civil Engineers*, 1904.
28. JORGENSEN, L. R. "The Constant-Angle Arch Dam", *Engineering News* 1912, part II, and *Transactions American Society of Civil Engineers*, 1915.
29. RITTER, H. "Die Berechnung von bogenförmigen Staumauern", J. Langs Buchdruckerei, Karlsruhe, Germany, 1913.
30. STUCKY, A. "Etude sur les barrages arqués", *Bulletin technique de la Suisse romande*, 1922.
31. "Trial Load Method of Analyzing Arch Dams", US Bureau of Reclamation, Denver, USA, 1938.
32. "Report on Arch Dam Investigation", *Proceedings American Society of Civil Engineers*, May 1928, part 3.
33. "Two Arch Dams Fail Through Undermining of Abutments", *Engineering News-Record*, 1926.
34. "Le dighe di ritenuta degli impianti idroelettrici italiani", *Associazione nazionale imprese produttrici e distributrici di energia elettrica*, Milan, Italy, 1961, Volume I.
35. SEMENZA, C. "Arch Dams Development in Italy", *Journal of The Power Division of the American Society of Civil Engineers*, June 1956; and Symposium on Arch Dams, *American Society of Civil Engineers*, New York, USA, 1957.
36. COYNE, A. "Arch Dams, their Philosophy", *Journal of the Power Division of the American Society of Civil Engineers*, April 1956; and Symposium on Arch Dams, *American Society of Civil Engineers*, New York, USA, 1957.
37. MARY, M. "Barrages-voûtes; historique, accidents et incidents", Dunod, Paris, France, 1968.
38. GICOT, H. "Conceptions et techniques de quelques barrages-voûtes suisses", *Wasser- und Energiewirtschaft* 1961.

6

The evolution of buttress dams

N.J. Schnitter

PART ONE

After discussing historic buttress gravity dams and true buttress dams, the rapid evolution of this type of structure at the beginning of the 20th century is retraced. The evolution of reinforced concrete construction and the development of two distinct types of the multiple-arch and the flat-slab buttress dams are mentioned. This article forms part of the International Commission for Irrigation and Drainage's *History of Dams*, being prepared by the German National Committee of that Commission.

BUTTRESS DAMS are, in today's understanding, derived from the massive gravity types by the introduction of intermediate spaces, which allow water seepage through the foundation and dam body to discharge, not only downstream, but also sidewards and upwards into these spaces, thus greatly reducing the uplift pressures (Fig. 1). For the classical triangular gravity profile with a vertical upstream face, the total elimination of uplift pressures would theoretically permit a reduction of the dam's mass by 40 per cent without reducing its stability. However, the real reduction is considerably smaller, as the intermediate spaces must be closed upstream by slabs, arches or a thickening of the buttress heads to make them contiguous. Still smaller is the saving in costs, since buttress dams require more formwork of a more complicated shape and often a higher cement content in the less easily placed concrete. On the other hand, the heat produced during the setting of the cement is dissipated more readily in a buttress dam, and therefore elaborate temperature control measures are seldom required.

Larger savings are possible in the absence of uplift if the upstream face is also inclined, thus mobilizing the vertical water load on it for sliding stability. If the latter, in terms of the inclination of the resultant of the forces, is maintained the same as with the vertical upstream face, a buttress dam with 50 per cent (horizontal to vertical) slopes on both faces permits a theoretical reduction of the dam's mass of 64 per cent. At 100 per cent upstream and 33 per cent downstream slopes the reduction is as much as 89 per cent. Also the eccentricity of the resultant is greatly improved which moves from the downstream edge of the "middle third" in case of the vertical upstream face almost to the centre of the section with the 100 per cent upstream slope. Correspondingly, the vertical stresses are almost uniform over this section, although fives times larger. Except for high dams, this is welcome because it permits a better use of the strength of the construction materials.

* Motor-Columbus Consulting Engineers, Baden, Switzerland.

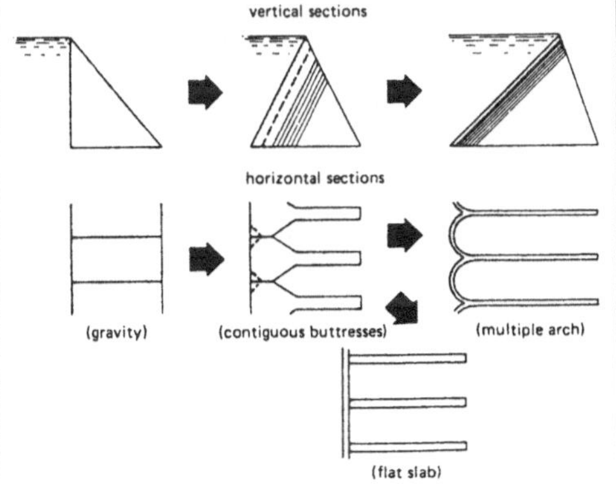

Fig. 1. Evolution of buttress dam types from the massive gravity dam.

Forerunners

These basic considerations were, of course, unknown to the Roman engineers who introduced for the first time not only the arch[1] but also the buttress dam. They may have developed its concept from buttressed earth retaining walls. In fact, some of their buttress dams, like the ones in the Wadi Megenin (Sidi Gelani)*, 40 km south of Tripoli, Libya[2], and at Olisipo near Lisbon, Portugal, are only partially and irregularly propped up rectangular walls which are rather slim. Closer to true buttress dams are those of Iturranduz, 30 km southwest of Pamplona in northern Spain[3] and of Esparragalejo, 10 km northwest of Merida in southwestern Spain[4]. While the water retaining element in the former consists of vertical walls placed between the buttresses at about their upstream thirds, these are interconnected by upstream arches at Esparragalejo, which is thus the oldest known multiple-arch dam. The flat vertical arches show only on the downstream side, whereas the steeply stepped upstream face is straight. The structure was built of rubble concrete-faced with masonry and was carefully reconstructed in 1959.

After the disruption of the Roman empire and its cultural and economic ties, dam building came to a standstill in the western world. Although it continued to flourish in various parts of Asia, the buttress dam concept was practically lost. An exception appears to be the Ereruisk dam, 40 km south of Leninakan in Soviet Armenia, which has double tuff walls with a loam core in between, propped up on the downstream side by small buttresses[5]. The structure dates from the early fifth century when Armenia, although caught between the Byzantine and Sassanid empires, experienced a cultural floresence.

The real re-emergence of the buttress dam concept, however, occurred more than one thousand years later in Spain, when that country attained the peak of its political and economic power under emperor Charles V (1516-1556) and his son King Philip II (1556-1598), who also initiated the construction of several dams. Among these were two irregularly buttressed gravity dams of rectangular cross-section.

One is the low Ontigola dam built by the King's architect and engineer Juan B. Herrera (1530-1597), 48 km south of Madrid, to irrigate the royal gardens at Aranjuez[6]. The other, higher, structure was begun in 1572 in Trujillo, 72 km northeast of Merida, to serve a mill further downstream[7]. Its construction was directed by Francisco Becerra, who later became a famous engineer and architect in Spain's American colonies. At

* For structural details, see Table.

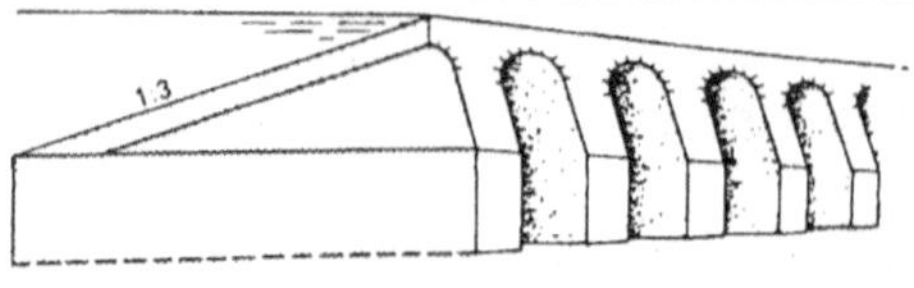

Fig. 2. Retraced drawing of proposed multiple-arch dam from the codex attributed to Juanelo Turriano (1560s).

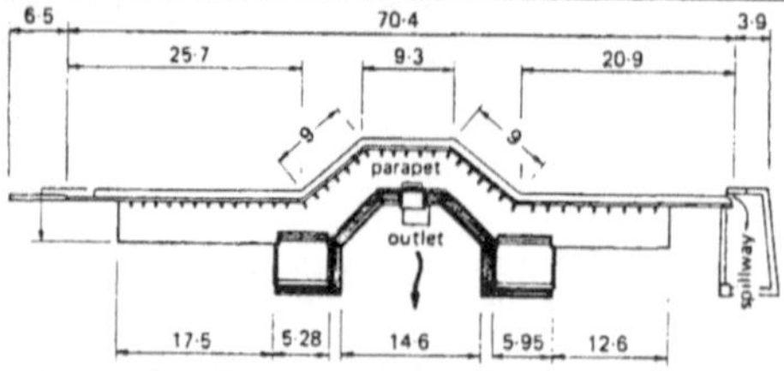

Fig. 3. Plan of the 18th century Topuzlu dam north of Istanbul (from Large Dams in Turkey[11]*).*

about the same time, the codex attributed to Juanelo Turriano (1500-1585), clock-maker and hydraulician to the King, claimed for its author the invention of the multiple-arch dam[8]. The author also proposed a sloping upstream face, which, he correctly stated, would help ensure stability, because of the favourable influence of the weight of water acting on it (Fig. 2).

This important insight was put to practical use about 150 years later in four low multiple-arch dams east of Bilbao in northwestern Spain, the most important of which is that of Bedia[9]. It consists of five flat arches, which, like the Roman dam at Esparragalejo, appear only on the vertical downstream side, while the sloping upstream face is straight. Thus the thickness of the arches increases horizontally from crown to buttresses and vertically from crest to base. The dams are solidly built of ashlar masonry, so as to permit spilling of the floods over their crests. The water diverted by them drove mills and forges, some of which were possibly owned or operated by the designer of the dams, Pedro B. Villareal de Berriz (1670-1740). A Basque nobleman, active in the iron industry and commerce and taking a lively interest in the great scientific progress of his time, Villareal developed empirical rules for the design of his multiple-arch dams, which he published in 1736[10].

Barely a decade after Villareal's exploits in northwestern Spain, the construction of buttress dams was resumed on a large scale in the southwest of the country to provide water storage and head for mills placed as a rule at the toes of the dams[7]. Their design did not, however, take advantage of the insights of the Turriano codex and Villareal, but reverted to the local pattern set by the Trujillo dam, that is, that of a rectangular wall propped up, more often than not unsystematically, by buttresses. Well over a dozen of such, usually low, structures were built by the end of the 18th century. The most important of them is the Feria dam[11], 47 km south of Merida, completed in 1747 to a height of 20 m and raised, in stages, by 1950 to 23.5 m. The spaces between its seven buttresses are closed downstream, roofed over and house two mills and a chapel, the latter a reminder that the dam's construction had been instigated by a bishop.

The clergy was also instrumental in transferring Spanish hydraulic techniques to Mexico, among which

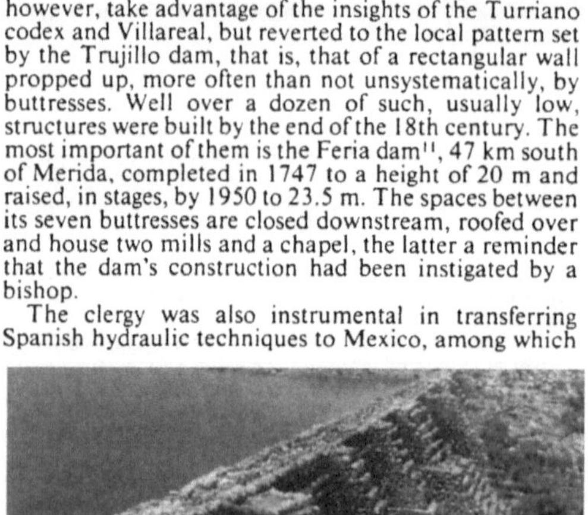

The 18th century Natillas buttress dam north of Aguascalientes in central Mexico (photo by the author).

was the type of buttressed gravity dam described earlier. The first was the Olla dam erected from 1742 to 1749 to protect the mining town of Guanajuato from floods. However, the many structures built west and north of the city of Aguascalientes in central Mexico[12], to form irrigation reservoirs and mill ponds (both purposes often being combined), are especially noteworthy. The largest of them is the San Blas dam near Pabellon 33 km north of Aguascalientes, while the lower Natillas dam some 20 km farther north is remarkable for its unusual cobble-masonry.

On the subject of unusual structures, it is appropriate to mention here some of the dams built under the Ottoman sultans in the Belgrade forest north of Istanbul for the water supply of that city[13,14].

While the Topuz dam, completed in 1620, is a lightly buttressed gravity type with a sturdy rectangular cross-section, the similarly designed Topuzlu and Valide dams of 1750 and 1796, respectively, are not straight in plan but have a polygonal upstream protrusion in their central part, which abuts against two massive buttresses, a queer mixture of buttressed gravity and multiple-arch dam (Fig. 3).

More unusual still is the Gasco dam, begun in 1787 about 30 km northwest of Madrid to the design of the French engineer and brigadier of the Spanish army Charles Lemaur and his sons [15]. The Gasco dam was intended to reach a maximum height of 87 m, which at that time was extraordinary and would have made it the highest dam in the world until early in the 20th century. It was designed as a buttress dam with sloping decks closing the intermediate spaces on both faces (Fig. 4). To

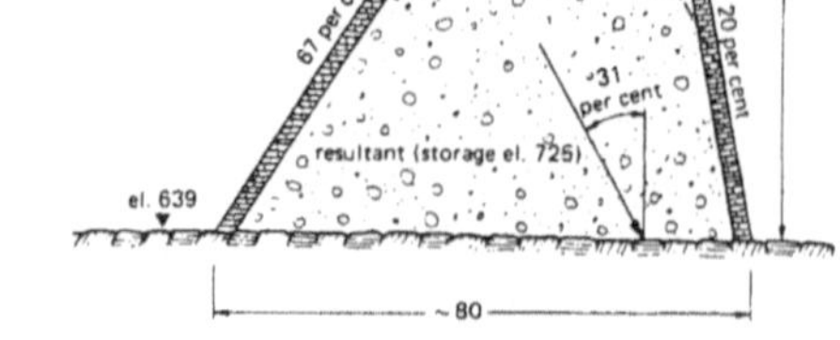

Fig. 4. Vertical and horizontal sections of the Gasco dam near Madrid.

Name	Country	Period of construction	Height[1] (m)	Length (m)	Type[2]	Span[3] (m)	Relative thickness (per cent)[4] Slabs/ arches	Buttresses	Slope (per cent)[5] u.s.	d.s.	Relative base[6] width (per cent)
Historic structures[7]											
Iturranduz	Spain	Roman	3	80	FS	4-5	17	43	0	0	140
Esparragalejo	Spain	1st century	6	312	MA	6-10	25	36	0	0	90
Olisipo	Portugal	Roman	7	50	BG	5	n.a.	40	0	0	155
Sidi Gelani (Megenin)	Libya	2nd century	5	91	BG	10-19	n.a.	8	0	?	110
Ereruisk	USSR (Armenia)	5th century	5	160	BG	5	n.a.	6	0	0	100
Ontigola	Spain	1568	6	280	BG	varies	n.a.	varies	0	varies	220
Trujillo	Spain	1572-	11	175	BG	15-16	n.a.	18-27	0	varies	110
Topuz	Turkey	1620	10	65	BG	8-18	n.a.	19-44	0	15	74
Bedia	Spain	1734	4	53	MA	10-12	30	24	53	0	270
Lancho	Spain	1744	7	142							
Feria	Spain	1747/1950[8]	24	170	BG	8.7	n.a.	37	0	0	93
Olla	Mexico	1742-49									
Topuzlu	Turkey	1750/1786[8]	16	81	BG	23	n.a.	30	0	17	74
Arce de Arriba	Spain	18th century	7	262	FS	5-11	23-53	17-48	0	0	80
S. Blas (Pabellon)	Mexico	18th century	24	177	BG	?	n.a.	40	0	varies	57
Arquitos (Arcos)	Mexico	18th century	18	219	BG	?	n.a.	50	0	varies	66
S. Jose (Guadalupe)	Mexico	18th century	11	?	FS	8.8	21	17	0	129	105
Natillas	Mexico	1760	12	100	BG	varies	n.a.				
Casillas I	Spain	18th century	10	182	BG	16.5	n.a.	30	0	117	150
Lampy	France	1777-82	19	130	BG	11.5	n.a.	32	0	varies	78
Valide	Turkey	1796	14	104	BG	20	n.a.	33	0	15	67
Gasco	Spain	1787-99[9]	87[9]	230	FS	9.8	29	17	67	20	92
Zalamea	Spain	1800/1955[9]	17	113	BG		n.a.		0	0	
Mir Alam	India	1804	12	1067	MA	29-52	5-9	14-25	0	varies	108
Tillot	France	1830-34	15	199	FS	28	25	18	0	22	90
Chazilly	France	1830-37	23	536	BG		n.a.				
Grosbois I	France	1830-38/42	29	550	BG	40-62	n.a.	16-25	varies	23	96
Pont	France	1878-82	22	151	BG	21.1	n.a.	16	varies	varies	95
Junction (Belubula)	Australia	1897	19	100	MA	8.5	10	30	58	varies	85
Steel (Ashfork)	USA (Arizona)	1898	14	241	MA	2.4	steel structure		100	0	100
Innovative and record-breaking structures											
Theresa	USA (New York)	1903	3	37	FS	1.8	8	17	167	10	200
Schuylerville	USA (New York)	1904	9	75	FS	2.5	9	15[10]	100	varies	204
Ricketts	USA (Pennsylvania)	1906	16	183	FS						
Ellsworth	USA (Maine)	1907	20	152	FS	4.6	14	10[10]	100	58	177
Prele (Douglas)	USA (Wyoming)	1908-09	41	99	FS	5.5	15	13[10]	119	9	131
Hume (Ten Mile)	USA (California)	1908	20	199	MA	15.2	5	9	63	42	120
(New) Bear Valley	USA (California)	1910-11	28	110	MA	9.8	5	9[10]	75	50	120
Gem	USA (California)	1915-16	34	210	MA	12.2	6	7[10]	84	10	
Combamala	Italy	1915-16	42	96	FS	5.5	16	15[10]	70	29	110
Hodges	USA (California)	1918	41	188	MA	7.3	7	12[10]	100	25	115
S. Chiara (Tirso)	Italy	1918-24	70	283	MA	15	7	35[10]	65	35	115
Pleasant (Waddell)	USA (Arizona)	1926-27	78	602	MA	18.3	6	27[11]	100	15	114
Carranza (Martin)	Mexico	1927-28	35	234	CB	9	n.a.	22	46	varies	91
Dixence	Switzerland	1931-35	87	459	CB	13	n.a.	42	4	78	82
Lampasas	USA (Texas)	1935	51		FS						
Rodriguez	Mexico	1937	73	579	FS	6.7	17	16[10]	100	20	121
Bartlett	USA (Arizona)	1936-39	87	244	MA	18.3	8	28[11]	90	36	125
Escaba	Argentina	1948	83	240	FS	12.2		22[10]	100	27	124
S. Giacomo Fraele	Italy	1940-50	96	970	CB	15	n.a.		3	62	67
Ancipa	Italy	1949-53	105	253	CB	11	n.a.	37	45	45	88
Albigna	Switzerland	1956-59	115	810	CB	20	n.a.	75	0	varies	73
Grandval	France	1959	88	400	MA	50	7	11	70	40	95
Hatanagi I	Japan	1962	125	269	CB	11/16	n.a.	36/37	50	50	100
Johnson (Manic. 5)	Canada	1961-68	214	1314	MA	76/162	7/10	17/20	60	60/65	95
Alcantara	Spain	1969	135	570	CB	11	n.a.	40	45	45	89
Itaipu	Brazil/Paraguay	1983	196	2050	CB	17	n.a.	35	58	46	104

Table title: Buttress dams throughout the world

1 = maximum height above lowest foundation; 2 = BG = buttressed gravity, CB = contiguous buttresses, FS = flat-slab and MA = multiple-arch; 3 = between buttress centrelines; 4 = average over dam height in per cent of span; 5 = upstream (u.s.) and downstream (d.s.) face slopes of buttress (horizontal in per cent of vertical); 6 = in percentage of height (buttresses); 7 = until 1899 (after 1750 only over 10 m high structures); 8 = second date indicates subsequent heightening to which structural data are referred; 9 = construction interrupted in 1799 at 56 m height; 10 = with struts or braces between buttresses; 11 = double walled, hollow buttresses with internal bracing.

provide additional weight, these spaces were filled with earth, a provision which entailed the ultimate failure of the bold enterprise. After a heavy rainfall in May 1799 the wet earthfill pushed out the downstream slab over four intermediate spaces, whereupon the already flagging construction came to a standstill at a dam height of 56 m.

With this failure, the catastrophic (1000 dead) break of the Puentes gravity dam near Lorca in southeastern Spain in 1802 and the silting up of the nearby Valdeinfierno curved gravity dam a few years after its completion in 1806, Spanish dam engineering reached a low point, a reflection of the political and economic downfall of the country, which culminated in the Napoleonic wars[16]. The leadership in the field of dam engineering passed over to France, where the first large dams were built, probably not by coincidence, near the Spanish border, for the water supply of the epochal Canal du Midi[17] (initial construction 1666-1681).

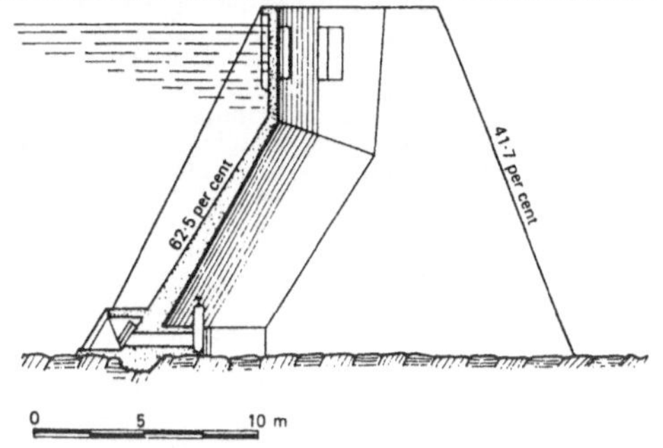

Fig. 5. Cross-section of the Hume (Lake) multiple-arch dam completed in 1908 in California[22].

Among them is the massive, buttressed gravity dam of Lampy, completed in 1782, about 60 km southeast of Toulouse. Closer to a true buttress type is the Tillot dam completed in 1838 about 35 km southwest of Dijon for the water supply of the Canal de Bourgogne, while the nearby larger Grosbois I and Chazilly gravity dams, completed at the same time and for the same purpose, were reinforced with buttresses after they began to slide on their foundations during impounding[18].

In contrast to these rather unsatisfactory, conventional structures, the French engineer Russle, serving in the Madras Engineering Corps, is credited with the design of the most remarkable multiple-arch dam of Mir Alam[19] (or Meer Allum), constructed in 1804 for the water supply of Hyderabad in southern central India. Built of solid masonry it is up to 12 m high and consists of 21 semicircular, vertical arches of constant thickness but with variable spans of up to 51 m. This figure was only attained again more than 120 years later at the Coolidge dam in Arizona.

The next major multi-arch dam was built almost a century later in 1896/97 across the Belubula river at Junction Point Reefs, 210 km west of Sydney in Australia[20]. It is higher than the Mir Alam dam, but much shorter and consists of just five arches with small spans. These are, however, inclined upstream and of elliptical shape. Moreover, the arches, built of brick like the buttresses, were covered with concrete on the upstream face, so that it became a straight sloping plane. The dam was designed by Oscar Schulze, who analyzed it like a bridge turned through 90°.

Prime of multiple arches

It may have been coincidence that while the Belubula dam was still being built, a similar design was under study for a 31 m-high structure near Ogden, 56 km north of Salt Lake City, Utah[21]. However, as in the case of single-span arch dams[1], some kind of transpacific technology transfer may have taken place. Both regions, Australia as well as the West of the United States of America, were at that time undergoing similar pioneering developments, which favoured solutions requiring a minimum of imports and transport.

An alternative for the never built Ogden[21] dam, even envisaged a rather light steel structure of concave (downstream curved) sheets resting on trusses or struts spaced 3 m apart and braced together for its upper half. Similar designs were realized in 1898 at the 14 m-high Steel (or Ashfork) dam 70 km west of Flagstaff, Arizona; in 1901 at the 23 m-high Redridge dam 20 km north of Houghton, Michigan, and in 1907 at the 25 m-high Hauser (Lake) dam 24 km north of Helena, Montana[22]. The central portion of the latter rested on gravel, which despite an 11 m-deep cut-off of sheet-piles, washed out on 14 April 1908, causing the collapse of the steel structure. It was replaced in 1909-1919 by a concrete gravity dam founded on rock throughout.

In the meantime, the first multiple-arch dam of reinforced concrete had been completed in 1908 in only 114 working days. It impounded the Hume Lake logging and fluming reservoir on the Ten Mile Creek 80 km east of Fresno in the Californian Sierra Nevada mountains. The dam was designed and supervised during its construction, by direct labour, by John S. Eastwood (1857-1924), who immediately after his graduation from the University of Minnesota as a civil engineer[23], went to work in the western USA. There he became chief engineer for a 1020 kW powerplant on the San Joaquin river 60 km north of Fresno. Its head of 430 m was the highest in the world when it started operation in 1896. Afterwards he worked on the design of the first phase of the later famous Big Creek project in the same area, until he set up his own consulting practice in 1910.

Mir Alam multiple-arch dam built in 1804 in Hyderabad, India (photo by the author).

The 20 m-high Hume (Lake) dam[22] is made up of twelve circular arches of 15.2 m span and a central angle at the crest of 124°. In the upper 5 m of the dam the arches are vertical and some of them have spilling openings, while below they are inclined 63 per cent upstream (Fig. 5). The arches as well as the buttresses are very thin, so that the whole dam required just 1700 m³ of concrete or barely 10 per cent of an equivalent gravity dam. The reinforcement consists of old logging cables and rail-road scrap iron. Placed mostly in the arches, it serves principally to resist the temperature stresses, while those produced by the hydrostatic load are almost exclusively compressive. Eastwood still used the simple "cylinder formula" for the analysis of the arches, whereby arch stress equals pressure times radius divided by thickness. He designed the buttresses for zero eccentricity of the resultant of those forces.

Shortly after completion of the Hume (Lake) dam, Eastwood designed the replacement of the extremely slender (Big) Bear Valley single-arch dam, built in 1883/84 by Frank E. Brown (1856-1914) in the San Bernardino mountains about 130 km east of Los Angeles, California[1]. The new 28 m-high dam, moreover, was made approximately 9 m higher than the old one, so that it could store three times as much water[24]. Interestingly the total cost of both dams was about the same, which attested once more to the low investment involved with Eastwood's new design. This was modified in the (Big) Bear Valley dam only by the addition of heavily reinforced, horizontal strut-ties between the buttresses near the crest and at two and one-third of their height in view of their lateral stability in case of seismic disturbances or vibration of any kind.

Overall a dozen multiple-arch dams were built in the following decade and a half to Eastwood's design, among which, in 1918, was the record-breaking 41 m-high Hodges dam 37 km north of San Diego, California. The concept was quickly taken up by such other renowned dam designers in the western United States, as Lars R. Jorgensen (1876-1938), better known for his constant-angle, single-arch designs[1], and by Fred A. Noetzli (1887-1933). The latter, an immigrant Swiss engineer, in 1924 proposed double-walled, hollow buttresses with interior diaphragm bracings to give them adequate lateral stability for still greater dam heights and arch spans[25]. This improvement was first used at the 78 m-high (Lake) Pleasant or Waddell dam, completed in 1927, 40 km north of Phoenix, Arizona, but with dubious success, since extensive reinforcements became necessary after the first filling of the reservoir[26]. The application of the double-walled buttresses was wholly successful twenty years later at the 87 m-high Bartlett dam 60 km northeast of Phoenix, Arizona, built by the Bureau of Reclamation. Despite their record heights,

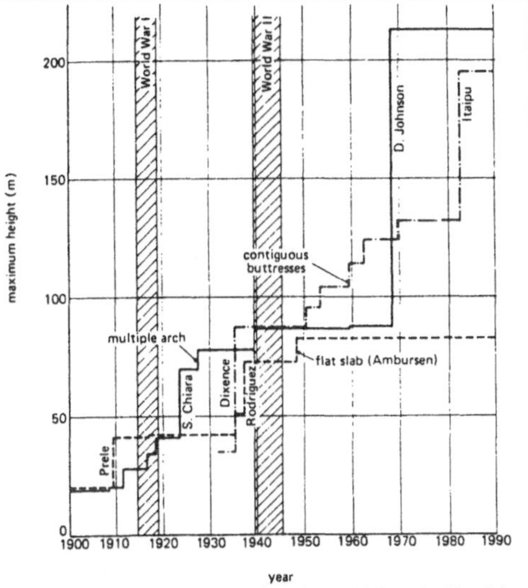

Fig. 6. Maximum heights reached by multiple-arch, flat-slab, and contiguous buttress dams since 1900.

however, both dams still have relatively modest arch spans of 18.3 m.

A decidedly larger arch span of 54.9 m was achieved in 1928, as mentioned before, at the 76 m-high Coolidge dam of the Bureau of Indian Affairs 140 km east of Phoenix, Arizona. The three thin (1.2 to 6.1 m-thick) arches are also curved vertically. This was the first application of a double-curved dome-structure in dam engineering apart from the 27 m-high single-span arch dam designed for Ithaca, New York by Gardner S. Williams (1866-1931), professor of hydraulic engineering at Cornell University in the same city, the construction of which was, however, interrupted in 1903 at one-third of the planned height[27].

Large-span multiple-arch dams were pursued further by the prominent French dam designer André Coyne (1891-1960) after World War II. In pre-war Europe, multiple-arch dams had attained a certain importance only in Italy during the 1920s. After the war, several were built in France, some of them sizeable structures. However, the most important project of Coyne, and the apogee of the multiple-arch dams, was built in 1961-1968 in Northern Canada on the Manicouagan river 470 km northeast of Quebec to form the D. Johnson reservoir. Its 214 m-high central arch has a span of 162 m (Fig. 6).

PART TWO

The history of multiple-arch dams was discussed in Part One last month, and this month the history of the flat slab type of structure is outlined, both types playing a part in the development of buttress dams.

Heyday of flat-slab buttresses

In the enthusiasm for the still relatively new, but fast spreading use of reinforced concrete, which at the beginning of the 20th century also influenced quite a few dam builders, the most powerful competitor to the multiple-arch dam was the flat-slab buttress type developed and patented by Nils F. Ambursen. Born in Norway, in 1876 he emigrated to the United States immediately after his graduation from the Porsgrund engineering college as mechanical engineer at the age of 18. Barely five years later he set up the firm of Ambursen & Sayles in Watertown in northern New York State, which in 1903 became the Ambursen Hydraulic Construction Co of Boston, Massachussetts.

In the same year the new firm completed within only 18 working days, its first, rather modest, dam of the new design at Theresa, 30 km north of Watertown[28]. Interestingly its main competitor was not the gravity dam, as in the case of the multiple-arch dams, but a timber structure. And in fact, Ambursen was inspired in his design by the traditional wooden mill dams[29] as used for a long time in Europe and the United States (Fig. 7). Like many of these, Ambursen's design takes full advantage of the stabilizing effect of the vertical water load on a strongly inclined (usually 100 per cent) upstream face, which requires a minimal buttress thickness per unit of dam length. This conforms well with the rather narrow buttress spacing dictated by the limit on the bending stresses in the upstream slab. The latter is articulated by expansion joints along its interfaces with the buttresses, which support the slab on protruding haunches. From the second project, three times higher,

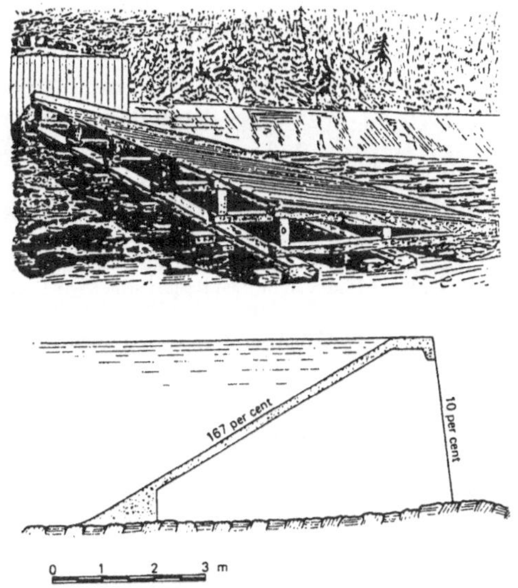

Fig. 7. Cross sections of traditional timber dam[29] (top), and of Ambursen flat-slab buttress dam built in 1903 at Theresa, New York, USA[28].

at Schuylerville 50 km north of Albany, New York, and subsequent ones, the lateral stability of the buttresses was assured by horizontal brace beams between them, as in multiple-arch dams[22].

The Ambursen dam type soon became quite popular and by the end of 1910 almost 60 such dams had been built, among which, in 1909, was the record-breaking 41 m-high La Prele (or Douglas) irrigation dam 60 km east of Casper, Wyoming. It required 17 200 m^3 of concrete or 43 per cent of that of an equivalent gravity dam. The average reinforcement content is 23 kg/m^3 of concrete. In 1977/79 the dam[30] had to be provided with a new upstream slab, since the original one had disintegrated over about 20 per cent of its thickness in the rough mountain climate at an elevation of 1600 m. At the same time, underseepage was reduced to about one-third or 280 l/s (still substantial) by grouting below the upstream cut-off wall and into the alluvium left in place below the foundation slab across the valley floor.

The provision of a foundation slab seems to defeat the very principle on which the stability of buttress dams rests and the slab's foundation must therefore be thoroughly and permanently drained. Nevertheless, quite a few Ambursen dams were built on alluvium and this was even claimed as one of their additional advantages after the 13 m-high Ashley (or Pittsfield) dam 55 km northwest of Springfield, Massachussetts survived undamaged the wash-out of its alluvial foundation on 19 January 1909. Less lucky, however, were a 5 m-high dam in Dansville, New York, a 3 m-high one in Janesville, Wisconsin, and a 16 m-high one in Stony, West Virginia. They failed on 5 December 1909, January 1912, and 15 January 1914 respectively, following piping in their alluvial foundations[26].

Nevertheless Ambursen-type dams continued to proliferate and by the end of the 1920s more than 200 had been constructed, outnumbering multiple-arch dams by far. Outside the United States the flat-slab buttress type was adopted, especially after World War II, for about 50 dams more than 15 m high in Ambursen's native Norway, because many sites there, were of difficult access and required only moderately high structures[31]. To improve the dams' resistance to the severe winter climate, the Norwegians usually added a thin, vertical insulating wall between the downstream part of the buttresses. Moreover, they reduced the number of expansion joints in the upstream slab by making it continuous over two or three buttresses. Almost no flat-slab buttress dams were built in the rest of Europe and only a few in Latin America. There the highest of them all was completed in 1948 to a height of 83 m at Escaba in the northwestern Argentine province of Tucuman (Fig. 6).

Survival of contiguous buttresses

Despite the temporal popularity of reinforced concrete dams, especially of Ambursen's flat-slab type, economic factors, that is, the rise of labour costs in relation to prices of materials and the ensuing mechanization, worked strongly against them (Fig. 8). Moreover, several of them deteriorated unexpectedly quickly under the severe climatic conditions they were sometimes exposed to, or developed cracks and leaks. The wisdom of using thin concrete members in dams at all was being increasingly questioned. Thus some designers reverted to dam types in which the arches or slabs were replaced by enlargements at the upstream heads of thicker and closer spaced buttresses and which were more amenable to modern mass concrete techniques.

Such solutions were used by Noetzli for the 35 m-high spillway section of the Venustiano Carranza

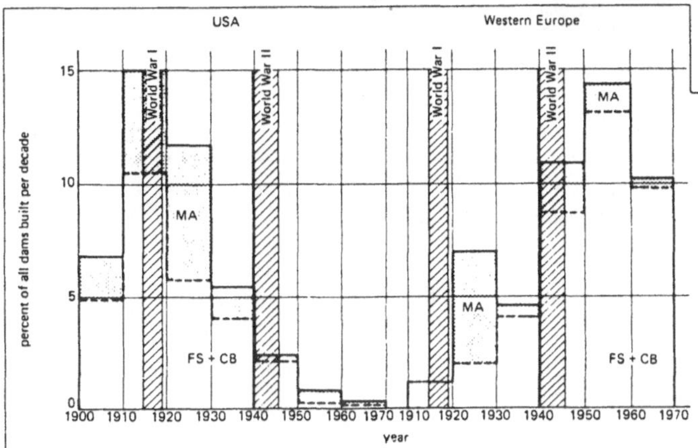

Fig. 8. Development of the share of buttress dams in all dams of more than 15 m height built in the United States and Western Europe (FS + CB = flat-slab and contiguous buttress types; MA = multiple-arch dams).

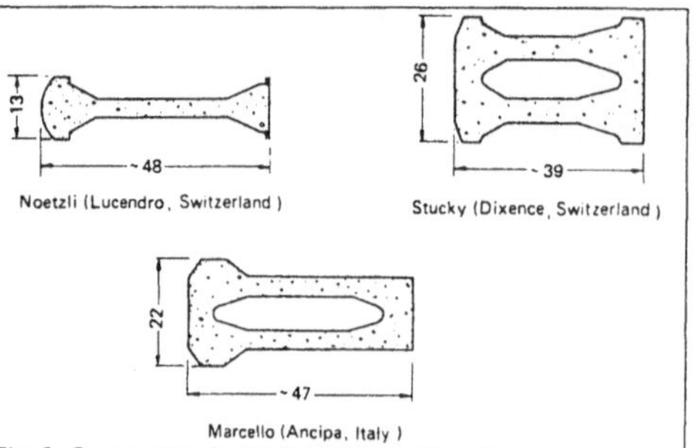

Fig. 9. Comparative horizontal sections of contiguous buttress dams 50 m below storage water level.

(formerly Don Martin) dam built in 1927/28 200 km north of Monterrey in Mexico[32] and by the Spanish engineer Federico Cantero (1872-1946) at the 26 m-high Burgomillodo dam, completed in 1929, 100 km north of Madrid in Spain[33]. While the latter's upstream face has a slope of only 21 per cent, Noetzli kept to a 46 per cent inclination, on average, to take advantage of the stabilizing effect of the water load on it and thus get as wide a spacing of the buttresses as the bending stresses in their heads would permit. To reduce the latter, he made the upstream face of the heads convex, from which the name of the new design became "round-head buttress". Similarly, the polygonal approximation of the round-head became known as the "diamond-head buttress-dam". Moreover, Noetzli introduced T-shaped enlargements at the downstream ends of the buttresses for a better distribution of the stresses, which are highest there, and/or to reduce the clear span of overflow or insulating decks (top of Fig. 9).

In deference to the high altitude, about 2200 m, the downstream T's were made contiguous at the Dixence dam 17 km south of Sion in Switzerland[34]. Moreover, contraction joints were provided for at only every second intermediate space, thus resulting in double I-shaped buttresses (middle of Fig. 9). The upstream face of the Dixence dam is almost vertical, but in plan it is curved for additional stability. The maximum height of 87 m was a world record at the time of its completion in 1935, which was maintained until after World War II. The dam was designed by Alfred Stucky (1892-1969), professor of applied hydraulics at the Institute of Technology, Lausanne, Switzerland, and later a renowned arch-dam designer[1]. The Dixence dam was submerged in 1957 in the reservoir of the 284 m-high Grande Dixence gravity dam.

The Dixence design was adopted during World War II by Claudio Marcello (1901-1969), chief engineer for hydropower plants at the Edison company of Milan, for the 54 m-high Trona and the 41 m-high Inferno dams 70 km north of Milan, Italy, in which he omitted the downstream T's in the intermediate spaces having contraction joints[35]. Moreover, in the many buttress dams Marcello designed thereafter, he used similar slopes of 45 to 50 per cent on both faces and carefully standardized the shapes for buttress heads and intermediated spaces (bottom of Fig. 9).

In fact, most of the buttress dams built in the last decades followed either Marcello's double-buttress or Noetzli's single-buttress design, sometimes with rather trivial variations. In general, however, both dam types became relatively less important for the same economic reasons, which had already rendered the multiple-arch and flat-slab buttress dams obsolete before World War II (Fig. 8). Nevertheless, the largest hydropower plant ever built so far, that is, the 12 600 MW Itaipu scheme on the Brazilian-Paraguayan border has a double-buttress main dam with an unprecedented height of 196 m and a long wingwall of single buttresses.

References

1. Schnitter, N. J., "The Evolution of the Arch Dam", *Water Power & Dam Construction*; October and November, 1976.
2. Vita, C. and Brogan, O., "Roman Dams on the Wadi Megenin", Libya Antiqua; 1965.
3. Saenz, F., Wais, F. and Garcia, J. A., "La presa romana de Iturranduz", *Revista de Obras Publicas*; January, March and June, 1973.
4. Diaz, D. M. and Garcia, J. A., "Comentario al articulo 'Don Pedro Bernardo Villareal de Berriz y sus presas de contrafuertes' y contestacion del autor", *Revista de Obras Publicas**, March, 1972.
5. Zonn, I. S., "A History of Ancient Irrigation in the Soviet Union". International Commission on Irrigation & Drainage Bulletin; July, 1981.
6. Smith, N. A. F., "The Heritage of Spanish Dams", Spanish National Committee on Large Dams, Madrid; 1970.
7. Garcia, J. A., "Old Dams in Extremadura", History of Technology, Mansell Ltd London, England; 1977.
8. Garcia, J. A., "The Chapter on Weirs in the Codex of Juanelo Turriano", *Technology and Culture*, No. 2; 1976.
9. Garcia, J. A., "Don Pedro Bernardo Villareal de Berriz y sus presas de contrafuertes", *Revista de Obras Publicas*; August, 1971.
10. Villareal, P. B., "Maquinaz hidraulicas de molinos y herrerias y goviemo de los arboles y montes de Vizcaya", Published by A. Marin, Madrid 1736 (Reprinted by Sociedad Guipuzcoana, San Sebastian, Spain; 1973).
11. Lazaro, J., "La presa de la Albuera de Feria (Badajoz)", *Revista de Obras Publicas*; 1936.
12. Hinds, J., "200-Year-Old Masonry Dams in Use in Mexico", Engineering News-Record; 1 September, 1932.
13. Ural, O. M. and Ungan, U., "Large Dams in Turkey", General Directorate of State Hydraulic Works, Ankara, Turkey; 1967.
14. Ozis, U., "The Ancient Dams of Istanbul", *Water Power & Dam Construction*, July and August, 1977.
15. Garrandes, E., "La presa de El Gasco sobre el rio Guadarrama" *Boletin de informacion del ministerio de obras publicas*; November and December, 1963.
16. Schnitter, N., "Alte Talsperren in Spanien", *Wasser-und Energiewirtschaft* No. 5, 1973.
17. Rolt, L. T. C., "From Sea to Sea", A. Lane, London, England; 1973.
18. Ziegler, P., "Der Talsperrenbau", First published by A. Seydel, Berlin 1900 and subsequently by W. Ernst & Sohn, Berlin; 1911/1925.
19. "An Unusual Arched Dam", *The Engineering Record*, Vol 47, No. 2; 1903.
20. Schulze, O., "Notes on the Belubula Dam", *Transactions Australian Institute of Mining Engineers*; 1897. (Summarized in *Engineering News*, 8 September, 1898.)
21. Goldmark, H., "The Power Plant, Pipe Line and Dam of the Pioneer Electric Power Company at Ogden/Utah", Transactions, American Society of Civil Engineers; December, 1897.
22. Wegmann, E., "The Design and Construction of Dams", J. Wiley & Sons, New York; 1888/1927.
23. Jackson, D. C., "John S. Eastwood and the Mountain Dell Dam", Journal Society Industrial Archaeology; 1979.
24. Eastwood, J. S., "The New Big Bear Valley Dam", *Western Engineering*, No. 6, 1913.
25. Noetzli, F. A., "Improved Type of Multiple-Arch Dam", Transactions, American Society of Civil Engineers; 1924.
26. "Lessons from Dam Incidents", ICOLD, Paris; 1974.
27. Williams, G. S., "Discussion on Lake Cheesman Dam and Reservoir", Transactions, American Society of Civil Engineers; 1904.
28. Ambursen, N. F. and Sayles, E. W., "A Hollow Concrete-Steel Dam at Theresa, NY", Engineering News; November 5, 1903.
29. Leffel, J., "Construction of Mill Dams", Springfield, Ohio, 1881 (Noyes, Park Ridge/New Jersey; 1972.)
30. Tolle, J. M., Simard, P. W. and Brown, L. A., "Modern Engineering Saves Troubled Dam", *Civil Engineering* (ASCE); June, 1979.
31. Berdal, A. B., "Slab and Buttress Type Dams", *Concrete Dams in Norway*, Norconsult AS, Oslo, Norway; 1968.
32. Noetzli, F. A., "Don Martin Dam, Mexico", *Western Construction News*; 1928.
33. Gomez-Navarro, J. L. and Aracil, J. J. "Presas de Embalse", Tipografia Artistica, Madrid, Spain; 1958.
34. Stucky, A., "Le barrage de la Dixence", *Bulletin Technique de la Suisse Romande*; 1946.
35. Marcello, C., "Considérations sur les exemples réalisés d'un type de barrage à éléments évidés", V Congress of ICOLD, Vol. IV; Paris, 1955.

7
The evolution of British dams

Geoffrey Binnie

I will start by explaining briefly the difference between a weir and a dam. A weir raises the water level in a stream or river to a more or less fixed level. On the Thames, for example, there are several weirs including the one at Teddington. In the old days weirs were frequently built to divert water to a water mill or for fish breeding purposes. Very often a weir served both purposes and depending on whether the stream was navigable or not, it may or may not have had lock gates in it. A structure may also be built across a stream for the purpose of storing water. It has to be designed so as to allow a large variation in water level, the larger the variation the greater the amount of water stored. These structures are referred to as dams, the lakes so formed reservoirs.

Anyone who has flown on a clear day over the Pennines on the way to and from Scotland must have been struck by the extraordinarily large number of reservoirs nestling in these hills. There is probably a greater density of dams in Great Britain than anywhere else in the world.

As compared with dams in Switzerland, three of which range in height from 230 m to 280 m, and even higher dams under construction in Russia, dams in Great Britain are very small, the highest being Lynne Brianne, completed in 1972, which is 91 m high. In an urbanised country such as ours even a dam only 15 m high is an engineering project which demands the most careful investigation, design and construction because of the appalling potential consequences of failure. There are more than 400 dams 15 or more metres high in Great Britain.

British dams, although modest in size, deserve a place in the history of civil engineering. I intend to outline the part that they have played in our culture and to depict some of the developments in their design and construction.

It is convenient to classify this period into the eras of the water wheel, river navigation, canals, ornamental lakes and water supply. These eras began at different times but, with the exception of the water supply era, they can all be said to have ended about 1850. The water supply era started in the nineteenth century and has continued to the present day.

THE WATER WHEEL ERA

The earliest authentic reference to an Anglo Saxon water mill appears in a document of A.D. 762 and Domesday Book of A.D.[1] 1086 records at least 5,000 water mills.

> 'See you our little mill that clacks,
> So busy by the brook?
> She has ground her corn and paid her tax,
> Ever since Domesday Book'.[2]

Fig. 1 shows the Chester weir which was built across the Dee by Hugh Lupus, First Earl of Chester (1071–1101), shortly after the Norman conquest and which, in spite of opposition from navigation interests throughout the centuries and, more recently, from agricultural interests, still remains as the oldest surviving milldam in Great Britain.[3]

It was used for corn mills and fulling mills, fulling being a process for cleaning and thickening short haired wool. Today it still serves a useful purpose in that it protects the water supplies of Merseyside from salt water.

By the beginning of the 15th century, and possibly earlier, water power was used to work bellows for smelting. There were many 'iron mills' in the Weald, for example. Fig. 2 shows an old mill dam at Cowden on the Kent-Sussex border which was formerly a centre of the iron industry. The small lake, named Furnace Pond, created by this earthern embankment, is now used for fishing.

The long reign of the horse drawn wagon coming to the mill with sacks of grain faded away after the repeal of the corn laws in the eighteen forties.

After a period lasting from the fourteenth to the seventeenth century when the woollen industry flourished on the upper reaches of rivers and tributary valleys in the west country, the West Riding of Yorkshire became in the eighteenth century the chief centre of the woollen industry. History repeated itself in that it was the banks of swift Pennine streams and not the towns that were sought as sites for fulling mills.

Fig. 1. Chester Weir.

Fig. 2. Cowden mill dam. *Photo by author.*

In the early nineteenth century, Arkwright's machines for carding and spinning cotton and Cartwright's power loom, invented 1785, required power; in consequence spinning mills were erected along the steeply graded Pennine valleys of Lancashire where water power could be exploited. Eastern Lancashire, which already had a share in the woollen industry, became the great manufacturing centre for the cotton industry.

Whilst Watt's famous partnership with Matthew Boulton had begun at the Soho Manufactory in Birmingham in 1775 and the first of the improved steam engines started appearing on the industrial scene shortly afterwards, even by the turn of the century the use of steam engines in mills were comparatively rare and improvements in the reliability of water supply for power purposes contined to be made. As late as 1845–48 the 17 m high Kentmere Head dam was built for the millers of

Kendal but after the turn of the century the steam engine became dominant and water power began to decay.[4]

RIVER NAVIGATION ERA

In the past the greatest obstacles to river navigation were the weirs which served two, often linked, purposes, firstly for the breeding and capture of fish, and secondly for raising the water level for the use of water mills. On the other hand, it must not be supposed that these weirs served no useful navigation purpose for they banked up water which could on occasions be released and it was on these 'flashes' of water that barges passed over shallows; in other words they also served as flash locks. However, the bargemasters were entirely at the mercy of the mill owners who might refuse flashes from their weirs, either because they wanted the water themselves or merely to extort excessive charges for the passage of boats.

Very often considerable delays occurred. In due course, these difficulties were overcome by substituting the modern form of pound locks which raises and lowers boats within a confined space, for these weirs or flash locks. Whilst pound locks were developed in the 10th and 11th centuries under the Sung dynasty in China[5] it was not until several centuries later that they began to appear in Europe, the first in this country being built by Trew on the Exeter canal in the 16th century.

During the years 1662 to 1665 numerous Acts for improving navigation on rivers were passed. In 1663 we find the Speaker of the Commons presenting Bills to the House of Lords and commencing his address in the following terms:[6]

> Cosmographers do agree that this Island is incomparably furnished with pleasant Rivers, like veins in the natural body, which conveys the Blood into all the Parts, whereby the whole is nourished, and made useful, but the Poet tells us he acts best, *qui mascuit utile dulci.* Therefore we have prepared some bills for making small rivers navigable.

Parliamentary activity continued and more Acts for improving inland water navigations were passed, particularly during the years 1697 to 1700 and 1718 to 1721.

The coastal trade was closely linked to inland water navigation. Between the year 1600 and the years 1724 and 1727 when Defoe toured England, several thousand square miles of previously inaccessible country came within one days' carriage on land of navigable water, a very important factor when roads were still so bad. 'The improvement of the Aire and Calder and the Mersey had opened up the industrial regions of the north and broken almost in two the 'backbone' of England, which had hitherto remained inaccessible to water carriage'.[7]

Sometimes the locks depended for their operation upon a supply of water in excess of the natural flow. One example is the series of flash locks constructed at the end of the twelth century by Godfrey de Lucy, Bishop of Winchester, who in 1189 decided to make the entire length of the Itchen navigable from Southampton Water to Alresford. To overcome this difficulty an impounding reservoir with a dam about 6 m high and 76 m long was constructed near Alresford. This dam, which can still be seen, is one of the oldest surviving earth dams in England.[8]

Another example is the series of pound locks for the Exeter canal already referred to. In this case a masonry dam to divert water to the canal was constructed on the river Exe. The original dam can no longer be seen but there still exists today a dam constructed on nearly the same site in 1676, which is named Trew's Weir.

Whilst not at so great a pace as between the years 1600 and 1750, improvements continued to be made to river navigations until the end of the canal era.

THE CANAL ERA

During the seventeen fifties, the 3rd Duke of Bridgewater decided to develop his estates at Worsley. There was coal at Worsley but the problem was to transport it to Manchester and so he conceived the idea of making a canal. The Duke instructed Brindley to make a survey of the proposed line from Worsley to Manchester and so the fine collaboration began which resulted in the historic Canal Act of 1759,[9] the first of its kind. On Brindley's recommendation, the canal was carried over the river Irwell on an aqueduct instead of locking down the slopes on either side, thus an unbroken level was obtained throughout the 8½ mile length of the canal.

Where the canal was built on embankments it was lined with clay-puddle which confounded the sceptics who foretold that such embankments could never hold water. The first boatload of coal sailed along the canal on 17 July, 1761.

According to Samuel Smiles, when the Duke of Bridgewater's Bill to extend his canal to Runcorn was before Parliament in 1762, Brindley gave a practical demonstration of the properties of clay puddle to an astonished committee of the House of Commons which Smiles described as follows:[10]

> Brindley caused a mass of clay to be brought into the Committee room, and, moulding it in its raw untempered state into the form of a trough, he poured into it some water, which speedily ran through and disappeared. He then worked up the clay with water to imitate the process of puddling, and again forming it into a trough, filled it with water, which was now held without a particle of leakage.

After a lapse of two centuries, the water in our canals is still held in its bed by this simple principle which Brindley demonstrated. Thus began the great canal era. Between 1763 and 1800, over 80 different canal schemes were authorised by Parliament and most of them were completed, the peak period of construction being the last ten years of the century. Smeaton (in the earlier phase), Whitworth, Henshall, Outram, Jessop, Rennie and Telford were the foremost engineers of the network of canals which followed the success of Brindley's work. In time the major barriers, such as the Pennines, were crossed by flights of locks with usually a tunnel or a deep cutting at the summit. The water requirements of such a canal have been described by Rolt as follows:[11]

> If the canal does not (as frequently is the case) supply water for industrial use, and there is no serious leakage, then the only 'flow' or consumption of water will be occasioned by the passage of a boat through the water way. Assuming that the canal passess over a summit level on its course between two river valleys, then the boat will draw the capacity of two locks away from that summit level, one as it ascends and the other as it descends. Adequate water resources must therefore be provided on this summit level to replenish the water so consumed, and the extent of the supplies available will determine the amount of traffic that can pass through the canal.

For maintaining water supplies in these canals, particularly at their summits, storage reservoirs were usually constructed. Whilst not in all cases still fulfilling their original function, and in many cases being used for recreational purposes, a large number of these reservoirs still exist. Half a dozen of them are created by some of the oldest major dams in existence. By a major dam I mean one which is 15 or more metres high. As a result mainly of the coming of the railways, the canal era can be said to have come to an end about the same time as the waterwheel era, that is to say about 1850.

THE ORNAMENTAL LAKE ERA

At Grimsthorpe in Lincolnshire the 4th Duke of Ancaster employed John Grundy of Spalding (1718–83) at different times between 1745 and 1767 on various alterations and improvements to his estate. Fig. 3, which is reproduced from a manuscript in the library of the Institution of Civil Engineers, shows a cross section and part of an estimate prepared by Grundy for a proposed dam, the particular items in the estimate being those referring to the proposed puddle clay trench and the proposed puddle clay corewall. The ancient unit *Floor* used in the estimates was equivalent to 400 cubic feet. The stage construction indicated by FGHI and KLMN on the cross section was for the purpose of testing the watertightness of the proposed lake bottom before bringing the dam up to full height.

It is remarkable that Grundy is here advocating a form of construction which became accepted as being the best practice in the nineteenth century and continued in the Pennines until about ten years ago. Thus, my grandfather, who was water engineer for Bradford from 1875 to 1890, wrote a hundred years ago:[12]

> 'The water which is retained in the reservoir by the embankment may, and often does, find its way through fissures in the rock or porous beds beneath its base, to prevent which it is usual to sink what is called the puddle trench. This is a trench sunk on the centre line of the embankment and extending up each side of the valley to a point above top water level, where it should tie into some watertight strata; this trench is afterwards filled with puddle, concrete or other watertight materials.
>
> Earthern embankments, as constructed in this country are generally formed with a backbone of tough impervious clay, called the puddle wall, and which is a prolongation upwards in the same plane of the filling of the puddle trench; this wall is supported on each side by material selected on account of its retentive character, the remainder of the embankment being made up of all the harder and less watertight earths, the inner slopes being covered with stone pitching to prevent the wash of the waves injuring the earthwork'.

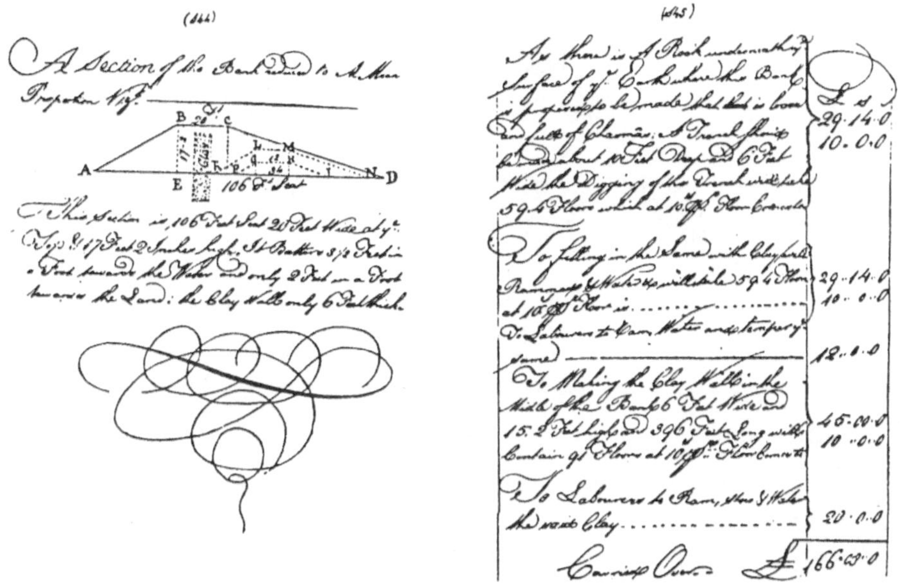

Fig. 3. John Grundy's Cross Section and Estimate (part only). *By courtesy of the Institution of Civil Engineers.*

Fig. 4 from a map by Grundy shows the lakes at Grimsthorpe as they existed in 1763 and as they still exist almost unchanged to this day. There are two lakes, one formed by the Red Bridge dam and the other by the Great Water dam which is 130 m long and has a maximum height of approximately 5.4 m. In Grundy's day the difference in water level between the two lakes was 1.60 m whereas today it is 1.40 m. In addition can be seen a proposed extension of the Great Water. The cross section and estimate in Fig. 3 were for this third dam which was never built, although traces of stages FGHI and KLMN remain. Grundy was responsible for the Great Water dam constructed with a clay corewall and completed in 1748. Fig. 5 shows the great Water dam as it exists today.[13]

A Kip drawing dated 1674 shows a structure on the alignment of the Red Bridge dam which was probably a bridge but at some later date it was converted to a dam. Traces of the old stone arches of the bridge which was buried in the dam can still be seen on the upstream side. This conversion from a bridge to a dam probably took place prior to 1745 but Grundy appears to have raised the dam. As shown on Fig. 4, Grundy proposed to make the dam 'a bridge of deception' by constructing sham arches on the downstream side but this proposal was not carried out.

Fig. 5. Great Water Dam.

THE TWELFTH DICKINSON MEMORIAL LECTURE, THE EVOLUTION OF BRITISH DAMS

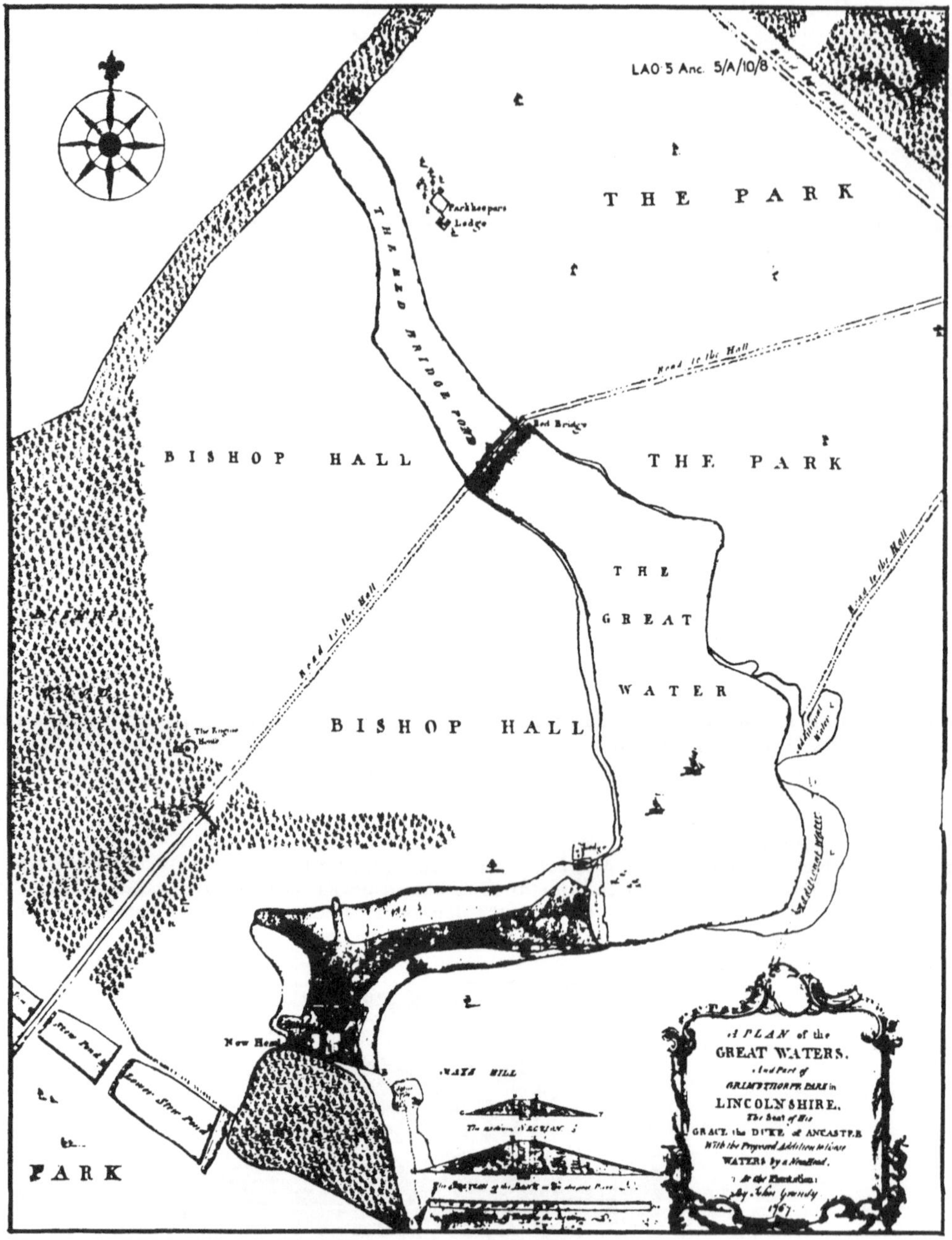

Fig. 4. A Plan of the Great Waters, by John Grundy 1767. *By courtesy of the Earl of Ancaster.*

Grundy is best known as being one of the founder members in 1771 of the Smeatonian Society[14] and for his work in the Fens but he also deserves to be known as being the first engineer to design and build earth dams on relatively modern lines.

Lancelot 'Capability' Brown (1716–1783), Humphry Repton (1752–1818) and others constructed a large number of dams for creating ornamental lakes of which many still exist, but with the exception of Grundy's embankments, all such dams built before 1850 appear to be of homogenous construction without corewalls, the fill material probably being obtained from excavation within the reservoir area[15]. The largest dam built for purely landscaping purposes appears to have been one constructed by 'Capability' Brown during the seventeen sixties for transforming two streams into a lake at Blenheim. This dam has a maximum height of 9 m.

WATER SUPPLY IN THE NINETEENTH CENTURY

As a consequence of the acceleration of the Industrial Revolution at the commencement of the nineteenth century there was a migration from the country into towns. For example between 1821 and 1831, the population of Manchester and Salford jointly grew by 47%, Bradford by 78%, West Bromwich by 60% and Duckinfield's population nearly trebled, So great was the movement that the labouring classes lived in over-crowded and grossly insanitary conditions.

In his monumental report of 1842 Chadwick recommended the disposal of night soil by suspension in water to be conveyed in salt-glazed circular bore drains instead of by horse and cart. This had a profound effect on the health of urban populations. It also enormously increased the demand for abundant fresh water.

The growth of the population[16] and the growth in the number of major dams built from 1800 to 1974[17] is shown in Fig. 6. Also shown is the upper limit in the heights of the dams built during this time. From the graphs it can be seen that the population increased from approximately 10½ million in 1801 to approximately 22 million in 1851 and 38 million in 1901 while the number of major dams increased from 3 in 1800 to 27 in 1850 and 168 in 1900. This big increase in the number of major dams in the latter half of the century was not only due to the increase in population and industry but also to the introduction in urban areas of water-borne sanitation.

It can also be seen that the upper limits in the heights of dams constructed increased from 18 m in 1800 to 36 m in 1850 and 42 m in 1900. Two or three of the major dams were originally constructed to improve the reliability of supply to mills but, with the exception of half a dozen built for the canals, all these major dams are now used either for water supply for domestic and industrial purposes or, in a few cases, to form compensation reservoirs to enable water to be abstracted elsewhere. All except four of the major dams were earthen embankments. In addition to the major dams, an unknown number of other dams of lesser height were also constructed.

The vast majority of the early water supply works were in the North of England and the same sources in the Pennines and the Lake District which had formerly powered water wheels for industrial purposes were now exploited for water supply for the rapidly growing industrial cities.

WATERWORKS PIONEERS

The chief pioneers in modern waterworks engineering were James Simpson (1799–1869), Thomas Hawksley FRS (1807–93) and John F. Le Trobe Bateman FRS (1810–89). Simpson's outstanding contribution was the introduction at the Chelsea Waterworks in 1827 of the slow sand filter which revolutionised the treatment and purification of water. Simpson was also responsible for the Barrow No. 1 and Barrow No. 2 dams, both less than 15 m high, of the Bristol Waterworks Company and the 18 m high Horse Coppice dam for Stockport in Cheshire.

Hawksley & Bateman between them were responsible for the construction in whole or in part of the waterworks of nearly all the towns in the manufacturing districts of the north of England and the Midlands. Those for which Hawksley was the consulting engineer included Nottingham, Liverpool, Sheffield after the Dale Dyke disaster and Leicester.

Hawksley and Deacon, the water engineer to the Liverpool Corporation, were jointly responsible for the design of the Vyrnwy dam, a gravity dam built in masonry, the drawings for which were nearly all prepared in Hawksley's office. Until Hawksley resigned when the Works were half com-

THE TWELFTH DICKINSON MEMORIAL LECTURE, THE EVOLUTION OF BRITISH DAMS

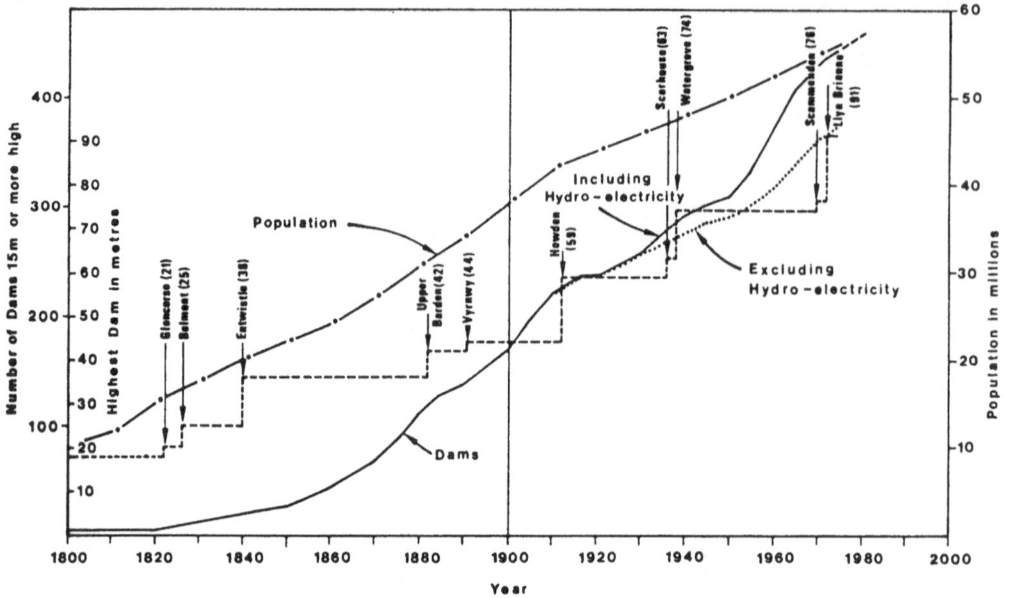

GROWTH OF MAJOR DAMS AND POPULATION IN GREAT BRITAIN 1800–1974

Fig. 6. Growth of Major Dams and Population in Great Britain 1800–1974.

pleted, he was also jointly responsible with Deacon for construction. The dam was completed in 1891. Hawksley was also responsible for at least six major earth dams on two of which, Cowm completed in 1877 and Tunstall completed in 1879, he used, for what is believed to have been the first time anywhere in the world, cement injections to seal fissured rock. The firm which Thomas Hawksley founded still exists and until 1972, when his great grandson died, it had preserved a professoinal continuity from father to son through four generations.

Bateman constructed in whole or in part the waterworks of at least 25 different towns and cities varying in size between 7,000 and 600,000 inhabitants including Manchester and Glasgow. He was the most prolific earth dam builder this country has ever known, having 30 major dams to his credit. For two-thirds of these dams, the undivided responsibility fell on Bateman alone and for the other third he shared the responsibility with others. One of his last works was the inception, with a masonry dam of gravity type as the impounding structure, of the supply from Thirlmere in the Lake District for Manchester, construction of which was begun in 1885 and completed in 1894 by G. H. Hill, who had been his assistant for many years and was his partner from 1881 to 1885. In 1885 Hill established his own firm which still exists.

MASONRY DAMS

Of the 168 major dams built during the nineteenth century, 164 were earthen embankments. The other four were Abbeystead, a 20 m high masonry dam with concrete backing in two arches designed by James Mansergh and completed in 1881, Vyrnwy and Thirlmere which I have already mentioned and the Burrator masonry dam of gravity type designed by Edward Sandeman for the water supply of Plymouth and completed in 1898.

In a short survey such as this, space does not permit discussion of these masonry dams but for a good historical review on the theory of masonry and gravity dams, the Unwin lecture by the late Professor Pippard FRS, published by the Institution of Civil Engineers in 1949, is recommended[18]

EARTH DAMS

An early example of a major earth dam is the 21 m high Glencorse dam designed by Telford and Jardine and completed in 1822. The upper part of an old drawing in the possession of the Lothian Regional Authority is shown in Fig. 7. The lower part is inscribed with specification and 3 endorsements as follows:

> The head to be of the dimensions represented and figured in the cross sections with a good puddle in its middle. The loose earth to be all cleared off the foundation, and ends of the Head down to the solid earth or rock or partly both as the case may be, and the puddle sunk into the ground until it meets with a substance sufficiently impervious to water. The tunnel to be made elliptical, the vertical axis being 6 feet and horizontal 4½ feet, the walls to be constructed with squared Ashlar of good durable sandstone 18 inches thick well set in good mortar mixed with burnt ironstone Purroland, the Tunnel to be properly secured with good well wrought puddle one yard in thickness.
>
> The cast iron pipe which is to be laid through the puddle, in the Head, between upper and lower tunnels to be 3 feet interior diameter, and 2 inches thick of metal, with proper flanges to keep it in its situation. The interior of the shaft to be cylindrical and the wall to be built with good sandstone Ashlar backed with good crossed rubble masonry in thickness equal to that of the Ashlar. The wall of the shaft to be made 3 feet thick at the bottom and to taper regularly to 2 feet thick at the top. The said shaft to be surrounded with a puddle wall 1 yard in thickness and to have a stone stair within. The waste weir to be constructed of square sandstone Ashlar at least 18 inches deep well set in good lime mortar and pointed with Parker's cement.
>
> I have examined and do approve of the above Sections and Specification.
>
> Signed Thos. Telford
> Edinb. 19th August, 1819
>
> This is the Plan referred to in my report regarding the sufficiency of Head of the proposed Reservoir on 'Glencorse Burn'.
>
> signed John Rennie
> Edinb. July 7th, 1819
>
> (signed) Ja: Jardine.

Telford was acting on behalf of the water company. Rennie was acting on behalf of the landowners and millers.

For their outlet works Telford and Jardine, who had few precedents to guide them, admitted water to the upstream side of the clay corewall through a culvert, across the clay core through a single 36 inch dia. cast iron pipe 2 inches thick and below the downstream shoulder through twin free standing 12 inch dia. cast iron pipes, not shown on the drawing, inside a masonry culvert of sufficient size to admit a man. In addition, there was a valve shaft immediately downstream of the corewall.

The construction of the Glencorse dam was much more difficult than Telford and Jardine had anticipated. Sixteen metres of alluvial material had to be excavated before an impervious bottom for the clay filled cut-off trench could be reached and twice during construction the steep sides slipped which considerably delayed completion and put a great strain on the financing of the project but after five years their efforts were rewarded and they demonstrated conclusively that clay could be used to obtain a watertight dam.

Puddle clay cut-off trenches and puddle clay cores for making dams watertight became established practice. However, either through the short-sighted policy of not persisting with the excavation of the cut-off trench until it reached a solid watertight bottom, or due to slight leakage along rock joints where the puddle clay came into contact with the bottom of the trench, or to unequal settlement, or to a stream of water being permitted to flow across the trench during the progress of the work, the clay was liable to be damaged. One example, given by Deacon in the tenth edition of the *Encyclopedia Britannica* (1902), is shown in Fig. 8.

Once slight movement of water sufficient to cause erosion commences, the finer particles of clay are washed away while the sandy particles, which nearly all clays contain, are left behind and form a constantly enlarging porous vein of sand rising vertically in the puddle clay. Percolation through the sand is thus added to the original leakage and, having passed through the puddle clay, the leaking water sometimes rises to the surface as a visibly turbid spring thus giving the first warning of trouble. A common feature of such internal erosion is a depression of the embankment above it as shown on the drawing. In the case shown in the figure the reservoir was worked for 17 years after the leakage was discovered and the larger vein rose 13.5 m before the corewall was reconstructed.

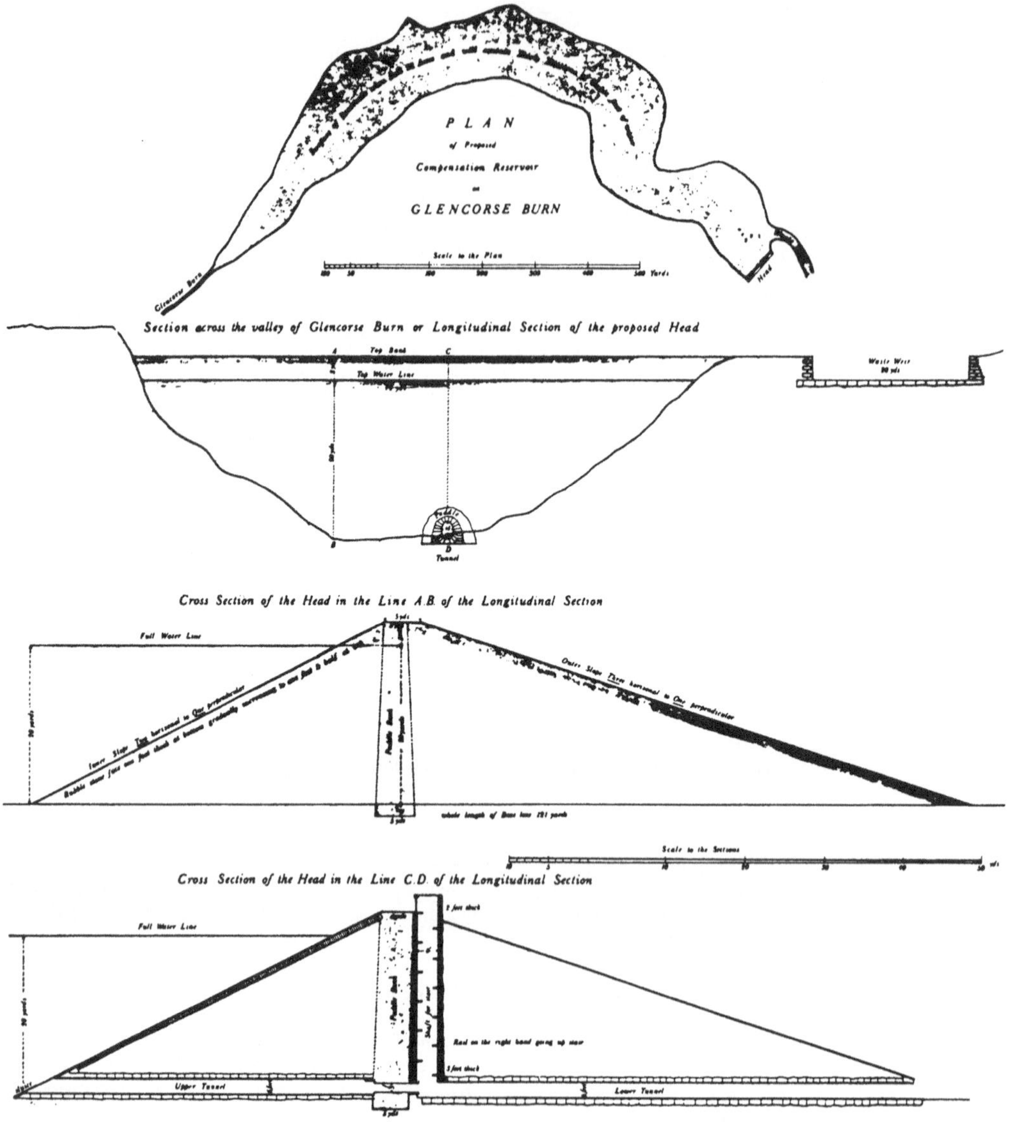

Fig. 7. Plan and Sections of Glencorse Dam 1819. *By courtesy of the Lothian Regional Council.*

BILBERRY DAM FAILURE

Bearing in mind the old adage that there is more to be learnt from one failure than several successes, I will now refer to the Bilberry reservoir dam which failed in 1852. This dam failure has not attracted as much attention as the Dale Dyke dam disaster, which occurred twelve years later and caused the death of nearly 250 people besides much destruction of property in and near Sheffield, but nevertheless it was a great disaster which killed 81 people and destroyed a large amount of property at

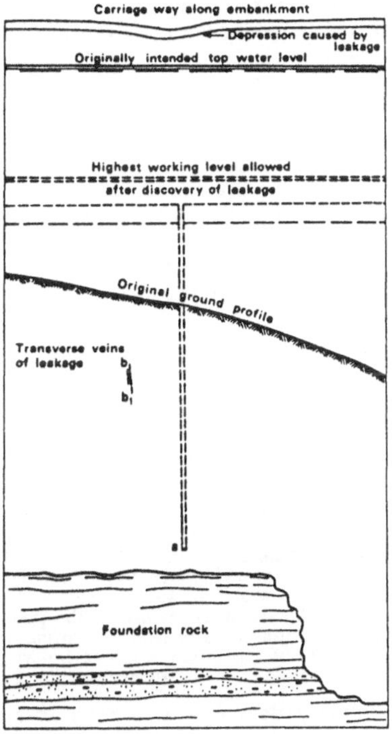

Fig. 8. Deacon's Longitudinal Section of Puddle Clay Trench.

Holmfirth near Huddersfield. Few accidents in history have made such a deep and lasting impression on the British public as the 1879 Tay Bridge collapse but the half-forgotten Bilberry accident killed more people and caused more damage to private property.

An ironic coincident was that George Leather, the engineer from Leeds who prepared the designs for the Bilberry dam, was the uncle of John Towlerton Leather, who was the consulting engineer responsible for the designs and specification for the Dale Dyke dam.

In essence, the story of the Bilberry dam failure is one in which the cost of the works exceeded the funds available for constructing them and where vitally necessary repairs were allowed to go by default because there was no money available for executing them. There was also very little appreciation of the need for skilled and competent supervision of construction. In the letter inviting Leather to become the engineer, the commissioners wrote 'you will only be required to come over now and then as circumstances may require' because they would employ a supervisor[1] to attend to the works'.[19]

The Bilberry dam had a maximum height of 20 m and was about 90 m long at crest level. The design included a puddle clay corewall 5 m wide at the bottom and 2½ m wide at the top.

For the removal of surplus water during floods, a spillway or waste pit in the form of a circular chimney 4.5 m in diameter which communicated with the tunnel by-passing the dam was constructed. The normal method of releasing water from the reservoir was through a trap door at the bottom of the chimney.[20]

> 'In the event of the trap-door being insufficient to carry the surplus water during extraordinary supplies, the water on rising to the level (of the top) of the chimney or bye-wash, would meet with a source of escape presumed to be adequate for all contingencies'

In October, 1838 the first tender for the construction of the dam was accepted by the Commissioners. A dispute arose with the first contractor and his contract was terminated in 1843. A second contractor was engaged on the dam during the years 1843–45. Leather, who was not being paid his fees and who appears seldom, if ever, to have visited the works previously, visited the dam site twice in 1844. After giving some instructions during his first visit in July and finding on his second visit in September that, due to lack of funds, his instructions had been countermanded, Leather quarrelled with the Commissioners and refused to give them any more help. He considered that he had no further connection with them.

During the evening of 4 February, 1852, the clouds moved away after a day of heavy rain and the moon shone. The water in the reservoir began to rise at a rate of about 18 inches an hour. The trap-door was blocked with stones and heather and very little water could escape through it. About midnight water started flowing over the embankment and an hour later it failed. One witness who saw the embankment give way described the scene as the rising of an immense sheet of mist accompanied by a rumbling sound like thunder.

At the coroner's inquest the governent inspector, Moody, referred to the waste pit and said: 'This waste pit is designed to carry off the waste or flood water, but the top of the embankment having sunk below the top of the pit, *and being suffered to remain so,* the flood waters had no proper or sufficient escape, but went over the dam, which, as a necessary consequence, gave way'.[21]

After criticising the construction of the dam, Moody commented on a spring stated to have been discovered in the trench under the embankment. He referred to the fact, proved in evidence, that it was not led away by any of the usual and necessary modes, but:

> That very objectionable plans were resorted to in the hope of choking it up, or 'weighting it down'. But it was not to be 'weighted down', it rose as the work rose, materially infusing the lower portion of the puddle, making it weak and bad, of a nature easily to be worked away with the water of the spring, as the latter forced itself through the outer part of the embankment like a little rill of water issuing from the foot. At times this rill was clear, and at times muddy and yellow. The muddiness varied with the head of water in the reservoir. To the weak nature of the puddle at the base, and the washing away from time to time by the continuous run of water from the spring under the bottom of it, the great settlement of the puddle in the centre is to be attributed, a settlement which continued to go on during the construction, and after the dam had been raised to the height required in the specification.

The Bilberry dam case is a good example of the susceptibility of puddle clay to erosion. On account of this susceptibility, it became common practice to fill at least the bottom of the cut-off trench with concrete as, for example, in Fig. 9 which is a cross section of the Bottoms dam designed by Bateman for the water supply of Manchester and completed in 1872. Quite frequently the whole cut-off trench was filled with concrete but puddle clay was still retained for the corewall above ground level.

THE LLUEST WEN SUBSIDENCE

Another accident which might have had serious consequences if prompt action had not been taken occurred on the Lluest Wen dam. Except that the valve shaft is sited on the upstream side instead of the downstream side of the corewall, the outlet works of this dam are rather similar to the outlet works of the Glencorse dam. Construction of the dam was completed in 1892.

More than 75 years later, a horseman riding across the dam on Christmas Eve 1969 suddenly fell with his horse into a large swallow hole which had remained concealed until the horse stepped on it. A State of Emergency was declared and immediate action was taken to lower the reservoir water level 9 m. The internal subsidence was found to be due to water from the reservoir escaping through a crack in a 15 in dia. drain pipe which crossed from the valve shaft below the puddle clay core to a man-sized culvert under the downstream shoulder of the dam and to clay in suspension being removed by the water.[22]

THE TWENTIETH CENTURY

During this century, the demand for water for domestic and industrial purposes has continued unabated until the 1973–76 financial crisis. In addition, water has been exploited for power extensively in Scotland and to a lesser extent in Wales. Referring again to Fig. 8, it can be seen that the population increased from approximately 38 million in 1901 to 56 million in 1973 and that the number of major dams built increased from 168 in 1900 to 440 in 1974. Of these dams 369 were for

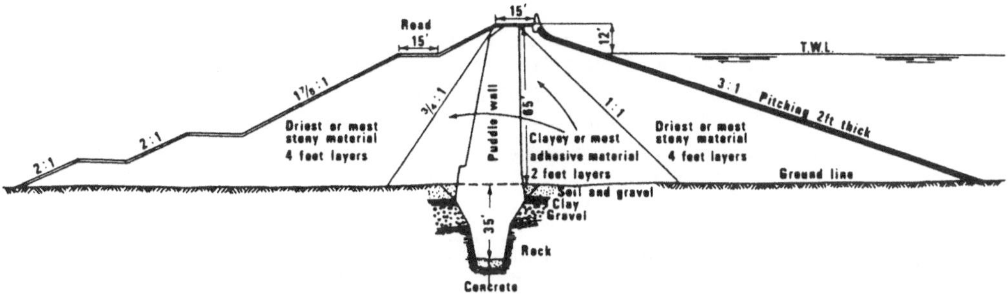

Fig. 9. Bottom's Dam 1872–based on drawing in *History & Description of the Manchester Waterworks, 1884*, by J. F. Le T. Bateman.

water supply and inland navigation and 71 were for hydroelectric power. The upper limit in the heights of the dams increased from 42 m in 1900 to 91 m in 1976.

From the beginning of this century up to the end of the second world war, the design and construction of earth dams followed traditional lines based on more than a century of recorded practical experience. Almost complete watertightness continued to be aimed at and this resulted in works of considerable magnitude and expense. As examples can be quoted the Silent Valley reservoir (1933) in Northern Ireland and the Ladybower reservoir in Derbyshire (1945) where the respective maximum depths of the cut-off trenches were 65 m and 76 m. To ensure watertightness at the ends of the dam, wing trenches of the Howden dam (1912) were 800 m and 400 m long.

After the war, some dams with clay corewalls were still being constructed but the last one, namely the Dove Stone embankment, was completed in 1967. The popularity of this type of design over so long a period has been partly due to the availability in many parts of the country of boulder clay from which excellent puddle clay could be made, but it does not lend itself particularly well to mechanical construction and it is doubtful whether any more dams with puddle clay cores will now be built. The puddle clay core has been superseded by the much wider rolled-fill clay core which can be constructed by modern earth moving and consolidating equipment.

The stability of an earth dam is very dependent upon the foundations on which it is to be constructed and on the cross-sectional profile of the dam. Up to the outbreak of the second world war, the complex factors controlling the mechanical properties of clays and other soils presented such intractable problems that the upstream and downstream slopes of earth dams were decided empirically on the basis of the designer's own judgement and experience.

However, design methods based on more clearly defined concepts of soil properties were gradually being developed on the Continent by Terzaghi. In 1938 the Metropolitan Water Board had a new four sided reservoir, the Chingford reservoir, under construction and a major slip occurred in one of the embankments. Terzaghi was called in to advise on the necessary measures and he also gave a memorable lecture at the Institution of Civil Engineers in 1939. After the war Terzaghi's concepts were welcomed with enthusiasm by the post-war generation of British engineers and he was recognised as being the founder of a new science called 'Soil Mechanics'. Whilst experience and judgement still remain very important factors in the investigation and design of dams, the empirical methods of the past have now been superseded by methods based on impartial analysis and scientific logic. For the last two or three decades we have been fortunate in having in this country several excellent schools of soil mechanics such as the one at Imperial College formerly under our Vice-President, Professor Skempton, and now under Professor Bishop.

An example of modern methods of design and construction is shown on Fig. 10 which is a cross section of the recently completed Empingham embankment for which T. & C. Hawksley are the consulting engineers. Although only 40 m high, this embankment is 810 m wide at the base. This large width is dictated by the strength properties of the lias clay forming the foundations of the dam. The dam consists basically of a sloping rolled clay core supported by specially drained fill material.[23]

During the last 100 years about 300 major dams for water supply have been constructed but it is not likely that so many dams will be built in future on dry land. Whilst the current reduction in the

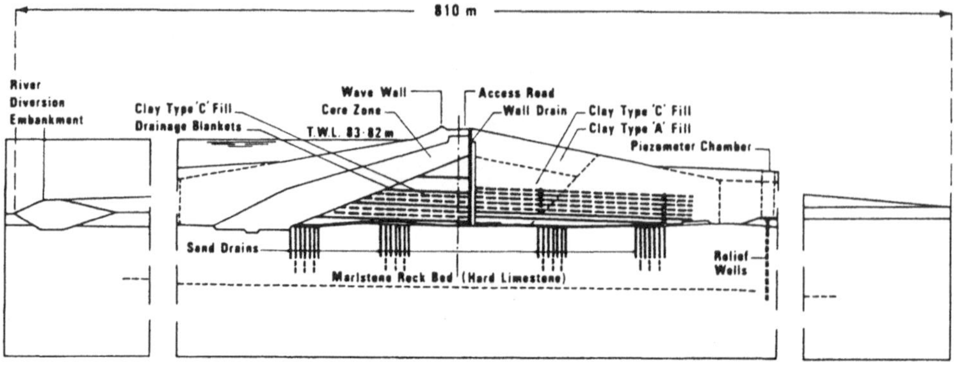

Fig. 10. Empingham Dam 1976 designed by Messrs. T. & C. Hawksley. *By courtesy of the Anglian Water Authority.*

growth of demand for potable water is a consequence of the present economic crisis and it will no doubt increase again after the crisis is over, it seems likely that in the next century water supply requirements will be met mainly by estuary schemes such as those for the Wash and the Dee.

HYDRO-ELECTRICITY

Whilst the earliest hydro-electric scheme in Scotland was the Foyers project for the British Aluminium Company completed in 1896, it was not until the same company built the Kinlochleven scheme at a later date that the first major dam for exploiting water power was constructed. Subsequently three other major dams were built by the company, the last one being completed during the last war. Four major dams were built in Wales before the war of which the most notable is the 31 m high Maentwrog arch dam. In addition, nine other major dams were constructed in Scotland for exploiting water power which, with one exception, now belong to the South of Scotland Hydro-electric Board.

The ways in which, as I have tried to show, the evolution of British dams has influenced and reflected social and economic changes is exemplified by the projects undertaken by the North of Scotland Hydro-electric Board which was founded under Tom Johnston as its Chairman in 1943. One of the main purposes for which the Board was set up was to improve the quality of life for a poor and sparsely populated region and to provide the power needed for economic development and expansion. No less than 50 major dams, many of them of the concrete buttress type, were constructed after the war by the Board. In addition four major dams completed in 1961–62, were constructed in Wales by the Central Electricity Generating Board. Two of these dams were for the upper and lower reservoirs for the Ffestiniog pumped storage scheme.

Having regard to economic constraints, the water power resources of Scotland and Wales have now been more or less fully exploited. It is not likely that any more hydro-electric schemes to develop new sources of water will be constructed in Great Britain but pumped storage schemes which enable energy supplies to be tailored to meet peak demand appear to have a future. The Central Electricity Generating Board has a very large pumped storage scheme named the Dinorwic project under construction near Caernavon. In Northern Ireland, construction of the Camlough pumped storage scheme was suspended in 1973.

THE DOLGARROG DISASTER

It is now necessary in my survey to refer once more to a flood disaster, this time to the dam burst flood which drowned sixteen people and swamped the village of Dolgarrog in 1925. This flood arose from two quite distinct events. The first was a local under-mining or blow out which occurred beneath the base of a small concrete dam built in 1911 on clay foundations while the second and major incident was the consequential failure of an earth dam about 3 kilometres downstream. The discharge of water released by the upper dam was much in excess of the capacity of the spillway of the lower dam with the result that the lower dam failed rapidly after being overtopped and released very suddenly an enormous quantity of water.

RESERVOIR SAFETY ACTS

As a result of the Dolgarrog disaster, the Reservoir (Safety Provisions) Act of 1930 came on the statute book. In 1863 an Act had been passed whereby any person who wished to complain about the safety of a reservoir could do so but the 1930 Act went much further. Under this Act, there is a Panel of engineers responsible for inspecting reservoirs to which only civil engineers considered competent to do so can belong and every reservoir with a capacity exceeding 25,000 cu. m. has to be inspected by a Panel engineer at intervals not exceeding ten years. This Act is still in force but it is expected to be superseded very shortly by a new Act which is the same in principle but contains some amendments considered desirable as a result of 45 years experience of the present Act.

FLOODS AND SPILLWAY CAPACITY

World wide experience has shown that about a quarter of recorded dam disasters have been due to spillways being too small when the exceptionally large flood, which is bound to come sooner or later, occurs. In the past, this has frequently been due to difficulties in estimating in each case how large a flood should be allowed for without being unnecessarily extravagant when considering spillway capacity, also what margin of the level of the top of the dam above the effective storage water level, referred to as freeboard, should be allowed to protect a dam against erosion due to wave action during such a flood. An example of wave action is shown in Figs. 11 and 12 of the dam at Haweswater taking a battering during an autumn gale. No damage is being done to the concrete but it would be a different story if Haweswater dam was an earthern embankment.

Fig. 11. Waves on Haweswater Reservoir. *Copyright Ivor Nicholas.*

A document was published in 1975 which makes the analysis of floods a less empirical task. It considers the meteorological and flood producing characteristics of the country and is the work of a team based largely on the Institute of Hydrology and the Meteorological Office.[24] This report makes it possible at any site to determine a design flood which might be exceeded only once in a hundred years or any other stated return period limited only by the probable maximum flood which is determined by the purely physical factors involved.

Fig. 12. Waves breaking over Haweswater Dam. *Copyright Ivor Nicholas.*

Including those less than 15 m high, there are perhaps as many as 2,000 existing dams which are subject to reservoir safety legislation. On some of the older dams in particular the spillway capacities are inadequate to cope with the theoretical maximum probable flood or even that which may occur once in 10,000 years. Safety is a purchaseable commodity and it would be possible to enlarge the spillways of these reservoirs so as to make all these dams safe under every conceivable circumstance but such a policy would be exceedingly expensive and would make excessive demands on the Nation's limited resources which, insofar as the sancity of human life is concerned, could be allocated to greater effect on the roads and elsewhere. Depending upon where the dam is sited, one has to also consider whether or not a breach would endanger the lives of a community.

Fig. 13. Todd Brook Dam 1968. *Photo by courtesy of the North West Water Authority.*

To study, not only the application of the *Floods Studies Report* to reservoir safety but also flood protection standards for different types of dam in different situations, the Institution of Civil Engineers formed a Working Group under my chairmanship in 1971 and plans to publish in accord-

Fig. 14. Todd Brook Dam Main Spillway. *Photo by author.*

ance with our recommendations a guide on floods and reservoir safety soon which will supersede an interim report first published in 1933 by a committee of which my father was chairman.

TODD BROOK DAM

An interesting example of increasing spillway capacity on an old dam is the work done on the Todd Brook dam in the South-West Pennines which is 22 m high and was completed in 1831 to supply additional water to the Peak Forest Canal. Fig. 13 shows a photograph of the dam taken in the nineteen sixties. Fig. 14 shows a photograph of the original old spillway. Evidently during the nineteen sixties the Panel engineer was not satisfied that the spillway capacity was adequate and Fig. 15 shows the dam as it is today with an additional large concrete lined auxiliary spillway constructed directly on top of it in 1969.

Fig. 15. Todd Brook Dam 1976, with auxiliary Spillway. *Photo by author.*

COOMBS DAM

A warning on the dangers which may arise occurred in 1976 on the 16 m high Coombs dam which was authorised by an Act of Parliament dated 1794 and was built about the beginning of the nineteenth century. It also supplies water to the Peak Forest canal. During the last 10 days of January, 1976, many parts of the country were hit by tremendous gales. On 29 January a slip occurred on the Coombs dam which is shown on Figs. 16 and 17. Thus even after standing safely for more than 150 years a dangerous situation arose on this old major dam. The situation was saved by the prompt action taken by the British Waterways Board in lowering the reservoir water level by pumping and by cutting a deep cut in the spillway. By 7 February when the photographs shown on Figs. 16 and 17 were taken the reservior water level had been drawn down about 6 m.

In recent years, Britain probably has a better record than any other country on the safety of dams due in large measure to the success of the Reservoir (Safety Provisions) Act 1930 but, as the Lluest Wen and Coombs incidents show, constant vigilance is necessary.

Fig. 16. Twentyninth January 1976. Slip on Coombs Dam. *Photo by author.*

THE TWELFTH DICKINSON MEMORIAL LECTURE, THE EVOLUTION OF BRITISH DAMS

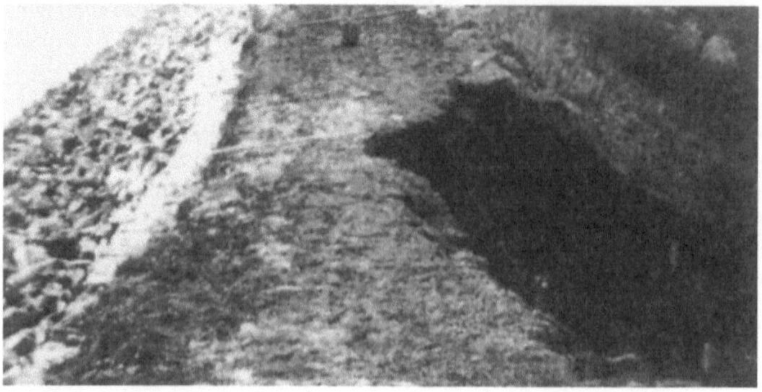

Fig. 17. Coombs Dam Slip, close up of crest. *Photo by author.*

NOTES

1. R. Bennett and J. Elton *History of Corn Milling Watermills and Windmills*, Vol. 2, 1898, p. 97.

2. Rudyard Kipling, *'Puck of Pook's Hill.*

3. Hemmingway, *History of the City of Chester from the foundation to the present time, 1831*, J. H. E. Bennett 'The Causeway Mills in Handbridge', *Cheshire Sheaf, XLIX*, 1954, Information supplied by Miss A. M. Kennet, City Archivist, Chester City Record Office.

4. P. N. Wilson, 'Kendal Reservoirs', *Trans. Cumberland & Westmorland Antiquarian & Archeological Society*, LXXXIII 1973.

5. J. Needham FRS, 'China & the Invention of the Pound Lock', *Trans. Newcomen Society*, XXXVI 1963–64.

6. Quoted in T. S. Willan, *River Navigation in England 1600–1750*, 1936, p. 29.

7. Ibid p. 133

8. N. Smith, A History of Dams, 1971, pp. 164–165.

9. *An Act to enable the Most Noble Francis, Duke of Bridgewater to make a navigable Cut or Canal from a certain place in the township of Salford to or near Hollin Ferry in the country palatine of Lancaster.*

10. S. Smiles, *Lives of the Engineers*, 1874, Vol. V, p. 196.

11. L. T. C. Rolt, *The Inland Waterways of England*, 1966, pp. 106 & 107.

12. A. R. Binnie, *Lecture on Water Supply 1877* – 2nd. ed. 1887 p. 38.

13. The Great Water dam built in 1748 at Grimsthorpe has traditionally been attributed to Capability Brown but on p. 161 of the 1975 edition of her book *Capability Brown* (first published 1950) Miss Dorothy Stroud now confirms that there is no evidence of Brown having worked at Grimsthorpe before 1770 and that John Grundy was employed during the two previous decades.

14. A. W. Skempton & Esther C. Wright 'Early Members of the Smeatonian Society of Engineers' *Trans. Newcomen Society* XLIV p. 27.

15. M. F. Kennard, *Eighteenth Century Dams in England: BNCOLD/University of Tyne Symposium 24/25 Sept. 1975.* 'Inspection Operation & Improvement of Existing Dams'. Paper 5.1 p. 5, published by University of Newcastle upon Tyne, edited by A. I. B. Moffat.

16. Based on the Registrar General's *Statistical Review of England & Wales for the year 1973*, Part II Tables, Population.

17. Based on *World Register of Dams.* International Commission on Large Dams (ICOLD) updated to 31 December 1974. pp. 182–203 with dams less than 15 m high excluded, Bureau Central: 51, Rue Saint Georges, Paris (9°)

18. A. J. S. Pippard, Unwin Lecture 'The Functions of Engineering Research in the University, *Journ. Inst. Civ. Eng.* vol. 33, 1949–50, pp. 265–86.

19. Report of the Inquest on the Bilberry disaster victims from the *Huddersfield Weekly Chronicle*, February, 1825.

20. *The Holmfirth Flood caused by the Bursting of the Bilberry Reservoir on Thursday morning, February 5th, 1852*, Holmfirth, Eli, Collins & Co. Ltd. Print 'Express' office 1910.

21. Ibid p. 63.

22. A. L. Little, *Maintenance & Remedial Works at Some Embankment Dams: BNCOLD/University of Tyne Symposium*, 24/25 Sept. 1975, 'Inspection & Operation & Improvement of Existing Dams'. Paper 5.7.

23. Information & cross section supplied by courtesy of the Anglian Water Authority & Messers. T. & C. Hawksley.

24. *Flood Studies Report*, 5 vol., 1975, Natural Environment Research Council, 27, Charing Cross Road, London.

8
The Jones Falls Dam on the Rideau Canal, Ontario, Canada

Robert F. Legget

That a Canadian should have been asked to present a Paper to the Newcomen Society must be a matter of surprise to some members, so young a country is Canada still regarded. The completion of the Canadian Pacific Railway through the western mountains in 1885 has been so well publicised that it is natural that many should think of this great feat as the first major engineering undertaking of the Dominion. Long before that, however, the building of the Grand Trunk line linking Montreal with Chicago, now a main line of Canadian National Railways, had involved impressive civil engineering works. And before this early railway construction came the first commercial canal building, the successors of the first little St. Lawrence canals being especially newsworthy today in view of the pending completion of the modern St. Lawrence Seaway which will now supersede them. Earlier still were the military works considered necessary for the proper settlement and defence of the young colony after the victory of Wolfe in 1759 and the ensuing Treaty of Paris, which really marks the start of the history of Canada as a part of the Commonwealth. To begin with these were fortifications, notably the great bastion of Quebec. Further to the west, the fortress of Kingston (at the eastern end of Lake Ontario) was strengthened and enlarged, all these works being pioneer endeavours of the Royal Engineers, always important units in the British garrisons overseas. When, in 1812, war broke out between Great Britain and the fledgling United States of America, Kingston came into its own, playing an important part not only as a fortress but also as the site of an important naval dockyard. Much of the war of 1812 was fought on the waters of the Great Lakes. Sir James Yeo, the British commodore, flew his pennant in a three-decker vessel on Lake Ontario more powerful than Nelson's flagship at Trafalgar.

The war finished in 1814, the Treaty of Ghent being signed in that year, since when the international border has never again been the scene of martial dispute. But hard feelings persisted long after the fighting stopped; contemporary records show how strong was the fear of further conflict.

THE JONES FALLS DAM ON THE RIDEAU CANAL, ONTARIO, CANADA

This sense of alarm was communicated to the British Government in London. Army strategists were concerned at the vulnerable position of Kingston, should hostilities recommence. A glance at Fig. 1 will show that in order to reach Kingston from Montreal, the full length of the international section of the St. Lawrence has to be negotiated. All supplies for Kingston and the little settlements to the west, including all naval stores and equipment, and even—on one occasion—two small gun-boats made in England and dismantled for transhipment, had to be brought up the rapids of this famous waterway in small river boats. Why the Americans did not ambush this vital supply line during the 1812 war is almost a mystery even today. Evidence has been found, however, which showed that they intended to take this step if war had again broken out. The British, therefore, had to find an alternative water route from Montreal to Kingston, there being naturally no convenient land communication between these points at that time because of the thick virgin forests. Fortunately, an alternative route was already known, one long used by the Indians and shown by them to early explorers. A survey of the route was made by a young officer of the Royal Engineers (Lieutenant French) as early as 1783, as a part of the early exploration of the new colony. He had proceeded 120 miles up the Ottawa River to the great rapids already known as the Chaudière, thence up a river that entered the Ottawa on its south bank until he reached a group of lakes. Shown the way through these interlocking bodies of water, and crossing the old Indian portages, he had come to the headwaters of the Gananoque River, descending which he came to the St. Lawrence River and sailed down this back to Montreal. Later explorers followed the same route as far as the Rideau Lakes but then descended the Cataraqui River, which also had its source in the Lakes, finding that this led them directly to Kingston. It was therefore decided by the British authorities that the possibility of using this route as an alternative military supply line between Montreal and Kingston should be investigated. Various estimates of cost were prepared, all roughly approximate. Discussions were held with the government of the colony of Upper Canada but no agreement as to the division of cost could be reached. The British Government finally decided to construct the waterway itself, with no financial contribution from the local government, almost certainly at the direct orders of the Duke of Wellington.[1]

The Rideau Canal

The Rideau Canal is not a canal in the normally accepted engineering sense of the word, but a canalised river system, since it follows the Rideau River from the Ottawa River to the Rideau Lakes and the Cataraqui River from the Lakes down to Kingston on Lake Ontario. Its location is shown in Fig. 1 upon which is also indicated the outline of the Precambrian Shield. This is the name given to the great mass of Precambrian rock which makes up about one-half of the land area of Canada, stretching from Labrador in the east to the Mackenzie River in the west. The map shows the extreme southern part of the shield. It will be seen that a narrow neck (the "Frontenac Axis" to geologists) crosses the St. Lawrence River and forms the Adirondack Mountains of northern New York State. The Rideau Lakes are flooded depressions in the glaciated surface of this ancient rock massif, which rarely has more than a very thin cover of soil; in this they are similar to the tens of thousands of lakes which so distinguish the Canadian northland. The two rivers drain from this lake system, the Rideau River dropping about 270 ft. to its confluence with the Ottawa, which is made over a fine waterfall, its course of about 60 miles passing through flat country, now agricultural but originally virgin hardwood forest. The Cataraqui drops through 160 ft. but does so in less than 40 miles. Originally a turbulent rushing stream with many falls and rapids, it is today submerged

[1] Robert Legget, *Rideau Waterway*, University of Toronto Press, 1955, 249 pp.

beneath the waters of the land flooded by the dams necessary for forming the canal, for all but its last few miles in Cataraqui Bay at the level of Lake Ontario.

The Canal was built with locks 134 ft. by 33 ft., with a 5 ft. 6 in. draft over the lock sills. In order to achieve clear sailing from the one end to the other, it was necessary to construct forty-seven locks. Thirty-two of these are between the Ottawa River and Lower Rideau Lake; two are between

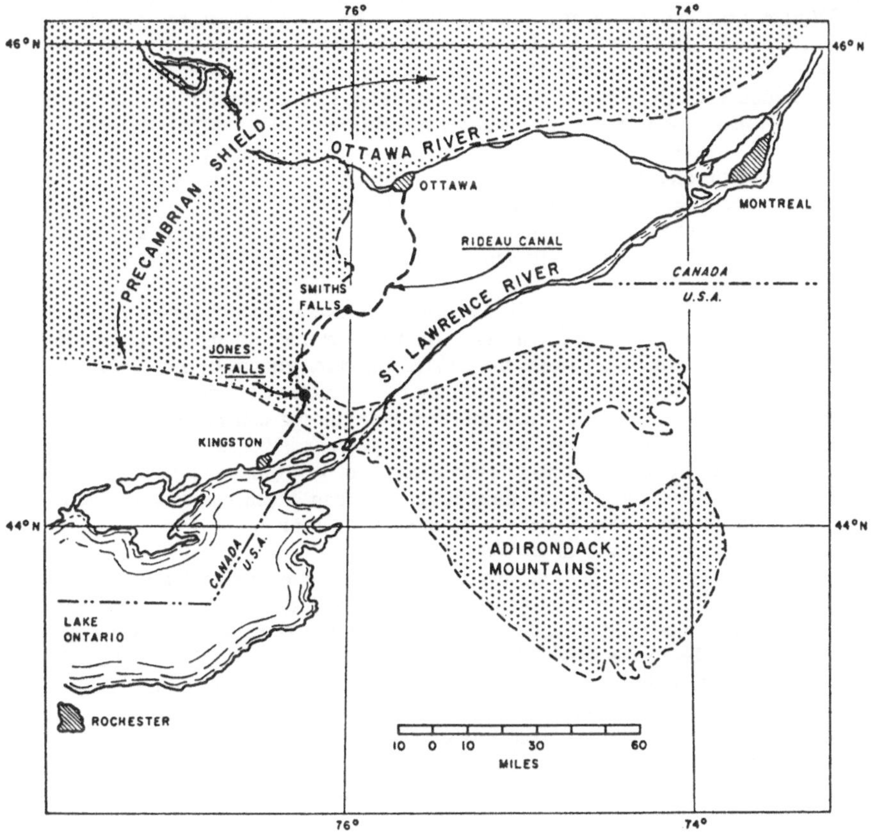

Fig. 1. Location of Rideau Canal

the lakes that were linked together to form part of the waterway; and thirteen lie between Indian Lake and Cataraqui Bay. In order to raise the natural water level of the rivers and lakes the requisite distance to provide for navigation between and into the locks, fifty-two dams had to be constructed. These range all the way from simple earth embankments a few feet high to several massive masonry structures, the largest of which is the subject of this Paper. The most troublesome of all the dams to build was the Hog's Back Dam, now within the limits of the city of Ottawa, which was first built

THE JONES FALLS DAM ON THE RIDEAU CANAL, ONTARIO, CANADA

as an earth fill dam but which had to be rebuilt as a rockfilled timber crib after a serious failure of the initial structure. That this was the only failure of any sort during construction is but one of the many features of the work which testify to the professional skill and ingenuity of the builders. Lieutenant-Colonel John By, Royal Engineers, was selected for the great task of directing the work, once the decision to build had been made. He had served for 11 years in Canada as a junior officer but was brought from retirement (almost certainly by the Duke of Wellington) to return to Canada at the age of 52 in order to construct this canal through a forest wilderness. He arrived in Canada in May 1826; the first sod was turned in September of the same year in the presence of the Earl of Dalhousie, Governor in Chief. Colonel By sailed through the completed canal, from Kingston to Bytown (now Ottawa) in early May 1832. Almost immediately, he was recalled to England, not to receive the honours which he had so surely earned, but to face an inquiry initiated by the Lords of His Majesty's Treasury with regard to alleged expenditure of about £22,000 without authority. By was exonerated but retired from public life and just 4 years later died at Shernfold Park, Frant, Sussex, worn out by his exertions in the forests and swamps of Canada and certainly not encouraged by the reception he received on his return to his native land.

Although the general route had been selected on the basis of earlier surveys and reports, Colonel By had to select the final route to be followed, decide upon all water levels, design all the dams and locks, and direct and supervise all construction. He was assisted by a group of unusually able young assistants (at least three of whom eventually became full Generals in the British Army) and by two companies of the Royal Sappers and Miners. He was fortunate in having the services of four reliable contractors for his masonry work, his satisfaction being shown by the silver cups which he presented to each firm at the end of the work, theirs by the modest fortunes which they established as a result of their efforts. By was much less fortunate in the contractors he obtained (always through public tenders) for his excavation work. The popular belief, at that time, that excavation was a simple job led to innumerable troubles as a result of low bids and thorough incompetence. Despite all such difficulties, the job was finished in less than 4 years and at a total cost of only £800,000; the figure is mentioned if only to counter any wrong impression that may have been created by the reference to contractors' profits.

The full significance of Colonel By's achievement can only be appreciated when the setting of the canal building is realised. The only settlement of any kind on the entire route was Kingston, and it was little more than a small military and naval fortress. There were no more than a dozen settlers' cabins along the 130 miles of waterway, the shores of all rivers and lakes being thick forest to the water's edge. Two trails only came through the forest as far as the canal route from the shores of the St. Lawrence upon which one or two small settlements had been established. Apart from the few settlers, the only other inhabitants were Indians, fortunately friendly and helpful. The climate ranged from sub-zero weather during the winters, when work had to stop because of the intense cold and deep snow, to the tropical heat and humidity of high summer. Surprisingly, the heat of summer caused Colonel By far more trouble than the winter cold.

All construction materials had to be obtained locally, with the exception of the ironwork for sluices and gate fittings. Timber was cut from the surrounding forests. Stone was quarried at sites as close to the locks and dams as possible. Cement, for grouting masonry was made near the Ottawa end of the canal, by burning limestone from a quarry which is still in use today; it now provides raw material for one of the main plants of the Canada Cement Company Ltd. Supplies, including most of the necessary food, had to be brought from Montreal either up the Ottawa River to Bytown, or up the St. Lawrence to Kingston, thence by canoe to the work camps along the waterway. Even the pay for the workers had to be carried in this way (usually as silver coin in kegs) since the local Paymaster General objected to Colonel By's very reasonable suggestion that the men should be paid by cheque.

THE JONES FALLS DAM ON THE RIDEAU CANAL, ONTARIO, CANADA

Of construction equipment as it is known today, there was none. All the power used was man-power, except in a few locations where oxen were used for long haulage. Horses seem to have been used only for riding; Colonel By on his black charger became almost a legendary figure around Bytown. (This was the name given, almost jokingly at first, to the small construction camp on the Ottawa where Colonel By made his headquarters; thus was Ottawa, the nation's capital, founded.) Earth excavation was by pick and shovel, earth transport by wheelbarrow. Rock drilling was by hand; blasting by means of "black powder." Stone cutting was similarly all the hand work of masons, Scottish and French Canadian; stone handling was done by means of wooden "stone-boats" and hand derricks. By such simple means were built, in the midst of the virgin forests of Canada, masonry dams that exceeded in size all that had hitherto been constructed in the New World and most of those in the Old. Of the Rideau dams, that at Jones Falls was and still is pre-eminent.

The Dam at Jones Falls

Before the building of the Canal, water drained from Clear Lake down to Whitefish Lake (about 35 miles north-west of Kingston) through a rocky gorge. There is available a description of this, the most difficult part to be canalised of the entire Rideau route in an account written by John McTaggart, published in London in 1829. McTaggart served as Colonel By's first Clerk of Works; some of his writing seems to have been tinged by his singularly vivid imagination, but his description of this natural feature of the Canal route can be accepted as reasonably accurate.

> "These [Jones's Falls] are the greatest in the least distance that are met with in the whole route, rolling down a narrow ravine scarcely a mile in length, and having a 60-foot fall. The banks of this narrow and crooked ravine are lofty, averaging 90 feet in height; and on the west side there are deep bogs, surrounded by high land. . . . It is also an objection that as the ravine is crooked, and cannot be straightened, from the nature of its steep freestone banks, the locks can not be placed in such a manner as not to have their entrances awkwardly set for boats to get in and out of them."[1]

This was the natural barrier that had to be circumvented. McTaggart advanced several ideas to his chief as to how this could be done but they were generally impracticable. Following his own examination of the site, Colonel By, with his fellow engineer officers, devised the scheme which they carried successfully to completion but which must have seemed foolhardy at the time. This was to dam the gorge right up to the level of Clear Lake and to use "Macdonald's Gully," a natural depression in the general rock profile about 3,000 ft. to the south, with some additional excavation, in order to build the four locks necessary to get boats up and down the difference in water level of 58·5 ft. between the two lakes (Fig. 2). The locks themselves are fine examples of masonry work, arranged as a group of three with the fourth (upstream) lock separated from the chain of three by a large basin. They are similar in design to the other locks on the Canal. Attention must be concentrated upon the dam which made their operation possible and which is still a unique feature of the Waterway.

The downstream face of the dam is arched in plan, concave downstream, so that it appears to be a true arched dam. It is formed of hand-hewn sandstone blocks, each about 6 ft. by 4 ft. by about 18 in. The blocks are arranged in vertical, not in horizontal courses (Plate XXI, Fig. 1). The joints are so tight that it is impossible to state whether they are mortar joints or not. There is certainly some mortar in the masonry work, however, as became evident when the dam had to be cut into some years ago. The masonry was carried right up to the crest of the dam which measures 350 ft. from abutment to abutment. The masonry is built solidly into the bedrock that forms the slopes of the

[1] John McTaggart, *Three Years in Canada etc.*, H. Colburn, London, 1829, Vol. I, p. 148.

THE JONES FALLS DAM ON THE RIDEAU CANAL, ONTARIO, CANADA

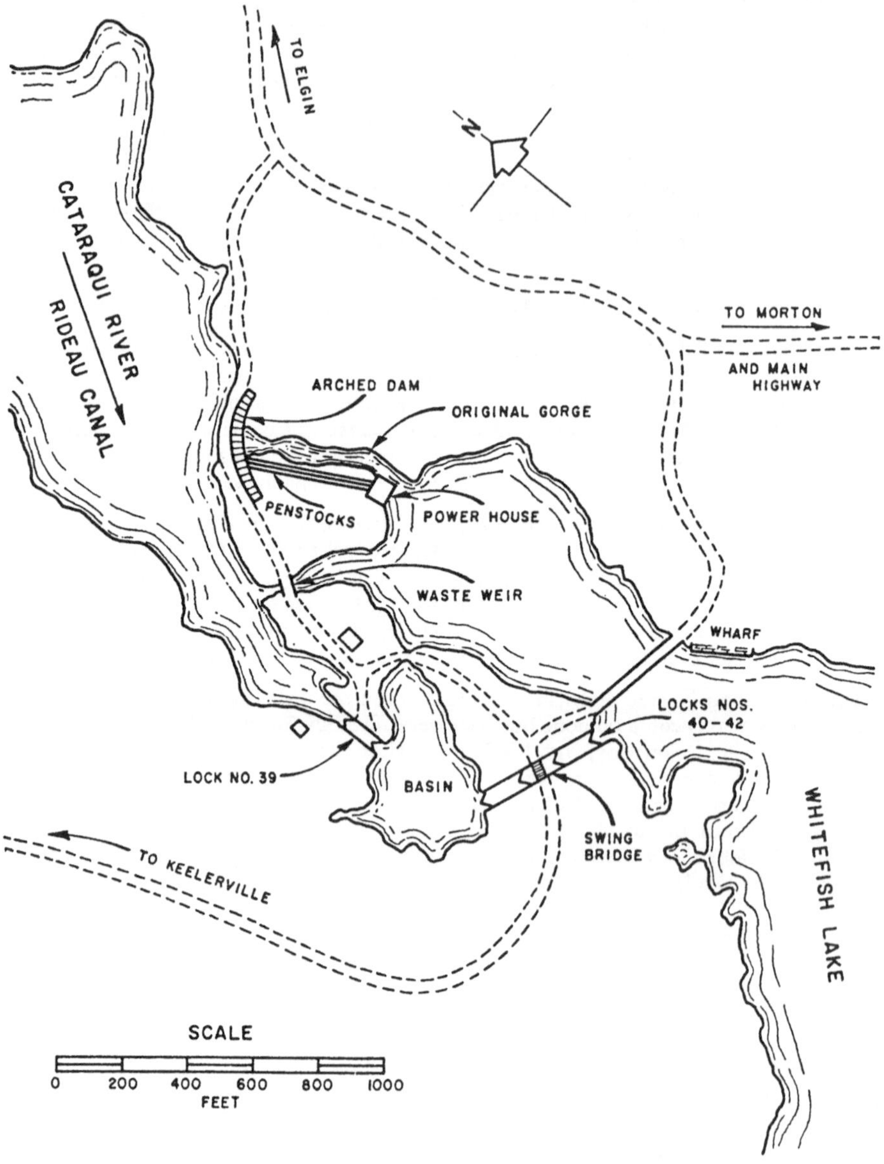

Fig. 2. LOCATION OF DAM AND LOCKS AT JONES FALLS

THE JONES FALLS DAM ON THE RIDEAU CANAL, ONTARIO, CANADA

gorge so that the structure is, in fact as well as in appearance, a true arched dam. Analytical design methods were not available to Colonel By when he had to design this structure and so he chose not to rely on the masonry entirely but to use it as a means of retaining an earth dam upstream. Although it cannot be seen today, nor has it been seen since the dam was built, this earth dam is shown in an old sketch of the structure which is all that has yet been discovered in the way of any record drawing of the dam as constructed. The dam consists of the masonry structure, backed by the earth fill, and between them a puddle core (Fig. 3). At its maximum cross-section the masonry is 27 ft. 6 in. wide at rock level, decreasing in width to 21 ft. 6 in. at water level, which is 56 ft. 6 in. above lowest rock level. With freeboard of about 5 ft., this gives a maximum height for the dam of between 61 ft. and 62 ft. No record has yet been found of any dam standing in North America in 1832 which approached this height. The construction of this dam at Jones Falls antedates by a few years that of the famous dam near Aix-en-Provence (1843) designed by Zola and frequently referred to as a pioneer structure. At its maximum section the earth fill extended upstream 127 ft. out from the masonry structure. It was finished off at an underwater slope of 2½ to 1. Along the crest of the dam the earth is level for a width of 25 ft. and stands about 3 ft. higher than the crest of the masonry. The separation of the masonry and the earth fill by a clay puddle core is a feature probably drawn from British engineering practice of the time. Scattered references in contemporary accounts suggest that it was not possible to found the masonry solidly on the bedrock in the bed of the gorge so that the puddle core was almost certainly intended to ensure the watertightness of the completed structure. During the course of construction, the clay was replaced by broken stone grouted with cement mortar, as explained in this note which appears on the sketch to which reference has been made:

> "The broken stone grouted was an afterthought and therefore does not actually reach the bottom everywhere; however, the Puddle was taken out wherever it could be replaced with grouted stone; the back or the Key Work was also jointed with mortar. As the grouted stone does not *extend quite* to the bottom everywhere, it is certainly a disadvantage as it will probably prevent the Clay from settling properly close to the base."[1]

The concern of the writer of this note was dissipated, it may be hoped, before the officers of the Royal Engineers left the Canal works, for the dam has performed quite successfully through all the intervening years. The crest of the earthwork is today used as part of a local road which crosses the dam. The masonry crest remains exposed, the sandstone exhibiting no signs of deterioration. With its crest curved to a radius of about 175 ft., the masonry blends almost imperceptibly into the bedrock at the sides of the gorge; the dam stands today as a singularly graceful structure in its woodland setting.

Building the Dam

The initial reaction of civil engineers when they see the dam at Jones Falls for the first time, is one of admiration for the structure itself. When the date of its construction is known, initial reactions are displaced by surprise at the fact that such a structure was built so many years ago in a narrow and confined gorge with no obvious possibility for the necessary control of the river which flowed through the gorge, river control being always so critical a part of dam construction. Here are no large diversion tunnels, no convenient spillway opening to be gradually closed up as the dam rises to its full height. Fortunately, the way in which this construction operation was carried out can be described in the words of one of the most remarkable of Colonel By's assistants, Lieutenant William

[1] Lieut. W. Denison, "Rideau dams" in *Papers on Subjects connected with the Duties of the Corps of Royal Engineers*, J. Crane, London, 1838, Vol. II, pp. 120–1.

THE JONES FALLS DAM ON THE RIDEAU CANAL, ONTARIO, CANADA

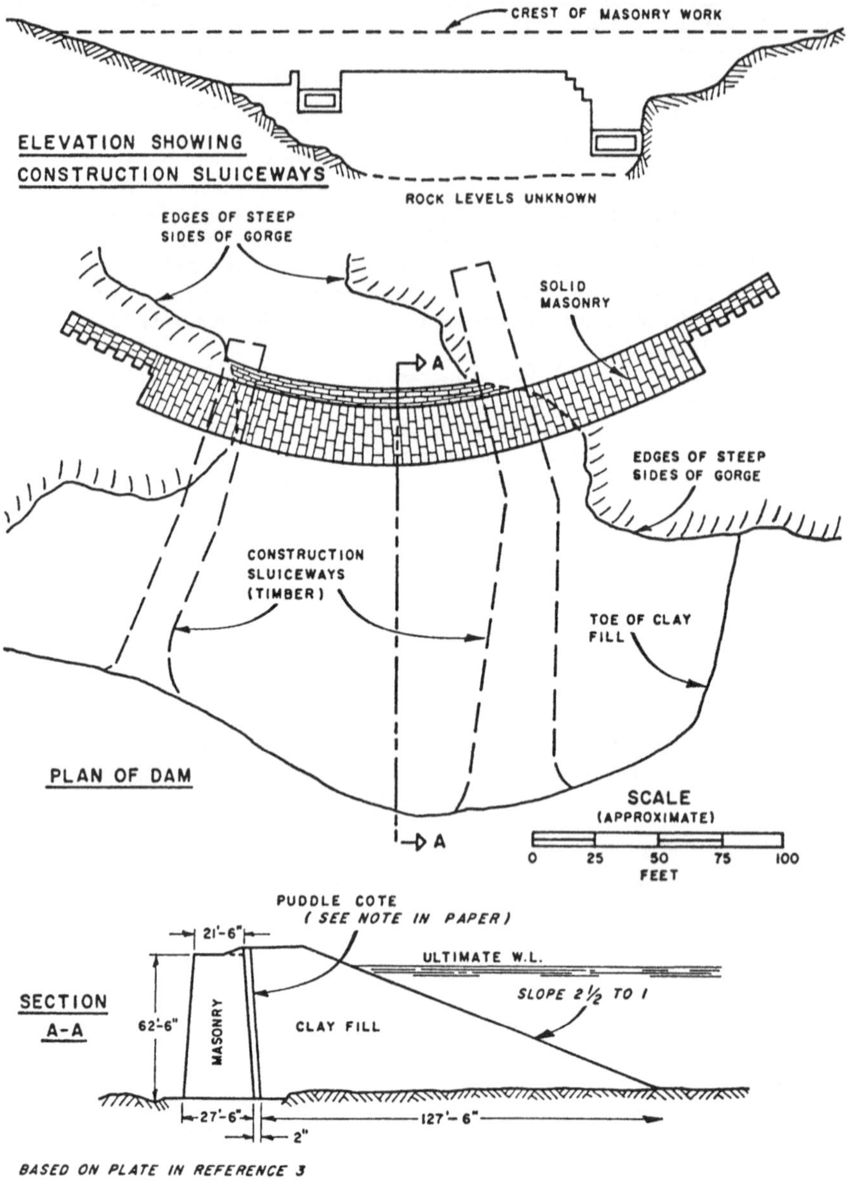

Fig. 3. Sketches of Arched Dam at Jones Falls

THE JONES FALLS DAM ON THE RIDEAU CANAL, ONTARIO, CANADA

Denison who, although not the resident officer at Jones Falls, must have visited the works during the course of construction while on some of the special missions he undertook for his chief. He recorded that—

> "The work was commenced by erecting a wooden temporary dam at the mouth of the gully, by which, when necessary, the outlet of the lake could be stopped, and the works kept free from water for about a week in the dry season, without any fear of raising the water in the lake so high as to do any mischief. A temporary wooden sluice-way . . . was then constructed close to one side of the ravine, at the spot where the dam was to be constructed; and this being made of sufficient width and depth to carry off all the water that usually passed down the gully, the remainder of the passage was blocked up with earth, and rubbish, behind which the key-work was built to the height of about eighteen or twenty feet. At this height, on the opposite side of the dam, a fresh sluice-way was made, the part through the key-work being framed of wood, and the remainder of masonry; sheeting piles were driven in several places, to prevent the water working its way beneath the floor of this sluice, which was carefully planked. As soon as this sluice-way was completed, and the key-work and remainder of the dam raised some height above it, the outlet of the lake was stopped, and every exertion made to close the old sluice-way filling in the opening with rubbish, and building the key-work behind. The water being then allowed to pass down the ravine, rose in front of the dam, and discharged itself by the second sluice-way. After this, the work went on as before, and a permanent waste-channel having been excavated in the rocky bank of the ravine, provided with sluices, and everything requisite to regulate the height of water when this second temporary sluice-way was closed, the water was turned into this permanent channel, and the dam was then completed to the requisite height without any further trouble."[1]

As thus described, the operation (known to construction men as "leap frogging") appears to be deceptively simple, but it requires little reflection to appreciate that it must have been the work of a master builder. Colonel By was fortunate in being able to award the contract for this work to an experienced firm of masonry contractors who had already carried out important masonry contracts in the Montreal area. Thomas McKay and John Redpath were partners; together, they carried out some of the most important of the canal works. By agreement, John Redpath took the responsibility for the dam at Jones Falls; the records show that he gave to it his continuous personal attention. It has been the writer's privilege to examine at McGill University (where the name of Redpath appears on several notable buildings) the old account books of John Redpath. Many are the human touches there revealed (such as the repetitive item "Expenses to see the men off at Lachine," men for the job at Jones Falls about to proceed up river by canoe from the famous wharf at Lachine). These records also show the total volume of masonry in the dam, for which the partners were paid at the rate of 1s. per cubic foot, to have been £426,161.70.

The primitive setting for the dam works has already been sketched—the virgin forest, with no settlements of any kind within 25 miles. Accordingly, a camp had first to be built in forest clearings; this was ready by the early summer of 1828, accommodation being provided for 200 men. So simple were the tools to be used that there would be no trouble in bringing them up to the site of the work by canoe, or even on men's backs along the forest trail from the tiny settlement of Brockville on the St. Lawrence. Wood would be cut from the forests around for the hand cranes which appear to have been the only "mechanical" equipment used in any way on the work.

Perhaps the most surprising aspect of the building operations is the fact that the stone for the dam was not taken from the vicinity of the works, as was the clay for the puddle and earth fill, but from a quarry about 6 miles from the dam. The dam site lies on Precambrian rock of extreme hardness,

[1] Lieut. W. Denison, *op. cit.*

THE JONES FALLS DAM ON THE RIDEAU CANAL, ONTARIO, CANADA

not easily "workable" by masons, so that it would not have been a suitable stone for building. It so happens that the boundary between the Precambrian of the Shield and the palaeozoic rocks of the St. Lawrence lies about 5 miles to the east of the Falls. The boundary is not normally exposed and there is only one major exposure of the Beekmantown sandstone to be seen within convenient distance of the site. How the builders of the dam at Jones Falls knew that such stone could be found, and how they found the one exposure which must have been in the midst of unbroken forest when the work commenced are questions the answers to which can only be left to imagination. The sandstone was found, however, and all the stone for the locks and the dam was obtained from this quarry (now located between the little towns of Elgin and Philipsville) hauled by oxen over an improvised track through the forest, unloaded on to scows and floated down Clear Lake to the site of the work. The quarry may still be seen today, with some blocks of stone in semi-finished state, just as they were left by the canal builders. Since little masonry building has subsequently been done in the district, the quarry has never been worked for other purposes.

Even though all construction operations involve some hardships, mention must finally be made of the hardships encountered on this work. Not only had they to contend with the rigours of winter, without the advantage of modern heating, but far worse was the heat of summer with the fetid and humid atmosphere generated from the swamps on the canal route and the accompanying "swamp fever." It is reported that so bad was the fever at Jones Falls that at one time "no one . . . was able to carry water to a sick friend, because all were overcome, including the doctors." Mortality among the Irish labourers, and the Scottish and French Canadian masons, was heavy; for many years a small graveyard nearby bore mute testimony to the toll in human lives taken by this construction job. By the winter of 1831 the work was complete. A watchman was left to guard the work until it was officially taken over in 1832; some of his letters testify to the extreme loneliness of the location.

The Dam Today

Unlike so many older engineering structures of historic significance, the dam at Jones Falls stands today in perfect condition, still performing its intended function despite the changes that have come to the land that is now Canada, and the way in which the course of time has taken the history of the Rideau Canal almost through full circle. Constructed as a defence measure in view of fears in the young colony of invasion by American troops, the Canal is today well used every summer by a steadily increasing invasion of American visitors, who come to enjoy the beauties of the Rideau Waterway (as it is known to some), the delights of the Rideau Lakes, and the excellence of the local fishing. The Canal was used, for more than two decades from the time of its completion, for the conveyance of troops, military supplies, and even of small naval vessels from Montreal to Kingston. Gradually, however, the fear of war disappeared, concurrently with the steady opening up and development of the colony of Upper Canada. As military traffic diminished, civilian traffic on the Canal increased. Until the opening of the final section of the St. Lawrence canals in 1855, the Ottawa River and Rideau Canal together constituted the first St. Lawrence Seaway, for small ocean-going vessels did use the Canal. As through traffic from Montreal diminished, local traffic in coal, timber, cheese and other farm products increased, as also did passenger traffic along the Canal. The coming of the railways, of highways and the automobile, eventually led to the disappearance of all this traffic; use of the Canal reached its nadir in the 1930's. In recent years, however, summer traffic has shown a surprising increase but entirely of pleasure craft, Canadian and American; the increase continues. Vessels that come through connecting American inland waterways from as far away as Florida and New Orleans are now not uncommon on Rideau waters. Through all these vicissitudes, the great dam has continued to maintain the water level necessary for the approach to the flight of locks at Jones

THE JONES FALLS DAM ON THE RIDEAU CANAL, ONTARIO, CANADA

Falls. Today, the "narrow and crooked ravine" of John McTaggart is filled with large trees, its undergrowth almost impassable because of dead wood and thick brush. It is, therefore, difficult to obtain any photographs which give any real impression of the size and quality of the structure; those which accompany this Paper are about as good as can be obtained (Plate XXI). Standing remote even from the little side road that connects the locks with the main highway two miles away, the dam is seen only by the few. Most of those who now sail the Canal do not even realise its existence since all they see is its grassy crest as a break in the wooded shore line which they pass as they turn into or out of the narrow channel past "The Quarters," so called since at this location the resident Royal Engineer officers were quartered.

Throughout its 130 years of service, the dam has required practically no maintenance. There is today a small leak through its base, but this appears to have been constant for many years. The masonry is just as good as it was when first placed. In one way it is really better, the mortar having taken on a really rock-like consistency. This was demonstrated in 1947, when a strangely modern addition was made to the ancient structure. The Gananoque Electric Light & Water Supply Company Ltd. obtained permission from the Federal Government of Canada (which owns and administers the Canal through its Department of Transport) to use the normal flow of water past the Jones Falls dam and locks for generating water power to be fed into its small local supply system. The dam was therefore breached close to its western end, and a small concrete intake structure was erected; as many as possible of the masonry blocks were restored to their original position in order to preserve the appearance of the dam. A wood stave penstock was connected to the intake, led down the side of the old gorge and into a small concrete power-house located on the edge of the pool at the foot of the locks. The head of approximately 60 ft. was thus made available for power purposes. The one water turbine and electrical generator initially installed were soon supplemented by three more units fed by two additional penstocks, so that the little station now has an installed capacity of 4,000 kW. The power-house is remotely controlled from the station at Kingston Mills, 25 miles away. Near to the little building is a television aerial. What contrast these modern features provide to the nearby dam will be apparent to all who have any feeling for history. Standing on the crest of the dam, however, they can easily be forgotten as the significance of the old structure is again appreciated and silent tribute paid to the men of the Royal Engineers, and their contractors, who gave this early work, and so much more, to assist in the founding of the land that is now Canada.

Discussion

Professor A. W. Skempton recalled with delight the day he and Mrs. Skempton had spent with Mr. Legget seeing the Rideau Canal. There could be no doubt that it was a master work, and that John By was a great engineer.

The Jones Falls dam, however, could be claimed neither as the first arch dam, nor as the highest masonry dam of its period. A gravity dam at Elche, in Spain, built by the Moors in the tenth century, had been replaced in 1590 by an arch dam, the dimensions of which were given in the table. Only 4 years later a gravity dam with a height of 135 ft. was completed at Alicante, also in Spain. Nevertheless, the Jones Falls dam is a most notable structure of extreme importance in the history of civil engineering in America.

Mr. Legget writes of Jones Falls that "it appears to be a true arched dam." Since the ratio of base width to height is only 0·44, whilst for gravity dams this ratio is never less than 0·75 and is usually about 0·90, there is no need for caution in accepting its true function as an arch dam. The clay bank was necessary in construction and also served to increase the watertightness of the base of the dam. Its presence slightly increases the total pressure which the masonry arch has to withstand, above the pressure due to the water alone.

THE JONES FALLS DAM ON THE RIDEAU CANAL, ONTARIO, CANADA

EARLY ARCH DAMS

Dam	Completion date	Height ft.	Base / Height	Length / Height	Radius ft.	Engineer
Elche, Spain	1590	76	0·51	3·0	205	—
Jones Falls, Canada.. ..	1831	62	0·44	3·6	175	John By
Zola, France	1843	120	0·35	1·7	158	M. Zola
Bear Valley,* U.S.A. ..	1884	64	0·31	4·7	335	F. E. Brown
Sweetwater, U.S.A... ..	1888	90	0·51	4·2	222	J. D. Schuyler
Barossa, Australia	1903	94	0·36	5·0	200	Alex Moncrieff

* Considered dangerous, and replaced 1910.

Dr. A. P. THURSTON remarked that his brother-in-law still owned land round one of the locks and that he himself still actually had money from it. They would probably remember that Napoleon offered Louisiana to the fledgling United States if they would pay £3 million, and they could have Canada if they liked to take it. That was in 1803, but in 1805 Nelson won the battle of Trafalgar, and that changed things. In 1812 Napoleon took Moscow and the War Hawks thought the old country was done and they crossed the Niagara and St. Lawrence rivers and invaded Canada; but we had a loyal Canada and the invasion was repulsed. Subsequently we crossed the border, took Detroit, New York, Washington and Philadelphia and burned The White House: then Napoleon fell and we had peace.

The Rideau Canal originally was strategically important for it enabled the rapids on the St. Lawrence to be by-passed and gave easy access to Kingston from Montreal. It was its military value as a line of communication with an important town which led to its being built.

When the railways were first introduced to Canada they were built round the portages on the rivers. One of them, the Carillon–Grenville Railway, was built in 1843. It was 14 miles long, the second oldest in Canada and continued to run until 1910. The railways followed the canals and were ultimately linked together to form a complete system. The whole question of transport in Canada wanted to be thoroughly thrashed out. Much of it was greatly to the honour of the Royal Engineers to whose excellent work there were but few and fragmentary monuments. There was one at the cross roads in Hawksbury near the Perley Bridge, which carried an inscription: "Designed and constructed by the Royal Engineers: commenced in 1819; completed in 1833."

Much of the early exploration of means of communication and transport had been done by the Indians and the early pioneers. On one occasion he had visited an Indian church, far and far away from civilisation, and had found that the communion plate they were using there had been a gift from Queen Anne. It was great gentlemen like Colonel By of Frant, near Tunbridge Wells, whose devotion had helped to make Canada and the United States.

Mr. A. STOWERS recommended members to read a book, edited by Frank N. Walker and called *Daylight through the Mountain*, published in 1957 and reviewed in *The Newcomen Bulletin*, June 1959. In it about 400 letters between two brothers describe a great deal of pioneering work in Canada from 1836 onwards. They were two engineers, Walter Shanly (1817–99) and Francis Shanly (1820–82). The title of the book refers to the Hoosac tunnel, 5 miles long, through which was established the direct railway route between Boston, Mass., and Lake Ontario. Mr. Stowers asked Mr. Legget to convey to the Engineering Institute of Canada the congratulations of the Newcomen Society for its good work in starting to publish accounts of the pioneer Canadian engineers and their civil engineering works.

THE JONES FALLS DAM ON THE RIDEAU CANAL, ONTARIO, CANADA

Mr. H. CLAUSEN said he was struck by the fact that that dam had been conceived merely as a piece of utilitarian engineering and with no idea of impressing anyone as an achievement over the forces of nature or anything else. It had a close parallel in that respect in the Roman Aqueduct at Pont du Gard, built about 40 B.C. simply as a piece of utilitarian engineering to carry water across a valley.

The tools used and the means of lifting the stone blocks into position must have been very much the same on the Falls dam as in building the Pont du Gard. The problem of cutting pieces square that measured 6 ft. by 2 ft. by 4 ft. was a formidable one in itself. He would like to know how Colonel By and his party did it.

Professor A. F. BURSTALL said he would like to know how Colonel By felt when the water first went into his dam. How did he know that he had designed to the right dimensions; but apparently he had. Was the structure ever designed? If so, what data did he go on, seeing that that was the first real dam of its kind? How did he know his abutments were secure? Did he make calculations based on theory, or did he work to empirical rules which were well known to civil engineers of that day? He (Professor Burstall) was not a civil engineer. To him it seemed most important to know to what extent Colonel By could have had confidence in his work when the thing was finished. He was sure that many novelists would be able to write a good story on it; the anxious Colonel standing on top of the dam as the water slowly rose; just not quite certain or absolutely sure.

Mr. J. FOSTER PETREE referred to the question which had been asked concerning the use of winches by the Romans and mentioned that Professor F. M. Feldhaus, in his book, *Die Maschine im Leben der Völker*, gave an illustration from a monument which showed builders using something very similar to a winch, operated by a tread-wheel.

The PRESIDENT (Mr. L. E. HARRIS) said what had been to him so remarkable was that a lecture on civil engineering achievements had produced so much interesting evidence about the country of Canada itself. It was also very gratifying to see that so much interest was being taken in the history of engineering in Canada itself and in the preservation of ancient buildings, as had been particularly exemplified in the slide which Mr. Legget had shown of one of the locks on the Rideau Canal.

Following the President's remarks, a film was shown which had been made expressly for Canadian Television on the period during which Colonel By was working in Canada.

Mr. LEGGET, replying to the points raised, said the tunnel to which some speakers had referred, that built by the Shanlys, those two most remarkable engineers, was still in use as a main railway tunnel. He had visited it about 6 months ago. It was the longest tunnel to be built in North America at the time and for many years after. Many American firms had tried to do it and given up in disgust before the Shanlys took it over and completed it.

The comparison of the Roman work at the Pont du Gard and the Rideau Canal fascinated him, for he had never thought of it in that way. As far as he could see there had been no significant difference at all between the Roman construction methods and those of Colonel By, apart from his use of black powder. He did not know whether the Romans used hand winches, but they knew from sketches that Colonel By's stones were moved by means of heavy winches made from tree trunks cut from the forest. He hesitated to say more, but one of the baffling things to him was that the Romans did not use circular units in their construction except as pipes.

Finally, a conjectural answer to the most interesting question about Colonel By's design. Most of the letters he wrote (and he wrote many) suggested that he did not make any calculations as such in his designs, but based them on his own engineering knowledge and skill. Being the man he was, he had almost certainly packed away at the back of his mind experience and memories of other works which he had visited. It was the same with his other structures—locks, for example. The lock walls were built up with masonry, at first without any grouting in between the stones; when they put the water in they found the leaks. By then found suitable limestone and started to make his own cement for the grouting of the lock walls.

In another place, he built an earth dam with which he was not satisfied. He had not realised that the clay which was used had been frozen before it was consolidated. The dam was about 30 ft. high and performed quite satisfactorily throughout the winter, but in the spring when the snow began to melt, the clay unfroze and the dam began to leak. There was a very interesting letter in which he described his experience on the top of the dam when he saw it was going to fail; indeed, he only just escaped being carried out on the flood water.

PLATE XXI

Fig. 1. SEEN FROM THE RAVINE

Fig. 2. SEEN FROM ABOVE

THE DAM AT JONES FALLS, THE RIDEAU CANAL, ONTARIO

9

Nineteenth-century hydropower: design and construction of the Lawrence Dam, 1845–1848

Peter M. Molloy

FROM THE EARLIEST colonial days until well after the Civil War, American iron-workers, sawyers, fullers, and mill operators relied on waterpower to drive bellows, saws, fulling mills, machine tools, cards, mules, and looms. During the seventeenth and eighteenth centuries most towns grew along the fall line of the Eastern seaboard, where the small rivers and streams dropped a few feet in elevation. Abrupt waterfalls were preferable to a series of rapids. A fall of 10–15 feet was sufficient to power the overshot and breast waterwheels of the period, and the quantity of water needed to operate a mill was not large. In numerous locations the millwright was satisfied to install a wooden headrace or to dig a shallow ditch from the pond upstream of a waterfall to the mill's wheelpit. At other sites millwrights built small dams to ensure a supply of water during dry periods. Such dams were small affairs of wooden cribs, frames, or even branches and tree trunks, seldom exceeding 5 feet in height (figs. 1, 2, 3, 4, 5.[1] If a freshet breached or carried away such a dam, as often occurred, repair or replacement was fairly simple.

Colonial millwrights avoided large rivers as sites for mills because of the difficulty in building a dam across a wide stream that was subject to heavy flooding. The power requirements of a colonial mill did not justify such an effort. Where there was no alternative to the large rivers, millwrights either built headraces from naturally ponded water above the falls or installed wing dams to channel part of the river's current into a headrace. The earliest waterpower systems at the sites of Lowell, Massachusetts, and Manchester, New Hampshire, utilized such devices to obtain a few cubic feet per second from the flow of the mighty Merrimack.[2]

During the third decade of the nineteenth century, the nation's power requirements began a radical swing upward. The growth of such cities as Philadelphia, Boston, and New York led to a demand for water-pumping stations. The fledgling textile, machinery, and locomotive building industries and other new technologies needed thousands of horsepower units rather than a few dozen. The ready availability of waterpower sites in America meant that hydraulic and not costly steam power would dominate the American industrial scene, although steam was always used to some extent even at the largest waterpower sites.

As early as the 1790s, developers attempted to establish industrial cities near the falls of large rivers. In Paterson, New Jersey, attempts to build a large waterpower system at the Great Falls of the Passaic were only partially successful; other smaller developments on the Blackstone River at Pawtucket, Rhode Island, and on the Charles River at Waltham, Massachusetts, enjoyed greater success. Not until the 1820s did a combination of millowners, merchants, and engineers succeed in establishing a large industrial center along one of New England's largest rivers. A group of entrepreneurs calling themselves the Proprietors of Locks and Canals built a series of mills, boarding-houses, and related structures at Pawtucket Falls and harnessed the power of the Merrimack. They

The author acknowledges the generous assistance of Helena Wright, librarian at the Merrimack Valley Textile Museum, North Andover, Massachusetts.

[1] For excellent descriptions of such dams, see James Leffel, *Leffel's Construction of Mill Dams* (Springfield, Ohio: By the author, 1881). Although this book was published late in the nineteenth century, the dams described are similar to those of the colonial and early national periods.

[2] U.S., Department of the Interior, Census Office, *Reports on the Water Power of the United States*, in *Tenth Census*, 22 vols. (Washington, D.C.: Government Printing Office, 1885) (hereafter cited as *Tenth Census*), 16:78, 83–84.

Fig. 1. Brush, stone, and gravel dam. From James Leffel, *Leffel's Construction of Mill Dams* (Springfield, Ohio: By the author, 1881), p. 9.

Fig. 2. Hollow-frame dam. From James Leffel, *Leffel's Construction of Mill Dams* (Springfield, Ohio: By the author, 1881), p. 16.

Fig. 3. Economical dam. From James Leffel, *Leffel's Construction of Mill Dams* (Springfield, Ohio: By the author, 1881). p. 19.

Fig. 4. Log dam for soft or sandy bottoms. From James Leffel, *Leffel's Construction of Mill Dams* (Springfield, Ohio: By the author, 1881), p. 22.

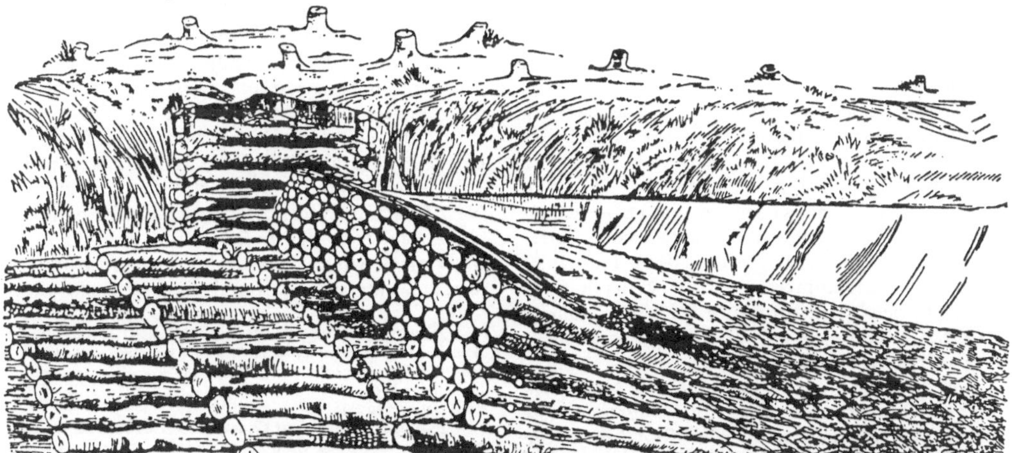

Fig. 5. Crib dam. From James Leffel, *Leffel's Construction of Mill Dams* (Springfield, Ohio: By the author, 1881), p. 45.

named their city after Francis Cabot Lowell, a leading textile mill owner at nearby Waltham, Massachusetts. By the 1830s a crib dam, averaging 8 feet in height, spanned the entire river above the falls, and water flowed through a complex system of power canals to the textile mills, printworks, and machine shops. From the power canals water flowed into the mills' penstocks, fell from 15 to 30 feet into the buckets of one or two breast wheels, and exited through tailraces into the river below the falls.[3]

The average flow of the Merrimack at Lowell was 6,000 cubic feet per second; the power canals of the spindle city delivered close to 10,000 horsepower to the mills as early as the 1840s. The canals were much more impressive achievements than the dam. The height of Pawtucket Falls, almost 30 feet, required a dam only a few feet high to create a reservoir sufficient for dry spells.

Other towns in New England and New York developed at similar sites. Manchester, New Hampshire, capitalized on a 40-foot fall in the Merrimack; Holyoke, Massachusetts, relied on a 20-foot fall in the Connecticut; Cohoes, New York, developed where the Mohawk dropped over 130 feet into the Hudson. All of these sites needed only low dams which were, for the most part, larger and better constructed versions of colonial dams. Almost uniformly of wooden construction, these dams required constant repair and failed with alarming regularity.[4]

By the 1840s developers had claimed the best of these natural falls on the larger rivers in the East. Consequently, the backers of yet another New England waterpower development company chose a site for a new mill town at a location on the Merrimack which had been considered and rejected twenty years before.[5] This new site was at

[3] *Tenth Census*, 16:78–83. Peter M. Molloy, *Historic American Engineering Record Inventory of the Lower Merrimack Valley* (Washington, D.C.: Government Printing Office, 1975), pp. 69–81.

[4] *Tenth Census*, 16:83–84, 221, 364. See also Louis C. Hunter, *Waterpower in the Century of the Steam Engine*, vol. 2, *A History of Industrial Power in the United States, 1780–1930* (Charlottesville: University Press of Virginia for the Eleutherian Mills–Hagley Foundation, 1979).

[5] Jeremiah Spofford to Charles Storrow, January 7, 1846, ms. group 306, Essex Company Collection, Merrimack Valley Textile Museum (hereafter cited as MVTM). Museum librarian Helena Wright has catalogued all of the Essex Company records, which include sketches of the dam under construction, the calculations of the company engineers, and the correspondence of chief engineer Storrow.

Bodwell's Falls, about 12 miles downstream from Lowell and equidistant from Methuen and Andover. A few miles downstream lay Haverhill and a few miles beyond that the seaport of Newburyport.

The site had advantages not even enjoyed by Lowell. The banks of the river were high enough to accommodate mill islands parallel to the direction of the Merrimack, making short, straight power canals and short tailraces possible. These canals could waste their excess water into small streams which flowed into the Merrimack on either side of the river. Most of Lowell's power canals dropped only about 15 feet, partly because of the site's topography but also to avoid the necessity of constructing unwieldy breast waterwheels to capture the entire fall of more than 30 feet. By 1845 the reaction turbine was on the scene. Turbines could easily utilize large falls and did not need huge wheelpits. They made it feasible for a millsite to utilize the total available head in a single drop. The topography of Bodwell's Falls favored this type of installation. Where the designers of Lowell found it necessary to build over 8 miles of canals to obtain 10,000 horsepower, at Bodwell's Falls a single canal 1 mile in length delivered approximately the same power.

Bodwell's Falls possessed major disadvantages: the river fell only about 5 feet and did so in a series of rapids.[6] A pond at the proper height to supply the mill island with water required a dam of considerable size. For these reasons potential developers had passed up the site in previous years.

The developers of the new site had extensive experience in waterpower engineering. Most of the principal shareholders of the Essex Company, as these developers called themselves, owned mills or shares of the waterpower systems at Lowell and at Manchester. The principal investors were brothers, Abbott and Samuel Lawrence. Samuel, owner of Middlesex Mill, Lowell's only large woolen company, led the drive to establish Essex Company and planned to build the region's largest woolen mill at Bodwell's Falls. Abbott Lawrence, for whom the town was named, was primarily an investor and provided most of the financial support for the company, whose treasury contained $1 million for development. The developers earmarked half of this sum for construction of a dam, mill island, and canal on the north side of the river. The budget for the dam alone was $250,000. The magnitude of this figure comes more into focus when one considers that the combined costs of the dams at Lowell, Manchester, and Holyoke, which were all completed during the 1840s, came to less than $100,000. The directors of Essex Company realistically anticipated the formidable nature of a dam at Bodwell's Falls.[7]

It might be useful at this point to consider some of the difficulties in the design and construction of overflow weirs in large rivers, such as the Lawrence Dam. An overflow weir is a dam whose entire crest serves as a spillway for excess water. In contrast, a storage dam, such as an earthen embankment, employs a relatively narrow spillway and waste gates to accommodate overflow; should the embankment be overtopped, its destruction follows shortly. The failure of Johnstown Dam in 1889 is a good example of this kind of failure. Similarly, high masonry storage dams, profiled to retain water no higher than their crests, also employ spillways and waste gates. An overflow weir is designed to withstand steam flows high enough to obliterate the dam from view. In 1852, for example, the Merrimack, flowing at over 150,000 cubic feet per second, overtopped Lowell Dam by almost 20 feet. The problems of designing a structure to withstand such forces are immense. A statement by the noted English engineer John Smeaton regarding the difficulties of overflow-weir design is relevant: "There is not a more difficult or hazardous piece of work within the compass of civil engineering than the establishment of a high dam upon a rapid river that is liable to great and sudden floods."[8] Significantly, Smeaton was referring to his design of an 8-foot-high weir across the Coquet, an English river whose flow was but a small fraction of the Merrimack's.

By the middle of the nineteenth century, engineers such as Sazilly and Delocre had developed theories of masonry storage-dam design which by 1900 resulted in the construction of dams over 200 feet high. None of these developments really applied to the design of overflow weirs, the difficulties of which Daniel Mead, professor of hydraulic engineering at the University of Wisconsin and author of an excellent text on waterpower engineering, aptly summarized in 1908: "Where [a dam] is constructed to retain a definite head of water without overflow, as in the case of reservoir

[6] H. K. Curtis to Storrow, December 15, 1845, MVTM.

[7] William R. Bagnall, *The Textile Industries of the United States* (Cambridge, Mass.: Riverside Press, 1893), vol. 1. The manuscript of vol. 2, which contains considerable information concerning Lawrence, Lowell, and Manchester, is held by MVTM. See also S. N. D. North, *The New England Wool Manufacture* (Boston, 1897).

[8] *Reports of the Late John Smeaton*, 2 vols (2d ed.; London: M. Taylor, 1837), 2:63.

embankments, the problem becomes one largely of statics. . . . When a dam is constructed in a running stream and is subject to the passage of extensive floods of water over it, frequently accompanied by large masses of floating ice, logs or other materials which in many cases may strike the crest of the dam, and many unknown and violent strains, the problem becomes one largely of experience and judgment." As late as 1973 the author of the United States Department of the Interior's *Design of Small Dams* noted the impossibility of accurately determining such important factors as uplift and ice pressure on an overflow weir's crest.[9]

At Lawrence in the 1840s it was necessary to build a dam over 30 feet in height at a site where the Merrimack was about 750 feet wide. The river's average flow was 7,000 cubic feet per second; flows of three times higher were by no means unusual. No weir of comparable size across such a river as the Merrimack existed in either the United States or Europe.

In addition, there existed several theories of dam design that applied to overflow weirs, although some of these theories were not specifically developed for that purpose. Finally, a number of overflow weirs across large American rivers were already in place, although none were of any great height. There were, however, some fairly high dams across small rivers.

The directors of Essex Company nonetheless went ahead with the project. The time was ripe for such a venture. Earlier railroad and canal building booms had led to the growth of construction firms capable of undertaking large projects. By 1845 suppliers could provide thousands of cubic yards of dressed granite blocks, millions of board feet of timber, hundreds of thousands of bricks, tons of blasting powder, and hundreds of barrels of high-quality natural cement. There was also an adequate supply of highly trained and experienced civil engineers, surveyors, and artisans in the area. Essex Company's chief engineer, Charles Storer Storrow, had been trained in both the United States and Europe. His assistant, Charles Henry Bigelow, was a graduate of the United States Military Academy, the nation's leading producer of civil engineers since the 1820s. Most of the other engineers, surveyors, draftsmen, and artisans who were employed by the company had received on-the-job training at railroad, canal, and other waterpower projects.[10]

The Engineers

The dam's chief engineer, Charles Storrow, was one of the nation's best engineers.[11] The son of Thomas Wentworth Storrow, an English officer during the American Revolution who remained to become a prominent New England merchant, and Sarah Brown Storrow, a member of an old New England family, Charles was born in 1809 in New England. His family moved to France soon after, and his early education began in a Paris pension. As a teenager he returned to New England and attended Round Hill School in Northampton, Massachusetts. Harvard followed, and he graduated in 1829 at the head of his class. It was during his last two years at Harvard that Storrow began his formal education as an engineer. The elder Storrow was a close friend of Loammi Baldwin, Jr., a prominent civil engineer of the day. Charles obtained permission to study in Baldwin's library, which contained volumes collected by both Baldwin and his father Loammi Baldwin, designer of Middlesex Canal. As an engineering library, it was surpassed in the United States only by that of the military academy at West Point. During these years, the younger Baldwin was heavily involved in designing a water supply system for Boston, and he may have interested Storrow in hydraulic engineering. In 1829 Storrow sailed for France and enrolled as *élève étranger* at Ecole nationale des ponts et chaussées, Europe's finest school of civil engineering. During the next two years Storrow also attended lectures at the Ecole polytechnique, the preparatory school for all of France's chemical, civil, military, mining, and topographical engineers. In all likelihood, he did not graduate from the school of bridges and roads but rather audited courses, the common procedure for foreign stu-

[9] Daniel W. Mead, *Water Power Engineering* (New York: McGraw Publishing Co., 1908), p. 583; U.S., Department of the Interior, *Design of Small Dams* (Washington, D.C.: Government Printing Office, 1973), pp. 329–44.

[10] For discussions of the growth of civil engineering during the early national period, see Daniel H. Calhoun, *The American Civil Engineer: Origins and Conflict* (Cambridge, Mass.: Harvard University Press, 1960); Forest G. Hill, *Roads, Rails and Waterways: The Army Engineers and Early Transportation* (Norman: University of Oklahoma Press, 1957); and Peter M. Molloy, "Technical Education in the New Republic: West Point as America's Ecole Polytechnique, 1802–1833" (Ph.D. diss., Brown University, 1975).

[11] *Dictionary of American Biography*, 1933, s.v. "Storrow, Charles Storer." *Engineering News and Record* 28, no. 36 (September 1, 1892): 207; 29, no. 7 (February 16, 1893): 147; 50, no. 18 (May 5, 1904): 422. *A Biographical Dictionary of American Civil Engineers*, Historical Publication no. 2 (New York: American Society of Civil Engineers, 1972), pp. 112–13.

dents. It is also likely that Storrow examined a number of engineering projects upon which the students at the school practiced as apprentice engineers. While touring in 1832, he studied such projects as the transportation canals of France and the canal and railroad systems of Britain. He spent a considerable amount of time as an observer of the construction of the Liverpool and Manchester Railway, an experience which he put to early use in 1833 on his return to the United States, as an assistant engineer to Loammi Baldwin, Jr., on the Boston and Lowell Railway. On its completion, Storrow became general manager, a position he retained until 1845.

While engaged in the Boston and Lowell project, Storrow became a consulting engineer for the Proprietors of Locks and Canals at Lowell, where he helped to develop methods of accurate water measurement and where he recommended the appointment of James Bicheno Francis as the company's chief engineer.

In 1835 Storrow published *A Treatise on Water-Works,* a book he may have written in an attempt to repay Baldwin for the use of his library. Baldwin was still deeply involved in designing a water supply system for Boston. The treatise, a small book, contains no original research by Storrow; it is instead a condensation of European theories of canal design and construction, flow in pipes and flumes, the measurement of water flow, and related topics. Storrow paraphrased for his American readers the most useful theories of Jean Baptiste Belanger, Charles Bossut, Charles Coulomb, Pierre Louis Du Buat, Johann Albrecht Eytelwein, Claude Louis Navier, and Baron de Prony, most of whose writings had never appeared in English. For those engineers interested in English techniques, Storrow summarized the writings of Smeaton and Young which applied to water supply. The book does not contain a single reference to theories of dam design. Storrow may have considered this subject peripheral to the subject of water supply engineering. Nonetheless, it is quite likely that he was familiar with the dam design theories of those engineers he cited in the treatise.[12]

Storrow was a wealthy man by the 1840s, both from family resources and from his own efforts as an engineer and an investor. In 1844 he joined Essex Company, assuming the duties of treasurer and chief engineer, positions second in authority only to that of Abbott Lawrence. The minutes of the company's board meetings show that Storrow was given authority to plan the entire city, including the locations of streets, bridges, railroads, commercial buildings, boardinghouses, and residences, as well as the dam, canal, and mill island.[13] He was responsible for both designing and constructing those facilities to be owned and operated by the company, such as the dam and canal; canal bridges, locks, headgates, and wasteway; a sawmill; and Lawrence Machine Shop, in which the shareholders intended to produce textile machinery, water turbines, locomotives, boilers, steam engines, and machine tools. Storrow was also responsible for the design and construction of the boardinghouses and for the design of the mill buildings. Essex Company shareholders hoped to earn a substantial return for these services through the sale of waterpower and the leasing of millsites to manufacturers, as well as through the sale and lease of commercial and residential lots to businessmen and artisans. They also expected the machine shop to be profitable. At Lowell, similar ventures had all been lucrative.

For such a large project, Storrow needed a good staff led by a first-rate assistant engineer, since Storrow himself would be required to spend much time in Boston as the company's treasurer. During 1845 he relied on two men: H. K. Curtis, a business manager with little engineering experience; and Joseph Bennett, an excellent surveyor. Bennett laid out the sites for the dam, mill island, canal, and all of the city's main streets. He also prepared an accurate map of the riverbanks upstream from Hunt's Falls to the damsite, plotting those areas to be submerged by the millpond and which must consequently be purchased by the company. Bennett was a skilled engineer as well as a surveyor, since he edited an English translation of J. F. d'Aubuisson's *Traité d'hydraulique à l'usage des ingénieurs* in 1852.[14] Nothing is known of Bennett's work prior to his engagement by Storrow. His letters to Storrow during the project reveal that he supervised the company's junior surveyors poorly and usually failed to meet Storrow's deadlines. He blamed most of these shortcomings on his own and his wife's poor health, the lack of proper equipment, the incompetence of others, and so forth. Bennett remained with Essex Company at

[12] Charles S. Storrow, *A Treatise on Water-Works for Conveying and Distributing Supplies of Water* (Boston: Hilliard, Grey, 1835).

[13] Records of Essex Company board meetings, 1845–1940, MVTM. See also Abbott Lawrence to Storrow, May 15, 1846, MVTM.

[14] Bennett to Storrow, June 4, 19, and 25, 1845, and April 9, 1846, MVTM. J. F. d'Aubuisson, *A Treatise on Hydraulics, for the Use of Engineers,* trans. Joseph Bennett (Boston: Little, Brown, 1852).

least until completion of the dam and possibly for years afterward, but he was not the man to be assistant engineer.

Business manager Curtis was energetic and well organized but lacked any training as an engineer. Nonetheless, he supervised the drilling of test holes in the bed of the river at the damsite, the construction of cofferdams, the blasting of the cutoff trench in the riverbed, and the construction of boardinghouses, in addition to his activities as purchasing agent, paymaster, and general contractor. In 1846 Storrow was at last able to relieve Curtis of all engineering duties. A qualified assistant engineer was finally on the scene.[15]

The new man was Charles Bigelow, lately resigned captain in the United States Army Corps of Engineers. Bigelow was a native of Massachusetts who had graduated from West Point in 1835, ranking second in a class of fifty-six, and whose classmates included George Meade and Herman Haupt. Since the 1820s the Corps of Engineers had often assigned its officers of civilian firms engaged in the construction of railroads, canals, highways, and similar public improvements. In fact, a primary reason for the survival of the military academy during its early years was its ability to supply the nation with civil engineers. While on loan, army officers were paid at lucrative civilian pay scales far in excess of their regulation salaries. Consequently, many officers resigned to pursue full-time careers as engineers in the private sector.[16]

Bigelow followed the pattern and worked for several railroad construction companies until in 1841 he returned to active duty as assistant engineer for the construction of Fort Independence in Boston Harbor. Chief Engineer for the project was Col. Sylvanus Thayer, superintendent at West Point during Bigelow's early years at the academy and one of the nation's most distinguished civil engineers. Under Thayer, Bigelow received valuable experience in the principles of stonecutting, the use of concrete and hydraulic cement, and other useful skills of masonry construction. As early as 1845 Bigelow wrote Storrow about possible cement suppliers for the dam and described the construction of a huge overshot waterwheel which he had recently seen in Troy, New York. In early 1846 he resigned his commission and took charge of all construction projects at Lawrence.[17] In addition to supervising construction on the dam, canal, and mill island, Bigelow designed and built the Lawrence Machine Shop and many of the early mills at Lawrence, including the unfortunate Pemberton Mill, which collapsed in 1860 with considerable loss of life. The mill collapsed because of its faulty cast-iron columns rather than from poor design, but, as constructing engineer, Bigelow accepted full responsibility for the disaster, which may have contributed to his death at age forty-seven. Unfortunately, Bigelow has been remembered for his design of Pemberton Mill but not for Lawrence Machine Shop, which is still in use, for several of the other original mills at Lawrence, or for his construction of the dam and canal.

Under Bigelow were several junior engineers who trained as surveyors, draftsmen, and engineers on New England and New York railroads, or at mill towns such as Lowell. Thomas B. Coolidge, an assistant engineer, had worked as a surveyor for the Boston and Woburn Railway under Waldo Higginson. Another engineer, Phineas Stevens, was a self-trained millwright with extensive experience in the mill towns of New Hampshire and Massachusetts. The noted Uriah A. Boyden also served as a consulting engineer during the early stages of the dam and canal project at Lawrence. Of the junior engineers Storrow hired, George B. Russy's qualifications are fairly typical. In a letter of application to Storrow, he described himself as a self-trained engineer who began his career as a surveyor on New York's Black River Canal. His next position was that of a constructing engineer of the "State Lunatic Asylum" in Utica, and then he served as draftsman for the Proprietors of Locks and Canals in Lowell. In addition the company employed several master stonecutters, masons, and carpenters. The company's chief mason and stonecutter, a Mr. Folsom, had previously worked for the Proprietors of Locks and Canals in a similar capacity. Francis Hodgman, another Lowell stonecutter, was hired to operate Essex Company's quarry in Andover. The superintendent of Lawrence Machine shop was also a former employee of Locks and Canals Company Machine Shop.[18]

[15] Storrow to Curtis, January 1, 1846; Curtis to Storrow, June 25, August 26, September 5 and 15, October 9 and 18, and November 12, 1845, MVTM.

[16] *United States Military Academy Register of Graduates and Former Cadets* (18th ed.; West Point: West Point Alumni Foundation, 1963), pp. 220–21. See also George W. Cullum, *Biographical Register of Officers and Graduates of the United States Military Academy at West Point, New York*, 2 vols. (New York: D. Van Nostrand, 1868), 1:60. Molloy, "Technical Education," pp. 459–63.

[17] Bigelow to Storrow, July 7, 1845, and May 1, 1846, MVTM.

[18] Higginson to Storrow, August 10, 1847; A. A. Burleigh to Storrow, October 5, 1847; Bennett to Storrow, June 1845; Russy to Storrow, October 11, 1846; Samuel Lawrence to Storrow, June 13, 1845; Curtis to Storrow, May 12, 1846; Storrow to Curtis, April 7, 1846, MVTM.

James B. Francis's role in the design of the dam is unascertainable, but it is likely that he provided considerable advice. Francis, protégé of Storrow, had worked closely with Storrow at Lowell. Storrow tried to pirate Francis away from the Proprietors of Locks and Canals as his assistant in late 1845. Francis would accept the position only "if the Proprietors of Locks and Canals should completely wrap up their affairs." This never happened, and Francis remained at Lowell for another forty years. A number of letters from Francis to Storrow in the Essex Company collection show that the two men collaborated frequently in day-to-day construction problems at Lawrence. Francis loaned Essex Company some pumps and barges as well as promising young engineers such as Russy, for whom Francis wrote a flattering letter of introduction.[19] Francis may well have helped to design the crest profile for Lawrence Dam, since he duplicated its profile for a famous hydraulic experiment in which he attempted to derive a formula by which to calculate the flow of water over a large weir. Most of the experiments which appeared in Francis's *Lowell Hydraulic Experiments* (1855) took place during the design and construction phases of Lawrence Dam.[20] Also, the crest and overall profile of Lowell Dam, partially completed in 1848 and finished in 1876, closely resemble Lawrence Dam. Another similarity is the design of the coffer dams. The huge retaining wall of Lowell's northern canal, constructed in 1848, required cofferdams identical with those used at Lawrence. There is no doubt that Francis was consulted regarding the selection and installation of water turbines in the mills at Lawrence. He performed exhaustive studies of turbine efficiency at Lawrence in the 1850s and 1860s, and he was the leading turbine expert in the United States for many years.[21] In all probability, he collaborated with Storrow in the design phases of the dam, canal headgates, wasteways, and wheelpits for the mills.

With such skilled assistants and consultants, it is unlikely that Storrow overlooked any significant theories of dam design as he planned the weir across the Merrimack. Storrow's background in European engineering has already been described. Francis received his early training in his native England and apparently read French and German, since the library of the Proprietors of Locks and Canals contains an extensive number of European texts on hydraulics and civil and mechanical engineering. Bigelow, as a graduate of West Point, was quite familiar with the techniques of European, especially French, engineering. Bennett, too, was obviously aware of European theories. An examination of these theories is therefore very much in order.

Theories of Dam Design in 1845

Storrow and his associates could rely on no definitive source of information for designing overflow weirs. Not until the late nineteenth century did engineering texts deal even in a peripheral fashion with the subject. By way of illustration, Edward Wegmann did not include a chapter on overflow wiers in his authoritative *Design and Construction of Dams* until the sixth edition in 1911. It is possible, but unlikely, that Storrow had read Italian engineer Juanelo Turriano's sixteenth-century treatise on Spanish overflow weirs.[22] It is more likely that he was familiar with the work of French engineers Bernard F. Bélidor, Bossut, d'Aubuisson, and Coulomb, as well as that of Englishmen Smeaton and Henry F. Moseley. Storrow's 1835 *Treatise on Water-Works* indicates a familiarity with the work of all of these men save d'Aubuisson and Moseley. During the 1840s, Essex Company's library acquired texts written by both authors, so it is likely that Storrow gained knowledge of their theories as the project progressed. Joseph Bennett, well underway with his translation of d'Aubuisson's 1834 *Traité d'hydraulique,* probably brought the text to Storrow's attention. With the exceptions of Coulomb's and Moseley's theories, most of these studies would have been of only partial use to Storrow because their premises were faulty or because they contained only practical suggestions for the construction of low dams across steams of moderate flow.

Bélidor's *Architecture hydraulique,* published in 1750, mathematically computes the minimum thickness needed to prevent overturning in a dam of rectangular profile. The text would have been of marginal interest to Storrow, since he did not plan

[19] Francis to Storrow, October 14, 1845; Russy to Storrow, October 11, 1846; Francis to Storrow, September 9, 1845, MVTM.

[20] Edwin T. Layton, "Scientific Technology, 1845–1900: The Hydraulic Turbine and the Origin of American Industrial Research," *Technology and Culture* 20, no. 1 (January 1979): 64–89. James B. Francis, *Lowell Hydraulic Experiments* (New York, 1855).

[21] Richard Hale notebook, pp. 207–8, 102–4, MVTM; "Efficiency of Waterwheels at Lawrence by J. B. Francis, 1859–1863," MVTM.

[22] Edward Wegmann, *The Design and Construction of Dams including Masonry, Earth, Rock-Fill, Timber, and Steel Structures, also the Principal Types of Movable Dams* (7th ed., rev. and enlarged; New York: John Wiley & Sons, 1922), p. iii. J. A. Garcia-Diego, "The Chapter on Weirs in the Codex of Juanelo Turriano: A Question of Authorship," *Technology and Culture* 17, no. 2 (April 1976): 217–34.

a rectangular dam; moreover, the theories of Coulomb and Moseley showed that Bélidor's formula would not produce a safe dam. Bossut's work dealt only peripherally with dams and limited itself to the subject of spillway design and the prevention of scour at the downstream faces of dams.[23]

Of much greater interest to Storrow were the *Reports of the Late John Smeaton,* which first appeared in print during the 1820s. Smeaton's "Design for a Dam upon the River Coquet," dated 1776, was probably of most importance to Storrow because this dam, like Lawrence Dam, was founded on bedrock.[24]

Smeaton's design relied entirely upon experience; there are no calculations in his paper. Coquet Dam has an almost square profile, 8 feet on a side, and includes several features that Storrow employed on Lawrence Dam. Coquet Dam arched upstream with an outside radius of 178 feet and a chord length of 170 feet. Smeaton probably tried to strengthen the dam by arching it to concentrate the river's flow at the center of the dam, as far as possible from the two wing walls. These wing walls were massive in relation to the size of the dam and were 1 foot higher than the dam's crest, reflecting Smeaton's justifiable concern that the river might breach a wing wall and carry the dam away.

Smeaton also addressed the problem of water seeping through the dam, despite mortared joints of all exterior portions of the weir. To prevent this he recommended using an upstream embankment to provide an impervious barrier to the water in the pond. The dam itself was built of cemented stone with an ashlar exterior. The masonry was bonded with *civita vecchia pozzelana* ("hydraulic cement") and further secured by wooden dowels. Although excessively strong, Smeaton's design was sound; Coquet Dam stands today essentially unaltered.

Storrow employed many but not all of Smeaton's designs in Lawrence Dam. To apply Smeaton's profile to Lawrence Dam, which had an average height of 32 feet above bedrock and a length of over 900 feet, would have taxed even the resources of Essex Company. Fortunately, the theories of Coulomb and Moseley allowed Storrow to create a more economical profile for the dam.

During the 1770s Coulomb, in addition to his experiments with electricity, formulated a method that used statics to determine the forces exerted on retaining walls.[25] The theory, applicable to walls with rectangular or trapezoidal profiles, employed algebra and trigonometry to calculate the thickness of a wall given the density of the backing and wall and the height of the wall. Coulomb's theory tended to result in a sturdy retaining wall. When applied to dams, however, his theory produced a dangerously thin profile, because Coulomb considered only the pressure exerted by the portion of earth between the wall and the angle of repose of the hill which the wall retained. This assumption is invalid for a dam, because a dam receives pressure from the entire head of water that it ponds. According to the theory, for a dam with an average height of 32 feet and a rectangular section, the recommended thickness was only 11 feet. Such a dam would stand but would be perilously unstable.[26]

During the 1830s Moseley, an English mathematician, published several treatises which greatly refined Coulomb's theory for both retaining walls and dams. Moseley's largely graphic theory used vectors to determine the location and magnitude of the resultant force upon a wall or dam. This resultant force was the algebraic sum of the force of the water and the weight of the retaining structure. The structure's center of gravity determined its location. To prevent the buildup of tensile forces at any point in the structure, the theory required that the resultant force fall within the middle third of a wall's base thickness. (A masonry wall has little strength in tension.) The resultant force within the dam's middle third put the entire structure in compression, a condition necessary for stability.[27]

Moseley's theory allowed an engineer to design a dam's profile with precision. By varying the profile and lowering the center of gravity as much as possible, the resultant force could be made to pass through the dam's middle third. Since the size and direction of the resultant force vector changed as the water level in the pond increased, a safe, overflow-weir design required that, with the height of the pond at its highest theoretical level, the resultant force would still pass within the middle third of the dam's base (fig. 6).

[23] Bernard Forest de Bélidor, *Architecture hydraulique ou l'art de conduire, d'élever et de ménager les eaux pour les differens besoins de la vie,* 4 vols. (Paris, 1737–53), 1, pt. 2:80–83. Charles Bossut, *Traité théorique et expérimental d'hydrodynamique,* 2 vols. (1775; rev. ed., Paris, 1796).

[24] *Reports of the Late John Smeaton,* pp. 63–68.

[25] Charles Augustin de Coulomb, *Mémoire par divers savants étrangers* (Paris: Academy of Science, 1976).

[26] The theories of both Coulomb and Moseley are clearly described and compared in *The International Library of Technology* (Scranton, Pa.: International Textbook Co., 1907); see sec. 48, "Retaining Walls," and sec. 51, "Dams."

[27] Henry Moseley, *A Treatise on Hydrostatics and Hydrodynamics* (Cambridge: For T. Stevenson, 1830), and *The Mechanical Principles of Engineering and Architecture* (1840; 2d ed., New York: J. Wiley & Sons, 1869).

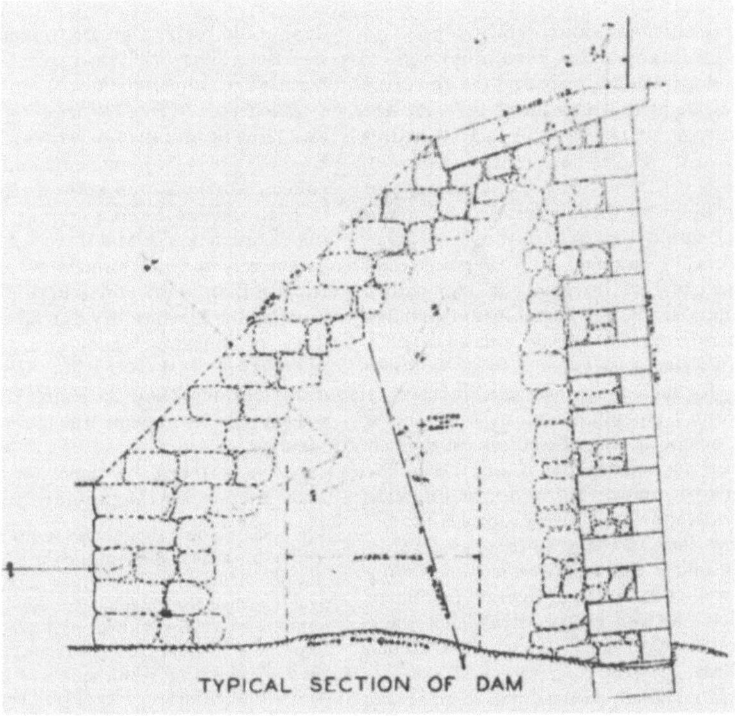

Fig. 6. A force diagram for Lawrence Dam which has been plotted with a pond height of 3 feet above the dam's crest. The resultant vector passes well within the dam's middle third. (Essex Company Records, Lawrence, Massachusetts.) Working drawings of the dam from the 1840s are missing. This diagram was carried out in 1948 for Essex Company by the engineering firm of Charles T. Main.

Moseley carefully stated the applicability of his theory to both retaining walls and dams: "It is apparent, that the pressure of a mass of earth upon a revetment wall is identical with that which it would produce if it were perfectly fluid, except that the pressure of the earth is not exerted on the wall in a direction perpendicular to its surface, as that of a fluid, but in a direction included to the perpendicular at a given angle, namely the limiting angle of resistance."[28]

Moseley's theory, with its elegant mathematical and graphic models, would have been of great value to Storrow. The only force that Moseley did not consider was uplift, the force exerted by water that flows from the pond into fissures beneath the base of the dam. But not until the twentieth century did dam engineers completely realize the magnitude of uplift forces.[29]

Another source available to Storrow was mining engineer d'Aubuisson's *Traité d'hydraulique,* which included a short chapter on "Considerations Relative to the Action of Water on Construction," in which he undertook to discuss the "Position and Forces of Dams." He noted the popularity of placing dams at an oblique angle to the flow of the river, on the theory that in a freshet the force of the water over a long crest would be less than that on the shorter crest of a dam built perpendicular to the flow of the stream. Another recommendation was to give the dam "a broken form, that of a rafter presenting a salient angle to the current, especially when it is intended to build mills at each end." The idea was to concentrate strong currents in the center of the river, away from the vulnerable wing walls.[30]

Like Bossut, d'Aubuisson devoted most of his essay to a discussion of a dam's downstream side

[28] Moseley, *Mechanical Principles,* p. 424.

[29] Norman Smith, *A History of Dams* (Secaucus, N.J.: Citadel Press, 1971), p. 205.

[30] D'Aubuisson, *Treatise on Hydraulics,* p. 197.

and the prevention of scour, which tended to undermine a dam. He recommended a downstream slope of one in six and the construction of stilling dams downstream of the weir. He also included suggested designs for such structures by engineers of the Corps des ponts et chaussées. Most of these recommended dams are almost square in profile, with sharply inclined or curved spillways. D'Aubuisson noted that the great eighteenth-century engineer J. R. Perronet had built for the canal of Burgogne a dam with a sinusoidally curved spillway. D'Aubuisson qualified his recommendation for curved spillways, however, by noting the great expense in stonecutting and masonry for such construction.[31]

D'Aubuisson's recommendations were sound enough but obviously were based almost entirely on experience and made little if any use of the principles of hydrostatics. Moreover, he intended the suggestions and illustrative examples for designers of low dams in the 5–12-foot range. Storrow did not follow the advice regarding stilling dams and spillways, but the overall profile of Lawrence Dam as well as the dam's siting and wing-wall construction conformed to the French engineer's ideas.

Dennis Hart Mahan, professor of engineering at West Point, also summarized French theories of dam design. Mahan's *An Elementary Course of Civil Engineering* first appeared in 1837 and was republished in updated form at intervals of four or five years for the next thirty years. The Proprietors of Locks and Canals library purchased an 1846 edition in 1847.[32] It is probable that Storrow was familiar with Mahan's text and a certainty that Bigelow had studied it while a cadet. Mahan, an admirer of French engineering, spent several years at French schools where he compiled extensive notes on all aspects of military and civil engineering. These notes, which he distributed to the cadets in lithographed form, were eventually published in the *Elementary Course.* Mahan periodically returned to Europe and updated his notes for revisions of his book. He also visited numerous American sites of canal, bridge, dam, railroad, and fortress construction.[33] Mahan recommended constructing overflow weirs of trapezoidal profile placed at an oblique angle to the river and arched, since such a dam would "offer a greater outlet . . . to the water that flows over [it]." Lawrence Dam conforms to Mahan's recommendations.[34]

Zachariah Allen's *The Science of Mechanics* might also have been useful to Storrow. As one of New England's best civil engineers and a noted mill architect, Allen was well known. His book appeared in 1829 and contained a short section entitled "On the Construction of Mill Dams." Allen had traveled extensively in Europe and admitted an admiration for English dam construction.[35] His recommendations for masonry overflow weirs paralleled those of Smeaton: water and air faces of well-dressed stone, doweled and mortared with hydraulic cement, and an embankment on the upstream side to prevent leakage of water through the dam.

Allen differed with Smeaton and agreed with Mahan regarding the profile of a weir:

> The wall of a dam should also be made much thicker at the bottom than at the top, and the whole slope or batter should be made on the lower face, the upper side, against which the embankment rests, being made perpendicular. . . . By mathematical calculation it has been found that a wall or bank [of a dam] will be sufficient to resist the pressure, if the square of its thickness at the base is to the square of its perpendicular height, as the weight of a given bulk of water, say a cubic foot, is to the weight of the same bulk of the material of which the bank is made.[36]

Allen's ratio is a variation of Coulomb's theory for the design of retaining walls. His formula allows for a thickness-to-height ratio of about 0.66 for a masonry dam, a marginal figure at best. (A ratio of about 1.0 is considered safe by modern engineers.) Lawrence Dam has a ratio of just over 1.0, but it conforms to Allen's recommendations as to shape and the use of an embankment.

Storrow and his colleagues probably gained some useful ideas from most of these authors. Smeaton and Allen had good advice on construction; Mahan and d'Aubuisson provided sound design theories. The theories of Coulomb and especially Moseley helped most. None of Storrow's calculations or sketches of the dam have survived, so it is impossible to know for sure whether he based his plan on mathematical theory or on empirical evidence. If the latter, then Storrow's design was extremely fortuitous; its profile is identical in di-

[31] D'Aubuisson, *Treatise on Hydraulics*, pp. 197–200.

[32] Dennis Hart Mahan, *An Elementary Course of Civil Engineering for the Use of Cadets of the United States' Military Academy* (new ed.; New York: Wiley & Putnam, 1846).

[33] Molloy, "Technical Education," pp. 405–7, 451–53. See also Thomas E. Griess, "Dennis Hart Mahan: West Point Professor and Advocate of Military Professionalism, 1830–1871" (Ph.D. diss., Duke University, 1968).

[34] Mahan, *Elementary Course*, p. 346.

[35] Zachariah Allen, *The Science of Mechanics as Applied to the Present Improvements in the Useful Arts in Europe, and in the United States of America* (Providence, R.I.: Hutchens & Cary, 1829), pp. 201–4.

[36] Allen, *Science of Mechanics*, p. 203.

mensions with mathematically designed weirs which date from the latter decades of the nineteenth century.

Existing Dams in 1845

Whether or not he used mathematical theories, Storrow inspected dams already in use and requested advice from other civil engineers of the period. Useful examples, however, were almost as scarce as helpful theories. Europe's numerous large storage dams bore no relationship to Storrow's large overflow-weir project. Europe had no weirs that exceeded 10 feet in height across large rivers, except in Spain, and there is no evidence that Storrow was familiar with Spanish dams.[37]

In the United States, however, there were several large weirs which were at least of partial use to Storrow. The oldest such structure was Fairmont Dam across the Schuylkill River, built between 1819 and 1821 by Ariel Cooley of Chicopee, Massachusetts, to provide water for the city of Philadelphia. The Schuylkill, with an average flow of about 600 cubic feet per second, is less than one-tenth the size of the Merrimack, but the dam was nonetheless a structure of impressive size which cost $150,000. The total length of the dam was almost 1,500 feet, of which 270 feet was an earthen mound dam. The remainder, constructed diagonally upstream, was a timber overflow weir made of "crib-work founded on the rock." The dam's height varied from 9 to 30 feet, depending on the contours of the river bottom, but averaged about 10 feet. The entire timber portion of the dam was replaced in 1842–43 because the original crib work had rotted, a lesson in durability that was probably not lost on Storrow. Fairmont Dam, the first large weir constructed in the United States, was a dramatic achievement. However, its earthen and timber construction offered little assistance; Storrow was designing a masonry dam for Lawrence.[38]

Another timber overflow weir which attracted Storrow's attention crossed the Kennebec at Augusta, Maine. In response to Storrow's May 1845 request, it was described by Robert Hallowell Gardiner, Jr., a friend who lived near Augusta. The 960-foot-long dam consisted of log cribs filled with stones and sheeted with planks on the upstream face to make it waterproof. Its profile was trapezoidal, with a base thickness of 127 feet, a crest thickness of 8 feet, and an average height of 16 feet. Gardiner described its engineer, William Boardman, as a "self-educated man" from Nashua, New Hampshire. Built by the Kennebec Locks and Canals Company, the dam spanned a large river, and, in 1838, just two years after its completion, it failed disastrously. A freshet carried away the wing wall on the mill island, completely destroying both dam and island.[39] The dam was rebuilt in 1842, apparently in an identical fashion, and was still in place at the time of Storrow's 1845 letter. A year later another freshet carried away 150 feet of the dam. Storrow had firsthand knowledge of two major defects with timber dams: vulnerability to rot and insufficient mass to withstand severe overtopping.

Storrow also received a letter from R. H. Gardiner, Sr., who since 1820 had built eight masonry dams on the Cobbosseecontee River, a tributary of the Kennebec. The senior Gardiner's father, a transplanted Rhode Islander, had built his first dam on the Cobbosseecontee in 1760. All of the Gardiner-built dams were small affairs across a small stream. Gardiner's advice resembled that of Smeaton and Allen: use dressed stone set in hydraulic cement, a trapezoidal profile, and an earthen embankment on the upstream side to prevent seepage.[40] The advice was sound.

There were three other masonry dams in North America which might have attracted Storrow's interest: one was in Windsor, Vermont; a second was on the Rideau Canal in Canada; a third was on the Croton River in New York.

In 1834, Ascutney Mill Dam Company completed a gravity arch dam across Mill Brook at Windsor, Vermont (fig. 7). Simeon Cobb of Westmoreland, New Hampshire, probably served as chief engineer for most of the project. The original dimensions are not known, but the dam was approximately 16 feet at the base and 40 feet in height, slightly arched, with a chord length of 360 feet. It used the high stone cliffs of Mill Brook valley as abutments. The air face was cut granite, doweled, and probably set in cement. The remainder of the dam was granite rubble, with a rel-

[37] An excellent summary of existing dams in Europe during the medieval and early modern periods is provided in Smith, *History of Dams*, chaps. 5 and 7.

[38] *Tenth Census*, 16:527, 618–21.

[39] R. H. Gardiner, Jr., to Storrow, May 15, 1845, MVTM. Henry D. Kingsbury and Simeon L. Deyo, eds., *Illustrated History of Kennebec County, Maine, 1629–1799–1892* (New York: H. W. Blake, 1892), pp. 417–18.

[40] R. H. Gardiner, Sr., to Storrow, May 2, 1845, MVTM. Kingsbury and Deyo, eds., *History of Kennebec County*, pp. 417–18; Walter Wells, *The Water-Power of Maine* (Augusta, 1869), pp. 288–90.

Fig. 7. Ascutney Dam, Windsor, Vermont. (Photo, Peter M. Molloy.)

atively tight water face made of wooden planks. Apparently recognizing the dam's dangerously low thickness-to-height ratio, the builders thickened the downstream side with a stone buttress 20 feet wide at the base and sloping in a stepped fashion to the dam's crest. The dam was designed for a small amount of flow over its crest; overflow sluices and a large scouring gallery could draw off the water contained in the small millpond in a short time. A 100-foot spillway provided an additional margin of safety. Although considerably altered and no longer used to supply water to power mills, the dam remains standing. Except for its height and the fact that it was a gravity dam intended to stand in place by virtue of its weight, the dam at Windsor would have been of marginal usefulness to Storrow. It ponded a very small stream and was designed more as a storage dam than as an overflow weir. It has not been ascertained whether Storrow or any of his associates were aware of the existence of Ascutney Mill Dam.[41]

[41] Edwin A. Battison, "Ascutney Gravity-Arch Mill Dam, Windsor, Vermont, 1834." *Journal for the Society of Industrial Archeology* 1, no. 1 (Summer 1975): 53–58. The dam was little known. No description of it appears in the compendious 1880 census of waterpower in the United States.

Storrow probably knew of Jones Falls Dam, built in 1832 as part of Canada's Rideau Canal system. The dam, which still exists, is of the storage variety, with a height of 62 feet and a length at the crest of 350 feet. Like Ascutney Mill Dam, it was built between two steep natural walls. It was a true gravity-arch dam, with a mean radius of 175 feet. Because of this arch, it has a very slim profile, with a thickness of 27 feet at the base and 21 feet at the crest. A large earthen embankment with a 127-foot-thick base protected the upstream face.[42] The use of an embankment would have interested Storrow, but the dam's acute arch springing from steep natural cliffs could not be duplicated at Lawrence. Nor was Jones Falls Dam designed with the possibility of severe flooding in mind.

The only other large masonry dam constructed in the United States ponded Croton River and created a reservoir for New York City (fig. 8). The designer was the well-known engineer John B. Jervis. The constructing engineer, Horatio Allen, was almost as experienced with canal and dam con-

[42] Smith, *History of Dams*, p. 178; R. F. Legget, "The Jones Falls Dam on the Rideau Canal, Ontario, Canada," *Transactions of the Newcomen Society* 31 (1957–59): 205–18.

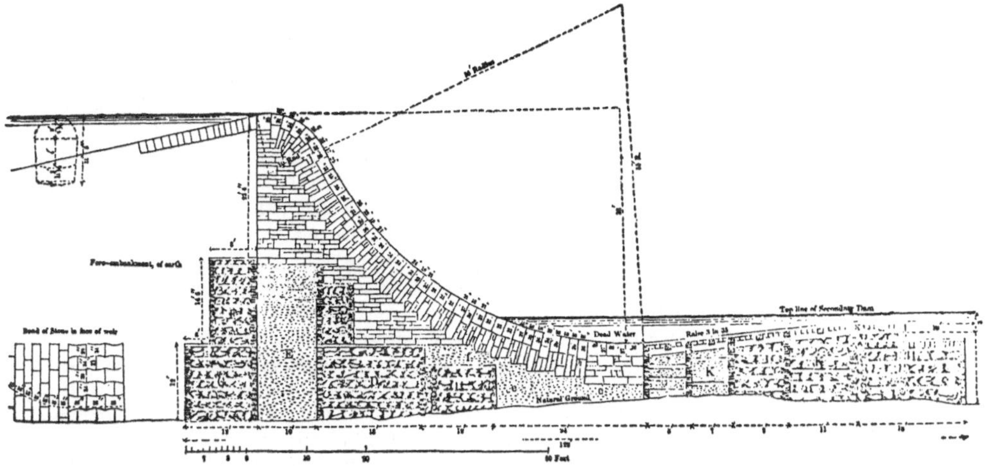

Fig. 8. Profile of Croton Dam. From U.S., Department of the Interior, Census Office, *Reports of the Water Power of the United States,* in *Tenth Census,* 22 vols. (Washington, D.C.: Government Printing Office, 1885), 18:523. This is the masonry section which replaced the original earthen embankment. The concrete caisson is at *E,* the stone-filled cribs, at *C, D, F, G,* and *H.* At *K* is the secondary dam intended to reduce scour in the riverbed. The curved spillway, made of carefully cut stone, also reduced the impact of overflowing water on the riverbed.

struction as Jervis and was also well acquainted with European engineering practices.[43]

Croton River had an average flow of about 400 cubic feet per second, but it was subject to considerable flooding. Since the dam's course did not extend entirely on bedrock, Jervis's original design called for a dam of composite construction. Over three-quarters of its 434-foot length was to be earthen embankment and the remaining 90-foot section a masonry overflow weir with a trapezoidal profile. This masonry section would function as a spillway for the earthen dam. The overall average height of the dam was 50 feet. Construction began in 1838. In 1841, with the dam almost complete, an unusually large freshet overtopped the masonry spillway and flooded the earthen section of the dam, carrying most of it away. Jervis then lengthened the spillway by adding a 194-foot masonry section. Together with the original stone portion, which had not been carried away, the new dam had a masonry length of 284 feet and a 150-foot earthen section. This new masonry section could not be built on bedrock, so Jervis designed an ingenious series of wooden cribs filled with rock and centered on a huge concrete bulkhead 27 feet high and 10 feet thick. The downstream side of the new section used a curved spillway as well as an apron to lessen the impact of the water on the alluvial soil in the riverbed. On the upstream side, huge earthen embankments backed both masonry sections. The dam's profile—a base of 54 feet, a crest of 12 feet, and a height of 50 feet—was compatible with the theories of Moseley. It allowed a considerable safety factor without being overly massive. The dam was completed in 1842 and served the city of New York until the pond of a much larger dam a few miles downstream submerged it in 1906.[44]

Since the dimensions of Lawrence Dam and the use of an earthen embankment on the upstream side closely resemble those at Croton Dam, it is possible that Jervis's design influenced Storrow. Undoubtedly, Storrow also recognized some of the negative aspects of Croton Dam. An earthen dam posed a danger even for a river as modest in size as the Croton, which had a flow of about one-fifteenth that of the Merrimack. Croton Dam's elaborate curved spillway contributed substantially to a cost of more than $250,000 for the 284-foot masonry section.[45] A similar dam at Lawrence

[43] *Biographical Dictionary of American Civil Engineers,* pp. 2–3, 69–70.

[44] *Tenth Census,* 17:521–24; Park Benjamin, ed., *Appleton's Cyclopaedia of Applied Mechanics: A Dictionary of Mechanical Engineering and the Mechanical Arts,* 2 vols. (New York: D. Appleton, 1880), 1:116–28; Lawrence Lankton, "Manhattan Lifeline: Engineering the Old Croton Aqueduct" (Ph.D. diss., University of Pennsylvania, 1977); Wegmann, *Design and Construction of Dams,* pp. 153–54.

[45] Lankton, "Manhattan Lifeline."

would have cost three times as much. Fortunately, at Lawrence Storrow had a foundation of igneous quartzite, an extremely hard rock. As Storrow set about his design, he also may have recognized a positive aspect. One reason why the original masonary section of Croton Dam withstood the freshet of 1841 was the strength of its wing wall, which was 12–15 feet above the crest of the dam.

In sum, none of the other dams offered solutions to Storrow's problems. Perhaps the most important findings were negative. Storrow knew that he did not want to build a dam of wood, despite initially low construction costs. Wood decayed too rapidly and was not strong enough to resist the floods of a large river. He rejected the idea of an earthen dam also. Although inexpensive to build and durable, it would be totally destroyed if overtopped by floodwaters. This posed too great a risk on the Merrimack, which had an average flow of 6,000 cubic feet per second and was subject to floods of up to 200,000 cubic feet per second. Storrow also eschewed costly curved spillways with their requirements for large numbers of skilled masons and stonecutters. In setting on a masonry dam, the only alternative to wood or earth, Storrow recognized that his design had to be right on the first try. Once built, the dam would be difficult to modify, since it would be impossible to lower the level of the pond by more than about half of its depth. Should the dam be broached, the cost of repairs might ruin Essex Company; masonry dams are not easily repaired. A broached dam would also spell financial disaster for the industries of Lawrence that relied on waterpower. With these thoughts in mind, Storrow set about the design of his dam.

Design of Lawrence Dam

Despite its shallow drop, Bodwell's Falls was an excellent site for an overflow weir. It had two very important advantages: a river bottom of extremely hard bedrock covered by a thin layer of gravel and loose rocks, and high rocky banks on either side of the river. The builders would be able to build strong wing walls and have an accessible, sturdy foundation for the dam. So hard was the riverbed at the damsite that Storrow entirely omitted plans for a sloped spillway. Eliminating the curved or inclined air face, elaborate aprons, and stilling dams saved a large amount of money. It was a courageous decision, and Storrow's judgment has proved to be sound. To this day, there is no sign of scour in the riverbed below the dam.

The dam could only be high enough to pond the river as far upstream as the foot of Hunt's Falls, which was a mile downstream from Lowell and 9 miles above Lawrence. Should the pond be allowed to extend beyond the falls, it would submerge the tailraces of many of the Lowell mills, causing a backwater and loss of power for the mills. If the water ponded by Lawrence Dam stood at 38 feet above the riverbed, then the foot of Hunt's Falls would be reached. Storrow decided to build the dam to a height of 32 feet above the riverbed and on flashboards to raise the level of the pond several feet higher. Flashboards, an inexpensive and prudent way to raise the level of a millpond, are simply planks set on a dam's crest and secured to iron rods set in the crest stones. Storrow thus saved several feet in the height of the dam and eliminated the need for hundreds of tons of additional rock and cement. In the event of a flood the iron supporting rods were designed to bend, allowing the boards to wash downstream and lessening the dynamic forces on the dam. In approving the height of Lawrence Dam, the proprietors of Locks and Canals at Lowell stipulated flashboards no higher than 3 feet. By 1890 Essex Company was using 5 feet of flashboards, but the pond still did not extend as far as the base of Hunt's Falls. The 10-mile-long pond covered 9,000 acre-feet without flashboards and 11,000 acre-feet with 3 feet of flashboards in place.[46]

Storrow designed a masonry gravity dam, slightly arched, at an angle with the flow of the river (fig. 9).[47] Extensive masonry wing walls anchored the dam to the high, rocky riverbanks. These wing walls also served as entranceways to the headgates of the power canals. The initial canal on the north side of the river ran between the mill island and the town of Lawrence. Storrow also made provisions for a canal on the south bank when the need for additional millsites arose. The dam's profile was trapezoidal in section, sloping upstream and inclining almost vertically from the riverbed on the downstream side (fig. 10).

Storrow had several reasons for skewing the dam across the Merrimack instead of building it at right angles to the riverbanks. In the first place, he wanted to build the dam across the most level grade, avoiding large boulders and holes in the riverbed if possible. In this Storrow largely suc-

46 Francis to Storrow, August 4, 1848; "Notes on Dam, 1869–1925," MVTM.

47 For descriptions of Lawrence Dam, see *Tenth Census*, vol. 16; Frizell, *Water Power;* and Richard Hale, "Lawrence Dam across Merrimack River," *Journal of the Association of Engineering Society* 7, no. 12 (December 1888): 494–500. See also "Notes in Regard to Dam, Canals, and Fishway, 1876–1929," MVTM.

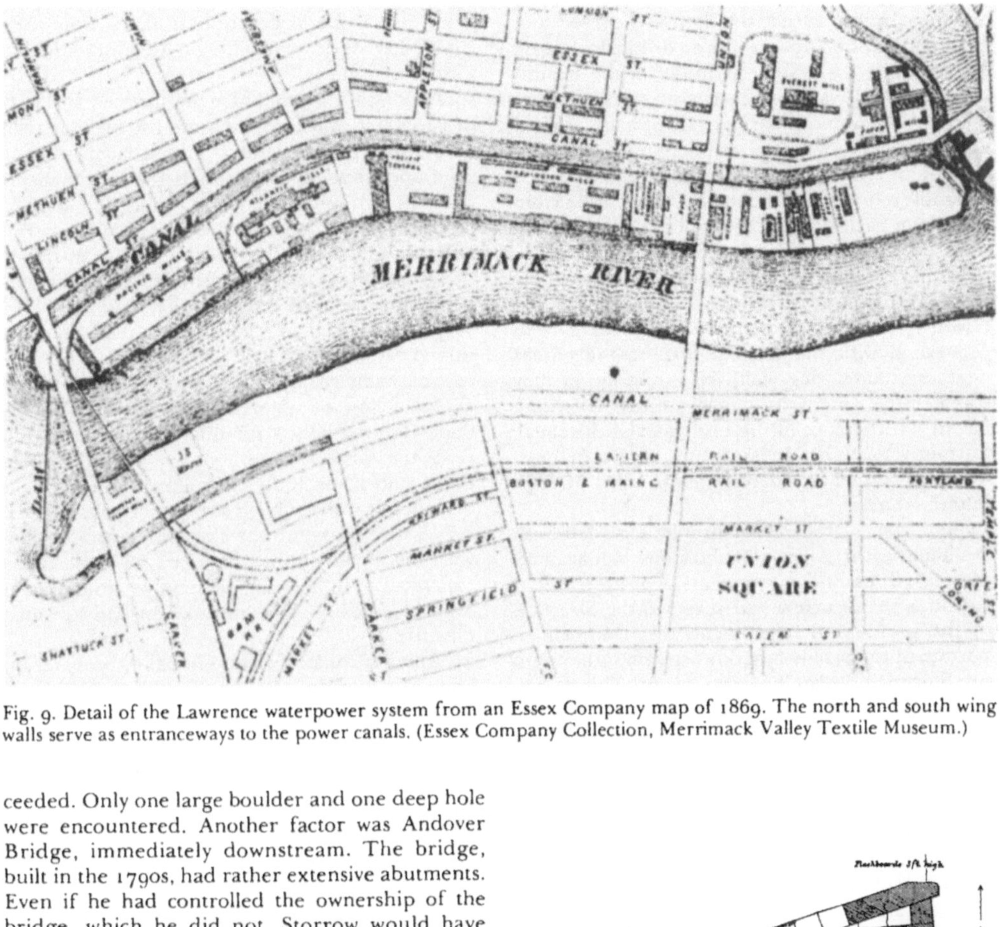

Fig. 9. Detail of the Lawrence waterpower system from an Essex Company map of 1869. The north and south wing walls serve as entranceways to the power canals. (Essex Company Collection, Merrimack Valley Textile Museum.)

ceeded. Only one large boulder and one deep hole were encountered. Another factor was Andover Bridge, immediately downstream. The bridge, built in the 1790s, had rather extensive abutments. Even if he had controlled the ownership of the bridge, which he did not, Storrow would have wanted to keep it intact since it connected Lawrence and Andover. Only by moving the south end of the dam upstream could he anchor it on a solid riverbank and avoid contact with the bridge abutments. Finally there was the advice of d'Aubuisson, Smeaton, Mahan, and most dam builders of the day. Park Benjamin, editor of *Appleton's Cyclopaedia of Applied Mechanics,* explained the theory: "In order to diminish the height and extent of backwater during floods, the crest of the dam is often made considerably longer than the breadth of the channel. This is effected by making it cross the channel obliquely."[48] At its skewed angle, the length of the dam was 943 feet, exclusive of wing walls.

It is unlikely that Storrow intended to strengthen the dam by designing it with a slight

[48] *Appleton's Cyclopaedia,* 1:116.

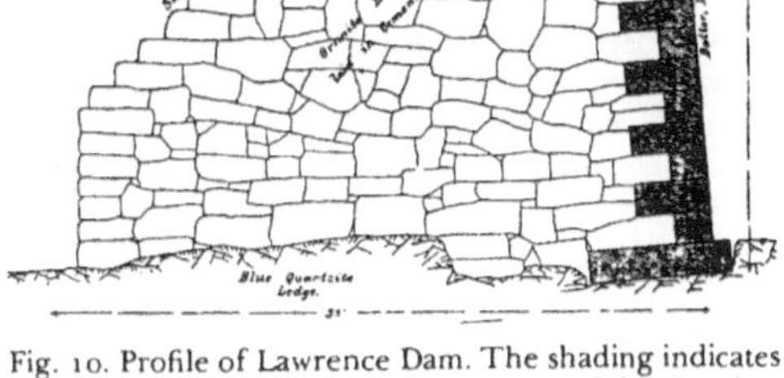

Fig. 10. Profile of Lawrence Dam. The shading indicates dressed stones. From U.S., Department of the Interior, Census Office, *Reports of the Water Power of the United States,* in *Tenth Census,* 22 vols. (Washington, D.C.: Government Printing Office, 1885), 17:72.

upstream arch. Indeed, the dam was curved to a mean radius of 7,433 feet with an ordinate of 14.95 feet.[49] While dams such as those on the Rideau Canal or the River Coquet derived strength from their nearly semicircular arches, Storrow arched Lawrence Dam for reasons well summarized by Benjamin: "For the protection of the banks . . . it is advisable so to form the dam that the cascade from the lateral parts of the crest shall be diverted from the banks and toward the center of the channel. This may be effected by making the dam slightly curved in plan, with the concavity at the downstream side." Storrow may well have been concerned with the formation of strong eddies near the dam's wing walls. Not only would they threaten the wing walls, but they might also cause harmful turbulence in the headgates of the canal, causing velocity head loss in the power canal system and funneling floating debris, such as logs, into the canal.[50]

The dam's trapezoidal profile accorded with the theories of Allen, d'Aubuisson, Mahan, and Moseley. Present-day engineers, however, might conclude that Storrow had reversed the direction of the profile and recommend that the inclined portion of the profile face downstream rather than up. Storrow's design, however, is entirely rational. Mathematically there is little difference in the location of the resultant force.[51] The resultant force of Lawrence Dam falls well within the middle third of the base length (fig. 6), until the level of the pond rises to almost 25 feet above the dam's crest.[52] Such a flood would leave most of eastern Massachusetts under water. Lawrence Dam, with an average base-to-height ratio of just over 1.0, is comparable with most modern overflow weirs, leaning slightly toward being overly massive.

By sloping the dam's profile upstream, Storrow effected two large savings. First, he needed far less earth to create an embankment on the dam's upstream face, the purpose of which will soon be discussed. Second, he avoided the need for an expensive dressed-stone inclined spillway, such as Jervis needed at Croton Dam. The inclined portion of Lawrence Dam, since it was protected by an earthen embankment, is formed of rough-cut stones of random size. The nearly vertical air face inclines 1 foot horizontally for every 12 feet vertically.

In his "General Description of the Dam to be Built Near Andover Bridge on Merrimack for the Essex Company," written in 1845, Storrow specified that the embankment, inclined at a slope of 1 foot in 6, must extend some 180 feet upstream of the dam's crest, with a bottom layer of "most solid and impervious materials for a thickness of five feet through the [entire 180 feet]."[53] Identical embankments had been constructed at Jones Falls and on the Croton River. Many European storage dams of the period used similar embankments. Although late nineteenth-century engineers such as Joseph Frizell scoffed at them, embankments served two vital functions, only one of which Storrow probably realized.[54] Storrow intended to prevent the flow of water through the body of the dam. Otherwise the dam, unable to use its mass to resist the force of the water, would tend to float; this would result in its destruction. Modern dams of concrete construction are impervious to seepage, but this was not the case with masonry dams with their many joints. In 1845, with the use of concrete it is infancy and with only natural cement available, the use of an embankment was a prudent measure.

Another cause of failure in dams, probably unknown to Storrow, was the force of uplift. Should water from the pond force its way underneath the dam through fissures in the bedrock or cracks in the base of the dam, the water would act like a hydraulic jack to lift the dam from the bedrock, again causing failure. Dam engineers did not fully understand the importance of uplift until the twentieth century. As late as 1945 a textbook recommended using an embankment to prevent uplift forces from becoming excessive in concrete gravity dams.[55] By using an embankment, Storrow unwittingly reduced the possibility of uplift. Excavation for the canal made large amounts of fill available and kept the cost of the embankment at a modest level.

To ensure a solid footing for the dam in the riverbed and to prevent the force of water in the pond from causing the dam to slide downstream, Storrow designed a cutoff trench along the base of the dam on its downstream side. A row of granite header stones laid in cement in the trench pro-

[49] Hale notebook, MVTM; Thomas B. Coolidge, "Levels and Courses of Dam," ca. 1846, MVTM. The dam's length would occupy 1/48 of a circle with a radius of 7,433 feet.

[50] *Appleton's Cyclopaedia*, 1:116. Excessive velocity head losses would result in less water for the mills' turbines.

[51] Frizell, *Water Power*, pp. 15–26.

[52] Author's calculations.

[53] The original contract has been lost; Richard Hale, an Essex Company engineer, made an exact copy in 1870 and it is part of ms. group 306 (hereafter cited as "Contract for Dam"), MVTM.

[54] Frizell, *Water Power*, pp. 101–4; "Contract for Dam," MVTM.

[55] Walter P. Creager, Joel D. Justin, and Julian Hinds, *Engineering for Dams*, 3 vols. (New York: John Wiley & Sons, 1945), vol. 1.

vided a secure base for the remainder of the downstream wall.[56]

Storrow designed the crest to withstand the pressures exerted by ice floes and to ease the passage of ice and large objects over the dam. For crest stones he chose granite headers, 9 feet long and 2 feet wide, hammered on three sides to a ⅜-inch joint; they projected 1 foot beyond the face of the dam. These directed the overflowing water well away from the base of the dam. Their upstream slope allowed floating objects to gently coast over them and fall to the riverbed below.[57]

Rounded wing walls of massive stones followed the contours of the riverbanks and anchored the dam on both sides of the river. Their trapezoidal sections closely adhered to the retaining-wall theories of Coulomb and Moseley. Their heights exceeded the dam's crest by 16 feet to prevent overtopping during floods because the loss of a wing wall would probably result in the dam's destruction. Whether Storrow estimated the maximum possible height of water over the dam during a flood is unknown. In 1852, however, he corresponded with James Francis on the subject, and the two men concluded that a very large flood might place as much as 14 feet of water over the dam's crest. A flood in 1852 had caused almost 10 feet of water to flow over the crest, and Francis claimed that a flood in 1789 had been even more severe. The flood of 1936 vindicated Storrow and Francis's calculations; it delivered water almost 14 feet over the crest and the dam held firm.[58]

A review of the dam's design demonstrates the rationality of Storrow's plans, considering the topography of the site and the materials at hand. In view of the problems experienced with earlier large timber dams in Augusta and Philadelphia, a masonry gravity dam was a logical preference. An abundant supply of granite blocks was also available from several quarries within a few miles of the damsite. The hardness of the riverbed made an inclined spillway unnecessary. The presence of large quantities of fill made the construction of an upstream embankment a good idea. Every other feature of the dam had similar logical reasons for inclusion in the plan. Modern engineers would omit only the slight arch, skewed plan, and embankment. They would reverse the profile, but probably not the dimensions, of the dam and include a curved spillway. Of course, they would use reinforced concrete, modern material handling equipment, and pressure grouting, advantages which Storrow did not enjoy. It should also be remembered that Storrow could not refer to any earlier dams as precedents, since none the size of Lawrence Dam existed. His design had to be flawless. A dam failure, especially during a flood, would not merely destroy the water and power supply for the city of Lawrence; it could result in the destruction of the Lawrence mills as well as most of Haverhill and Newburyport, two sizable towns located only a few miles downstream of the dam. Storrow designed a safe dam which would last for the indefinite future with a minimum need for maintenance at a reasonable cost. Including the wing walls, the dam was completed for just over $150 per linear foot. This compares favorably with the costs of Croton Dam, whose masonry section averaged over $1,300 per linear foot.

Construction of Lawrence Dam

On July 5, 1845, Essex Company let the contract for the construction of Lawrence Dam. The general contracting firm, Gilmore and Carpenter of Boston, agreed to "perform all the labor necessary to construct and finish in every respect in the most substantial and workmanlike manner the Dam . . . across Merrimack River." The firm was experienced in railroad construction, having built the Boston and Woburn Railway, but little else is known about it. It demonstrated its competence during the next three years and completed the dam at a cost of $250,000 "in the most substantial and workmanlike manner." Gilmore and Carpenter simultaneously contracted to build North Canal and the mill island, a project as large as the dam. The canal, 5,330 feet long, 12 feet deep, and narrowing from 100 to 60 feet wide, included a set of headgates, a waste dam, and two sets of locks for barge traffic. The canal project, finished slightly before the dam, was identical with the dam in construction costs. It is also likely that Gilmore and Carpenter built several of the early mills at Lawrence.[59]

The contract did not set a total cost for the dam. Instead, Storrow indicated the prices that would be paid for performing various jobs, such as excavation of sand for fill, trenching, and setting

[56] "Contract for Dam," MVTM. The use of a cutoff trench is highly recommended by twentieth-century engineers; see Mead, *Water Power Engineering*, p. 589.

[57] "Contract for Dam," MVTM.

[58] "Essex Company Structural Permanent Notes, 1869–1925," MVTM.

[59] "Contract for Dam," MVTM. Higginson to Storrow, August 10, 1847; Samuel Lawrence to Storrow, March 9, 1846, MVTM.

masonry. Engineers and master mechanics of Essex Company supervised all aspects of the project. All materials except cement were to be furnished by the contractor, and they were subject to inspection by Essex Company engineers. Storrow refused to leave the purchases of natural cement to the contractor's discretion. Most of the fill for the embankment, sand and gravel for mortar and concrete, and granite rubble for the dam's interior came from Essex Company property near the damsite. Gilmore and Carpenter were paid only for labor and transportation costs for these materials. The numerous bills for lumber, bricks, stone, and gunpowder in Storrow's correspondence reveal that Essex Company actually purchased most of the materials used to construct the dam. Gilmore and Carpenter supplied the labor and tools to carry out the project.

Essex Company performed all surveys and designs for the dam and canal. Probably the company's steam-powered sawmill, located at the dam's northern side, cut the lumber for the projects.

For the purchase of cement, Storrow relied upon suppliers in New York and New Jersey. Bigelow, well experienced in masonry construction at Fort Independence, recommended purchasing cement from "the business of Mr. White carried on by L. W. Mansfield of Cohoes, New York." Canvass White had founded this pioneer hydraulic cement company in 1820. Instead, the company purchased "Rosendale Cement" of similar quality, principally from John W. Lawrence of Eddyville, New York. The cement arrived in Boston on schooners and traveled by rail to Lowell, from where it was carted to the damsite. Well over 5,000 casks were used on the dam. An Essex Company engineer tested each cask for purity as it arrived at the site. He mixed a sample from each cask, allowed it to harden, immersed it in water, and then checked it for tensile strength. Cement that passed was used in the dam; marginal cement was earmarked for less critical projects, such as the wing walls; condemned casks were returned to Eddyville. Similarly, every cask of lime was tested for purity.[60]

Storrow purchased black powder and probably Bickford fuses for blasting from Oliver Whipple's mill in Lowell. The Proprietors of Locks and Canals in Lowell supplied many of the pumps and related machinery used to dry the riverbed; a few additional pieces of equipment were borrowed from Amoskeag Company of Manchester, New Hampshire.[61]

Essex Company crews first surveyed the course to be followed by the dam. Then they built cofferdams to hold back about a third of the flow of the river. After damming a section of the river, steam pumps were used to dry the river bottom.[62] in the first stage of construction, workers leveled the river bottom, removed loose rocks and gravel, and blasted where necessary. Next the cutoff trench, 18–24 inches deep and at least 8 feet wide, was hand drilled and blasted out. In this trench a row of granite headers were laid, each 18 inches on a side and 8 feet in length. These stones were dressed on four sides to a ⅜-inch joint, cemented to the bottom and to each other, and leveled. At deep points in the riverbed they stacked several layers of these headers, taking care to avoid a continuous line of vertical joints. Once the headers were laid in the cutoff trench and leveled, a row of stretchers, each approximately 4 feet long and 2 feet on a side, and hammered to a ⅜-inch joint, were cemented in place. One foot of the original header course projected beyond the stretcher course (fig. 11). As an additional precaution the stretchers were also doweled to the header course. Masons then ran up a portion of the air face, which consisted of headers and stretchers identical in size with the lowest course of stretchers. Where headers and stretchers met, the joints were cut so as to dovetail the stones together (figs. 12, 13). The forward edges of these stones were cut so as to impart a slight arch to the dam. Since a number of engineers' notebooks in Essex Company records specify the dimensions and angles of cut for each stone in the air face, crest, and wing walls, probably the stone used for these was rough cut at several quarries in the area and hammered to a smooth joint at the damsite by Essex Company stonecutters.[63]

Masons joined the main body of the dam to the air face as it rose, setting massive blocks of rough-cut granite of differing sizes in concrete, using smaller stones to fill large gaps. Storrow insisted that each stone be smeared with a thin mortar before being set in concrete, to insure enough mois-

[60] Bigelow to Storrow, July 7, 1845; John W. Lawrence to Storrow, June 30, July 23, and August 20, 1846; Thomas B. Coolidge, "Test of Cement, 11 Aug 1846 to 27 July 1848," MVTM.

[61] Whipple to Curtis, April 1846 and May 20, 1846; Francis to Storrow, September 9, 1845; Storrow to Curtis, January 18, 1845, MVTM.

[62] "Contract for Dam," MVTM.

[63] "Contract for Dam," MVTM; "Return of Granite on N. Bank of River, Dec. 9, 1847," MVTM. All dimensional stone came from quarries in Pelham, Rockport, Cohasset, and Quincy. Rough-cut stones came from ledges in Andover near the damsite.

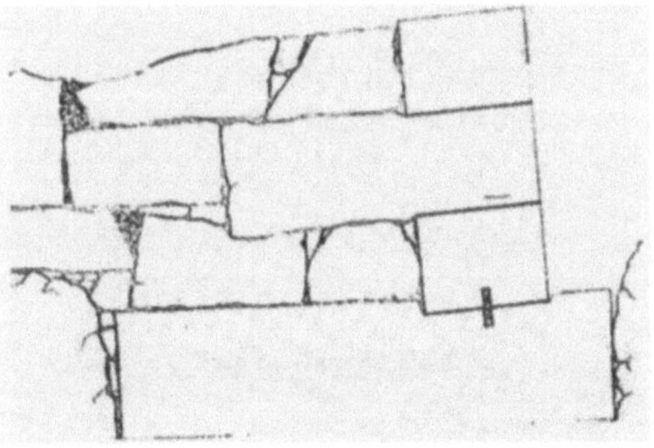

Fig. 11. Side elevation of the cutoff trench with foundation headers and first three layers of stone in place, Lawrence Dam. From a series of drawings by Thomas B. Coolidge, Essex Company engineer, 1849. (Essex Company Records, Lawrence, Massachusetts.)

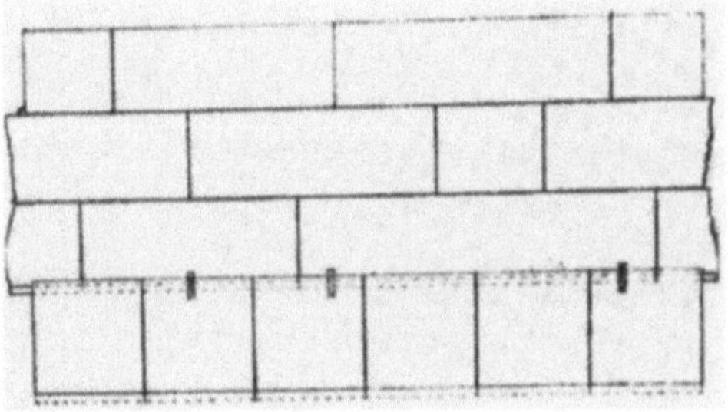

Fig. 12. Front elevation of the air face, showing the foundation headers and a typical plan of the arrangement of dressed-granite header and stretcher stones, Lawrence Dam. The row above the foundation consists entirely of stretchers doweled into the foundation to make up for the lack of headers, All remaining courses used headers at designated intervals. From a series of drawings by Thomas B. Coolidge, Essex Company engineer, 1849. (Essex Company Records, Lawrence, Massachusetts.)

ture for a good bond. He also specified the type of bonding to be used in each section of the dam. For example, a mixture of first-quality cement and sand bonded all exterior stones. Wing walls were generally constructed of stone set in second-class mortar made from cement, lime paste, and sand. "Cement-beton," a mixture of cement, sand, and broken mica slate from ledges on the south side of the dam, bonded the granite rubble within the dam. Wet places, fissures, and large gaps between the rubble were filled with lime and cement-beton—cement, lime paste, sand, and broken stone. Storrow determined the proportions for each recipe and insisted that an Essex Company engineer supervise all mixing of cement and the washing and screening of sand and gravel.[64]

The earthen embankment was laid in place at the same time as the masonry portion. For 5 feet behind the stonework, the embankment consisted of puddle, a mortar-like mixture of clay and gravel impervious to water. Puddle was commonly used in the cutoff walls of earthen dams at that time. The remainder of the embankment was fill from the wing-wall and canal excavations.[65]

Workers trammed materials for the embankment out to the damsite in wooden cars which traveled on wooden tracks atop the cofferdams (figs. 14, 15). These cofferdams, about 14 feet high, were filled with earth. They became part of the embankment as it rose from the riverbed. Since the rising embankment itself then acted as a cofferdam, a simple wooden bulkhead held back the river. As the embankment rose, fill was trammed out on crude log trestles. Trestles were also built to transport the stones to the damsite. A series of derricks, operated by manpower, lifted the stones into place (fig. 16).[66]

After raising a portion of the dam, planks were laid across it, a new series of cofferdams were erected on another third of the riverbed, and con-

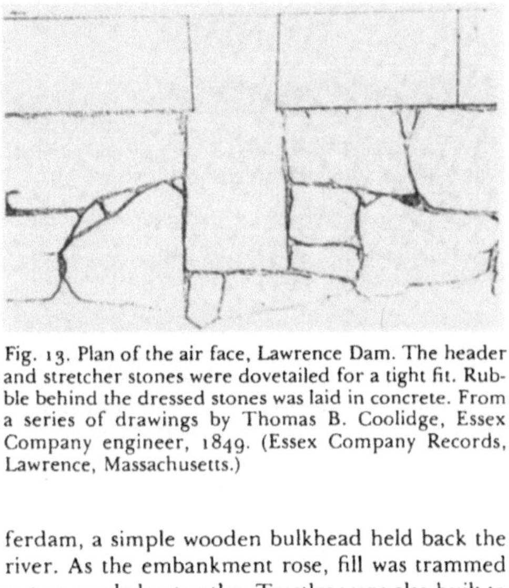

Fig. 13. Plan of the air face, Lawrence Dam. The header and stretcher stones were dovetailed for a tight fit. Rubble behind the dressed stones was laid in concrete. From a series of drawings by Thomas B. Coolidge, Essex Company engineer, 1849. (Essex Company Records, Lawrence, Massachusetts.)

[64] "Contract for Dam," MVTM; Hale notebook, "Details of Mortar and Beton used in the Construction of the Essex Company's Dam," July 11, 1874, MVTM.

[65] "Contract for Dam," MVTM.

[66] "Contract for Dam," MVTM; Hale notebook, "The Coffer Dam Erected along Guard Gates Northern Canal Lowell Mass," MVTM; Hale notes that this cofferdam was similar to those used in Lawrence Dam. Sketches of Lawrence cofferdams are held by Essex Company.

Fig. 14. Lawrence Dam, June 20, 1847, attributed to Marshall Tidd, Essex Company draftsman. Watercolor; H. 20.25 cm, W. 20.75 cm. (Essex Company Collection, Merrimack Valley Textile Museum.) The view is from the Lawrence side looking toward Andover. At far left are the piers for a railroad bridge. At the right is the dam, about a third completed. In the foreground oxcarts are delivering dressed stones and granite rubble to the site of the north wing wall (partially completed at far right). Workers are loading the stone onto flatcars and tramming it to work stations on the dam. When the stonework is raised a few feet higher than its level as shown, workers will add earth fill to the embankment to the rear of the stone portion. Derricks were operated by hand.

struction began anew. In the meantime, the river flowed over the newly completed section. Storrow took pains to insure that by the end of each building season, around October, the levels of each section of the dam were approximately even, to prevent the river from destroying isolated pieces of masonry.

Masons laid crest stones with particular care, since they would receive more abuse than any other portion of the dam. These stones were each laid as a row of headers, each 2 feet on a side and 9 feet long. The stones were cemented to the course below. As a final precaution they were also doweled to the lower course, the wrought-iron dowels being set in brimstone (liquid sulfur) to insure a tight fit. Stretcher stones were set in a row immediately behind the crest stones, cemented in place, and clamped to the crest stones with iron dogs. The tops of the crest stones were drilled to accommodate iron posts for the flashboards (fig. 17).[67]

The dam's height above the riverbed averages 32 feet, but at one deep part in the river it exceeds 40 feet. Its thickness at the crest is 12½ feet; that at the base is 35 feet (see fig. 10).[68]

The last crest stones were laid in place on September 19, 1848, three years to the day after the first header stone was lowered into the riverbed; North Canal (fig. 18) had been completed a few months earlier. Storrow had hoped to finish the dam in 1847, an unrealistic goal in light of the construction figures: over 1,700 cubic yards of rock excavated to prepare the foundation; 29,000 cubic yards of masonry used in the dam and wing walls; a surface area of 148,000 square feet of hammered granite.[69]

Only one serious accident occurred, and that was almost a year before the dam's completion. As

[67] "Contract for Dam," MVTM.

[68] *Tenth Census,* 16:73–74; Hale, "Lawrence Dam across Merrimack River."

[69] *Tenth Census,* 16:73–76; "Notes in Regard to Dam, Canals, and Fishway, 1876–1929," MVTM.

Fig. 15. *View of the Dam from the second Pier of the Bridge. May 22d 1847,* attributed to Marshall Tidd, Essex Company draftsman. Watercolor; H. 20.2 cm, W. 30 cm. (Essex Company Collection, Merrimack Valley Textile Museum.) The artist drew this scene from the middle of Andover Bridge. The Greek revival house at right was the construction-site headquarters, and Tidd sketched figure 14 from this house. It later became the home of North Canal's gatekeeper. At far left are some of the workers' shanties. In the center background is Essex Company's sawmill. Workers have completed the foundation course, a few of the upper courses, and a portion of the wing wall. The fencelike structures to the rear of the stonework are partially filled cofferdams. Pylons for a tramway are in position. Note the queen-post trusses employed on Andover Bridge at far right.

Captain Bigelow and a crew of fifteen Essex Company workers inspected a portion of the crest, a cofferdam gave way. The barge carrying the men was swept over the dam onto the rocks 30 feet below. Three men died and three, including Bigelow, were seriously injured. Of the remaining ten, five escaped with no injuries whatsoever and returned to work that day.[70]

Lawrence Dam since 1848

Even before the dam's completion, the mills of Lawrence were in production. By midsummer 1848 North Canal received enough water to power the turbines of Bay State Mills, Atlantic Mills, and Lawrence Machine Shop, all of which were partially complete (fig. 19). By 1880 the canal operated at peak capacity and supplied fourteen mills with from 10,000 to 13,000 horsepower, depending on the stage of the river. The largest mills had a capacity of 335,000 spindles and produced woolen, worsted, and cotton cloth. Even then the textile mills in this bastion of waterpower needed additional energy, which they met by using a steam boiler with a capacity of over 4,000 horsepower. They used most of this steam for heating, finishing, and dyeing, but a few of the larger mills used it to power steam engines.[71]

In 1865 Essex Company began a smaller canal on the south side of the river and extended it in stages until 1896. At completion it was 2,750 feet long and 10 feet deep. This canal generated a maximum of 2,000 horsepower. By 1880 a half-dozen small mills operated on its island. The limited power available from South Canal was ideal for these modest paper mills, tanneries, and bobbin manufactories.[72]

[70] Families of the three men who were killed received a month's pay as compensation; see "Men Injured on Dam," October 23, 1847, MVTM; Charles D. Appleton to Storrow, October 23, 1847, MVTM.

[71] *Tenth Census*, 16:74–75.

[72] Molloy, *Historic American Engineering Record Inventory*, p. 44.

Fig. 16. *View of the Dam,* August 12, 1847, attributed to Marshall Tidd, Essex Company draftsman. Watercolor; H. 20.35 cm, W. 30.2 cm. (Essex Company Collection, Merrimack Valley Textile Museum.) This scene shows the south side of Lawrence Dam. The north and south portions are about half complete, with the river flowing through a gap in the center. At the far right are two of the boardinghouses.

Fig. 17. Side elevation of the crest of Lawrence Dam. The header and stretcher stones were dovetailed for a tight fit. Rubble behind the dressed stones was laid in concrete. From a series of drawings by Thomas B. Coolidge, Essex Company engineer, 1849. (Essex Company Records, Lawrence, Massachusetts.)

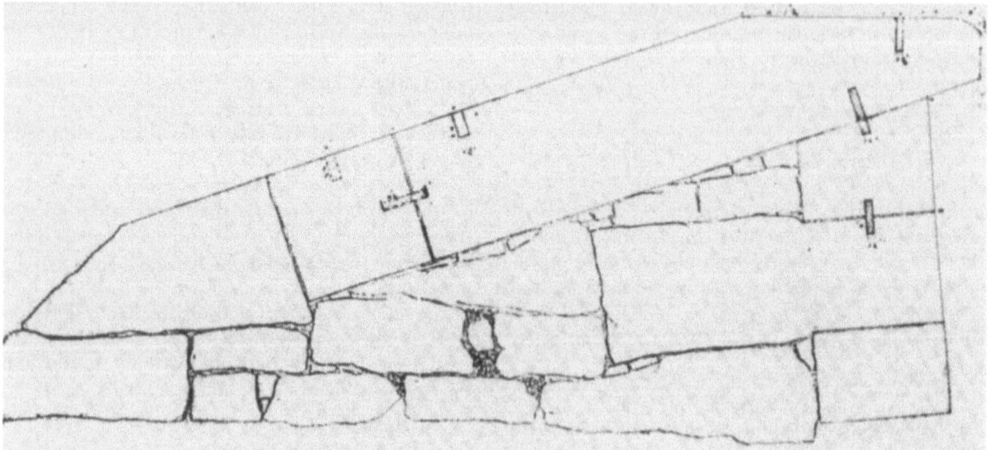

Fig. 18. *View of the Guard Locks from Blacksmith's Shop*, June 25, 1847, attributed to Marshall Tidd, Essex Company draftsman. Watercolor; H. 20.15 cm, W. 20.35 cm. (Essex Company Collection, Merrimack Valley Textile Museum.) The dam's wing wall is used as an anchorage for the headgates of North Canal. Wooden falsework is being erected to support the foundation for the gatehouse. The barrels contain cement and lime. The dam is on the other side of the construction headquarters building at right center.

By 1850 Essex Company also had a private fire-control system. Water was pumped from the millpond to a small reservoir on Prospect Hill, from which it then flowed to fire mains in the mill yards.[73] In 1875 the city of Lawrence installed a filter gallery upstream of the dam and a steam-operated pumping station near the north wing wall to supply several large water mains for the city's 49,000 residents. By 1880 the city was diverting some 700 million gallons per year from the millpond.[74]

New mills constructed in Lawrence after the 1880s were forced to use steam to power their machinery. Nonetheless, they placed considerable demands on the millpond for boiler feed and process water. For example, Wood Mill, completed in 1906, required a separate penstock from South Canal to supply water for its 10,000-horsepower steam engines and for its huge finishing department. The mill had 225,000 spindles and 1,500 broadlooms.

[73] MVTM.

[74] *Tenth Census*, 17:765.

Storrow and Francis had foreseen the need for a large continuous water supply. During the 1840s Essex Company and the Proprietors of Locks and Canals organized Lake Winnipiseogee Cotton and Woolen Manufacturing Company. Lake Company, however, did not manufacture woolens or cottons. Instead, it owned a portion of the water supply of Lake Winnipesaukee, Smith's Pond, Great Lake, Little Squam Lake, and several other New Hampshire ponds, all of which supplied the Merrimack with water. Lake Company constructed dams where the lakes flowed into the Merrimack and could lower the level of each pond by as much as 2½ feet as the need arose. This amounted to a reserve of almost 200,000 acre-feet of water for the mills of Manchester, Lowell, and Lawrence. Additionally, Lake Company could control to some ex-

Fig. 19. H. H. Bailey and J. C. Hazen, *Lawrence, Mass., 1876* (Merrimack Valley Textile Museum.) The dam is at the lower left. Just above the river is the upper mill island and North Canal. Excess water from that canal is wasted into Spicket River. South Canal, begun in 1865, is visible at the bottom of the lithograph.

tent severe flooding on the Merrimack by raising the dam levels during large rainfalls.[75]

The dam's maintenance costs, with the exception of flashboard replacements, have been less than $2,000 (fig. 20). After 1848 the wooden fishway was replaced several times, until, in 1917, the present concrete structure was built. The dam required modification only once. In July 1915, a freshet carried away a 285-foot portion of the crest of Lowell Dam. Since the crest designs of the dams at Lawrence and Lowell were almost identical, the Essex Company engineers decided to increase the number of dowels securing the crest stones on Lawrence Dam and drove 213 steel bars into the stones, anchoring them more securely to the interior of the dam. In 1945 small cracks were detected in two of the crest stones and repaired with steel plates. No other repairs have been made.[76]

The dam has withstood many floods, but the greatest test occurred in 1936, when the Merrimack, moving at a rate of over 200,000 cubic feet per second, overflowed the crest by almost 14 feet. At flood stage only a slight ripple marked the site of the dam (fig. 21). The dam survived the flood with no damage.[77]

Impact of Lawrence Dam on Other Weirs

Although expensive, Lawrence Dam's superiority was apparent from its first day of operation, November 19, 1848. The same day, Hadley Falls Company completed a huge timber dam across the Connecticut at Holyoke. The gates of the dam, which was 30 feet in height, were closed at 10 A.M. At 3:20 P.M., with the level of the pond 2 feet below the crest, the dam was swept downstream and totally destroyed. Hadley Falls Company remained undaunted and in 1849 completed a second timber dam. This weir, 35 feet high and

[75] *Tenth Census*, 16:77–78. The records of Lake Company are part of MVTM. See also Dwight Porter, "Water Power of the Merrimack River and of Rivers of Eastern Massachusetts between the Merrimack and the Blackstone," MVTM.

[76] "Notes on Dam"; Storrow to Essex County Commissioners, September 24, 1847; "Description of Fishway," MVTM.

[77] Essex Company scrapbook, MVTM.

Fig. 20. Air face, Lawrence Dam, ca. 1900. Note the slight arch. The workman is repairing flashboards. (Photo from original glass negative in the Essex Company Collection, Merrimack Valley Textile Museum.)

Fig. 21. The flood of 1936, Lawrence Massachusetts. (Newspaper photo, scrapbook, Essex Company Collection, Merrimack Valley Textile Museum.) The dam is marked by an arrow.

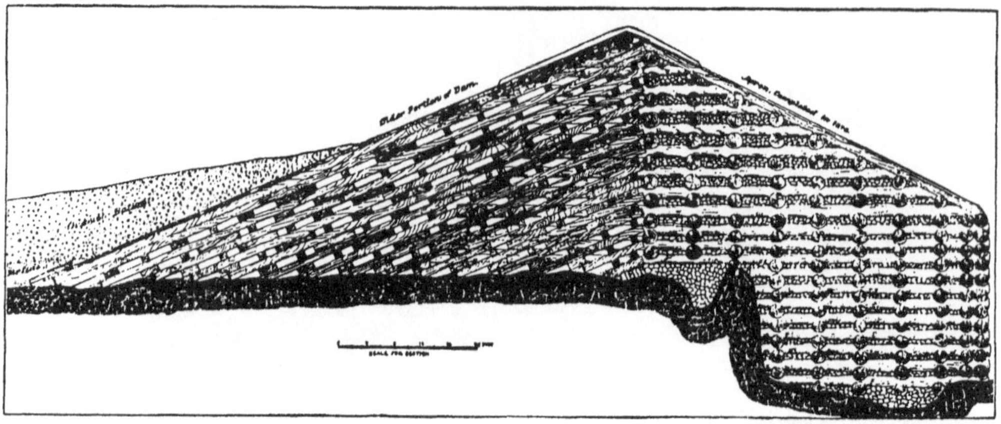

Fig. 22. Cross section of Holyoke Dam showing the 1870 apron. From U.S., Department of the Interior, Census Office, *Reports of the Water Power of the United States,* in *Tenth Census,* 22 vols. (Washington, D.C.: Government Printing Office, 1885), 17: facing p. 222.

Fig. 23. Plan and section of Housatonic Dam. From U.S., Department of the Interior, Census Office, *Reports of the Water Power of the United States,* in *Tenth Census,* 22 vols. (Washington, D.C.: Government Printing Office, 1885), 17: facing p. 311.

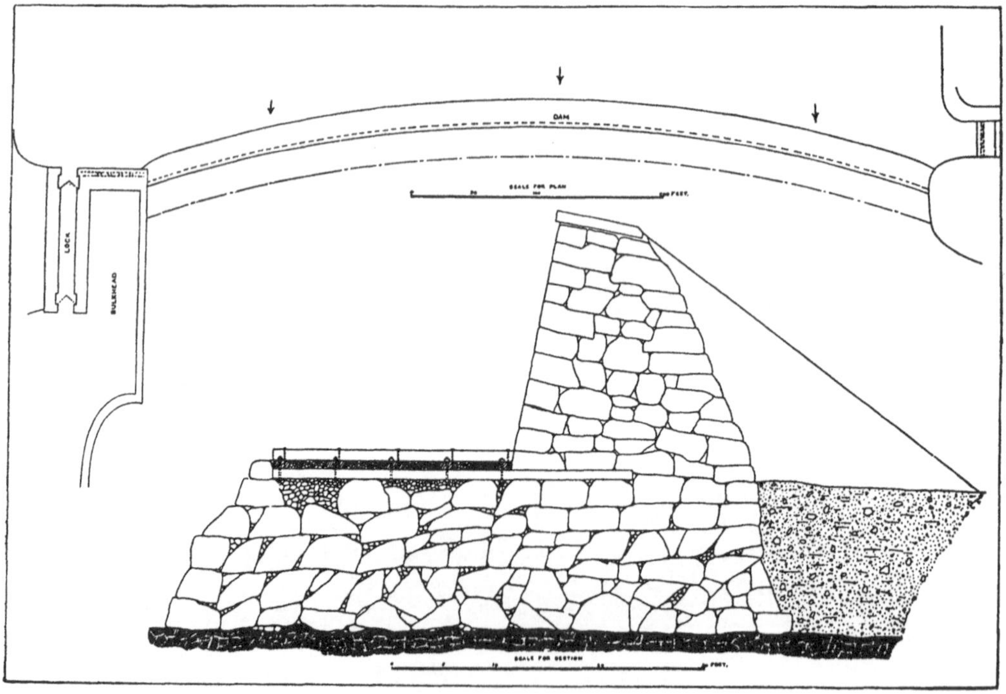

over 1,000 feet long, was without question the largest wooden dam in the world. It cost $150,000 to build. This dam remained intact until 1868, when the deep scour of the riverbed made it necessary to build an apron whose dimensions were actually greater than the dam itself. The apron cost $350,000 (fig. 22). Subsequent scour below the modified dam required extensive repairs in 1884 and 1885 that cost $65,000.[78] In 1899 the dam was replaced by a masonry weir with a curved spillway which cost close to $500,000. The total amount expended to pond the river at Holyoke between 1848 and 1899 totaled well over $1.1 million. By comparison, at $250,000 Lawrence Dam was a bargain.

Of the few overflow weirs exceeding 15 feet in height built in the United States after 1848, the majority bore a strong resemblance to Lawrence Dam. Most had an embankment on the upstream side, a slightly curved profile, and a steep downstream face—even on riverbeds which were subject to scour. High wing walls and crest designs similar to Lawrence Dam were also common. Lowell Dam, completed in two stages—1847 and 1875—is almost identical with that at Lawrence, although it is less than half as high. This is hardly surprising in view of the collaboration between Francis and Storrow. Other weirs that closely resemble Lawrence Dam were built on the Chicopee at Southworth Falls and Indian Orchard Falls, on the Shetucket at Greeneville, on the Hudson at Mechanicsville, and on the Housatonic at Birmingham, Connecticut.[79] The latter dam is remarkably similar to that at Lawrence: a slightly curved length of 636 feet, a trapezoidal profile with a width-to-height ratio of slightly more than 1.0, an earthen embankment, and a steep downstream face (fig. 23). It was designed in 1867 by Ezra Worthen, an engineer with previous experience at Lowell.

Joseph Frizell used the dams at Lowell and Lawrence as excellent examples of nineteenth-century overflow weirs, and, as late as 1902, *Engineering News and Record* favorably compared Lawrence Dam with more recent concrete dams.[80]

Perhaps John R. Freeman, a noted civil engineer who began his career with Essex Company, stated the case best in 1888:

> [The dam] was built in a day when there were few precedents to guide the engineer in designing such a structure, and it stands today, so far as I know or can learn, the most magnificent mill dam in the world.
>
> Though built forty years ago, the details of carrying out the work and the character of the supervision for securing the best of work form a good guide for the engineers of today. . . .
>
> [Storrow's] designs were all studied in the clear light of science, and as a result have stood the test of time. The water power whose development he superintended stands alone among our great waterpowers as one where the end and magnitude were clearly seen and continually kept in view.[81]

[78] *Tenth Census,* 16:221–25; Clemens Herschel, "On the Work Done for the Preservation of the Dam at Holyoke, Massachusetts in 1885, and on Some Studies for a New Dam for the Same Place," *Transactions of the American Society for Civil Engineering* 15, no. 1, paper no. 339 (1886): 543–80.

[79] *Tenth Census,* 16:350–53, 311.

[80] Frizell, *Water Power,* pp. 99–103; *Engineering News and Record* 44, no. 13 (September 27, 1900): 207–8, and 47, no. 9 (February 27, 1902): 176–77.

[81] John R. Freeman, "Discussion of R. A. Hale's Lawrence Dam across Merrimack River," *Journal of the Association of Engineering Society* 7, no. 12 (December 1888): 498–500.

10
The failure of the Bouzey Dam in 1895
Norman A.F. Smith

"Thank you for the copy of "Cassier" with your interesting article on the Bouzey Dam. It must have been an awfully jerry-built affair if the vertical bond was so wholly absent that it looked as if the mass filling the gap had slid and if the mortar joints, including the vertical bond, wouldn't stand such a paltry pull as 20 lbs. per sq. in. – about twice as much as a little boy's sucker!" (letter from Sir Benjamin Baker to Prof. W.C. Unwin in 1896)

Introduction

The Bouzey dam near Epinal in Eastern France failed almost exactly 100 years ago, on Saturday 27th April 1895 at 5-45 in the morning (Fig. 1). The ensuing flood-wave, pouring northwards along the valley of the Avière (Fig. 2) to the Moselle, drowned 85 people and extensively damaged canal works, railway structures, bridges, villages and farms. By the standards of dam disasters it was a serious accident although not so destructive of life as the Dale Dyke dam failure[1] near Sheffield in March 1864, which killed 244 people, and nothing like as terrible as the death-toll of over 2200 in Johnstown, Pennsylvania when the South Fork dam collapsed in May 1889[2]. For comparison, 75 people were killed in the famous Tay Bridge railway disaster of 1878.

There was another aspect of the gravity of the Bouzey failure, in the wider context of engineering safety and designer responsibility of great significance. The Bouzey dam, it was believed, had been designed rationally by the reasoned application of mathematics to a

Fig 1: A general view of the breach looking southwest and into the reservoir (from Louis Geisler, 15 Vues Photographiques de la Vallée de l'Avière après la Catastrophe de Bouzey, Roan-l'Etape, 1895)

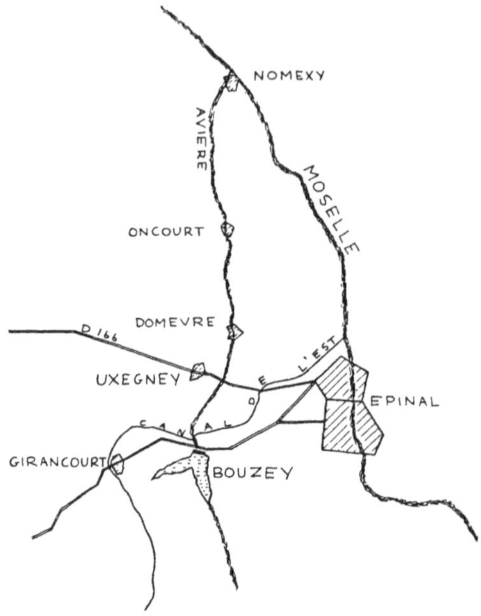

Fig 2: The Bouzey Dam and its environs

structural problem. Those engineers in France, Britain, Germany and the United States who had pioneered this first era of dam design were confident that new standards in safety of performance and economy of construction were now to hand. That they were astounded and alarmed when a dam of the new age went so badly wrong 100 years ago is hardly surprising.

This is not to be a "disaster" paper. It is about design and construction and why structures fail. Therefore, to begin with, it is necessary to examine, at least briefly, the background to the Bouzey dam's short life.

The Background

Masonry dams – gravity, arch, arched and buttressed – have been built from early times while the European tradition has clearly defined and well attested origins in the Roman period[3]. Some Roman dams are still in use as are Medieval structures, albeit with the aid, in most cases, of considerable repairs and reconstructions. There are big dams of the sixteenth century and later which, so far as trouble-free service is concerned, have much better records, indeed enviable. Some Spanish dams of the sixteenth and seventeenth centuries are a particularly impressive testimony to what could be achieved, and there are numerous examples of thoroughly successful dams built in France, Italy, Germany and Central Europe throughout the whole of the early modern period, the beginnings of a tradition carried on well into the nineteenth century.

I mention all of this to make the point, which if space permitted would bear considerable elaboration, that even in an era before dams could be designed rationally, they could still be designed very well. The old methods (pre-1850) of proportioning dams, whatever those methods comprised, were not unsafe even though they were undoubtedly uneconomic.

Even in the eighteenth century there was an emerging interest in calculating the ability of a masonry wall to resist overturning when loaded by some horizontal force. Charles Coulomb had applied himself to the particular problem of retaining walls bearing earth pressure, while B.F.de Belidor addressed the essence of the dam problem in his consideration of the behaviour of the walls of navigation locks.[4] Belidor examines the stability of a rectangular wall of height **b** and width **y**, to be determined, when resisting a depth of water **a**, less than **b** (Fig. 3). It is assumed that masonry is 5/3 times as heavy as water. By equating the overturning moment due to the water pressure, **P**, and the restoring moment developed by the weight of the wall, **W**, Belidor shows that $\mathbf{y}^2$ must be at least $\mathbf{a}^3/5\mathbf{b}$ and in addition, to achieve a safety factor of 1.5, he shows that $\mathbf{y}^2$ has to be $3\mathbf{a}^3/10\mathbf{b}$.

How far such rudimentary calculations were applied to practical dam-building in the late eighteenth and early nineteenth centuries has so far been impossible to establish. Serge Leliavsky[5] maintained that dams of the period *were* designed by Belidor's concepts, to which was added the requirement that the ratio of pressure to weight must not exceed the coefficient of friction appropriate to the masonry itself and the connection between the dam and its foundation. What *can* be said is that a number of French canal dams of the period – e.g. Lampy on the Canal du Midi; Vioreau, Bosméléac and Glomel on the Canal de Nantes à Brest; and Chazilly and Grosbois on the Canal de Bourgogne, ranging in date from the 1780s to the 1840s[6] – have profiles which are certainly not at odds with the "no-overturning" rule but that may be no more than coincidental.

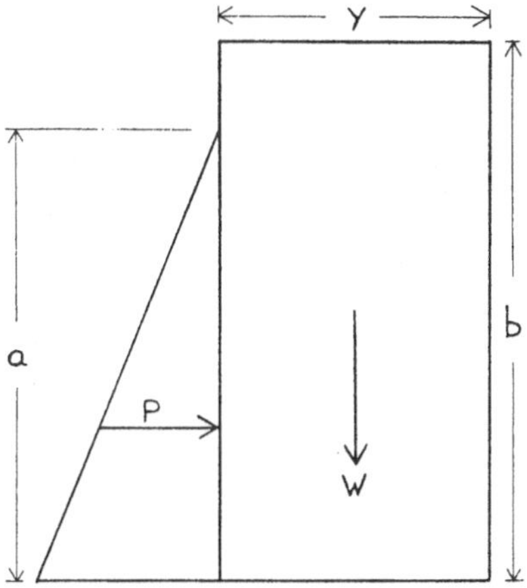

Fig 3: Belidor's problem

In 1853 came a decisive step. J.A.T.de Sazilly published his pioneering study of dam design, "Sur un profile d'égale résistance proposé pour les murs de réservoirs d'eau".[7] This long and complex paper was a further French contribution to the study of retaining walls generally and was based on previous work in the strength of materials by, for example, C.L.M.H. Navier and J.B. Bélanger. De Sazilly's ideas can be followed, in their essentials, by referring to Fig. 4.

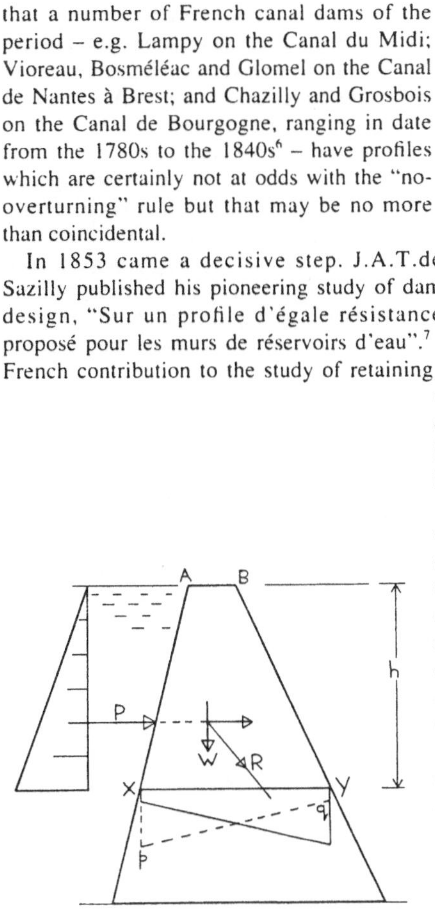

Fig 4: De Sailly's profile of equal resistance'; the vertical component of water pressure is ignored

It was assumed that a straight gravity dam could be visualised as a series of separate slices of unit thickness, and that any one could be treated independently of all the others. The limitations to these assumptions are considerable and obvious. The height of a dam varies across a valley and, particularly near the abutments, slices interact with each other and provide additional resistance. However, the most critical case can be covered by analysing the tallest slice and assuming it to be unsupported by its neighbours.

De Sazilly stated in full the two requirements mentioned above: that it should be impossible for the section of dam above **XY** to slide relative to the part below and that the line of thrust representing **W** and **P** should be contained within **XY** so that the piece **ABYX** could not overturn. A huge

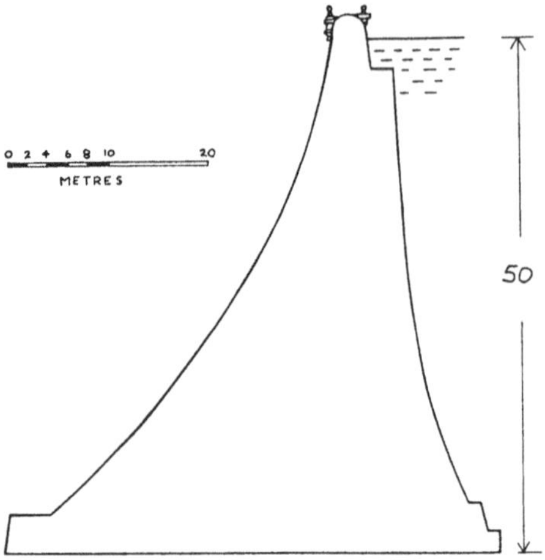

Fig 5: The Furens Dam, 1866

advance was de Sazilly's consideration of stresses. In effect he was extending the ideas of Navier and Bélanger to dams by treating each slice as a heavy vertical cantilever and assuming a linear distribution of stress on any horizontal section. When the reservoir is empty the self-weight **W** is taken to induce a trapezoidal compressive stress diagram **XYqp** whose maximum is at the water-face. With the dam loaded the maximum stress will occur at the air-face since the imposition of **P** induces bending stresses across **XY** to add to the direct stress due to **W**. In other words, the line of thrust moves from left to right as **P** increases.

The calculation of these states of stress was, for de Sazilly, perfectly straightforward and in order to maximise the masonry's performance and economise on construction costs he set a condition. He specified that throughout the dam on every section, at the point **X** reservoir empty, and at the point **Y** reservoir full, the maximum compressive stress that the material could withstand should be realised. Hence the term "profile of equal resistance" used in his title. It means, of course, that the two compressive stress diagrams, reservoir full and reservoir empty, are mirror images of each other. It also means that in Fig. 4 every horizontal section is subjected only to compression, never to tension; the shape of the profile and de Sazilly's conditions see to that.

De Sazilly's propositions were applied immediately. The Furens dam near St. Etienne, generally reckoned to be the first "modern" dam, was designed and built between 1858 and 1866 (Fig. 5). Its designer, F.X.P. Emile Delocre, in his discussion[8] of dam design follows de Sazilly exactly and their concepts were adopted immediately as the basis of European and American practice for half a century.

The next important advance in the theory of dams came from Professor W.J.M. Rankine. In 1870 he was consulted over the design of the Tansa dam in India. In a famous paper of 1872 – one of the last and one of the best of his publications[9] – he found in favour of the basic de Sazillian ideas but added new and crucial considerations. Two were particularly critical and for brevity we will confine ourselves to them here.

Rankine stated firmly that nowhere on any horizontal section of a dam should tension be allowed to develop, either at the air-face, reservoir empty, or at the water-face, reservoir full. It is essential to re-emphasise here that in de Sazilly's profile of equal resistance, for the style of cross-section shown in Figure 4, the "no-tension" condition is, in fact, met. However, neither de Sazilly nor his disciples positively spell out the need to avoid tension. They recognised that it could occur and indeed, following the earlier works of Navier, Méry and Bélanger, knew full well that according to their theories, if the force **R**, the resultant of **P** and **W**, cut the section **XY** outside the "middle-third" – i.e. acted with an eccentricity of greater than **XY/6** – then tension was inevitable.

In de Sazilly's paper there is a remark, at one and the same time chilling and portentous, which states, apropos tensile stresses, that it is allowable to neglect "the force of cohesion in the mortar, which is unfavourable to resistance".[10] The opinion is repeated verbatim by Delocre. In short, the de Sazilly-Delocre view of tension in a masonry dam was not that it should be avoided, but rather that its contribution to structural strength *should simply be discounted.* Rankine knew better. And so it was that he brought into gravity dam design a new imperative: that there must be a strict observance of the so-called "middle-third rule".

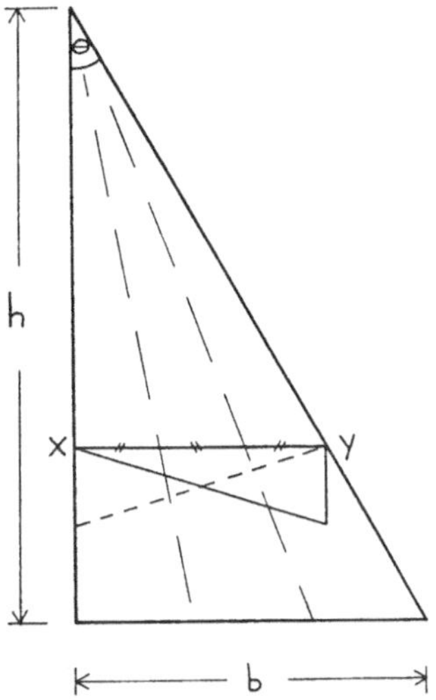

Fig 6: The basic triangular profile

The other of Rankine's observations challenged the assumption that maximum stresses did, in fact, occur on horizontal planes. Rankine puts it thus:

"The direction in which the pressure is exerted amongst the particles close to either face of the masonry is necessarily that of a tangent to that face; and, unless the face is vertical, the vertical pressure found by means of the ordinary formula is not the whole pressure, but only its vertical component; and the whole pressure exceeds the vertical pressure in a ratio which becomes the greater, the greater the "batter", or deviation of the face from the vertical".[11]

True, and quite correctly sensed here is that maximum *principal* stresses occur in directions parallel to those of zero shear. The face of a dam, by definition, is not subject to any shearing stress and therefore the *maximum direct* stress at the face of a dam must be on a plane at right-angles to that face. Moreover, its magnitude must exceed that of the associated vertical stress. Following Rankine's paper, a number of contributors to the theory of dams introduced variations on his criteria and by degrees the complexities of stress within a dam, including the role of shear, began to be assessed.[12]

As background to the Bouzey dam's design, which began in 1876, to its demise 20 years later, we can simplify the essential design concepts prevailing by considering the most basic of all dam profiles, the triangular one shown in Fig. 6. It is defined by the relationship $\mathbf{h^2=b^2w}$, where **w** is the specific gravity of the masonry.

It is elementary to establish a number of properties of this particular form of profile. For example, it can be shown that on the typical section **XY**, reservoir empty, the compressive stress diagram is itself triangular. If the same maximum compressive vertical stress is developed at **Y**, when the reservoir is full, then the same triangular stress distribution, reversed, will obtain, as shown. This will be true at every level, and in a sense what we have here is *a* profile of equal resistance. However, and this is crucial, it is not, and cannot be, *the* profile of equal resistance because the *maximum* working stress of the masonry cannot be realised at every level. Approaching the crest of a dam, up to and above maximum water level, stresses decrease progressively so that de Sazilly's theoretical profile is, to that extent, unattainable. On the other hand, the triangular profile automatically meets Rankine's middle-third condition because the lines of pressure, full and empty, pass through the two mid-third points, respectively, at every level. Thus we have here *the* most straightforward

The Failure of the Bouzey Dam in 1895

Maximum Principal Stress kg/cm^2	Specific Gravity of Masonry		
	2.0	2.25	2.5
	H E I G H T - metres		
5.00	16.7	15.4	14.3
10.00	33.3	30.8	28.6
15.00	50.0	46.2	42.9
20.00	66.7	61.5	57.1

Fig 7: Limiting heights of triangular dams for various specific gravities and different maximum principal stresses

profile for a gravity dam. For the air-face, maximum principal stresses are easily calculated as being equal to $\mathbf{p}\sec^2\theta$ where **p** is the maximum vertical compressive stress and θ the angle of the face. Now this basic shape, and its behaviour when loaded, are susceptible to variations in at least four parameters, all of which can have a bearing on the design process and a dam's safe operation.

Firstly, the density of masonry is a variable. Generally it was not usual to use a stone whose specific gravity was less than 2, while in the case of a really heavy specimen, it could be as high as 2.5 or a shade more; the masonry of the Vyrnwy dam in Wales was particularly dense at 2.6. Interestingly enough, between these extremes of density, a triangular profile is not required to change all that much, only from about 32 to 35 in the angle of the dam.

Secondly, and much more significant, is the masonry's compressive strength, be it measured or estimated. The latter was a real enough approach. To get started on designs French engineers, such as Auguste Graeff, applied their analyses to historic Spanish dams and determined the prevailing maximum stresses.[13] The answers were pretty mixed; from as low as $6.5kg/cm^2$ to as high as $14.5kg/cm^2$. In the early days, and to be safe, the lower limits were chosen; the Furens dam is stressed to $6.5kg/cm^2$. There was a bonus to this cautious approach to maximum stresses. Because it leads to a relatively wide dam profile – and not a particularly economic one – the middle-third rule is, fortuitously, likely to be met. However, in their determination to build more highly stressed dams, and hence ones which made more economic use of material, as well as satisfying a certain ambition to achieve "proper" design, European engineers generally were inclined, over the years, to allow higher compressive stresses in

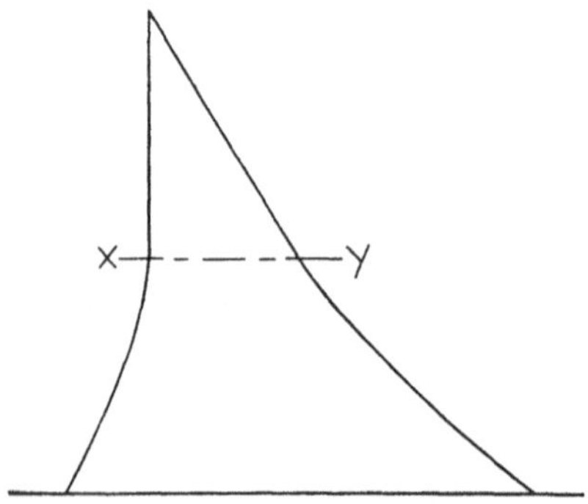

Fig 8: Below XY the profile must widen to restrict stress levels

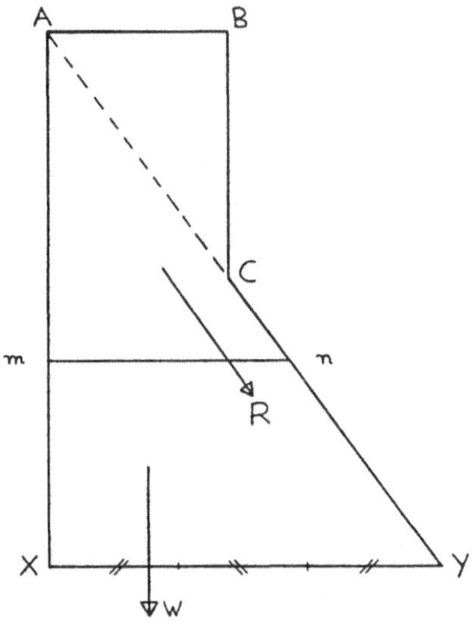

Fig 9: Dam with widened crest

increasingly slender profiles. Hence the risk of tension was increased; tension, remember, which in some quarters there was a willingness to discount.

The maximum stresses allowed have a marked effect on a dam's profile and the height to which it can retain the simplified triangular shape we have been visualising. It reaches its limits according to the range of numbers given in the Table in Fig. 7. To have a higher dam not stressed, front of back, beyond the prescribed limits requires a base of increasing width, and thus the more familiar shape of a high dam begins to emerge, as in Fig. 8.

This need to modify the basic triangular profile is compounded by another requirement, very relevant to Bouzey as we shall see shortly. A dam with a pointed crest is not a very practical proposition: it is difficult to build, vulnerable in use, and in any case fails to provide a roadway for maintenance, access or traffic across the structure.

As soon as the crest of gravity dam has some useful width, lines of thrust move and middle-third requirements are affected. Consider a width of road added to the top of a triangular profile as in Fig. 9. It is evident that when the reservoir is empty the mass of the piece **ABCYX**, will, below a certain level, **XY** say, shift the line of thrust *forward* and violate the middle-third rule *on the air-face*. This is very unlikely to be a problem but it is a fact. On the other hand, a very real issue arises when the reservoir is full. The fact that the top of the dam now has width means that above a section such as **mn**, the centroid of the profile is in any case much nearer the air-face. Therefore the possibility of generating tension at the water-face, at a point such as **m**, is greatly increased when the reservoir is full. In other words, on the section mn, the line of thrust, **R**, is forced to the right by the addition of **ABC**.

So it emerges that an unexpected property of a *practical* profile is that it is particularly susceptible to unfavourable stress distributions *near the top*. In the old days it was too easy to be unaware of this, the not unnatural assumption being that dams are vulnerable only in the depths where pressures are high. For a number of nineteenth century gravity dams – and Bouzey was a tragically outstanding example – it is elementary to show that the middle-third rule was nearly or *actually violated only near the crest*. It was a dangerous situation all too readily compounded by the fourth factor we are considering.

Inevitably designers and users alike a^sume that a reservoir's maximum water-level is known, predictable and controllable. However, a potential danger for any dam is that insufficient spillway capacity, or exceptionally rapid reservoir filling, or both, can produce an overload. It was especially true a century or so ago. Reservoir behaviour was, in its way, as problematical as dam behaviour. Indeed, the rates at which reservoirs filled, and overflowed, were as revealing of rainfall statistics and catchment area runoff as vice versa. So it is not difficult to appreciate that many a spillway was seriously under designed, and that a dam which was already stressed to the limit near its crest at *normal* maximum level,

The Failure of the Bouzey Dam in 1895

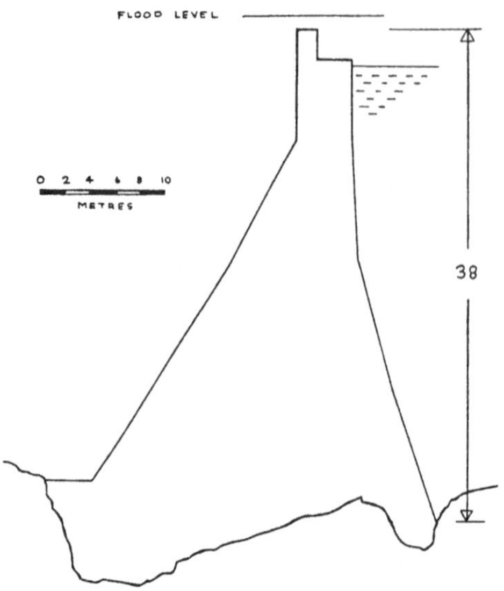

Fig 10: The Habre Dam

as discussed, was in a dire condition if an inadequate spillway allowed the water level to rise even higher.

A famous failure of this type befell the Habra dam (Fig. 10) in Algeria in 1881[14]. Massive runoff after storms and huge flood flows produced an overflow at the dam of, at its maximum, as much as $5000m^3$/sec. Not only was the spillway, such as it was, overwhelmed, but so too was the dam itself. The crest was overtopped to a depth of nearly 1m and the normal operational reservoir level was exceeded by about 4m. Proportionately, the increase in water pressure was not all that significant near the base of the dam, but it most certainly was at the top. At a depth of some 10m below the crest the horizontal section was being subjected to *twice* the designed thrust. Consequently, tensile stresses must have developed at the water-face and the masonry presumed to have cracked. Such cracking would have allowed the penetration of water as a result of which a new force came into action. It was an uplift force acting vertically within the crack. In effect, it augmented the horizontal water pressure in overturning the dam section above crack level.

As we shall see, the mechanism of failure at Bouzey was to be precisely the same, and to that extent Habra should have been a warning to be heeded. But apparently it was not. In fact, there seems to be no evidence that the Habra failure was considered even in the deliberations which *followed* the Bouzey accident, never mind those that went before.

There are three general observations to be made about the theory of dams in the period of Bouzey's original design, and during the years of maintenance and extensive reconstruction which preceded the final failure.

1). Following the work of de Sazilly, Delocre, Rankine and others, the sound design and construction of high, masonry, gravity dams was perfectly possible. It was achieved regularly in several countries, and the structures created continue to serve well to this day.

2). French engineers, pursuing rational design in the interests of economy, adopted maximum compressive stresses and the associated concept of the "profile of equal resistance" as the essentials of design. But "no tension" was not a design imperative. Observance of Rankine's "middle-third rule" was very often achieved but only as a consequence of other decisions, not as an object in itself.

3). Any theory of dams can be applied to two quite different problems and proper distinctions must be drawn. It is one thing – and relatively straightforward – to analyse an existing profile for stability and stress levels; it is quite a different matter to design a safe profile from scratch for a given situation. De Sazilly and Delocre attempted a generalised mathematical solution to the problem but had to give best to some prodigious equations; a close reading of their work is not for the faint-hearted or the mathematically inept. Rankine, by making concessions and being realistic, was more successful in realising a general solution, but ultimately it was relatively crude step-by-step design procedures, judiciously modified in the light of experience and common sense, which won the day. They usually do.

The Bouzey Dam

France continued to undertake major canal building long after the railways had won the argument in Great Britain. The Canal de l'Est was a particularly important project which was begun in 1872 following France's loss of north-south navigation along the Moselle-Rhine system as a consequence of the Franco-Prussian War. The Canal de l'Est consisted of two distinct parts: a northern section, essentially a canalised River Meuse, from Givet to Troussey; and a southern section from Toul to Corre on the River SÉone. The two were connected west of Toul by utilising 20km of the already built Canal de la Marne au Rhin. The summit-level of the southern section of the Canal de l'Est is situated just to the west of Epinal; it runs for 10km between the D166 road bridge and the village of Girancourt (Fig. 2). It was to supply this summit that the Bouzey reservoir was created by damming the River Avière.[15]

Originally, in 1876, the intended dam was to have held about 20m of water. Reservoirs, however, are subject to a tempting law of increasing returns. A nominal increase in the height of a dam can often realise a huge increase in reservoir capacity. The Canal de l'Est's engineers recognised this possibility at Bouzey and in September 1880 permission was obtained to add 2m to the dam's final height. This decision by the Ministre des Travaux Publics did not have the approval of the Conseil Général des Ponts et Chaussées which was unconvinced that canal traffic, at least at that stage, needed so much water, or that the dam, at least initially, should be subjected to so much pressure, as events were to prove, it is the case that the extra 2m of water did nothing to enhance the security of what turned out to be a very poor design indeed.

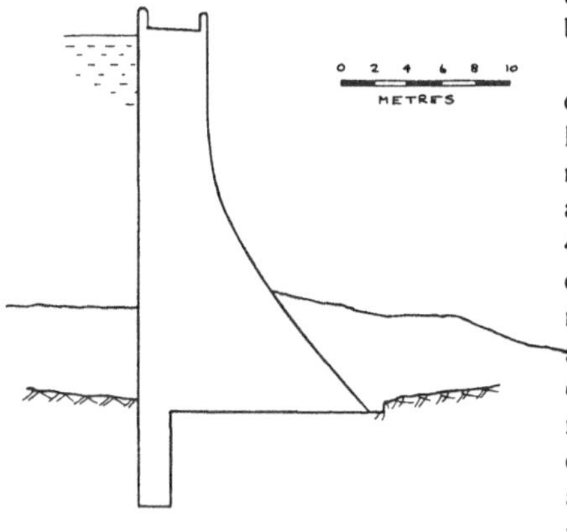

Fig 11: The Bouzey Dam as built

The Bouzey dam was completed right at the end of 1880 and filling of the reservoir began late in 1881. The reservoir's capacity was 7 million cubic metres, it covered 128 hectares, and could feed something of the order of $45{,}000m^3$ of water to the canal each day. The dam itself was straight, quite long at 525m, but not exceptionally high at 22.7m. The masonry and mortar used to build the dam had an overall specific gravity of 2, a low value but not of itself a cause for concern. In order to counter the fissured nature of the foundations, a cut-off wall was built reaching down as much as 5m below the base of the dam on the upstream side. As built the Bouzey dam's cross-section was the one shown in Fig. 11.

The Failure of the Bouzey Dam in 1895

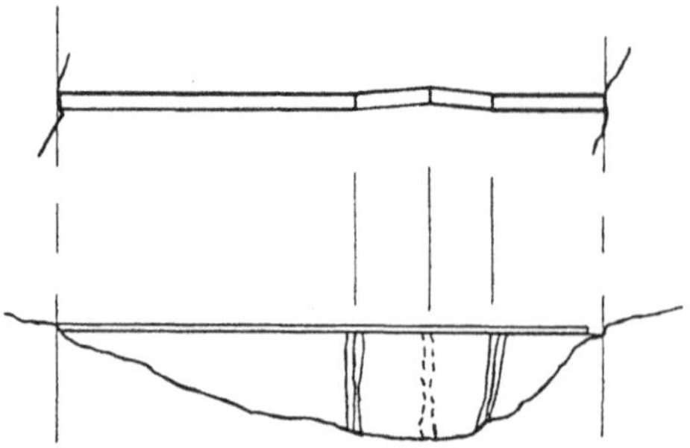

Fig 12: Plan and elevation of the failure of 1884 (not to scale)

Top water level was at 371.5m above datum or 22.1m above the foundation level.

As the reservoir began to fill, so the dam began to leak; even at water depth of less than 15m the leakage was up to 65 l/sec. In December of 1882 two fissures materialised to add to the leakage. They were put down to longitudinal contraction of the structure in cold weather. Come December 1883 the reservoir level had reached 15.5m, that is to say about two-thirds the maximum, and at this moment the water-supply to the reservoir was augmented very substantially by the completion of a 43km-long aqueduct bringing Moselle water from Remiremont. By March 1884 the reservoir was 18.6m deep; the leaks had grown only a fraction to 70 l/sec. The level rose a further 80cm. And then, on the 14th of March, just before noon, 135m of the dam slipped forward into the V-formation indicated in Fig. 12. The maximum displacement was 35cm and the break occurred at foundation level (Fig. 13).[16]

By any reckoning this was a serious situation and yet the astonishing thing is that not only were no precautions taken, the reservoir level was actually allowed to rise a further 20cm, to 19.6m of depth. In this condition, for over 18 months, the Bouzey dam continued in use with water pouring through cracks at the rate of 232 l/sec.

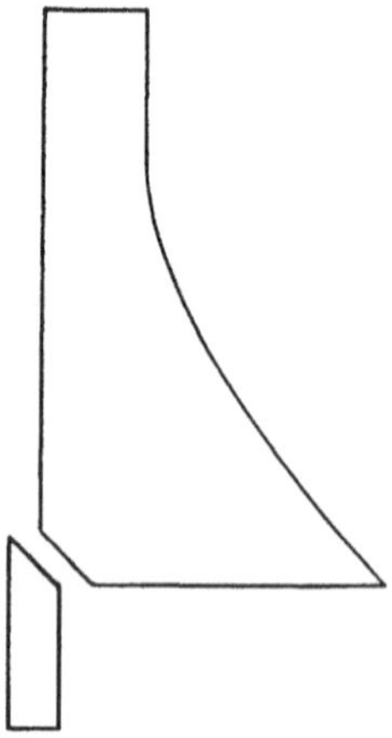

Fig 13: The nature of the fracture

The reservoir was drained in the autumn of 1885. Not surprisingly groups of vertical cracks were found at each end of the displaced section and at its centre. They, quite evidently, were the result of bending fractures, tension cracks in a broken "horizontal beam". Underneath the dam it was found that the cut-off wall had no means protected the poor rock immediately under the structure. Fissured and permeable beds reaching well into the reservoir area had not been sealed by the cut-off wall, and the formations under the dam were dislocated and saturated to a depth of 2-3m. Something which all the pioneer analysts of dams had thought about but never really believed could happen, in fact had. A dam had slid on its foundations, an effect probably exacerbated by a considerable uplift pressure under the dam, although that factor was not fully debated until a later date.

Several features of the 1884 accident and its aftermath are intriguing. That the broken dam was expected to carry on for 18 months cannot, surely, have been a decision of engineers. It must reflect the Canal de l'Est's paramount need for water to sustain traffic.[17] And yet once the reservoir was drained, progress was very leisurely. Exploration of the foundations occupied the years 1885 and 1886; reports were prepared and discussed in 1887 and 1889. There were conflicting opinions as to what had happened and what to do next. Oddly, the

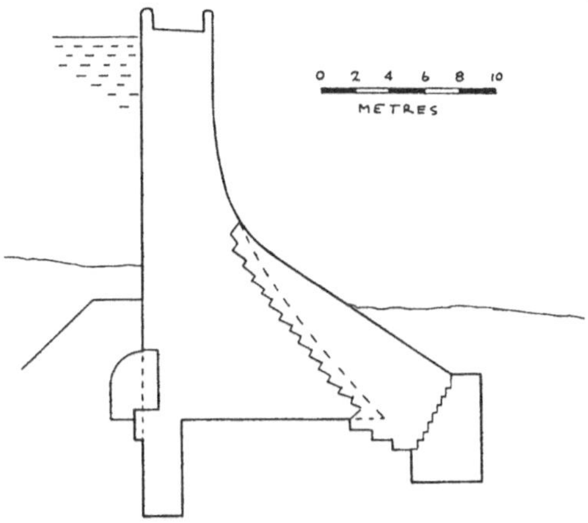

Fig 14: The Bouzey Dam rebuilt (the old dam is shown by the broken line)

one thing which was never agreed – surprisingly given how much time elapsed – was to demolish this ill-starred structure and start again.[18] The best chance to ensure the dam's future was missed. One detects here, though the issue is never broached in the relevant literature, profound economic problems. It was a post-war period and the Canal de l'Est was new and expensive. The urge to use the navigation and make it pay must have been powerful. Therefore the Bouzey dam was repaired.

On the upstream side at the base masonry was used to seal the gap between the body of the dam and the cut-off wall, the whole repair being generously covered with puddled-clay. For the downstream side there were several proposals. The most extraordinary was to construct an earthen bank to the full height of the dam; the best was for a series of tall buttresses in the manner of the measures taken at the dams of Grosbois and Chazilly.[19] In view of what was to happen in 1895 the decision *not* to build "contreforts" was another chance missed. In the event, the chosen solution was a massive buttress "toothed" into the original masonry and bearing against a deep level abutment. The intention was to prevent further sliding. But no strengthening was provided for the upper half of the dam and that was where the ultimate disaster was lurking. Drains were installed at the base of the dam to take away any water percolating underneath. The vertical cracks were filled with cement mortar or grout. The bore-hole used to investigate conditions on the lower side of the original dam was retained as a basis for monitoring water-pressures beneath the modified structure. The rebuilt cross-section is shown in Fig. 14.

On the 18th November 1889 the Bouzey reservoir began to refill. The level passed 19.4m, the depth at which the accident of 1884 occurred, in February 1890 and subsequently the water level was relentlessly increased. On 15th May it achieved a depth of 21.6m, just 50cm short of the maximum. In subsequent years the maximum level was always achieved. For 53 days in 1893 and for 167 in 1894 the water level fluctuated between 21.6m and 22.1m.

For the first half of the 1890s it must have seemed that all was well. The dam retained a high to full water level without mishap. Theodolite measurements[20] along the dam indicated a maximum deflection of 15mm; leakage was of the order of 70 l/sec., significant but not exceptional, especially by Bouzey standards; the piezometric levels were about 10m above foundation level. These various observations were not, as the years went by, maintained. The deflection measurements were given up in 1893 and replaced, totally unsatisfactorily, by visually sighting pegs. The piezometer proved itself very easily damaged by ice so that after the savage winter of 1894-5 it was removed. This same fearsome winter is supposed to have significantly opened vertical fissures in the dam as the wall contracted, a crack width of as much as 7mm being recorded.

The Failure of the Bouzey Dam in 1895

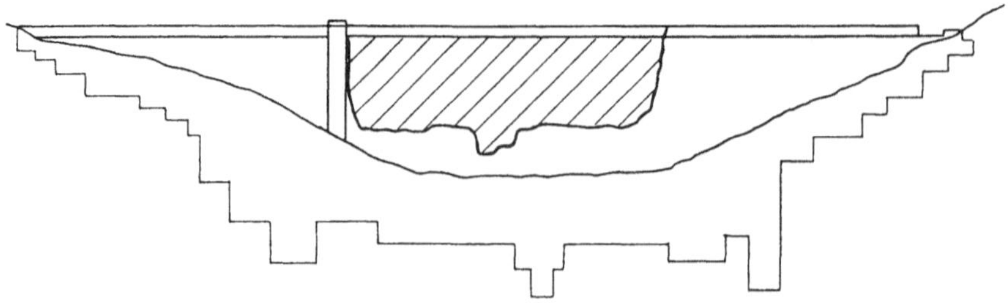

Fig 15: The failure of 1895, viewed from upstream. The vertical scale is 5x the horizontal; the stepped line shows, approximately, the bottom of the cut-off wall

To what extent accurate and frequent measurement might have predicted the final collapse will never be known. Several individuals claimed that towards the end a pronounced downstream curve of the dam developed which could be detected by the naked eye. If such could strike passers-by it surely would have registered officially had it been measured accurately. But it was not and the Bouzey dam failed.

The Failure

There were three witnesses to the collapse on the 27th April 1895. Early that morning a workman called Thiriat was crossing the dam and he stopped, 160m short of the spillway at the eastern end, the right bank, to talk to a mason working below. As they talked the dam began to break up. M. Thiriat had to jump over a 10cm crack in his dash to safety. The mason he had been chatting with was the first victim of the disaster. The third eyewitness was the innkeeper M. Gihin who was watering his horse in the village of Bouzey, 300m downstream. He was able to provide a useful account of the failure sequence which comprised two phases: an initial break near the crest in the middle of the structure over a length of 20m, followed by the main collapse of 180m of the dam to a depth of about 10m. Both Thiriat and Gihin commented on the noise. The failed dam is shown diagrammatically in Fig. 15.

Reactions to the disaster came at several levels. On behalf of an outraged public the popular press clamoured for explanation, redress and the names of those responsible. Louis Geisler was quickly into print with his "15 Vues Photographiques de la Vallée de l'Avière après la Catastrophe de Bouzey", a photographic collection in the style of the even better one covering the ruins of the Dale Dyke dam put together by the Manchester photographer, James Mudd. There was huge international coverage in the engineering press in France, Germany, Great Britain and the United States.[21] Of course there was an official enquiry, conducted by French engineers, and tours of inspection by other experts two of whom were Professor W.C. Unwin of Imperial College and Dr. G.F. Deacon, designer of the Vyrnwy dam in Wales. The latters' observations are among the most useful and objective certainly in English, that the historian can consult.[22] Not the least interesting feature of the deliberations which took place is the degree of disagreement as to what exactly had happened and how to explain it. It certainly was not enough to observe that the design of the dam was bad and that it had bad foundations.[23] Moreover, agreed conclusions were paramount because, as pointed out at the beginning of this account, Bouzey was supposed to have been *designed*. And if dam design was to be trusted, it was vital to comprehend how and for what reasons it could go wrong with an unsafe structure the result.

Fig 16: A close-up view of the break (from a photograph by Profession W.C. Unwin)

The failed dam itself is the place to begin a brief account of what happened. Naturally there was a readiness to associate the final collapse with the accident of 1884 and the repairs of 1888-9. In reality no such connection could usefully be made. The great abutment had not yielded and appeared to have prevented all further movement of the base of the dam. And in any case the failure of 1895 did occur *entirely above* the level of the remedial works of six years before. In addition, the extremes of the failure of 1895 did not coincide with the vertical cracks at the limits of the movement of 1884 while the central group of fissures in the earlier accident were contained completely by the final collapse. As Unwin put it, "The fracture of 1895 seemed to avoid, rather than to depend on, the fissures of 1884."[24]

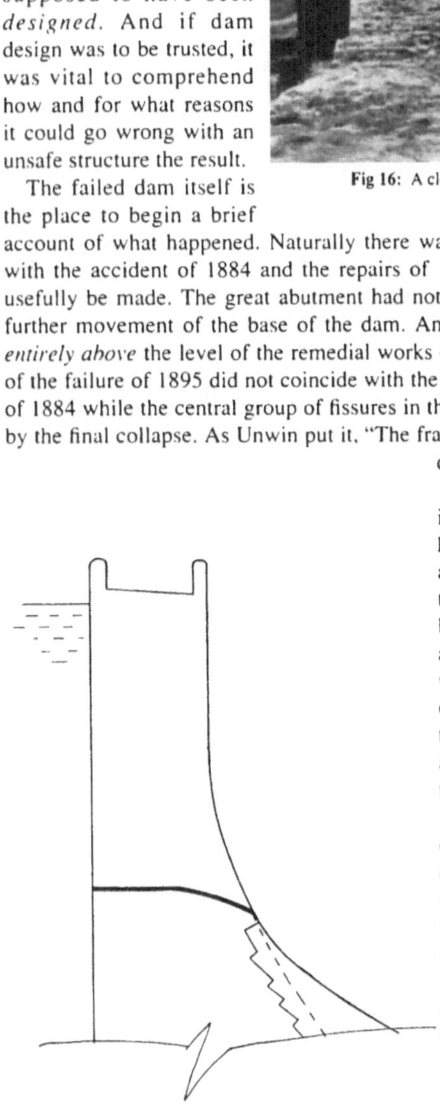

Fig 17: The typical shape and position of the failed section

The cross-section of the failed dam is instructive (Figs 15–17). The break runs horizontally through about half the section and then drops away. Unwin was struck by the fact that there was no evidence of failure by crushing, that "The mass was torn out along a nearly level plane of fracture ..... without any obvious cracking or splintering of the stones along the down-stream edge of the fracture. The mass filling the gap appeared rather to have slid or overturned."[25]

One school of thought was quite convinced of a shear failure. Of the three experts appointed to the tribunal of enquiry, two advanced this view and they were supported by the expert witness of the eminent M. Maurice Lévy.[26] In a sense they were probably correct. The mechanism of a dam's failure is a very complex affair, a whole series of effects occur in quick succession, and the one which starts the chain is not necessarily the one associated with the final moments of collapse. At Bouzey the final stage very likely was a shear

The Failure of the Bouzey Dam in 1895

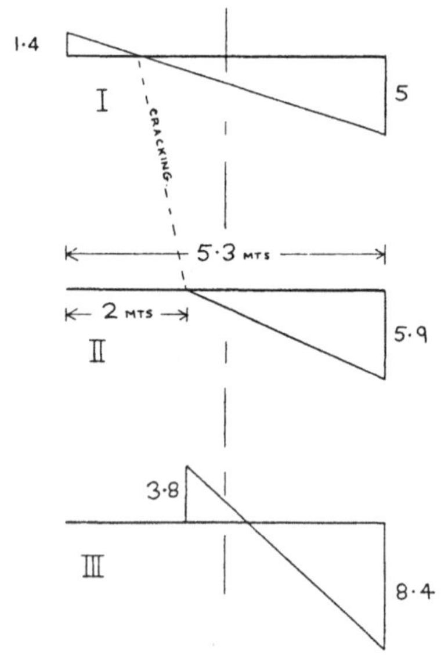

Fig 18: Successful sates of stress (in kg/cm²) and cracking across the critical section

failure on a surface sloping down towards the air-face. But the key question remains: what initiated the failure in the first place?

By applying the analytical techniques described earlier, the ones that were available to the engineers of the day, it is easily shown how completely the middle-third rule was violated in the upper part of the dam, and that the greatest deviation was at the level of the fracture. Actually the compressive stresses induced at the air-face were not destructive. The maximum vertical stress was about 5 kg/cm^2, the corresponding principal stress parallel to the air-face being of the order 6 kg/cm^2. Such compressions were well within the strength of the masonry used and relatively low compared with many other dams. By contrast the situation at the water-face, reservoir full, at a depth of 10m was fatal. Here the vertical *tensile* stress was about 1.4 kg/cm^2; a metre higher it was still 1.1 kg/cm^2.[27] It seems very probable that both the masonry and the mortar could have taken these stresses even though good design should not have required them to, particularly since neither was of first class quality. The crucial defect was that the bond between the masonry and mortar was very poor and should not have been relied upon in any case.[28] It was quite incapable of resisting the tension which developed with a full reservoir and so the Bouzey dam cracked.

Referring to the stress diagram I in Fig 18 we can show that if the tension of 1.4 kg/cm^2 was relieved by cracking, the crack would penetrate 2m and the next stress diagram, II, would then apply, the minimum stress now becoming zero and the new maximum being a compression of 5.9 kg/cm^2. So far the dam is still not over-stressed, exactly as de Sazilly, Delocre and their followers were willing to allow. However, as soon as water pressure penetrated the tension crack a new vertical force came into play, the uplift force we were considering in the case of the Habra dam, cracked in that case by a chronic hydraulic overload. The uplift **U** within the cracked Bouzey dam significantly increased the overturning moment and the magnitude of the effect is readily calculated (Fig 19).

At a depth of 10 m of water we can now construct a third stress diagram, III (Fig 18); it applies to the surviving 3.3m length of the section and, of course, it includes a new element of tension – 3.8 kg/cm^2 at maximum – which, if it is relieved by cracking, allows an even larger uplift force and leaves a decreased length of section intact. It is hardly necessary to emphasise the progressive nature of the failure now underway, except to point out again that at the final stage, overturning about the downstream edge would most likely give way, if that is the phrase, to shear fracture. Such a fracture would be along a surface that falls away towards the air-face, as indeed the ruins of the dam confirmed (Fig 17). A similar surface of failure had been observed in the Habra dam.

Such then is the explanation as to why the Bouzey dam failed. And yet, given that its profile was so poor that tension *did* occur, one wonders not so much why it failed but how on earth it lasted so long.[29] The explanation has a number of elements. Although the basic theory treats a dam as a series of thin vertical "slices", in practice it is not; the slices interact one beside the other in resisting deformation and so a crack has to extend to a considerable depth before *all* the resistance is finally overcome. The point is emphasised by the great length which gave way at Bouzey. Also, because Bouzey was a "leaker", its face was from time to time treated to a waterproof coating. Presumably, therefore, the percolating water was inhibited at intervals and sufficiently to prevent uplift pressures equal to the full head of water. And then again the reservoir was not always full. A canal which was busy in the summer presumably allowed the dam some seasonal respite.

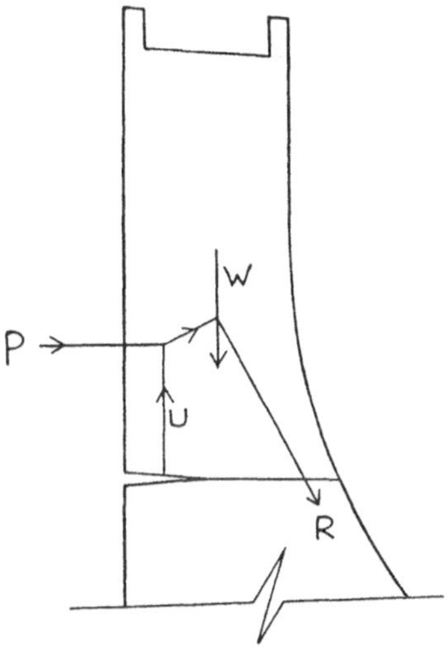

Fig 19: System of forces with uplift

The Puzzle

The question has not been answered. Why *did* the Bouzey dam have such inadequate shape and one that allowed so much tension? After all, the basic mechanics of dam design are straightforward and by 1876, when the project was begun, all the essential elements of the theory were in place including the middle-third rule.

To my knowledge the designer of the Bouzey dam is never named. That he could produce a profile which simply *looks* so appallingly inadequate is interesting in itself.[30] Earlier I noticed that it was easy to overlook the vulnerability of the gravity dam near its crest when the reservoir is full or, as Habra was, overfull. Conceivably Bouzey's proportions, long but not very high, were too readily regarded as ordinary, elementary to design and posing no significant risk in practice. The reason why French engineers pursued the design and construction of masonry dam so assiduously was because, to quote M.I. Auguste Graeff apropos the 50-m Furens dam, "We need not in this case consider earthen dams which are of very doubtful security at a height of 20 metres; at 50 metres they would, of course, be quite out of the question."[31] Always of concern was this question of height. But Bouzey was not a high dam and therefore if, for that very reason, the dam's designer was already complacent and if, in addition, he was prepared, like de Sazilly and Delocre, to allow tensile stresses and then discount them, either because he believed the material could withstand them or because it was assumed permissible to settle for a cracked section, as Fig 18, then Bouzey's design is explained. There is a telling paragraph in Langlois relating to the "calculs des ingénieurs de 1876". He says "No observations have come to light on the subject of upstream tension, of which there is no consideration, neither in the calculations of 1876 nor in those which preceded the reconstruction of 1888-9."[32]

There is one other feature of Bouzey's ill-judged design to consider. The dam was proportioned, apparently, by the method of M.Bouvier. His procedure was much in vogue at the end of the nineteenth century and in a French ministerial circular of 1897 was even specified as mandatory. But in reality Bouvier's method was never more than a device and its use at Bouzey may explain a good deal. Consider Fig 20. Bouvier, in an effort to allow that maximum stresses are not vertical ones, considers the resultant of **P** and **W**, the force **R**,

The Failure of the Bouzey Dam in 1895

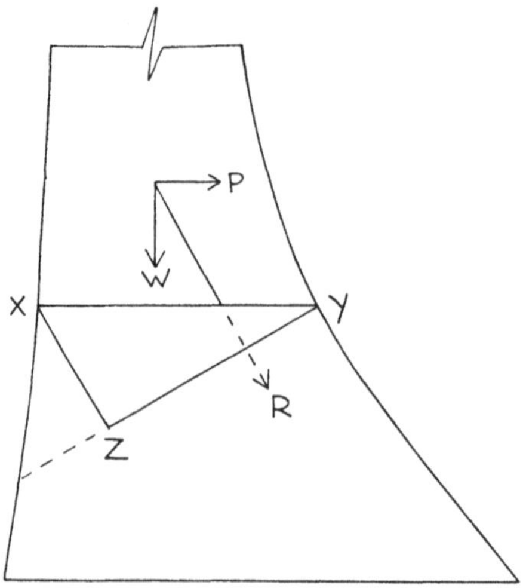

Fig 20: Bouvier's method

as acting on the plane **ZY** rather than **XY**. The stress distribution on **ZY**, at right angles to **R**, is assumed linear, as usual, but the calculated stresses are higher than those obtained by considering **W** acting on **XY**. It is a device which is plausible but without rationale. But what is crucial is that whereas Bouvier's method gives an augmented compressive stress value for the air-face point **Y**, as intended, it gives no stress value for the water-face at all, the point **Z** being in effect nowhere. As Langlois complained of Bouvier's method , "it does not determine anything about what happens on the air-face."[34] In short a designer relying on Bouvier's method will almost certainly design an adequate air-face at the same time as he ignores the water-face completely. The method is intrinsically a technique for determining a peak stress at a point, not a stress distribution between two points.

The Aftermath

Bouzey failed at a crucial time and strange though it may seem to say so, it was not altogether a bad thing: much good came out of this catastrophe.[35] For nearly half-a- century there had developed a growing confidence in dam design but considerable variations in how to do it, what assumptions could and should be made, and what criteria to apply. And this remember was in a period when dams were being built in increasing numbers world-wide, for public water-supply, irrigation and the newly developing application to hydro-electric power (the Bouzey dam's use for a canal was actually rather old fashioned). Moreover, dam-building was by no means the monopoly of European countries. Much, and often the most significant, was happening in North Africa, India, the United States and Australia. That is where one finds the very big dams, e.g. Aswan, and the emerging fashion for the thin arch dams, e.g. in California and New South Wales.

The need to fully understand the events at Bouzey was therefore of great importance and the audience awaiting enlightment was large and international. Close scrutiny of the failure and its full discussion disposed of a number of erroneous or unreliable propositions and established positively that for safe dam design (a) the no-tension condition was an imperative, (b) the middle-third rule was a correct concept, essential to observe, (c) a maximum compressive stress must not be exceeded and (d) a dam must not be able to slide. Not far behind came some instructive lessons to do with construction techniques and the quality of materials.

The Bouzey failure not only summed up the essential developments so far. It also helped to confirm the direction of future work. New design problems and attitudes to them, already it is true under consideration anyway, were given fresh illumination and a much sharper focus.

There was, for example, the matter of uplift and the difficulty of comprehending its behaviour in various circumstances. Some attention had already been given to uplift pressures in the design of such dams as Vyrnwy, in Wales, and Alfeld, in Alsace.[36] Avoiding uplift altogether was one possibility and in some quarters, following a much misunderstood French ministerial circular of 1897,[37] this led to dams being profiled so that the compressive stress at the water-face was greater than the water pressure, the belief being that the infiltration of cracks then became impossible. One result was some very broad profiles indeed.

In fact the 1897 instructions were *intended* to allow for uplift by reducing the effective density of the masonry. They also drew attention to the need to determine maximum compressive stresses on oblique joints, in effect maximum principal stresses, and to investigate the magnitude and danger of shear stresses, something about which the Bouzey failure had been instrumental in promoting discussion, as we have seen. Thus, and this is the point, Bouzey was central to the developing interest in the stress analysis of dams beyond the elementary concepts of de Sazilly and Rankine. In France, Maurice Lévy was prominent in pursing these increasingly sophisticated concepts while in England an equally mathematical approach succeeded in triggering a first class controversy in which the main protagonists were Professors Karl Pearson and W.C.Unwin.[38] And then, in attempt to throw light on these intractable problems, recourse was made to models utilising, in the early experiments, such rather unsuitable materials as plasticine, gelatine and India-rubber. Nevertheless, for all the crudeness of the techniques, it was the start of a very significant, and at times a crucial, trend in twentieth century dam design.[39]

The failure of the Bouzey dam, when all is said and done, stands at a cross-roads where the first phase of rational dam design ended, first of all in controversy but then in clarification. Most branches of civil engineering have their land-marks; for dam-building the Bouzey failure was truly a key one.[40]

Let me quickly finish the story of the dam itself. It was rebuilt twice more. The first reconstruction was a makeshift affair of 1901-2. According to Bellet[41] it was only 6.4m in height forming a reservoir of 1.5 million m^3 , useful to the Canal de l'Est no doubt, but a great reduction in capacity all the same. The present dam dates from 1939 and is of rock- fill construction, 27m high and 504m long. It occupies exactly the site of its ill-fated predecessor of which the only surviving part is the spillway.

References

1. Technical details are in G.M.Binnie, *Early Victorian Water Engineers* (1981), pp.263-277. A popular account is G. Amey, *The Collapse of the Dale Dyke Dam 1864* (1974).
2. A recent account is W. Frank, "The Cause of the Johnstown Flood", *Civil Engineering*, 58 (May 1988), pp.63-66.
3. Two general histories of dams are: N.A.F.Smith, *A History of Dams* (1971) and N. J. Schnitter, *A History of Dams* (Rotterdam, 1994).
4. B. F. de Belidor, *Architecture Hydraulique*, Vol.1, Part 2 (Paris 1750), pp.77-83.
5. In, for example , S. Leliavsky, *Irrigation and Hydraulic Design*, Vol. 3 (London 1960), pp.375-78.
6. These and other early French canal dams are covered in E. Wegmann, *The Design and*

Construction of Dams (New York 1907) and H. Bellet, *Barrages en Maçonnerie et Murs de Réservoirs* (Paris 1907).

7. J.A.T.de Sazilly, "Sur un type de profil d'égale résistance proposé pour les murs de réservoirs d'eau", *Annales des Ponts et Chaussées*, 2 (1853), pp.191-222.
8. E. Delocre, "Sur la forme du profil à adopter pour les grands barrages en maáonnerie des rèservoirs", *Annales des Ponts et Chaussées*, 2e sem. (1866), pp.212-272.
9. W. J. M.Rankine, "Report on the Design and Construction of Masonry Dams", *The Engineer*, 33 (January, 1872), pp.1-2.
10. J.A.T.de Sazilly "Sur un type de profil d'égale résistance proposé pour les murs de réservoirs d'eau", p.193.
11. W.J.M. Rankine, "Design and Construction of Masonry Dams", p.1.
12. These developments are reviewed by E. Wegman, *Design and Construction of Dams* Chaps. II-IV and H. Bellet, *Barrages en Maçonnerie*, Chaps.V-VI.
13. A. Graeff, "Sur la forme et le mode de construction du barrage du gouffre d'Enfer, sur le Furens, et des grands barrages en général", *Annales des Ponts et Chaussées*, 2e sem. (1866), pp.207-211.
14. "La Rupture du Barrage de l'Habra, près de Perrégaux", *Le Génie Civil*, XCII (1928), No.11, pp.256-9; No.12, pp.283-5: the author – denoted by "A.C." – may be André Coyne. See also E. Wegmann, *Design and Construction of Dams* pp.94-8 and H. Bellet, *Barrages en Maçonnerie*, pp.54-6.
15. The most complete general account of the Bouzey dam is Léon Langlois, *Rupture du Barrage de Bouzey* (Paris 1898).
16. These and other details of the Bouzey dam's career are taken from, in particular, L.Langlois and the report of a special commission of the Corps des Ponts et Chaussées which was published in 1895. There is a copy in the library of the Institution of Civil Engineers. It was abstracted by A.W.Binnie, "Failure of the Bouzey Dam", *Proc. Inst. Civ. Eng.*, CXXV (1895-6), pp.461-466. This same report appears to be the basis of "Failure of the Great Dam at Bouzey", *The Engineer*, 17 January 1896, pp.50-1.
17. The pint is noticed in "Historic Accidents and Disasters: No.XVIII, The Bouzey Dam", *The Engineer*, 4 December 1942, p.452. Langlois, *Rupture du Barrage de Bouzey*, pp.93-4 deals with canal consumption and there is a lengthy comparative study of canals' water requirements in Maurice Lévy, *Observations sur la Catastrophe de Bouzey*, (Paris n.d, circa 1896).
18. *The Engineer*, 4 December 1942, p.453 notes this oversight but does not reflect on the practicality of rebuilding. The dam had after all only been displaced, not destroyed.
19. H.Bellet, *Barrages en Maconnerie*, pp.37-9. The Grosbois dam is the subject of M.Guenot, "La Stabilité des Digues du Réservoir et du Contre-Réservoir de Grosbois alimentant le Canal de Bourgogne", Annales des Ponts et Chaussées, .4 (1949), pp.467-91.
20. So far as I can determine the Bouzey dam is the earliest to be instrumented in this way.
21. The bibliography of the Bouzey failure is enormous, far too much to record here. Serge Leliavsky once referred to the many articles written as being "comparable to a sea of ink"; see his discussion to E.Gruner, "Dam Disasters", *Proc. Inst. Civ. Eng.*, 27 (1964), p.362. For an introduction to the bibliography of the failure there is ICOLD, *Lessons From Dam Incidents* (Paris, 1973), p.29 and A.O. Babb and T.W.Mermel, *Catalog of Dam Disasters, Failures and Accidents*, (Washington D.C., 1968), pp.31-2.
22. W.C.Unwin, "The Failure of the Masonry Dam at Bouzey, France" *Cassier's Magazine*, XI (1896-7), pp.9-19. G.F.Deacon , "Water Supply", *The Encyclopedia Britannica*, 13th Edition, Vol.28, p.403. Professor W.C.Unwin frequently raised the Bouzey dam issue at

meetings of the Civils. See, for example, his discussion to G.F.Deacon, "The Vyrnwy Works for the Water-Supply of Liverpool", *Proc. Inst. Civ. Eng.*, CXXVI (1895-6), pp.93-7 and then to Sir John Ottley and A.W.Brightmore, "Experimental Investigations of the Stresses in Masonry Dams subjected to Water-Pressure", Proc. Inst. Civ. Eng., CLXXII (1907-8), pp.70-78.

23. Unwin is worth quoting on this: "It had been said that the design of the dam was bad, though it was not said in what respect; and also that it had a bad foundation. That was something like the mayor of a town in France who had seventeen reasons for not firing a royal salute, and the first was that there was no powder". W.C.Unwin discussion to Deacon, p.93.
24. W.C.Unwin, *Cassier's Magazine*, p.18.
25. W.C.Unwin, *Cassier's Magazine*, pp.18.
26. H.Bellet, *Barrages en Maçonnerie*, p.68. and Lévy, *Observations*, pp.10-23.
27. And it can be calculated that a zone of tension covered a height of about 6 mts above and below the facture.
28. Analyses of the accident are replete with discussions about the strength of the materials used: see Langlois, Rupture du Barrage de bouzey, and Lévy, Observations, pp.7-22. The author of "Historic Accidents and Disasters", *Engineer*, 4 December 1942, p.455 makes much of the use of lime rather than cement mortar.
29. Philip A.Morley Parker puts it in an interesting way when he says: "Nevertheless, these facts create a feeling not so much of astonishment at the disaster, as of confidence in a well designed dam. We have a dam violating in every possible manner all the present-day principles of sound construction. The foundation is bad, and the dam is not carried down far enough; yet it only partially fails." – Philip A.Morley Parker, *The Control of Water*, (1913), p.393.
30. Although one must be wary of hindsight here. In 1876, or thereabouts, the basis for appreciating a well proportioned dam was limited to only a decade or so's experience.
31. A.Graeff, "Sur la Forme et le mode de construction ....". p.196.
32. L.Langlois, *Rupture du Barrage de Bouzey*,. p.57.
33. M.Bouvier, "Calculs de résistance des grands barrages en maáonnerie", *Annales des Ponts et Chaussées*, 2 (1875), pp.173-205. Bouvier's method is assessed by H.Bellet, Barrages en Maçonnerie, pp.46-9.
34. L.Langlois, *Rupture du barrage de Bouzey*, p.56.
35. S.Leliavsky, discussion to E.Gruner, "Dam Disasters", pp.362-3.
36. For Vyrnwy see I.Davidson, "George Deacon (1843-1909) and the Vyrnwy Works", *Transactions of the Newcomen Society*, 59 (1987-88), pp.85-6. For Alfeld see S.Leliavsky, *Uplift in Gravity Dams*, (1958), pp.1-2.
37. "Circulaire du Ministère de l'Agriculture du 15 Juin 1897" is reproduced in H.Bellet, Barrages en Maconnerie, pp.293-302.
38. A.J.S.Pippard, "The Functions of Engineering Research in the University", *Jour. Inst. Civ. Eng.*, 33. (Feb.1950), pp.265-284.
39. And interestingly enough, the Bouzey dam itself was treated to an experimental investigation long after the events described; see V. de Belaevsky, "Note sur la rupture du barrage du Bouzey", *Le Génie Civil*, LXXXVI (March 1925), pp.239-40.
40. This is the theme of "Lessons Learnt from Dam Disasters", *Engineering*, 15 May 1964, p.683.
41. H.Bellet, *Barrages en Maçonnerie*, p.70.

11

George Deacon (1843–1909) and the Vrynwy works

I. Davidson

George Frederick Deacon was born at Bridgwater in July 1843. He was the eldest son of Frederick Deacon, who practised as a solicitor first in Bridgwater and later in Preston. The son was educated at Heversham Grammar School, and was apprenticed to Robert Napier & Sons, Glasgow, in 1860. He entered Glasgow University in 1863, where he studied under Professor Rankine and under William Thomson (later Lord Kelvin), with the latter of whom he formed a lasting friendship. These studies were prematurely concluded when Deacon sailed on the *Great Eastern* during the abortive attempt to lay a second Atlantic cable in 1865.

In that same year Deacon set up as a consulting engineer in Liverpool at the early age of 22, and apparently found enough work in both civil and mechanical fields. In 1870 he was appointed a lecturer in civil engineering and mechanics at Queen's College, Liverpool, and the following year, although only 28 years old, he was appointed Borough and Water Engineer.

LIVERPOOL WATER SUPPLY

In his new post Deacon was much occupied with sewerage and street pavements, but his chief concern was with water supply. The Rivington Works had been opened in 1857, having been constructed under the direction of Thomas Hawksley. The latter had disappointed the citizens of Liverpool, because he had estimated the maintainable yield of the Rivington Works as 22 million gallons per day (mgd), whereas experience had shown that it was scarcely more than 16 mgd.[1]

The shortage of water in 1872 meant that only an intermittent supply could be maintained, and this provoked constant anxiety.[2] It was realized that the trouble was due largely to excessive leakage and wastage. Deacon's ingenuity found the solution by the invention, in 1873, of his waste-water meter. This was a simple and robust, but portable, device which drew a chart of the rate at which water passed through it. The method of use was to insert the meter into a main feeding a district of the city. During the night, when flow should be a minimum, all other feeders into the district were shut off, together with all except one of the street mains in the district. By manipulating the street valves and stopcocks to premises it was possible to record the flow into each house, in turn. The resulting chart indicated the location of significant leakage and wastage.

The Deacon waste-water meter proved to be a great success, and by its use water consumption in Liverpool was reduced by 37%, even when a constant supply was reintroduced.[3] The meter was also applied by many other undertakings both at home and abroad, and is still in use. However, the continued growth of the population indicated that further supplies of water would be required, and between 1874 and 1877 various proposals were considered. In 1877 Deacon was instructed to report on Haweswater and Vyrnwy as possible sources of supply.

His reports are both dated November 1877. As regards Haweswater, Deacon considered that the water was of good quality, and that by raising the lake 29 ft above its normal level of 725 ft OD, (Ordnance Datum) a reliable supply of 26 mgd could be obtained. He estimated that the total cost for a supply of 13 mgd would be £1560,000.[4]

As regards Vyrnwy, Deacon mentioned that this supply from the headwaters of the River Vyrnwy—a tributary of the Severn—had been proposed by Bateman in 1865 as a possible supply for London. Deacon suggested an impounding reservoir with its water level at 817 ft OD, which would give a reliable supply of more than 50 mgd. The estimated total cost was £1280,000 for a supply of 13 mgd. Deacon strongly recommended Vyrnwy above all other proposals.[5]

In a supplemental report dated August 1878, Deacon said that shafts had now been sunk to rock at the proposed Vyrnwy Dam site, and he recommended that the dam should have a height of 79 ft above the bed of the valley. The revised estimate was £1181,700 for a supply of 13 mgd. Three different routes were proposed for the aqueduct, none of which was that finally chosen.[6] Hawksley was invited to comment on Deacon's reports and he stated that he believed that the Vyrnwy scheme was perfectly feasible, that the works were simple and safe, that the water was good and abundant, and that, considering the distance, the cost was small. Bateman concurred in this recommendation.[7]

Nevertheless the Water Committee had great difficulty in reaching a conclusion. The Chairman of the committee was J. H. Wilson until November 1881, when A. Bower took over.[8] In particular they were much concerned over the question of peat in the Vyrnwy water, in regard to which Hawksley offered a typically pungent opinion:[9]

> Allow me at the end of my 50 years of experience, to suggest that the great problem of the water supply of a million people is not to be solved by minute considerations. The microscopic and immaterial points raised by theorists are innumerable, so that if weight and importance are to be fancifully attributed to these, a practical solution will never be arrived at.

There was also a last minute suggestion that, in the short term, a supply should be taken from Thirlmere, in conjunction with Manchester who had now decided to develop this source.[10] However, after a deputation to Manchester, the third report of the Water Committee concerning the new water supply was finally issued on 17 June 1879.[11] It recommended Vyrnwy. This report was approved by the Council on 29 July 1879.[12]

VYRNWY WORKS: EARLY STAGES

Deacon was instructed to complete the surveys and to prepare Parliamentary plans for the Vyrnwy scheme. He had only five months for this as they had to be deposited by 30 November. The Council appreciated that they urgently needed a consultant of adequate seniority to promote the passage of their bill through Parliament, and Bower came to an agreement with Thomas Hawksley in Paris.[13] According to his memorandum dated 26 September 1879,[14] Hawksley and Deacon were to be joint engineers and both were to sign the Parliamentary plans; if the Act were obtained Hawksley was to be joint engineer, and was to prepare working drawings, for which he was to receive a fee of 2.5% of the cost of the works. Deacon was to undertake general and constant supervision in consultation with Hawksley. It is surprising that Hawksley was prepared to act jointly with Deacon, and at half his usual fee. However, the scheme was a prestigious one, Deacon had prepared the Parliamentary plans, and the Corporation would pay his salary. On 8 September 1879 the Council accepted Deacon's resignation from the posts of Borough Engineer, Gas Inspector and Building Surveyor, and appointed him Water Engineer at a salary of £1200 p.a.[15]

Hawksley and Deacon jointly prepared the Parliamentary estimates with Sir Frederick Bramwell,[16] and the Act was obtained on 6 August 1880 at the expense of an unexpectedly large amount of compensation water. In his preliminary estimates for Haweswater and Vyrnwy, Deacon had allowed for an earth embankment at both sites. That at Vyrnwy was intended to have an overflow weir at each end,[17] but Deacon received information about a cloudburst which had occurred over part of the gathering ground shortly before he knew the locality. He estimated that the discharge had certainly exceeded 1000 cusecs (cubic feet per second) per 1000 acres.[18] It may well be that this indication that a lengthy overflow weir would be advisable, plus the knowledge he now had that a continuous and good rock foundation was available at the dam site, caused Deacon to consider the advisability of substituting a masonry dam (the whole crest of which could be used as an overflow) for the intended earth bank. He mentioned this possibility to the Chairman of the Water Committee and, by a curious coincidence, Hawksley had done so too a few days earlier.[19]

Moreover the fact that the compensation water requirement was greater than expected, suggested to the engineers that the top water level of the reservoir should be raised, and the draw-off water level lowered, by the maximum permitted by the limits of deviation of the Act, in other words by 5 ft in each case. All these variations on the original proposals were discussed and agreed with the Chairman of the Water Committee,[20] but no formal instructions were given. The completed drawings of the masonry dam, signed by Hawksley and Deacon as joint engineers, were agreed, without comment, by the City Council on 1 June 1881.[21]

These activities did not pass without friction, for the characters of the joint engineers had little in common. Hawksley was aged 74, a self-taught water engineer and an outstanding expert witness with immense experience, who preferred not to concern himself with minute details and who was capable of personal dislikes verging on hatred.[22] Deacon was only 38, with an outstanding theoretical and technological education and an extreme aptitude to concern himself with details; he was self-confident and assertive, dour and unsmiling.[23] One of his engineers on the Vyrnwy works said of him, 'Mr. Deacon's fertility of resource and quickness in coping with difficulties have rendered service under him both interesting and stimulating'.[24]

By December 1880, when the Town Clerk was trying to draw up formal contracts with Hawksley and Deacon, the former refused to continue to use the term 'joint engineer'.[25] The Chairman maintained that the only reason why Hawksley did so was because the other half of his 5% commission was retained by the Corporation to pay Deacon. Hawksley's objection was that Deacon remained a servant of the Corporation, and was not independent.[26]

There were prolonged negotiations until a legal contract between Hawksley and the Corporation was signed on 2 March 1881 in which Hawksley was appointed Engineer-in-Chief, with the assistance of Deacon, who was to supervise the works, in consultation with Hawksley, and to act as Resident Engineer.[27] The Corporation exhibited some duplicity in the matter of these contracts, because only a fortnight later they signed a contract with Deacon which stated that he was to act as Engineer in conjunction with Hawksley.[28] These contracts were incompatible.

The early stages of their disagreement had some of the features of a Greek tragedy. On 15 September 1881 Deacon wrote a personal letter to Hawksley which was rather naïve—almost pathetic:[29]

> Until the Act I had believed, on the assurance of the Chairman, that I was to be appointed Engineer to carry out the works . . . Until after the Act I had no whisper of the Paris proposal that you and I should act as joint engineers. It came to me as a surprise and disappointment, but when I saw that the committee desired a joint engineership, I threw aside all previous expectations, and endeavoured to work up to the new position . . . I do not understand why the fact that I was not paid by commission should have made you wish to make my position even less important . . . My own agreement with the Corporation, appointing me engineer in conjunction with you, makes my position much better than your agreement leaves it . . .

Deacon entrusted this letter to Hawksley's son and partner, Charles, who was on a visit to Vyrnwy. Charles promised to give it to his father at the first opportunity. In fact, he left it in a drawer at their office, where it remained until 1885. The lack of any reply or acknowledgement must have made Deacon unwilling to co-operate in the manner he had promised.

The construction of the dam was by direct labour, under the supervision of Deacon, who was now living at Vyrnwy. After the first batch, all drawings were signed by Hawksley alone;[30] this proved to be to Deacon's advantage. The nature of the masonry was unique.[31] The rock produced from the local quarry was described as Lower Silurian clay-slate, but it lacks the regular cleavage planes of a roofing slate. Very large irregular stones—33% of them between 4 and 10 tons—were roughly dressed and incorporated into the masonry. The mortar was made from natural sand, crushed rock and Portland cement, and was mixed very dry. A 2 in layer of mortar was spread, and each large stone was beaten into it by a gang of men using two-handed mallets in unison. The spaces between the stones were filled with concrete, which was produced by placing successive layers of mortar, on to each of which small stones were scattered, and then beaten in to produce a surface like jelly.[32]

During these years Hawksley continued to complain: 'Deacon wants not only to conduct the band, but to play every instrument, and write the music too';[33] 'Deacon wearies and worries me so much that my labours are more than doubled. He is not of the slightest use to me';[34] 'The stones of the dam are in their shapelessness horrible. I cannot conceive how they are to bear the pressure'.[35] With these complaints, Hawksley also declared his wish to resign, 'The sooner the better'.[36]

THE QUARREL

In 1885 it was noticed that the costs of the Vyrnwy works showed signs of exceeding the estimates. In part this was because Hawksley had increased the thickness of the dam by 8 ft at the base, but this

GEORGE DEACON (1843-1909) AND THE VYRNWY WORKS

TABLE 1

Summary of estimated costs of Vyrnwy works

	Deacon's Preliminary Estimate	Parliamentary Estimate	1885 Estimates	
			Hawksley	Deacon
Dam etc.	Earth	Earth £509,400	Masonry £1038,150	Masonry £815,700
Aqueduct		£992,700	£1145,400	£957,800
Total	£1181,700	£1502,100	£2183,550	£1773,500

was not disclosed until years later,[37] and was only hinted at during the ensuing quarrel. In response to a request from the Water Committee, Hawksley submitted a return of costs to date and anticipated costs to completion (Table 1). He also wrote to explain that a Parliamentary estimate is made according to standing orders, and certain things, although important, are not included. Also there was the change in the nature and height of the dam. He concluded, 'The return will in my judgement be of no real use, and of course it could not be held to be binding or conclusive'.[38] Deacon did not immediately submit a return.

Hawksley's return and letter were considered by the City Council on 4 March 1885. They provoked an acrimonious debate which included the following opinions:[39]

> The Water Committee had been secretive and had been misled by their engineers; the Council had been grossly deceived . . . Hawksley was Chief Engineer and they were in his hands, until they dismissed him . . . No engineer of standing should be charged with fraud and deceit without evidence . . . We can put the work under another man . . . To swap horses in midstream is a most risky thing.

Hawksley was furious and wrote curtly to the Town Clerk on 6 March 1885, 'I shall be much obliged if you will communicate to the Water Committee that I decline to be Engineer-in-Chief from this date'.[40]

This was followed by further letters, including:[41]

> I have neither deceived nor been a party to deception. I cannot remain in office with comfort or self-respect . . . I am injured and offended, and I am not likely to be appeased by bland and courteous expressions . . . The wall was not suggested by me, nor by or through Mr. Deacon. It has been my duty to adapt the construction. Greater height, greater width. Do the critics wish the wall to be found at Gloucester?

Deacon submitted his return of costs on 26 March 1885 (Table 1) and added a statement which included:[42]

> The Parliamentary plans show generally more expensive works than I had contemplated in my original estimates, at Mr. Hawksley's instance. I have not always concurred as to the necessity for adopting so large a scale; but Mr. Hawksley is so greatly my senior that, in such matters, I have felt it incumbent upon me to bow to his decision.

The efforts of the Chairman to soothe Hawksley's feelings were not helped by this last document and, to make things worse, Deacon's contract was made public by Alderman Forwood.[43] This provoked another outburst from Hawksley in a long letter dated 2 June 1885, which included:[44]

> It is now sought to establish that by the phrase 'in conjunction' an irresponsible professional partner was imposed upon me. The performance of my duties has frequently been made extremely difficult and painful by what I now perceive to have been the covert operation of the 'dual control'. I therefore ask the Council to accept my resignation.

In the ensuing debate in the Council, strong views were expressed, but motions to set up an enquiry and to alter Deacon's contract were defeated.[45]

On 5 May 1885 Hawksley wrote to Deacon:[46]

> I have not been a party to your agreement and have never assented to the position you have endeavoured to assume. I am sure that you are doing yourself no good by these irritable and irritating self assertions.

GEORGE DEACON (1843-1909) AND THE VYRNWY WORKS

In reply Deacon wrote:[47]

When you wrote [your last letter] you must, I think, have forgotten my letter to you of 15 September 1881, in which I said 'My own agreement with the Corporation, appointing me engineer in conjunction with you, makes my position much better than your agreement leaves it'.

Hawksley replied that there was no such letter on his files, but luckily Deacon had kept a copy, and he forwarded one to Hawksley. The latter then found the original in a drawer, allegedly unopened, as already mentioned.[48]

The dispute dragged on, and in October 1885 the Chairman of the Water Committee published a 'Comprehensive Report on the Circumstances of the Engineers'[49] and Hawksley published 'Some Observations on Mr. Bower's Report'.[47] It is from these that the above account of the conflict has been condensed. The final break came in September 1885, when Hawksley refused to sign the August certificates.

The foregoing summary of this extraordinary episode shows the two Hawksleys in a poor light. The Committee went as far to placate Hawksley as was reasonable, even offering to withdraw Deacon (with his concurrence),[50] and there does not seem to have been any general vendetta, although individuals on either side were violently partisan. The Committee were probably well satisfied by 1885 that Deacon had then enough experience which, added to his intellectual qualities, meant that he could be relied on to complete the works. On 13 January 1886 Deacon was appointed to take charge of the Vyrnwy works, while remaining Water Engineer, for which he received a special allowance of £800 p.a. in addition to his salary of £1200 p.a.[51]

VYRNWY DAM

The reason why Hawksley increased the thickness of the dam, subsequent to the original jointly-signed and approved drawings, is of technical interest. Professor Rankine laid down three vital principles for dam construction in 1872: maximum compressive stresses should be limited, no tensile stresses should occur and there should be no risk of sliding.[52] In 1881—the year in which the first Vyrnwy Dam drawings were finished—the Habra Dam in French Algeria failed completely. This had been built before Rankine's theory was published, and the design was based only on acceptable compressive stresses. The failure occurred as a result of inadequate spillway capacity, when the dam was considerably overtopped.[53]

Under these conditions vertical tensions could occur on the water face; if these caused horizontal cracks, to which the impounded water had access, uplift forces could develop which could destroy the equilibrium of the section above the crack. This idea of uplift was novel at that time, and nothing had been published about it. It seems possible, although it is no better than speculation, that some unknown Frenchman had in fact identified this basic phenomenon at Habra, and that this came to the ears of Hawksley through his French connections. Certainly Deacon gave Hawksley, who had never built a masonry dam, the credit for insisting that possible uplift forces should be taken into account as one of the forces tending to overturn the dam,[54] but they differed in their solution of this problem.

Charles Hawksley explained in 1896 how the thickened section had been derived.[55] Taking the whole section of the dam on its foundation, an uplift pressure was assumed over the base, varying uniformly from the full reservoir head at the water face to the tailwater head at the air face. This was combined with the usual overturning moment of the water. Hawksley had then designed the section so that the total overturning moment should not exceed half the restoring moment due to the weight of the masonry.

Deacon applied the uplift forces in a more realistic manner. First he bonded the dam masonry as securely as he could to the bedrock for a distance of 40 ft from the water face. Over the rest of the foundations, almost to the air face, he provided a network of drains, which he vented by vertical risers to a drainage tunnel which ran the length of the dam, slightly above tailwater level. He then assumed an uplift pressure equal to the full reservoir head extending over 40 ft, and equal to the tailwater head over the remainder of the base. The section of the dam as ultimately constructed, when these uplift forces were applied, was such that the pressure of the water on the water face would have to be multiplied by a factor of 1.8 before the vertical stress at the bottom of the water face changed from compression to tension.[56] This printed report was the first ever to consider uplift, but it fell below present standards in that Deacon failed to calculate similar load factors throughout the height of the water face.

GEORGE DEACON (1843-1909) AND THE VYRNWY WORKS

Fig. 1. Vyrnwy Dam—the first masonry dam to be built on reasonable scientific principles. The dressed stone faces enclose a core of Cyclopean rubble concrete.
G. M. Binnie Collection, by Courtesy of the Institution of Civil Engineers

Fig. 2. The Vyrnwy Dam was built with eight steam locomotive cranes which ran on inclined tracks. Note the very early use of arc lights by which the crane tracks were moved at night.
G. M. Binnie Collection, by Courtesy of the Institution of Civil Engineers

Fig. 3. Vyrnwy Dam and Straining Tower. When completed in 1888 this was the largest artificial reservoir in Europe.
G. M. Binnie Collection, by Courtesy of the Institution of Civil Engineers

GEORGE DEACON (1843-1909) AND THE VYRNWY WORKS

In this comprehensive report[56] Deacon also gave full details of the geology of the site, of the construction of the dam in Cyclopean rubble concrete, and of his analysis of the stresses in the dam. At that time it was usual to consider the loads acting on horizontal sections, and it was possible to calculate the stresses only on the assumption that their distribution was linear. On that basis Deacon calculated that there were no tensions, and that the maximum compressive stress was 8.7 ton/ft^2 with the reservoir empty and 6.36 ton/ft^2 with the reservoir full. He showed how the real stresses would vary from these. Numerous specimens of concrete, rock and mortar had been tested and had shown that a safe compressive stress was 150 ton/ft^2. The stresses in the dam were calculated by computer by the author and it was found that there were no tensions, and that the maximum compression was 8.7 ton/ft^2 with the reservoir empty and 7.3 ton/ft^2 with the reservoir full.

The Vyrnwy Dam (Fig. 1) was quite a pioneer among masonry dams: it was the first to have underdrains, the first to be designed on reasonable scientific principles, the first high masonry dam in Great Britain and it impounded the largest artificial reservoir in Europe. The celebrated failure of the Bouzey Dam occurred later—in 1895—and it stimulated the first published theoretical studies, including work on uplift, of Levy in that same year.[57]

At the time when Deacon took over full responsibility for the Vyrnwy works, the dam had been constructed up to the top of the outlet tunnels,[58] and he began to reduce the section towards that originally agreed in 1881.[59] This provoked a further series of letters from Hawksley, who claimed that the dam would be unsafe.[60]

As a result of the doubts cast on the dam, the Council obtained four reports on it. The first was by Deacon and has already been discussed.[61] The second and third were by independent engineers: Major General Sir Andrew Clarke, Assoc.I.C.E.[62] and Russel Aitken, M.I.C.E.[63] A trial shaft had been sunk 22 ft deep from the drainage tunnel in the dam. Both reports commented on the very high standard of the masonry disclosed, samples of which gave a mean stress to failure of 289 ton/ft^2. Both considered that the section of the dam was very adequate. The fourth report was by Councillor Holme, Assoc.I.C.E.,[64] who was experienced as a contractor. He asserted that the people of Liverpool had in this dam a piece of masonry that could not be surpassed. His only complaint was that the masonry was costing £1 8*s* per cubic yard whereas a contractor could do it for £1 by increasing the proportion of small hearting stones, reducing the proportion of cement in the mortar and by using some lime with the cement. He reported that Deacon, whatever he may have been in the past, was then undoubtedly a most able engineer, thoroughly conversant with every detail.

The construction of the dam continued without further alarms; the river diversion valves were closed in 1888 and the level was 9 ft below the crest by the end of that year. Deacon reported that the leakage from the foundations and the dam was insignificant.[65] The author traversed the drainage tunnel in the 1920s, and remembers it to have been dry, although lime deposits gave evidence of earlier seepage.

AQUEDUCT

There are three structures on the aqueduct—all due to Deacon—which are worth mention: the Straining Tower, the Norton Tower and the Mersey Crossing.

Straining Tower. The Straining Tower (Fig.4) stands in deep water, some distance from the dam, and is the point of draw-off from the reservoir. Deacon had always been concerned about the impact of engineering on the environment,[66] and the Straining Tower, which is in the style of a Rhine Castle, is one of his most successful creations. It is built of concrete with stone facings, and it houses hydraulic machinery by which water may be drawn from any depth and by which the fine gauze strainers may be lifted from the outlet pipes and washed. The *tour de force* is that the tower is prestressed, at the springing of the dome which supports the gallery and roof, by plough-steel wire wound under tension into a groove in the concrete. To achieve this, Deacon designed a machine which was driven round the tower, and which carried a die through which the wire was drawn.[67] The same process was employed much later by Preload Ltd.[68] Reinforced concrete was not introduced until 1897.[68]

Norton Tower. Norton Hill is three miles south-east of Runcorn, but as the hydraulic gradient in the aqueduct is more than 100 ft higher than the top of the hill, the balancing tank had to be placed on a substantial tower—the Norton Tower (Fig. 5). Deacon succeeded again with a most impressive

GEORGE DEACON (1843–1909) AND THE VYRNWY WORKS

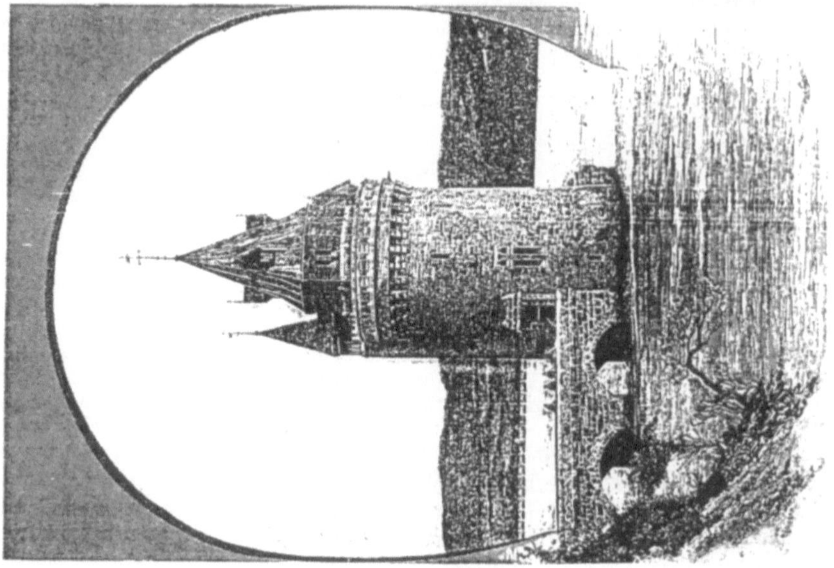

Fig. 4. The Vyrnwy Straining Tower is a good example of Deacon's concern to make his works suit their environment.

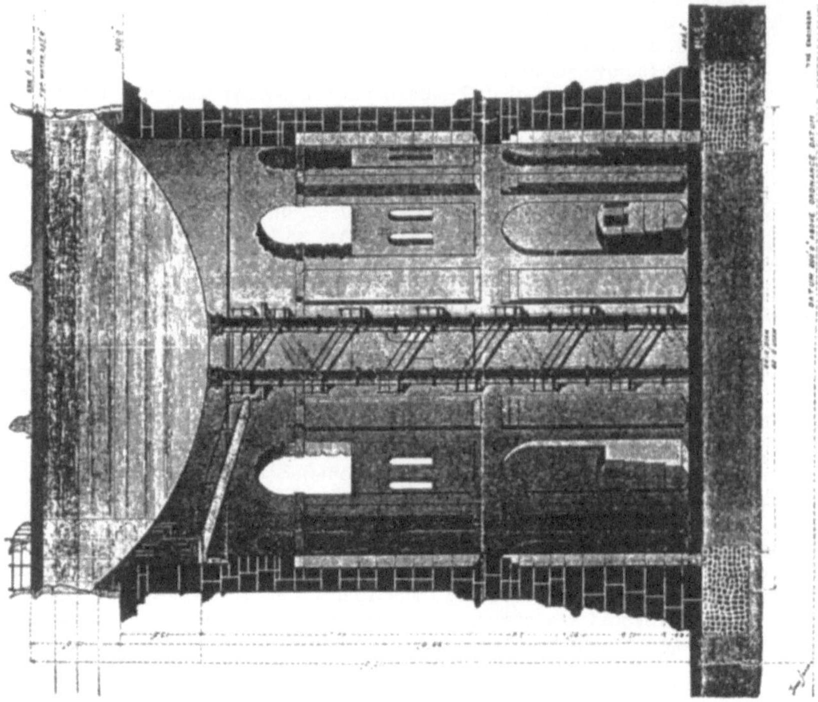

Fig. 5. The Norton Balancing Tank supports 2600 tons of water at a height of 110 ft. The bottom of the tank spans 76 ft, without intermediate support, yet it is only $\frac{5}{8}$ in thick.

structure, constructed in local red sandstone, to a Roman Doric design. The balancing tank which it supports is 80 ft in diameter by 29 ft deep in the centre; it holds 600,000 gallons. The upper cylindrical portion of the tank is 10 ft 6 in deep and is formed of flanged cast-iron plates. The bottom of the tank is suspended in pure tension from a hoop compression member formed of riveted steel plates, which is supported on rockers.

The remarkable feature is that the bottom of the tank, which spans 76 ft without intermediate support and which carries 2600 tons of water, is composed of riveted steel plates which are only ⅝ in thick.[70] Deacon achieved this apparently impossible feat by shaping the bottom to a carefully chosen curve. It is theoretically possible to do this so that both circumferential and azimuthal stresses are the same at every point, thus achieving an ideal tank of constant strength. The required curve is considerably more complex than a catenary; its theoretical derivation is given by Timoshenko.[71] In practice Deacon chose to make the bottom of the tank in the form of a segment of a sphere of radius 45 ft 9 in.[72] This radius, in conjunction with the span and the depth of water, gave a near approximation to the ideal. In the ideal case all the stresses would have been 1.9 ton/in^2 tension. As built, the stresses at the centre are both 2.4 ton/in^2 tension, and at the periphery the radial stress is 1.7 ton/in^2 tension and the hoop stress is zero. All these stresses are perfectly safe, even allowing for the riveted joints.

One does not know whether to admire Deacon more for his theoretical ability or for his supreme self-confidence in adopting a design which at first sight seems completely inadequate. The riveted steel bottom was fabricated at Warrington by Messrs. Pearson & Knowles.[73] After the design was complete and construction almost finished, it was found that a similar design had been studied in Germany.[74]

Mersey Crossing. The Mersey Crossing carries the aqueduct across the River Mersey between Warrington and Runcorn. It eventually proved to be the most difficult and expensive work on the whole aqueduct. The Parliamentary plans showed the pipes buried a few feet below the bed of the river, but the Board of Trade subsequently insisted that the pipes should not be higher than −20 ft OD, i.e. approximately 40 ft below spring tide level. The width of the river is about 800 ft. This change in plan meant that instead of the pipes being laid in a trench dredged in the river bed, it was necessary to drive a tunnel to contain them.

Deacon sank four boreholes, apparently on the south bank, from which he found that there was peat, sand and gravel down to about 60 ft, then 40 ft of boulder clay, with new red sandstone below.[75] This information was incomplete, and subsequent experience showed that on the north bank there was 80 ft of the soft clays, peats, etc. and then 33 ft of glacial sandy clay over the sandstone.[76]

Deacon thought it best to tunnel in the clay, and let a contract in April 1888 to Messrs. Monk & Newell of Bootle for completion in 12 months. The specification recommended a shield, but did not call for compressed air.[77] The contractors sank a shaft on the south side, 10 ft 9 in in diameter and lined with cast-iron segments to a depth of 84 ft 9 in. They were obliged to use dangerously high air pressures of 2.5-3 atmospheres to complete it. They managed to drive and line a tunnel 10 ft in diameter, without a shield, for a distance of 57 ft from this shaft. At this point the timber heading gave way, with an inrush of sand and water. This was in October 1889.[78] By this time the contractors had also sunk the north shaft to a depth of more than 82 ft 9 in. It would have been evident from the strata disclosed in the latter shaft that the tunnel must inevitably run out of the clay into the sand, as it in fact did. The contractors declined to continue, and the contract was terminated in December 1889.

The situation was serious. The Vyrnwy reservoir was full and the aqueduct substantially complete, apart from the Mersey Crossing, and there was a shortage of water in the city. In January 1890 a contract was signed with Messrs. Cochrane of London for £14,000 to drive the tunnel at a higher elevation, with its invert 52 ft below spring tide level.[79] The first task was to fill in the bottom portions of the existing shafts; then they were belled out at tunnel level to provide enough room for the insertion and removal of a shield, and special iron segments were inserted for the junction of the shaft and the tunnel.

The driving of the shield from the north shaft ran into immediate trouble as the previous contractors had left much debris, including iron tunnel segments and timber, around the shaft. The 10 ft dia. iron-lined tunnel was then in the alluvial deposits of silt, sand and gravel, and the shield proved to be too weak. After driving only about 182 ft, the upper part of the cutting edge buckled inwards for about a quarter of the circumference to a maximum extent of 1 ft 3 in and there was a longitudinal crack near the bottom of the skin, extending from the tail nearly to the diaphragm. The second contractor objected to continue the work in August 1891.[80]

The situation was now very serious, and some members of the City Council asserted that new powers would have to be obtained to cross the Mersey elsewhere. However, in October 1891 Sir Benjamin Baker recommended that the present shield should be repaired, and that a better shield should be designed in case it was required.[81] On the same date Deacon reported that he believed that the existing shield could suffice: 'Nothing remains but to push on with due deliberation, unslackened energy and full confidence.'[82]

A new agreement was signed in November 1891 with Messrs. Cochrane to continue the work at prime cost, under Deacon's direction. A. H. Cochrane continued in charge of men and materials, and A. W. Brightmore (Deacon's pupil) continued as Resident Engineer. It took about 7 weeks to repair the shield, which was strengthened by the insertion of a ring of cast-iron segments, hard up against the skin, just behind the diaphragm. The lower half of the diaphragm was cut away, and a bird fountain was provided.[83]

This tunnel was the first ever to be driven with a shield and compressed air in open water-bearing strata, but Deacon had correctly analysed the problems and the way to overcome them. The two main hazards are a 'blow', which is an uncontrolled exit of air at the top of the face; this will generally provoke a 'run', which is an uncontrolled entry of ground and water at the bottom of the face. The face is normally kept close-timbered with horizontal boards, only one being removed at a time in order to advance the face in steps of 1 ft 6 in (the width of the cast-iron lining segments).

The problem is that the ground-water pressure is naturally greater at the bottom than at the top of the face, while the air pressure is uniform. Deacon saw that the only way to succeed was to keep the air pressure approximately equal to the water pressure at the top (which was constantly varying in accordance with the tide), and to allow water to enter the tunnel at the invert, but to provide a restricted entry so that it could not bring too much ground with it. Then the frictional loss of head, due to the flow of the water through the ground, would approximately reduce its pressure to that of the air.[84] The water could be readily removed with a snorer.[85]

This technique succeeded, and the driving continued for 6 weeks, when it was found that the shield had an uncontrollable tendency to dive. The reason was that the existing longitudinal crack in the tail of the skin had extended and, just behind the diaphragm, a circumferential crack had developed. A portion of skin had turned down and was ploughing into the ground. The damaged portion of skin was 6 ft 6 in long by 9 ft in circumference. It was cut out and replaced by new plates bolted to the cast-iron ring girder. This took 12 days. Driving continued and the south shaft was reached on 22 March 1892.[86]

The author was engaged in driving the second identical Mersey Crossing in 1937, when the utmost care was required in the extremely difficult ground. The best rate of advance in the second tunnel was 9 ft per day. Deacon was prepared to take risks to complete the aqueduct as soon as possible; he achieved 12 ft 6 in per day at the expense of at least two runs, one of which flooded the tunnel, fortunately without loss of life.[87]

CONCLUSION

Deacon left the salaried service of Liverpool in December 1890, but was retained as Consulting Engineer.[88] He established an office at 16 Great George Street, Westminster,[89] where he built up a considerable practice. In 1903 he received the degree of LL.D. from the University of Glasgow of which he was extremely proud.

The iron determination which had served Deacon so well on the Vyrnwy works degenerated in old age into a tendency to fly into excessive rage when things went wrong.[90] He died very suddenly in his office on 17 June 1909.

Although Deacon had been friendly with Sir Alexander Binnie, the formal link did not occur until after his death, when his nephew Martin inherited the practice and went into partnership under the style Binnie, Son & Deacon.[91]

GEORGE DEACON (1843-1909) AND THE VYRNWY WORKS

ACKNOWLEDGEMENTS

The author is most grateful for assistance received from the late Mr. G. M. Binnie, F.R.S., formerly Senior Partner of Binnie & Partners, from Mr. R. J. Baty, M.I.C.E., Principal Engineer, North West Water Authority and from Dr. R. T. Smith of the Patent Office.

APPENDIX

An interesting postscript to the story of Deacon and the Vyrnwy works was started in 1904 by the publication of a memoir 'On Some Disregarded Points in the Stability of Masonry Dams' by L. W. Atcherley and Professor Karl Pearson.[92] This postulated—very reasonably—that, if engineers were accustomed to consider the forces acting on horizontal sections, then they should do the same on vertical sections. The memoir went on to show how a horizontal tension could theoretically occur near the air face, at the bottom of the dam. Some crude experiments with wooden models supported their thesis that, 'The current theory of the stability of dams is both theroretically and experimentally erroneous.'

This memoir caused intense alarm and distress among civil engineering circles, as was evidenced by the decision in March 1905 to postpone, if not abandon, the raising of the Aswan Dam.[93] A lively correspondence ensued in the columns of *Engineering* in which Professor Unwin played a leading part, but it seemed impossible to prove that Pearson was wrong. On 31 January 1908 three papers were presented for discussion at the Institution of Civil Engineers. The first, by Sir John Ottley and A. W. Brightmore, of whom the latter had been Deacon's pupil and Resident Engineer on the Mersey Crossing, described experiments on plasticine models.[94]

The second paper, by J. S. Wilson and W. Gore,[95] was an excellent account of experiments to determine stresses in dams by means of india-rubber models, which represented a section of the dam with its foundations. The rubber was 1 in thick, the height of the model dam was about 6.25 in and the depth of its foundation was about 4 in. The section of the model was divided into rectangular blocks, and lines defining these were accurately drawn on each side of the model. The rubber was carefully calibrated.

The thrust of the water was reproduced by plates pulled against the face by cords with weights, a strip of soft rubber separating the plates from the model. The self-weight was applied via a hard steel wire passing through the centre of each block, from which weights were hung by cords. The weights were chosen to magnify all the densities by a factor of 40, as it had been established that the resulting strains were less than the limit of proportionality of the rubber.

The strains were measured by photography on to glass plates 13 in by 8 in. The exposures were made by burning magnesium wire, which was kept moving to avoid shadows from the cords. After the model had been loaded for 24 hours a photograph was taken and the loads were removed; after a further 48 hours another photograph was taken. Lengths on the plates were measured to 1/1000 in to determine direct strains, and rotational strains were also measured; thus direct and shear stresses could be calculated.

The usual elementary equations relating stress to strain in an elastic body were applied to determine the principal stresses. The results were presented as ellipses of stress, in which the lengths of the major and minor axes represent the principal stresses to scale, and any other dimension represents the stress in that direction. The authors were well aware that Poisson's ratio for rubber is 0.5, as checked experimentally, whereas for masonry it is about 0.25, and they warned that the experimental results could not be applied directly to a real structure.

The conclusions were that tension was not observed in any part of the dam, but there was a horizontal tension in the foundation rock immediately upstream of the water face. It was shown that the maximum compressive stresses were not much different from those found by the old hypothesis of uniformly varying stress, but that their locations were different.

This paper received high praise from those present because it convincingly and finally removed all the doubts which Pearson had sown. Deacon drew attention to the remarkable ingenuity by which the virtual density of the model was increased. The use of ellipses of stress was, he believed, quite new. Of supreme importance was the tension at the upstream toe. It was necessary to consider the foundations with the dam.

GEORGE DEACON (1843-1909) AND THE VYRNWY WORKS

A remarkable fact—not, of course, recorded in the printed proceedings—was that Gore, joint author of this famous paper, worked in Deacon's office as his chief engineering assistant.[96, 97] One wonders to what extent Deacon's favourable comments, expressed in the discussion of the paper, stemmed, directly or indirectly, from his own intellect and from his own pocket. A comprehensive account of the whole controversy has been given by Professor Pippard.[98]

NOTES AND REFERENCES

1. G. M. Binnie, *Early Victorian Water Engineers* (Thomas Telford, London, 1981), p. 140.
2. *Porcupine*, Vol. **14**, 17 Aug. (1872), p. 307.
3. G. F. Deacon, "Report as to Water Supply", *Liverpool Water Supply Pamphlets*, Vol. **1** (1874), p. 8.
4. G. F. Deacon, "Report as to Supply of Water from Haweswater", ibid., Vol. **2** (1877).
5. G. F. Deacon, "Report as to Supply of Water from Vyrnwy", ibid., Vol. **2** (1877).
6. G. F. Deacon, "Supplemental Report on Vyrnwy", ibid., Vol. **2** (1878).
7. T. Hawksley and J. F. L. T. Bateman, ibid., Vol. **2** (1878).
8. *Liverpool Mercury*, 15 Nov. (1881).
9. T. Hawksley, Letter to Chairman, 1 Apr. (1879) op. cit. (3), Vol. **3**, Item 13.
10. Sir W. B. Forwood, "Draft Alternative Report", op. cit. (3), Vol. **3**, Item 14.
11. G. F. Deacon, op. cit. (3), Vol. **3**, Item 13.
12. *Proceedings of Liverpool City Council* 1878-79, p. 187.
13. A. Bower, op. cit. (8), 4 June (1885).
14. A. Bower, op. cit. (8), 19 Oct. (1885).
15. Op. cit. (12), 1878-79, pp. 226, 249.
16. G. F. Deacon, ibid., 1884-85, pp. 987-994.
17. G. F. Deacon, op. cit. (5), Vol. **2** (1877).
18. G. F. Deacon, "Impounding Reservoirs", *Proceedings of the Institution of Civil Engineers*, Vol. **115** (1893), p. 112.
19. A. Bower, op. cit. (8), 2 Apr. (1885).
20. Op. cit. (8), 2 Apr. (1885).
21. Op. cit. (8), 2 June (1881).
22. G. M. Binnie, op. cit. (1), p. 39.
23. P. G. Viall, *Binnie News*, Nov. (1971), p. 29.
24. T. Duncanson, "A Sketch of the Vyrnwy Works", *Proceedings of Liverpool Engineering Society*, (1892), p. 49.
25. A. Bower, op. cit. (8), 4 June (1885).
26. Op. cit. (8), 4 June (1885).
27. Ibid.
28. A. Bower, "Report on Circumstances of Engineers", op. cit. (8), 19 Oct. (1885).
29. Ibid.
30. G. M. Binnie, op. cit. (1), p. 147.
31. J. Mansergh, "The Vyrnwy Works for Liverpool", *Proceedings of the Institution of Civil Engineers*, Vol. **126** (1896), p. 69.
32. G. F. Deacon, ibid., pp. 31-36.
33. T. Hawksley, Letter to Bower, 3 Nov 1881, op. cit. (8), 21 Oct. (1885).
34. T. Hawksley, Letter to Bower, 1 Mar 1883, Ibid.
35. T. Hawksley, Letter to Deacon, 5 Aug 1884, Ibid.
36. T. Hawksley, Letter to Bower, 10 Dec 1881, op. cit. (8), 19 Oct. (1885).
37. G. F. Deacon, op. cit. (18).
38. Op. cit. (8), 12 Mar. (1885).
39. Op. cit. (8), 5 Mar. (1885).
40. Op. cit. (12), (1884-85), pp. 997-1009.
41. Ibid., pp. 997-1009.

GEORGE DEACON (1843-1909) AND THE VYRNWY WORKS

42. Ibid., pp. 987-994.
43. Op. cit. (8), 4 June (1885).
44. Ibid.
45. Ibid.
46. Ibid., 21 Oct. (1885).
47. Ibid., 21 Oct. (1885).
48. Ibid.
49. Ibid., 19 Oct. (1885).
50. Ibid., 19 Oct. (1885).
51. Op. cit. (12), (1885-86), p. 76.
52. W. J. M. Rankine, "Report on Masonry Dams", *Engineer*, Vol. **33** (1872), pp. 1, 2.
53. N. Smith, *History of Dams* (Citadel Press, New Jersey, 1972), p. 203.
54. G. F. Deacon, op. cit. (31), p. 30.
55. C. Hawksley, "The Vyrnwy Works for Liverpool", op. cit. (31), p. 74.
56. G. F. Deacon, "Report as to Masonry Dam", op. cit. (12), (1885-86), p. 695.
57. N. Smith, op. cit. (53), p. 205.
58. C. Hawksley, op. cit. (31), p. 79.
59. G. F. Deacon, op. cit. (31), p. 110.
60. Op. cit. (12), (1884-85), p. 312.
61. G. F. Deacon, op. cit. (56).
62. Sir A. Clarke, "Report on Vyrnwy Dam", op. cit. (12), (1885-86), p. 1195.
63. R. Aitkin, "Report on Vyrnwy Dam", op. cit. (12), (1885-86), p. 1233.
64 A. H. Holme, "Report on Vyrnwy Dam", op. cit. (12), (1885-86), p. 1259.
65. G. F. Deacon, "Progress Report on Vyrnwy Works", op. cit. (12), (1888-89), p. 1443.
66. G. F. Deacon, *The Aesthetics of Construction* (Liverpool Architectural and Archaeological Society, 1869).
67. G. F. Deacon, op. cit. (18), Vol. **126**, 1896, p. 49.
68. The early history of prestressing applied to concrete is poorly documented, but it is certain that Deacon was one of the first, if not the very first, to attempt it. It is generally considered that it was necessary to wait until the 1930s for the development of strong enough steels to make prestressing a practical proposition, as applied by Freyssinet. The plough steel wire used by Deacon probably had an ultimate strength of over 100 ton/sq. in but Deacon does not state at what tension he wrapped it round the concrete and so it is not possible to say whether or not it could still have much influence. Deacon's method of prestressing was first applied by the Preload Company in the 1930s; it is still used by their successors Prestensioned Structures Ltd.
69. P. Cusack, "François Hennebique". *Transactions of the Newcomen Society*, Vol. **56**, (1984-85), p. 75.
70. See figure bottom right. "Vyrnwy Aqueduct—the Norton Tower—details", *The Engineer*, 15 July (1892), p. 15.
71. S. P. Timoshenko, *Theory of Plates and Shells* (McGraw-Hill, 1959), p. 444.
72. G. F. Deacon, op. cit. (12), (1887-88), p. 1597.
73. G. Farren, op. cit. (24).
74. G. F. Deacon, op. cit. (67), p. 62.
75. G. F. Deacon, op. cit. (12), (1887-88), p. 1597.
76. R. J. Baty, *Redevelopment of Aqueduct Crossing of Mersey* (North West Water Authority), 1983.
77. G. F. Deacon, op. cit. (12), (1887-88), p. 1597.
78. G. F. Deacon, op. cit. (12), (1888-89), p. 1443.
79. G. F. Deacon, "City and S. London Railway", op. cit. (18), Vol **123** (1893), p. 100.
80. Ibid., p. 101.
81. Sir Benjamin Baker, "Report on Mersey Crossing", op. cit. (12), 1890-91.
82. G. F. Deacon, ibid.
83. A bird fountain is a subsidiary diaphragm about 2 ft behind the main one, extending from the bottom of the skin to a level higher than the top of the opening in the main diaphragm. If the air pressure is greater than the pressure of water in the ground at mid-height, then no more ground and water can enter than is sufficient to fill the bird fountain.

GEORGE DEACON (1843-1909) AND THE VYRNWY WORKS

84. G. F. Deacon, "Shield Tunnelling in Loose Ground", *Report British Association for the Advancement of Science* (1892), p. 532.

85. A *snorer* is a pipe with a small bore which runs from the face to the shaft, up through the air deck, and discharges above highest tide level. If the pipe is full of water the air pressure is not sufficient to lift the water up the shaft. To succeed the end of the pipe should be kept only half submerged, so that a mixture of air and water is discharged.

86. G. F. Deacon, op. cit. (18), Vol. **123** (1893), p. 104.

87. Ibid., p. 103.

88. Op. cit. (12), (1889-90), p. 146.

89. P. G. Viall, op. cit. (23), Nov. (1971), p. 29.

90. L. Pearson, op. cit. (23), May (1968), p. 34.

91. Ibid., p. 37.

92. L. W. Atcherley and Karl Pearson, *Drapers' Company Research Memoir*, Technical Series 2 (1904).

93. *Engineering*, Vol. **79** (1905), p. 414.

94. Sir J. Ottley and A. W. Brightmore, "Experimental Investigation of the Stresses in Masonry Dams", op. cit. (18), Vol. **172** (1907-08), Part 2, p. 89.

95. J. S. Wilson and W. Gore, "Stresses in Dams: an Experimental Investigation by means of India-Rubber Models", op. cit. (18), Vol. **172** (1907-08), Part 2, p. 107.

96. P. G. Viall, op. cit. (23), Nov. (1971), p. 29.

97. G. F. Deacon, "Report as to Vyrnwy Aqueduct", 1898. Liverpool Library, H 352-6.

98. A. J. S. Pippard, Unwin Memorial Lecture, *Journal of the Institution of Civil Engineers*, Vol. **33** (1949-50), pp. 265-286.

DISCUSSION

Mr. Dudbridge asked about the surveying work at Vyrnwy. **Mr. Davidson** said that no Ordnance Survey map was available for the route of the aqueduct. The whole 68 mile route had to be surveyed, as well as the site of the dam itself. Adequate instruments were available for the purpose, but no doubt the most difficult surveying task was in the tunnel—100 ft below the surface, down a narrow shaft obstructed by an impermeable air deck.

Dr. N. A. F. Smith pointed out that, contrary to some statements, he had never suggested that an influence on the design of Vyrnwy Dam had been the failure in 1881 of Habra Dam in Algeria. Of course it was possible. Habra Dam was well known and information about the failure might have reached Hawksley and Deacon in time to prompt reconsideration of the design of Vyrnwy Dam, in particular with regard to uplift. Nevertheless, the nature of the Habra failure needed to be borne in mind.

Fundamentally Habra Dam was destroyed by a massive hydraulic overload. French engineers in North Africa had seriously underestimated the water flows which could build up in the *sigs* and *wadis* after exceptionally heavy rain. It was said that on the night in question the total rainfall in the catchment area was about three times the capacity of the reservoir. In these conditions the spillway was totally inadequate and the dam was overtopped. The superelevation of the water level above that intended was nearly 4 m. In a dam that was not well proportioned in the first place this excessive depth of water would certainly have violated the middle-third rule near the top of the dam and, if tension cracks developed, water entering them would have added a vertical uplift force to the already excessive horizontal water pressure. Whether or not Hawksley and Deacon interpreted the Habra failure in this way, even assuming that they had heard of it, is not known. However, in view of the profile of Vyrnwy Dam it would be extraordinary if the engineers were thinking in terms of tension at the water face with consequent uplift forces developing in the cracks. What Hawksley and Deacon, and other designers, were worried about and anxious to quantify was uplift in and under dams due to the percolation of water under everyday conditions. When Deacon spoke to the Institution of Civil Engineers in 1896 and observed, 'In 1881 there was probably no high masonry dam in Europe so far watertight that an English engineer would take credit for its construction' he was probably referring to Gileppe Dam in Belgium which had been designed in the 1860s and finished in 1875. The especially interesting feature of Gileppe Dam was the attempt, however rudimentary, to design against uplift. The engineer—M. Bidaut—had assumed a specific gravity for masonry of 1.3 rather than 2.3 but, notwithstanding this allowance and the resulting massive profile, Gileppe Dam was notorious for leaking like a sieve. In 1877—four years before the design of Vyrnwy Dam was begun—a full account of Gileppe Dam had been published in French. Earlier in the 19th century Minard, in his *Cours de Construction* of 1841, discussed leakage through masonry dams including Lampy Dam built for the Canal du Midi in 1777-81. Percolation through Lampy Dam was also mentioned by A. F. Andreossy in 1804 who said that quantities of lime had been thrown into the reservoir close to the face of the dam in the hope that pores and cracks would be sealed. Indeed Andreossy was satisfied that the technique worked very well.

Evidently the observation that water could percolate through a masonry dam goes back at least a century before the design of Vyrnwy Dam. It was an obscure issue—and not to be confused with uplift contributing to a failure mechanism which has already been induced by other factors—which Hawksley and Deacon were ready to try to deal with. Some attempt had been made to take account of uplift in two earlier dams: Gileppe in Belgium and Alfeld in Germany, although details of the latter were not published until 1889.

The author replied that Dr. Smith was completely correct to distinguish percolation from uplift. There were indeed many leaky masonry dams, and these might be at risk either from leaching of cementitious material, or because the product of the pore surface area times the water pressure in the pores amounted to an overturning moment greater than could be resisted. Some early designers, such as those at Gileppe, simply deducted one from the specific gravity of the masonry. This was a rather confused notion, which assumed that the masonry was buoyed up by being immersed in water. This did not agree with reality, nor with the concept of uplift *per se*. Hawksley fell into a similar trap, but came out of it successfully after some unnecessary complications. Deacon saw the situation clearly and correctly.

In his report to the Liverpool Corporation, Deacon considered the case where both dam and foundations were impervious; then he considered the case where the dam was impervious, but the foundations were pervious (he could be sure of the dam, he said, because he had built it: he could not say as much of the foundations . . .); he did not consider in detail, as would be done in the present day, the forces exerted throughout the whole breadth and length of the dam by water in pores and crevices.

Dr. N. A. F. Smith demonstrated two scale models in slate of slices of Vyrnwy Dam. Made in about 1881, they appeared to be designed as aids to calculation.

Mr. G. M. Binnie said that Deacon was a true Victorian martinet; not only were his office staff required to sign in and out at all times: he even kept the washroom key in his own office and people who borrowed it for more than ten minutes were in trouble. To one unsolicited enquirer he had once responded, 'Two and two make four; and five make nine: you mind your own business, and I'll mind mine. I am, Sir, your obedient Servant . . .

Mr. R. J. M. Sutherland (President) asked whether what had been published about Deacon's prestressing work at Vyrnwy invalidated Messrs. Preload's patent. The author said that Preload never did get a patent, whether because of Deacon's prior work or not. Deacon's concrete, which was still visible, was still perfect.

Dr. D. P. Smith proposed a vote of thanks to the author which was carried by acclaim.

12

Engineering in the progressive era: a new look at Frederick Haynes Newell and the U.S. Reclamation Service

Donald C. Jackson

> The passage of the national irrigation law was one of the great steps not only in the forward progress of the United States but of all mankind. [THEODORE ROOSEVELT, 1903][1]

> The cumulative effect of the barrage of favorable publicity for reclamation has been the conditioning of the public mind to an unquestioning acceptance . . . that federal reclamation is a great and unmixed good, and that it is the major if not the exclusive factor in irrigated agriculture. [STANLEY DAVISON, 1952][2]

The American West constitutes an environment well suited to transformation through the agency of technology. As Walter Prescott Webb illustrated in his classic 1931 book *The Great Plains,* the expanse of America lying west of the ninety-eighth meridian (a line running through the center of Kansas) is largely defined by the lack of precipitation.[3] Water is a key factor in the economic and social development of the arid region, and water control technologies are

DR. JACKSON is assistant professor of history at Lafayette College. He gratefully acknowledges comments from Donald J. Pisani, Jack Cell, and several *Technology and Culture* referees that helped make this a better article. Some of the research underlying the article was completed as part of a contract between FraserDESIGN and the Bureau of Reclamation for documentation of historic resources affected by the Central Arizona Project; however, the opinions expressed are the author's and do not necessarily reflect the views of FraserDESIGN, the Bureau of Reclamation, or any agency employees.

[1]Telegram from Theodore Roosevelt to National Irrigation Congress, September 15, 1903, reprinted in F. H. Newell, comp., *Proceedings of the First Conference of Engineers of the Reclamation Service* (Washington, D.C., 1904), p. 121.

[2]Stanley Davison, *The Leadership of the Reclamation Movement, 1875–1902* (New York, 1979), pp. 32–33 (reprint of Ph.D. diss., University of California, 1952).

[3]Walter Prescott Webb, *The Great Plains* (Boston, 1931), pp. 3–44.

especially important because moisture is not evenly distributed over the landscape in either density or seasonality.[4] Torrential downpours might dump millions of gallons of water on the desert, but these storms can be spaced years apart during periods of severe drought. Most rivers depend on melting snowpack rather than rainfall to sustain their flow, and westerners have long recognized the importance of storage dams in capturing both torrential floods and spring "runoff" from snowcapped mountains.[5] Without dams, huge quantities of flood waters can be lost. If stored behind artificial barriers, however, this water can support sustained growth of agriculture, industry (e.g., mining and hydroelectric power), and municipalities.[6] Of these, irrigated agriculture (often called reclamation) historically constitutes the most widespread use of water in the West.[7]

Although irrigation in the West predates the era of European settlement, it started a new phase in 1847 when Mormon pioneers began diverting small creeks onto desert land near Utah's Great Salt Lake.[8] Other Anglo groups followed the Mormons and soon estab-

[4]Extensive references on western water history are provided in Lawrence B. Lee, *Reclaiming the American West: An Historiography and Guide* (Santa Barbara, Calif., 1980). Also see Donald J. Pisani, "Deep and Troubled Waters: A New Field of Western History," *New Mexico Historical Review* 63 (October 1988): 311–31; and Lawrence B. Lee, "Water Resource History: A New Field of Historiography?" *Pacific Historical Review* 57 (November 1988): 457–67. Notable recent books in the field include William L. Kahrl, *Water and Power: The Conflict over Los Angeles' Water Supply in the Owens Valley* (Berkeley, 1982); Donald J. Pisani, *From the Family Farm to Agribusiness: The Irrigation Crusade in California and the West, 1859–1931* (Berkeley, 1984); James Earl Sherow, *Watering the Valley: Development along the High Plains of the Arkansas River* (Topeka, Kans., 1990); Norris Hundley, *The Great Thirst: Californians and Water, 1770s–1990s* (Berkeley, 1992); and Donald J. Pisani, *To Reclaim a Divided West: Water, Law, and Public Policy, 1848–1902* (Albuquerque, N.M., 1992).

[5]The first major tract to discuss the importance of building storage dams in the West is John Wesley Powell's landmark *Report on the Lands of the Arid Region of the United States* (Washington, D.C., 1879). The importance of dam building in western economic growth is also reflected in James D. Schuyler, *Reservoirs for Irrigation, Water Power and Domestic Water Supply*, 2d ed. (New York, 1909).

[6]The pumping of subsurface water from underground aquifers also constitutes an important subtheme in western water history but will not be discussed in this article.

[7]In general, the words "irrigation" and "reclamation" are synonymous and refer to cultivation that depends on the artificial application of water to soil. Some authors use "reclamation" to connote large-scale projects, but in this article I have made no distinction between the two terms. The predominant use of western water for irrigation is noted in Fred Powledge, *Water: The Nature, Uses, and Future of Our Most Precious and Abused Resource* (New York, 1982), pp. 36–38, 114–15.

[8]For more on Mormon irrigation efforts, see Charles Brough, *Irrigation in Utah* (Baltimore, 1898); and George Thomas, *The Development of Institutions under Irrigation with Special Reference to Early Utah* (New York, 1920). Water use in the Spanish era is described in Michael C. Meyer, *Water in the Hispanic Southwest: A Social and Legal History,*

lished their own irrigation-based settlements.[9] In the late 1870s, John Wesley Powell, famed explorer of the Colorado River, submitted to Congress his *Report on the Lands of the Arid Region of the United States* and initiated what has become known as the "irrigation movement."[10] While eastern America grew ever more industrialized, supporters of this movement sought to open up arid public lands in the West to settlement by hard-working yeoman farmers.[11] In advocating irrigation-based colonies that could foster traditional rural values, the movement idealized the virtues of the owner-occupied family farm and, conversely, decried any commercial exploitation of the West's resources by "speculators" or wealthy ranch owners.

By the end of the 19th century irrigation enthusiasts confronted uncertainty over the amount of water and prime irrigable land that remained available in the West's public domain.[12] (See figs. 1, 2.) In response, Powell convinced Congress in 1888 to authorize an Irrigation Survey that would measure the streamflow of western rivers, survey potential reservoir sites, and assess what land was best suited for irrigation.[13] As director of the survey, Powell helped publicize the relationship between storage reservoirs and increased agricultural

1550–1850 (Tucson, Ariz., 1984). Early irrigation by Native Americans is described in R. Douglas Hurt, *Indian Agriculture in America: Prehistory to the Present* (Lawrence, Kans., 1987), pp. 19–26; while Hundley, pp. 16–24, discusses irrigation in California prior to Spanish settlement.

[9]For example, German farmers established the Anaheim Colony in southern California in the 1850s, while the Union Colony in Greeley, Colorado, was settled in the late 1860s. See Richard D. Bateman, "Anaheim Was an Oasis in the Wilderness," *Journal of the West* 4 (January 1965): 1–20; and David Boyd, *Greeley and the Union Colony* (Greeley, Colo., 1890). Early irrigation in California's Central Valley is described in Pisani, *From the Family Farm,* pp. 102–28.

[10]See Powell. This report was first submitted to Congress in 1878 (*H. Exec. Doc. 73,* 45th Cong., 2d sess.) and published for general distribution in 1879. During his career Powell was prominently associated with both the Smithsonian Institution and the U.S. Geological Survey. His life story is recounted in Wallace Stegner, *Beyond the Hundredth Meridian* (Boston, 1953).

[11]Analysis of this movement is provided in Davison; and Pisani, *To Reclaim a Divided West.* These discussions of Powell's work do not suffer from the hagiography of Stegner's account.

[12]Paul W. Gates, *History of Public Land Law Development* (Washington, D.C., 1968), provides an excellent review of federal laws and programs related to public land disbursement.

[13]For early treatments of this effort, see Everett Sterling, "The Powell Irrigation Survey, 1888–1893," *Mississippi Valley Historical Review* 27 (1940): 421–34; and John T. Ganoe, "The Origin of a National Reclamation Policy," *Mississippi Valley Historical Review* 18 (June 1931): 34–52. The best historical analysis of the Irrigation Survey and its demise in 1893 appears in Pisani, *To Reclaim a Divided West,* pp. 127–68.

FIG. 1.—Postcard view of irrigated California orange grove ca. 1900. Water from the laterals is released through small wooden gates into furrows that extend into the field. (Author's collection.)

production. In addition, he soon came to realize that practically all of the West's best irrigable land was already privately controlled. Powell revealed these findings at the 1893 International Irrigation Congress in Los Angeles, but, as Stanley Davison recounts, he received little thanks from congressional participants and was soon displaced from the movement's leadership by "the triumph of sentimentalism."[14]

Powell may have correctly understood patterns of land ownership in the West, but a vision of yeomen farmers gaining control of public land and upholding the values of American democracy proved too potent to die a quiet death. Its idealistic goals were perhaps rendered moot, but the irrigation movement's imagery and rhetoric survived. Within this cultural milieu, Frederick Haynes Newell came to advocate the creation of the U.S. Reclamation Service to direct the planning, financing, and construction of major reclamation projects in the seventeen western states. Authorized as

[14]Davison (n. 2 above), pp. 149–59, describes in detail Powell's treatment at the 1893 congress in Los Angeles; pp. 206–16 discuss his absence from the irrigation movement from 1894 until his death in 1902.

FIG. 2.—The basic elements of small-scale irrigation systems—a diversion dam and irrigation canal (center foreground) in Big Thompson Canyon, Colorado, ca. 1950. (Author's collection.)

part of the Department of the Interior in 1902, the agency represented a major Progressive Era program dedicated to solving the West's water supply problems.[15]

The term "progressive" and its relationship to early-20th-century reform is associated with myriad meanings that, at times, appear contradictory. But, as Daniel Rodgers explains, adherents of the loosely defined ideology spoke in "three languages of discontent," focusing on antimonopolism, social bonds and the social nature of human beings, and social efficiency.[16] As the Reclamation Service's first chief engineer (1902–7) and first director (1907–14), Newell combined elements of all three Progressive languages in advocating solutions to problems affecting western development. Above all, he drew on his background as an engineer to promote scientific analysis as a means of achieving efficiency in both technical and socioeconomic contexts. In this article, Newell's work as a reclamation engineer is examined within a historiographical

[15]An extensive bibliography of books, articles, and reports focusing on the early Reclamation Service is available in Institute for Government Research, *The U.S. Reclamation Service: Its History, Activities and Organization* (New York, 1919).

[16]Daniel Rodgers, "In Search of Progressivism," *Reviews in American History* 10 (December 1982): 113–32. This article provides a comprehensive overview of how historians have grappled with the issue of Progressivism.

framework and in the context of the Reclamation Service's reputation as a paragon of technological efficiency. As a result, the contributions of "progressive" engineering to western water development become clearer, while, conversely, the significance of Newell's rhetoric in advocating reform of the engineering profession becomes more problematic.

* * *

Born in Pennsylvania, Newell graduated in 1885 as a mining engineer from the Massachusetts Institute of Technology. Starting in 1888 he helped Powell administer the Irrigation Survey, and, after Congress discontinued the program in the early 1890s, Newell became the U.S. Geological Survey's chief hydrographer, with responsibility for recording the flow of America's rivers.[17] Allied with Gifford Pinchot, first director of the U.S. Forest Service, Newell became a leader in the self-styled "conservation movement" dedicated to protecting natural resources from crude corporate exploitation.[18] Proud of his association with America's scientific elite, Newell championed the virtues of research and planning and, in his own words, espoused "the need of careful and scientific study of natural resources, in the weighing of costs and benefits in utilizing these, viz., in the economics of their use."[19] Additionally, he advocated the scientific "study of hydro-economics" in which engineers would "explore far back into the fundamentals of many of the mathematical, physical and biological sciences. In their application in solving these problems engineering knowledge and skill are involved."[20] In *Conservation and the Gospel of Efficiency: The Progressive Conservation Movement, 1890–1920,* Samuel Hays comments on the larger significance of Newell and his views by noting, "Conservation above all was a scientific movement, and its role in history arises from the implications of science and technology in modern society . . . conservationists envisaged . . . a

[17]The best basic source of biographical information on Newell is his professional obituary: "Frederick Haynes Newell," *Transactions of the American Society of Civil Engineers* 98 (1933): 1597–1600.

[18]Newell's professional relationship with Gifford Pinchot is documented in Nelson McGeary, *Gifford Pinchot: Forester-Politician* (Princeton, 1960).

[19]Quote from Frederick Haynes Newell, *Water Resources: Present and Future Uses* (New Haven, Conn., 1920), p. 29. In the 1890s Newell served as the secretary of the National Geographic Society and the American Forestry Association. During this time he publicized western irrigation activities by authoring publications such as *Report of Agriculture by Irrigation in the Western Part of the United States at the Eleventh Census, 1890* (Washington, D.C., 1894), *Report of the Progress of the Division of Hydrography for the Calendar Years 1893 and 1894* (Washington, D.C., 1895), "Irrigation on the Great Plains," *Yearbook of the U.S. Department of Agriculture, 1896* (Washington, D.C., 1897): 176–96, and "Reservoir Surveys," *Irrigation Age* 15 (1899): 70–82.

[20]Newell, *Water Resources,* p. 34.

political system guided by the ideal of efficiency and dominated by the technicians who could best determine how to achieve it. . . . During the Roosevelt administration Newell became one of the architects of water policy and of the entire conservation movement."[21]

When Newell first became chief hydrographer for the Geological Survey, his mandate was limited and he could not aggressively promote a federal reclamation program. Within a few years, this began to change. Significantly, the disappointing efforts of men such as William Smythe (editor of the popular journal *Irrigation Age* and a leader of the irrigation movement in the post-Powell era) in sponsoring privately financed irrigation colonies during the 1890s gave impetus to government involvement in reclamation.[22] By the turn of the century, Newell was interacting with businessmen connected to western economic interests and proselytizing among the capital's political leadership.[23]

In 1893 Newell had been in Los Angeles when Powell reported on the paucity of prime irrigable land in the public domain; in fact, he

[21]Samuel Hays, *Conservation and the Gospel of Efficiency: The Progressive Conservation Movement, 1890–1920* (Cambridge, Mass., 1959), pp. 2–3, 5, 7. Hays further describes conservation as growing "out of the first hand experience of federal administrators and political leaders [who grappled with] problems of western economic growth, and more precisely with western water development."

[22]Smythe's career is discussed in Pisani, *To Reclaim a Divided West* (n. 4 above), pp. 235–51; and Davison (n. 2 above), pp. 136–204. Also see William E. Smythe, *The Conquest of Arid America,* 2d ed. (New York, 1905). In 1894 the U.S. Congress passed the Carey Act and authorized the transfer of up to a million acres of federal land to individual western states if they would make financial arrangements to irrigate this land. The Carey Act eventually fostered some significant projects (primarily in Idaho), but, in general terms, its impact was limited, and many perceived the program as a failure. Several good references to Carey Act projects are provided in Lee, *Reclaiming the American West* (n. 4 above).

[23]Aside from Pinchot, Newell's most valuable allies were George Maxwell, a former California lawyer who organized the National Irrigation Association, and Benjamin Fowler, a graduate of Yale who became a Phoenix rancher (presumably for health reasons) after a successful career as a New York publisher. Davison describes Maxwell as seeing "the need of extensive propaganda directed at members of Congress in behalf of favorable legislation, and for support he turned to the western industries [e.g., the railroads] which stood to benefit most from a boom in irrigated agriculture . . . from this time on [1897], Congress was under the pressure of this lobby to take some action in sponsoring reclamation" (p. 245). For more on Maxwell, see Pisani, *To Reclaim a Divided West,* pp. 285–94. Fowler's background is described in Karen Smith, *The Magnificent Experiment: Building the Salt River Reclamation Project, 1890–1917* (Tucson, Ariz., 1986), p. 16. In his diary entry for January 8, 1901, Newell reported that he gave an evening lecture at the home of Congressman Francis Newlands where he "showed lantern slides and talked irrigation." In attendance were sixteen guests including the secretary of the Department of Interior and several members of the U.S. Congress. See Frederick Haynes Newell Papers, Washington, D.C., Library of Congress, Manuscript Division.

even gave a speech to the Irrigation Congress echoing his mentor's remarks.[24] Despite his awareness of land ownership patterns in the West, in his 1902 book *Irrigation in the United States,* Newell projected a romantic image of yeoman farmers populating arid public lands: "Home-making is the aim of this book. . . . The dead and profitless deserts need only the magic touch of water to make arable lands that will afford farms and homes for the surplus people of our overcrowded Eastern cities, and for that endless procession of home-seekers filing through the Castle Garden. The national government, the owner of these arid lands, is the only power competent to carry this mighty enterprise to a successful conclusion."[25] Newell considered previously developed private and state-supported projects to be "strikingly meager" and "largely temporary in character" and asserted that "nearly all the land which is available for irrigation by individual effort or which affords profitable opportunities for comparatively small investments has been utilized."[26] As Lawrence Lee notes, "The predominant theme that ran through Newell's writings and Reclamation Service publications was that the West had given up on private enterprise and depended exclusively upon the federal government to build the large water storage and distribution works necessary for further irrigation advancement in arid America."[27]

In a quirk of history, plans for a federal reclamation program received a major boost after an assassin shot President McKinley in September 1901 and Theodore Roosevelt became president. Whereas McKinley possessed little interest in promoting government participation in internal economic affairs, Roosevelt embraced an activist approach to executive leadership. Pinchot and Roosevelt were good friends, and, as a result, Newell's and Pinchot's views concerning conservation and federal reclamation found explicit expression in Roosevelt's first message to Congress in December 1901.[28] In this

[24]Davison, pp. 150–51.

[25]Frederick Haynes Newell, *Irrigation in the United States* (New York, 1902), pp. 1, 406. The original edition of this book appeared before passage of the 1902 Reclamation Act; a second edition was published in 1906 that included a small amount of new material related to activities of the Reclamation Service.

[26]*First Annual Report of the Reclamation Service* (Washington, D.C., 1903), p. 60; and *Third Annual Report of the Reclamation Service* (Washington, D.C., 1905), p. 43.

[27]Lee, *Reclaiming the American West,* p. 17.

[28]See McGeary (n. 18 above), p. 54, for evidence that Roosevelt requested Pinchot and Newell to draft the conservation section of his first address to Congress. In later life Roosevelt declared "I first came in touch with him [Newell] when I was Governor [of New York], when I drew upon him for aid and advice in formulating the proper conservation policy for the State of New York. During the eight years that I was

manifesto, the President counseled legislators that "vast areas of public land [remain] which can be made available for homestead settlement, but only by reservoirs and mainline canals impracticable for private enterprise. These irrigation works should be built by the national government."[29]

Acting on Roosevelt's endorsement, in June 1902 Congress approved the National Reclamation Act and within weeks the Reclamation Service was organized as part of the U.S. Geological Survey.[30] The act placed proceeds from western federal land sales into a "Reclamation Fund" designed to finance construction of projects by the service; on completion of construction, farmers were to reimburse the federal government in ten annual payments that excluded interest costs. To prevent large landowners from receiving disproportional benefits, agency officials were to limit water disbursements to holdings of 160 acres or less (later expanded to 320 acres for husband and wife). Issues related to water rights would be determined by state and territorial law, but the Reclamation Service was given complete control over project construction, operation, and management.

Under Newell's leadership the early Reclamation Service built several large structures, including the Roosevelt Dam near Phoenix, Arizona, and planned several smaller irrigation developments, such as the Truckee-Carson Project in Nevada and the Garden City Project in Kansas. Some of these projects seemed to function successfully, while others, such as the Garden City Project, were abject failures.[31] However, practically all exceeded their original cost estimates and engendered enormous resentment among those responsible for repayment. After Teddy Roosevelt left the White House in early 1909, Newell lost political protection and the Reclamation Service entered a

---

President he was one of my right-hand men." Quotes taken from a December 22, 1914, letter partially reproduced in *Triangle Review* 5 (1915): 15. Available in the Newell Papers, Library of Congress.

[29]The portion of President Roosevelt's Message to Congress dated December 3, 1901, relating to federal reclamation is quoted in full in Newell, *Irrigation*, pp. 393–96.

[30]A complete transcript of the 1902 National Reclamation Act and all relevant legislative amendments through 1919 are included in Institute for Government Research (n. 15 above), pp. 105–24. The original act is legally referenced as "Act of June 17, 1902 (32 Stat., 388)." It is often called the Newlands Act in recognition of its cosponsor, Congressman Francis Newlands of Nevada. Political maneuvering involved in the act's passage is well described in Pisani, *To Reclaim a Divided West*, pp. 298–325.

[31]The failure of the Garden City Project is described in Sherow (n. 4 above), pp. 92–94.

prolonged period of public scrutiny.[32] In 1910 Congress provided $20 million to the Reclamation Fund to ward off insolvency and, not long afterward, Congress responded to western discontent by extending repayment schedules from ten to twenty years.[33]

The agency's fiscal shortcomings forced Newell's resignation in late 1914, but the fortunes of the service did not improve in the post-Newell era, and by 1923 it verged on complete collapse.[34] At that time, the agency was reorganized as the Bureau of Reclamation and project repayment obligations were allowed to extend for forty years or more.[35] Clearly, the chronic financial problems of both the Reclamation Service and the Bureau of Reclamation stand in stark contrast to Progressive ideals of efficiency associated with Newell. Even more striking, the notion of using federal resources to support family farms foundered as the bureau achieved reknown for supporting urban growth and corporate agribusiness.[36] Thus, in the wake of the New Deal, the bureau became famous for huge "multipurpose" projects such as Hoover Dam, a structure that generated hydroelectric power and provided water for greater Los Angeles and for large-scale irrigation tracts in California's Imperial Valley.[37] The ideal of the

[32]For example, see U.S. Congress, *Hearings before the Committee on Irrigation of Arid Lands of the House of Representatives*, 60th Cong., 2d sess. (1909).

[33]Repayment problems are discussed in Peggy Heim, "Financing the Federal Reclamation Program, 1902 to 1919: The Development of Repayment Policy" (Ph.D. diss., Columbia University, 1953); also see Ray P. Teele, *The Economics of Land Reclamation in the United States* (Chicago, 1927).

[34]The restructuring of the Reclamation Service into the Bureau of Reclamation is discussed in Donald Swain, *Federal Conservation Policy* (Berkeley, 1963), pp. 78–83.

[35]See William A. Warne, *The Bureau of Reclamation* (New York, 1973), pp. 62–68, for discussion of how the repayment system evolved from 1902 through the 1970s. Warne notes that costs for some bureau projects authorized in the 1920s are being reimbursed over a 104-year time span. Since 1939, most bureau projects stipulate forty-year repayment schedules that commence after a ten-year "grace period" following completion of construction. Early project repayment histories and future schedules are documented in U.S. Department of the Interior, *How Reclamation Pays: A Book of Project Repayment Histories and Repayment Schedules with History of Crop Production* (Washington, D.C., 1947).

[36]Warne, pp. 70–85. Written by a former Assistant Secretary for Water and Power in the Interior Department, this book sympathetically synopsizes the myriad problems associated with bureau enforcement of the 160-acre limitation.

[37]Ibid., p. 79. The bureau's 1933 contract with the Imperial Irrigation District in southern California stated that landowners within the district would not have to comply with the 160-acre limitation. In 1970 a federal court upheld this contract and "decided that the Imperial District would not have to enforce the 160-acre law on 800 landowners who had excess holdings." During the early 1980s, the Reagan administration obtained congressional approval to raise the landownership limit for bureau projects to over 900 acres.

family farm as primary beneficiary of federal reclamation faded into oblivion, but the reputation of the early Reclamation Service lived on in other ways.

Recently, many historians have learned about federal reclamation through polemics such as Marc Reisner's *Cadillac Desert* and Donald Worster's *Rivers of Empire*.[38] Though differing in tone and emphasis, both Reisner and Worster are concerned about the long-term environmental consequences of western water development and perceive the bureau as extremely significant in fostering regional growth. Concentrating on New Deal and post-World War II projects, the two authors readily accept the early Reclamation Service as being a technologically sophisticated dam-building organization that, as Worster ornately asserts, "astounded the world with its ingenuity, resourcefulness, mastery over nature, and grace of design."[39] Both Worster and Reisner are willing to lambast the service for social and financial failings but nonetheless believe that, in Reisner's words, "its dams were engineering marvels."[40]

Precedent for Reisner's and Worster's belief in the technical adeptness of the Reclamation Service can be found in Hunter Dupree's 1957 book *Science and the Federal Government*, which comments favorably on Newell and Pinchot in the context of early work by government scientists and engineers.[41] Following up on Dupree, Samuel Hays's 1959 *Conservation and the Gospel of Efficiency* projects an upbeat assessment of the Reclamation Service's accomplishments and suggests that contemporary criticisms of either Newell or the agency were generated by self-serving anticonservationists and/or greedy private landowners.[42] In a concluding chapter entitled "The West against Itself," Hays acknowledges that the Reclamation Service suffered

[38]Marc Reisner, *Cadillac Desert: The American West and Its Disappearing Water* (New York, 1986); and Donald Worster, *Rivers of Empire: Water, Aridity and the Growth of the American West* (New York, 1985). For historians interested in hydraulic engineering, there are several widely available publications focusing on the technical work of the service bureau. For example, see Ellis Armstrong et al., eds., *History of Public Works in the United States, 1776–1976* (Chicago, 1976); and Michael Robinson, *Water for the West* (Chicago, 1979). In terms of structural technology, the Reclamation Service and the Bureau of Reclamation receive prominent notice in Carl Condit, *American Building Art: The Twentieth Century* (New York, 1961).

[39]Worster, p. 173.

[40]Reisner, pp. 118–23. Worster, pp. 174–81, also criticizes the Reclamation Service's financial problems and ineffectiveness in enforcing the 160-acre limitation.

[41]Hunter Dupree, *Science in the Federal Government: A History of Policies and Activities to 1940* (Cambridge, Mass., 1957).

[42]Hays (n. 21 above), p. 249, avers that the Reclamation Service "thought in terms of maximum development of all western rivers, while the people of each local area considered their own welfare first."

financial difficulties but heralds Newell as a talented administrator who suffered at the hands of politicians once Roosevelt left the presidency.[43]

Hays established a form of "conservation orthodoxy" that Edwin Layton subsequently adopted in his 1971 book *The Revolt of the Engineers.* Layton is not primarily interested in Newell's work as a reclamation bureaucrat but instead examines his subsequent career as founder/president of the American Association of Engineers starting in 1915.[44] Perceiving Newell as politically significant in organizing engineers in the 1910s and early 1920s, Layton adopts Hays's idea that Newell's departure from the Reclamation Service was in large part politically motivated: "Newell, as director of the Reclamation Service, sought to preserve his independence by adroit and farsighted administrative practices. . . . It was, in part, Newell's professional independence that led to his dismissal as head of the Reclamation Service in December 1914. Newell wanted the settlers to pay the full cost of irrigation works . . . but the Westerners complained that the works were too expensive and refused to pay."[45] In *Technology and Culture,* Peter Meiksens focused on issues raised by Layton and further analyzed Newell's work with the American Association of Engineers.[46] While not mentioning Newell's experiences with the Reclamation Service (except to note blandly "his departure from government service in 1914"), Meiksens portrays Newell as a competent engineer who worked to "increase the engineer's role in civic affairs."[47] Both Meiksens and Layton assume that Newell possessed considerable technical abilities, and they implicitly use Newell's reputation as a Progressive engineer and conservationist to orient their historical interpretation of his later efforts as a spokesman/advocate for professional engineers.

In light of material referenced above, Newell's work in directing the U.S. Reclamation Service is widely considered to be scientifically sophisticated, and, while the agency's financial problems are gaining notice, its engineering record is accepted as a prominent example of Progressive-Era efficiency. For Newell himself, the technical reputation of the Reclamation Service is being used as a basis for interpreting his work with the American Association of Engineers.

[43]Ibid., pp. 241–60.

[44]Edwin T. Layton, *The Revolt of the Engineers: Social Responsibility and the American Engineering Profession* (Baltimore, 1971).

[45]Ibid., p. 119.

[46]Peter Meiksens, "The 'Revolt of the Engineers' Reconsidered," *Technology and Culture* 29 (April 1988): 219–46.

[47]Ibid., p. 230.

The questions now before us are, Does the early Reclamation Service warrant unquestioned adulation as a builder of "engineering marvels" and should Newell be considered a paragon of Progressive engineering? The answers are important both to our understanding of western water history and to our evaluation of Newell's work as an engineer and professional organizer. To help answer these questions, the remainder of this article examines some technical and administrative aspects of the early Reclamation Service with special attention given to the Salt River Project in central Arizona. The Salt River Project, featuring the massive Roosevelt Dam, is considered among the agency's most successful endeavors and provides a good focus for assessing the technological adroitness of the Reclamation Service.

* * *

Irrigation in the Phoenix area began in the 1860s when Jack Swilling, an ex-Confederate soldier, cleaned out an ancient canal originally built by the Hohokam Indians.[48] Swilling's Ditch (later called the Salt River Valley Canal) diverted water out of the Salt River and proved the viability of irrigating the valley's expansive desert lands. In the 1870s and 1880s several other privately owned canals began drawing water from the Salt, fostering concern that the river's natural flow could not serve all the competing ditches. Legal wrangling erupted over the priority of water claims, while other energy went into finding ways of increasing water supplies by capturing flood runoff.[49] In 1889 a survey expedition sponsored by the city of Phoenix and Maricopa County located a superb storage dam site about 60 miles east of Phoenix and near the confluence of Tonto Creek and the Salt River.[50] Known initially as the Tonto Dam site (and later renamed in

[48]The early history of the city is documented in Geoffrey P. Mawn, "Phoenix, Arizona: Central City of the Southwest, 1870–1920" (Ph.D. diss., Arizona State University, 1979).

[49]The valley's early irrigation canals are described in Arthur P. Davis, *Irrigation Near Phoenix, Arizona,* U.S. Geological Survey Water Supply and Irrigation Paper no. 2 (Washington, D.C., 1897). See Earl Zarbin, *Roosevelt Dam: A History to 1911* (Phoenix, 1984), pp. 19–28, for contemporary newspaper references documenting legal disputes over water supply in the 1870s and 1880s. Zarbin (p. 23) also quotes a letter in the *Arizona Gazette,* June 24, 1887, counseling that "now is the time to move in the matter of storing water for summer use and shortages in dry seasons. Let the citizens and farmers unite on both sides of the river in an association for the purpose of storing water."

[50]The 1889 reservoir survey led by Maricopa County Surveyor William Breakenridge is described in Mawn, pp. 220–21; also see James A. McClintock, *Arizona: Prehistoric-Aboriginal-Pioneer-Modern* (Phoenix, 1916), 2:431–32. This survey expedition occurred

honor of President Roosevelt), this reservoir location could store floodwater for use by all canals in the Salt River Valley. In the early 1890s the privately owned Hudson Reservoir and Canal Company filed an application to build a large dam at the Tonto site.[51] Although these efforts never progressed beyond the planning stage, the Hudson interests helped generate early interest in the Tonto reservoir project, and they continued to assert a claim to the site for the next decade.

At the start of the 20th century, Phoenix's civic leaders began lobbying for federal assistance through a local organization known as the Water Storage Committee.[52] In 1901 committee representatives went to Washington, seeking permission for Maricopa County to issue bonds to finance storage dam construction (Arizona was a territory until 1912, and such bonding required congressional approval). Not surprisingly, once Roosevelt became president and advocated direct federal involvement in reclamation development, the goal of Phoenix's leaders changed. Now they wanted to reap the benefits of a Reclamation Service–financed initiative that would erect a storage dam at Tonto and deliver water to the valley's existing irrigation systems. In 1903 their efforts proved successful when the Secretary of the Interior selected the Salt River Valley for one of the service's first major projects.[53]

Although financing was intended to come from the proceeds of public land sales, essentially *all* of the almost 200,000 acres of land to be served by the project was (as Powell had foreseen) in private hands. In addition, over 48 percent of this land was in holdings that exceeded 160 acres.[54] The Salt River Valley might seem an unusual

---

contemporaneously with Powell's Irrigation Survey but did not draw on federal sponsorship.

[51]Articles of incorporation for this company appear in the *Phoenix Herald,* February 18, 1893. The company's filings on Tonto Dam site are dated April 15, 1893 (and posted April 20, 1893) as recorded in "Secretary of State, Appropriation of Water, Dam and Reservoir Sites, 1893–1910," Phoenix, Arizona Department of Library, Archives and Public Records. I appreciate Richard Lynch's assistance in locating these references.

[52]Phoenix's efforts to obtain federal support for impoundment of the Salt River are detailed in Smith, *Magnificent Experiment* (n. 23 above), pp. 20–42.

[53]Karen Smith, "The Campaign for Water in Central Arizona, 1890–1903," *Arizona and the West* 23 (Summer 1981): 127–48. Although the Hudson interest's claims on the reservoir site were of questionable legal value, the Reclamation Service eventually paid the company $40,000 to quell any possible litigation.

[54]This statistic was calculated by Karen Smith from information relating to the original subscribers to the Salt River Valley Water Users' Association in the spring of 1903; although reflective of landownership patterns among early subscribers to the project, this figure may be a bit misleading because it excludes consideration of thousands of acres of land owned by Dwight Heard and others in the Tempe area that

place to attract the attention of an agency dedicated to nourishing the expansion of small farms on the public domain. But the Salt River Project offered the Reclamation Service an opportunity to develop one of the West's premier reservoir sites and serve one of the most agriculturally productive communities in the arid region. In this light, selection of the project is more understandable, especially for a nascent bureaucracy seeking to demonstrate its engineering skills.

By 1902 Newell had been a federal employee for almost 15 years and had achieved standing as a prominent irrigation advocate. However, his experiences were those of an administrator and bureaucrat rather than an engineer forced to cope with real problems associated with functioning water supply systems. Based in Washington, D.C., he had never been involved in the design, construction, or operation of any actual irrigation project, and the records of the Reclamation Service contain little evidence that Newell participated in the technological work of the agency. For example, in 1903 Newell directed his assistant chief engineer Arthur P. Davis to "concentrate your energies as far as practicable on the pushing forward of the Salt River Project . . . and laying out a scheme of further work, involving the purchase of lands, rights of way, etc., also [on] action which may be taken regarding the Arizona Canal and the construction of the new dam in the river. I should like to have, in brief, a consistent plan for future operations covering these important matters."[35] Such delegation can be interpreted as a sign of an administrator leaving details to a trusted subordinate, but it also suggests that Newell remained uninvolved in the agency's *engineering* work. Instead, Davis functioned as the agency's senior official responsible for technical matters. As John Wesley Powell's nephew, Davis attracted favorable attention, but he did not bring a great deal of expertise to his work as the agency's assistant chief engineer. After graduating from Columbian College in Washington, D.C., in 1888 he joined Powell's Irrigation

officially did not become a part of the Salt River Project until after World War I. The final legal determination of land to be served by the project stipulated that acreage which had been irrigated before 1903 (which was all privately owned) would have priority rights to water stored by Roosevelt Dam. See Smith, *Magnificent Experiment,* pp. 43–48, 125–35; and Earl Zarbin, "The Committee of Sixteen," *Journal of Arizona History* 25 (Summer 1984): 129–54.

[35]F. H. Newell to A. P. Davis, June 23, 1903, Washington, D.C., National Archives, Record Group 115 (Bureau of Reclamation), Entry 3, Box 490 (hereafter cited as NA RG 115, Entry 3, Box ____). This source refers to the General Administrative and Project Records (1902–19) for the Reclamation Service's Salt River Project retained in the National Archives.

Survey and spent the 1890s as a federal employee.[56] Although more active in field work than Newell, Davis possessed no direct experience in building or operating irrigation systems prior to formation of the Reclamation Service.

As a substitute for a strong engineering presence at the top of the service's organizational chart, Newell devised a system of consulting "engineering boards."[57] Usually consisting of three men (including full-time service employees and part-time consultants), these boards prepared, reviewed, and/or approved all major plans related to various projects. At times, such boards were used to adjudicate financial disputes concerning private claims against the service, but most work focused on technical issues.[58] On the surface, their use appeared to offer a good way to draw on the expertise of a wide range of skilled engineers. But problems arose with the management of at least one employee/consultant who did not place the interests of the Reclamation Service above his outside work; specifically, Joseph B. Lippincott provoked a major conflict-of-interest crisis in 1905 when he worked for both the service and the city of Los Angeles in assessing the water supply potential of California's Owens Valley.[59] The engineering board for the Salt River Project did not engender ethical scandal, but its membership highlighted a different kind of limitation.

When selecting an engineer to provide advice on large-scale dam design, Newell turned to George Y. Wisner, a 63-year-old hydraulic engineer from Detroit, Michigan, possessing no previous experience in dam design, irrigation, or western construction work. Prior to 1903, Wisner's professional interests focused on issues such as sanitary engineering, harbor development, and water transportation on the Great Lakes and the Mississippi River.[60] Despite his lack of

[56]Davis's career is described in his professional obituary: "Arthur Powell Davis," *Transactions of the American Society of Civil Engineers* 100 (1935): 1582–91. Columbian College later became George Washington University.

[57]The "engineering board" system of project review is described in the *Third Annual Report of the Reclamation Service, 1903–4* (Washington, D.C., 1905), p. 41.

[58]For example, the service used a special engineering board to establish a purchase price for the Hudson Reservoir and Canal Company's claims to the Tonto Dam site; see NA RG 115, Entry 3, Box 485.

[59]For more on the Lippincott controversy, see Kahrl, *Water and Power* (n. 4 above); and Abraham Hoffman, *Vision or Villainy: Origins of the Owens Valley—Los Angeles Water Controversy* (College Station, Tex., 1981). Simply stated, Lippincott used his position with the Reclamation Service to provide engineering data to Los Angeles authorities that otherwise would have been difficult for the municipality to obtain. Lippincott eventually resigned his position with the Reclamation Service to become full-time assistant chief engineer for the Los Angeles Aqueduct.

[60]For unknown reasons, no obituary (or "memoir") for Wisner was published in the *Transactions of the American Society of Civil Engineers.* His career is synopsized in Newell,

experience in irrigation or storage dam construction, Newell selected Wisner as senior member of the engineering board for the Salt River Project (as well as for several other projects), and the service also relied on him for expertise in mathematically analyzing dam designs.[61] Although Wisner's work did not precipitate overt controversy, the Detroit-based engineer (who died unexpectedly in July 1906) demonstrated little specialized insight that might have justified his role as a key technical consultant for the agency.[62]

Based on the publicity it received, the Salt River Project constituted the agency's most prominent early project and the Roosevelt Dam, with a reservoir capacity exceeding a million acre-feet,

---

comp., *Proceedings of the First Reclamation Service Conference* (n. 1 above), p. 350. He graduated as a civil engineer from the University of Michigan in 1865 and spent most of his professional life working on harbor and transportation projects involving the Great Lakes, the Illinois River, the Mississippi River, and ports on the Gulf of Mexico. He also served with the Lighthouse Service and the International Waterways Commission.

[61]Along with Wisner, the other members of the Salt River Project engineering board were A. P. Davis and W. H. Sanders. Sanders had previous experience with the Pomona Land and Water Company in southern California, but he is a minor figure in the history of western water development. See his professional obituary: "William Horatio Sanders," *Transactions of the American Society of Civil Engineers* 88 (1925): 1435–36. Wisner's role in designing and analyzing designs for Shoshone Dam and Pathfinder Dam is documented in the file labeled "Discussion Related to Dams," NA RG 115, Entry 3, Box 287. His role in reviewing the service's project plans for the lower Colorado River in California is noted in Pisani, *From the Family Farm* (n. 4 above), pp. 311–13. He also served on engineering boards for the Carlsbad Project (New Mexico), the Klamath Project (Oregon/California), the Milk River Project (Montana), the Minidoka Project (Idaho), and the Garden City Project (Kansas).

[62]In relying on the assistance of Wisner, Newell avoided associating with James D. Schuyler, the West's most prominent dam engineer at the turn of the 20th century. Schuyler's work was known to Newell as the U.S. Geological Survey had published his review of American dam building in 1897; see James D. Schuyler, "Water Storage and Construction of Dams," in *Eighteenth Annual Report of the United States Geological Survey* (Washington, D.C., 1897). For more on his work as a dam design engineer, see James D. Schuyler, "The Construction of the Sweetwater Dam," *Transactions of the American Society of Civil Engineers* 19 (1888): 201–32, and *Reservoirs for Irrigation, Water Power and Domestic Water Supply* (New York, 1901). Records of his numerous reports on western water projects are available in the James D. Schuyler Papers, Berkeley, Water Resources Center Archives. Perhaps Newell considered Schuyler too closely associated with privately financed water and power projects in the West to provide advice unclouded by the appearance of ethical conflict. However, utilizing Schuyler as a key consultant also would have belied the notion that earlier hydraulic developments in the West were inferior or "temporary" compared to works under construction by the Reclamation Service. In contrast, Wisner's lack of involvement in any previous western engineering projects could have been viewed as a positive attribute because his inexperience would insulate the agency from any potential conflict-of-interest charges.

represented the pride of the service.[63] In selecting a dam design for Roosevelt, the service did not attempt to develop a particularly innovative structure but instead adopted a conservative approach that harkened to tradition. Built using locally quarried sandstone, the Roosevelt Dam utilizes a gravity design that depends on its massive dimensions for stability. In addition, the structure is aligned along a pronounced upstream arch that appears to provide a supplemental source of strength. But appearances can be deceptive. In fact, the upstream curvature does little to increase the dam's stability beyond that provided by a similarly proportioned straight-crested gravity design. (See figs. 3, 4.)

The design employed at Roosevelt derived directly from the work of mid-19th-century French engineers who sought to "rationalize" the process of building masonry gravity dams.[64] This French technique (later refined by W. J. M. Rankine in Britain and Edward Wegmann in the United States) theoretically minimized the amount of material required for the cross-sectional profile of a gravity design.[65] Compared with thin arch dams (in which a structure's upstream curvature is absolutely necessary for stability), gravity designs required substantially more material and were thus much more costly.[66] In the late 19th and early 20th centuries, masonry gravity dams became the standard for metropolitan water systems such as those supplying New York (Croton Dam) and Boston (Wachusett Dam) as well as San Francisco (San Mateo Dam) and Denver (Cheeseman Dam).[67] In attempting to establish the prestige of the Reclamation Service, Newell considered it appropriate that the agency's most visible dam project an aura of permanence and stability similar to dams built by major municipalities. Newell and the Reclamation Service never explicitly stated that the Roosevelt

[63]An "acre-foot" is a term used to express water volumes in the West. It is equal to 1 acre of land covered with water 1 foot deep, or 43,560 cubic feet.

[64]For a description of how M. DeSazilly and M. Delocre "rationalized" the process of masonry gravity design in mid-19th-century France, see Norman Smith, *A History of Dams* (Secaucus, N.J., 1972), pp. 197–200.

[65]Detailed discussion of gravity dam design within the tradition established by DeSazilly, Delocre, and Rankine is presented in Edward Wegmann, *The Design and Construction of Dams* (New York, 1888), pp. 1–38.

[66]A history of dam design contrasting gravity and arch dam technologies appears in Donald C. Jackson, "A History of Water in the American West: John S. Eastwood and the 'Ultimate Dam' (1908–1924)" (Ph.D. diss., University of Pennsylvania, 1986), pp. 53–138.

[67]All of these structures are described in Edward Wegmann, *The Design and Construction of Dams,* 8th ed. (New York, 1927).

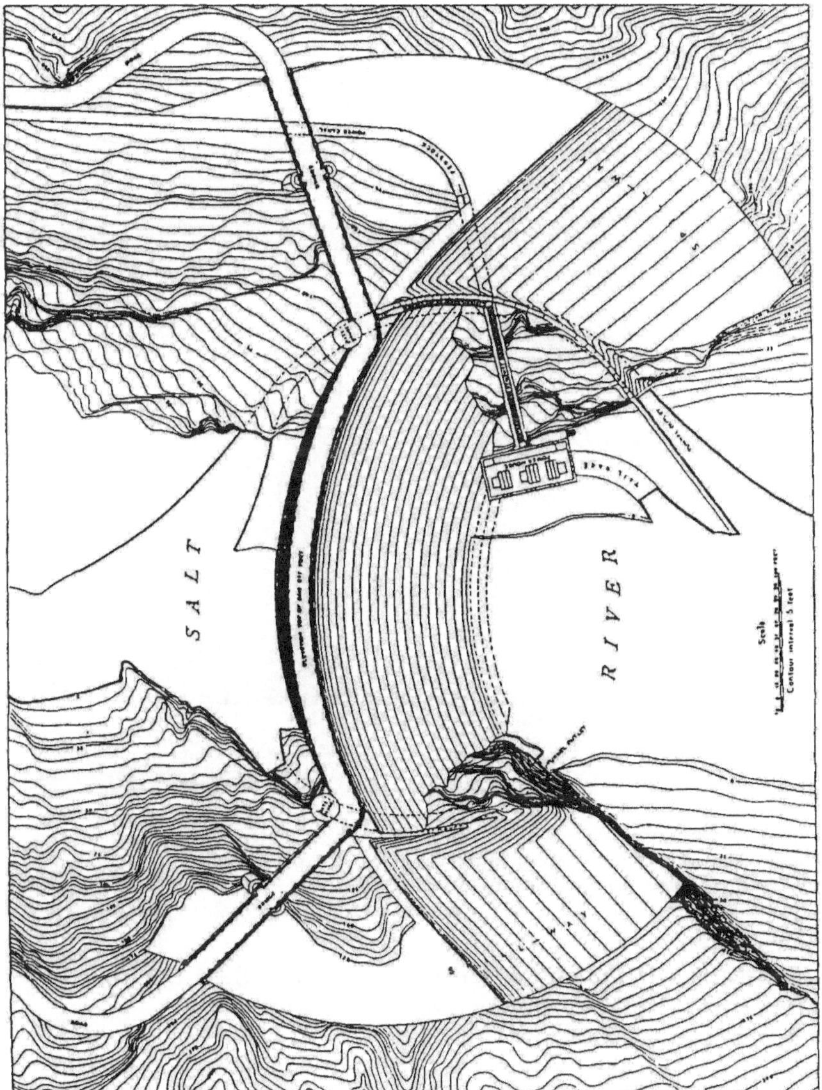

FIG. 3.—Plan view of the original design for Roosevelt Dam showing the structure's upstream curve. (*First Annual Report of the Reclamation Service* [Washington, D.C., 1903].)

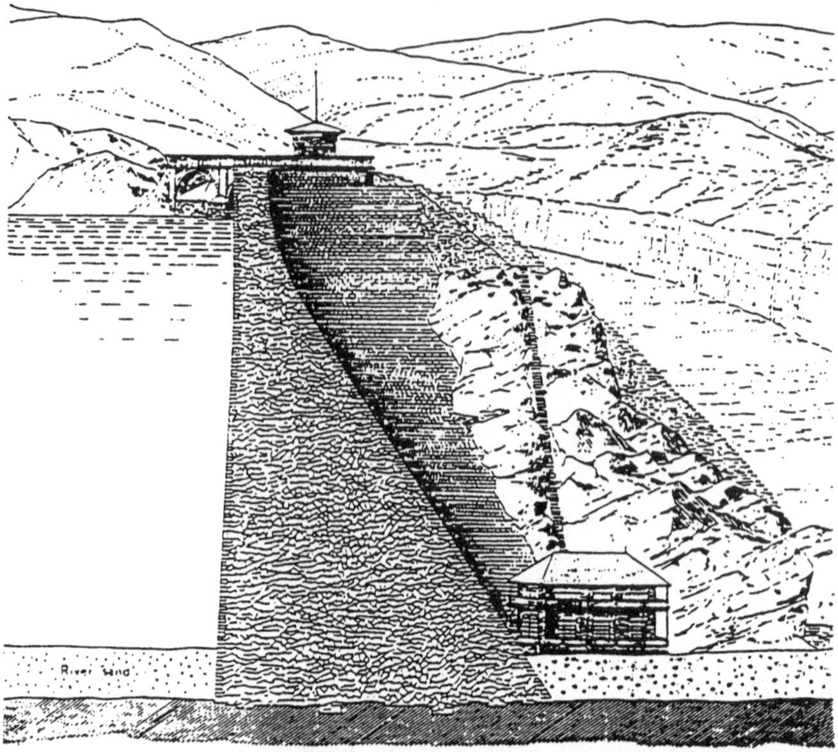

FIG. 4.—Cross-section of design for Roosevelt Dam showing the characteristic triangular shape of masonry gravity dams with vertical upstream faces. (*First Annual Report of the Reclamation Service* [Washington, D.C., 1903].)

Dam was to mimic major eastern structures, but little thought was given to anything other than a curved gravity design.[68]

[68]In the letter officially recommending selection of the Salt River Project for construction by the service, the director of the U.S. Geological Survey made explicit reference to recent work on the New York and Boston water-supply systems. The Salt River Project was the only initial service project intended to serve the region surrounding a burgeoning city. As such, it is not surprising that the Roosevelt Dam utilized a technology very similar to that employed for prominent eastern dams. See Charles Walcott to E. A. Hitchcock [Secretary of the Interior], March 7, 1903, NA RG 115, Entry 3, Box 492. Specific comparison of the proposed Salt River dam with the Croton, Wachusett, and Cheeseman dams is made in Arthur P. Davis, "Investigations in Arizona," in Newell, comp., *Proceedings of the First Reclamation Service Conference* (n. 1 above), p. 129.

In the 1890s, the Hudson Reservoir and Canal Company publicized a curved gravity dam design for the Roosevelt site.[69] In 1902 Davis proposed a structure that closely resembled the Hudson Company's plan, and the Salt River Project engineering board selected this design without seriously considering any alternatives.[70] Newell's attitude toward this process is well expressed in a letter he wrote near the end of his tenure with the agency: "We [the Reclamation Service] have been inclined to adhere to the older, more conservative type of solid dam, largely perhaps because of the desire not only to have the works substantial but to have them appear so and recognized by the public as in accordance with established practice."[71] Newell relished the idea of leading a scientific organization, but he was equally concerned with adhering to a status quo defined by "established practice." In this context, his concern with public perception and his interest in advocating "older, more conservative" technologies becomes evident. With storage dams, Newell's concern for appearance and its effect on public perception was particularly acute: "Plans for the construction of storage works, while they must be prepared with regard to reasonable economy, must be [undertaken] with a view to being not merely safe but looking safe. People must not merely be told that they are substantial, but when the plain citizen visits the works he must see for himself that there is every indication of the permanency and stability of a great storage dam . . . he must feel, to the very innermost

[69]The Hudson Reservoir and Canal Company's proposed curved gravity dam design is illustrated in Schuyler, "Water Storage and Construction of Dams," pp. 715–17.

[70]The first design description of Roosevelt Dam appeared in Arthur Powell Davis, *Water Storage on Salt River, Arizona* (Washington, D.C., 1903). This publication disseminated a report submitted by Davis in mid-1902 that was partially financed by Phoenix irrigation interests. The *First Annual Report of the Reclamation Service* (Washington, D.C., 1903) essentially repeats the description presented in Davis's 1902 report. As one of his first contributions to the Reclamation Service, Wisner visited the Tonto site in April 1903 and shortly afterward inspected the San Mateo Dam (a concrete curved gravity dam) in the company of Davis. In reporting on his trip, Wisner stated that "a curved masonry dam, such as designed by Mr. Davis, is particularly well adapted for the [Tonto] site." See George Y. Wisner to F. H. Newell, May 4, 1903, NA RG 115, Entry 3, Box 490. The engineering board was later consulted on plans that raised the Roosevelt design from a height of 240 feet to 280 feet. However, the enlarged gravity design was not modified except to proportionally increase the thickness of the base. For verification that Davis was primary designer of the Roosevelt Dam, see C. R. Olberg to A. P. Davis, March 6, 1911, NA RG 115, Entry 3, Box 509.

[71]F. H. Newell to A. H. Dimock, April 16, 1912, "Discussion Related to Dams," NA RG 115, Entry 3, Box 287.

recesses of his consciousness, that the structure is beyond question."[72] Unquestionably, Newell was justified in seeking to build dams that would not collapse, and, in this context, his concern for conservative design is well taken. Unfortunately, while the massive character of a gravity dam may have fostered a visual sense of safe design, the equation of greater mass with greater safety is not necessarily justified, no matter how the issue might appear to laypersons.

During the Reclamation Service era, an element of the hydraulic engineering community grew dissatisfied with gravity dam technology, and in 1913 George Holmes Moore published an article in *Engineering News* outlining these reservations. Moore complained that all gravity dam designs were "merely variations of a simple triangle, a triangle whose base dimension is two-thirds of its height" and noted that the wide, solid base of gravity dams made them vulnerable to sliding because of hydrostatic "uplift" forces acting across the foundations. In concluding his article, Moore challenged civil engineers to "face squarely and openly certain inherent weaknesses of [gravity dam] design, weaknesses long known to exist by those in the profession, but never honestly admitted to the public."[73] To Moore, Newell's interest in appealing to what the "plain citizen" considered to be a safe design would have been inconsistent with the ideals of good engineering. The selection of a gravity design for Roosevelt was not necessarily inappropriate, but Newell's overarching concern with visual appearance is at odds with what some engineers of the period might have considered a progressive approach to the science of dam technology.[74]

After selecting a design for Roosevelt, the Reclamation Service needed to develop a construction plan. Rather than take direct responsibility for hiring labor (often referred to as the "force account"

[72]Quote taken from F. H. Newell, "Irrigation: An Informal Discussion," *Transactions of the American Society of Civil Engineers* 62 (1909): 13. Newell's preoccupation with visual appearance is also reflected in his complaint that "the general appearance of the grounds at the Roosevelt headquarters could be improved . . . [because] our work is judged by the general public more by its outward appearance than by its real merit." See F. H. Newell to L. C. Hill, May 8, 1909, NA RG 115, Entry 3, Box 490.

[73]George Holmes Moore, "Neglected First Principles of Masonry Dam Design," *Engineering News* 70 (September 4, 1913): 442–45.

[74]In developing designs for the extremely narrow, hard-rock canyons that existed at the Shoshone and Pathfinder dam sites, the early Reclamation Service did utilize arch dams with profiles slimmer than required for a gravity design. Although more materially conserving than the gravity design used at Roosevelt, these structures were subject to criticism for their relatively bulky dimensions. For example, see John S. Eastwood, "An Arch Dam Design for the Site of the Shoshone Dam," *Engineering News* 63 (June 9, 1910): 678–80.

method), Newell and Davis sought to contract out as much work as possible. Davis explained:

> The Department and the Director, as well as most of the consulting engineers, are strongly in favor of doing work by contract wherever this is practicable, even if in advance it may not seem the most economical method. Where work is advertised and universal competition invited no criticism can be made against the award of a contract although the prices may be higher than it was expected the work would cost. If an equal amount were expended in the construction by force account, this would lead to criticism and charges of extravagence, bad management or corruption and it is not easy to answer such criticism.[75]

Thus for political reasons the ideals of efficiency were subordinated to more prosaic considerations of how the bureaucracy might fend off criticism of problems associated with its construction work.

At Roosevelt, the contract method proved disastrous in building the 19-mile-long power canal system designed to supply electric power during the construction process. Originally estimated to cost $215,260, this power canal (along with its diversion dam and associated hydroelectric power plant) eventually cost more than $1.4 million to complete.[76] The canal contracts were particularly troublesome because field conditions for the widely dispersed project presented many difficult construction problems.[77] Compared with other western hydroelectric power systems, however, the Roosevelt power canal system represented a small and unsophisticated project.[78] The Reclamation Service was a novice in this field, and its inexperience directly contributed to the increased cost of the power canal and hydroelectric plant.

Reliance on outside contracts also led to the selection of John M. O'Rourke and Company of Galveston, Texas, as the main contractor

[75]A. P. Davis to Louis C. Hill, February 27, 1906, NA RG 115, Entry 3, Box 490. Hill was supervising engineer for the Salt River Project.

[76]Initial estimates for the construction of the power canal and hydroelectric plant are given in the *First Annual Report of the Reclamation Service*, p. 97. Final costs are provided in the *Ninth Annual Report of the Reclamation Service* (Washington, D.C., 1911), p. 69.

[77]See David Introcaso, "Roosevelt Power Canal," Washington, D.C., Library of Congress, Prints and Photographs Division, Historic American Engineering Record (HAER), AZ-4. Further evidence of the poor work on the power canal can be found in G. Y. Wisner and W. H. Sanders to F. H. Newell, November 15, 1904, NA RG 115, Entry 3, Box 490.

[78]The technologically advanced nature of western hydroelectric power in the 1890s is described in Jackson (n. 66 above), pp. 171–269; and Thomas P. Hughes, *Networks of Power: Electrification in Western Society, 1880–1930* (Baltimore, 1984), pp. 262–84.

for the dam although the firm was new to Arizona and had no experience in dam construction.[79] In 1905, O'Rourke came in with a low bid of $1,147,000 to complete the main structure of the dam up to the 150-foot level in less than two years.[80] There is little reason to believe that O'Rourke's proposal should have been considered feasible. For example, when Boston's Metropolitan Water District constructed the Wachusett Dam, it took 49 months (June 1901–July 1905) to erect 273,000 cubic yards of masonry at a site directly accessible to a railroad and in the midst of a major industrial region.[81] In contrast, O'Rourke had offered to erect approximately the same quantity of masonry in only twenty-four months in a remote, harsh wilderness with a minimal supply of local skilled labor.

Because of heavy flooding, it may have been impossible for any company or organization to have built the Roosevelt Dam in a timely, cost-effective manner. But O'Rourke never came close to meeting the terms of the original contract, and his company received little acclaim for perseverance or hard work. Chester Smith, on-site supervisor for the Reclamation Service, characterized O'Rourke's early effort as "quite dilatory" and complained that the outfit worked "in accordance with their own ideas on the subject. They did not do as much as they should have."[82] By the spring of 1907, when the vast majority of the structure was to have been complete, O'Rourke had placed only 5 percent of the total masonry required.[83] In fact, the dam did not reach the 150-foot level until November 1909, more than two-and-a-half years past the deadline stipulated in the original contract. The structure was not completed until February 1911. With this performance, it is easy to understand why O'Rourke's company faded into obscurity. In structural terms, the masonry construction of the Roosevelt Dam has proven satisfactory. (See figs. 5, 6.) But historians should not believe that the dam represented a technological feat that only the Reclamation Service could have accomplished or administered.

The service encountered problems with other aspects of the Roosevelt project, including haulage of supplies and the operation of the hydraulic gates. This discussion concludes, however, by considering a significant task that involved a dramatic rejection of the contract

[79]O'Rourke's firm had recently completed the concrete seawall protecting Galveston's gulf coast shoreline, and this apparently qualified it for the Roosevelt contract.

[80]Construction bid data is published in the *Fourth Annual Report of the Reclamation Service* (Washington, D.C., 1906), pp. 69–71.

[81]Construction of Wachusett Dam is described in Wegmann (n. 67 above), pp. 185–94.

[82]Chester Smith to L. C. Hill, March 18, 1907, NA RG 115, Entry 3, Box 495.

[83]*Sixth Annual Report of the Reclamation Service* (Washington, D.C., 1908), p. 64.

FIG. 5.—A 1908 view of the Roosevelt Dam under construction showing the massive character of the masonry design. (Author's collection.)

Fig. 6.—A 1941 view of the completed Roosevelt Dam showing full reservoir. The Taintor gates controlling flow over the spillways were not part of the original design. (Author's collection.)

system. Roosevelt Dam primarily consists of large sandstone blocks bonded together by concrete requiring large quantities of cement. In his original 1902 design proposal, Davis estimated that purchasing cement and delivering it to the Roosevelt site would cost $9.00 per barrel; at the same time Davis reported that cement using local limestone and clay deposits could be produced for about $2.00 per barrel.[84] As a result, the service chose to build its own cement mill at Roosevelt despite subsequent offers by private companies to deliver cement on-site for as low as $4.51 per barrel.[85]

After letting construction contracts in March 1904, the service took a year to get the cement mill operating because "the inaccessibility of the dam site caused long delay in securing the necessary equipment for economical and rapid work."[86] After operations began, the need to fire the cement kilns with fuel oil proved to be the system's Achilles' heel. Fuel oil came to the Salt River Valley from California by railroad and was then hauled overland 60 miles to the cement plant. Because of a railroad tank car shortage and problems with the haulage contract, the cement plant operated sporadically (and inefficiently) for several years.[87] The plant ultimately worked, but the original cost estimates proved insufficient. Instead of $2.00 per barrel, the per unit cost was $3.14 per barrel, and this figure was substantially reduced because an additional 60,000 barrels (out of a total of 338,000) were produced for the Granite Reef Diversion Dam.[88] Despite honorable

[84]Davis, *Water Storage on the Salt River* (n. 70 above). These cost estimates for purchasing and making cement are repeated in the *First Annual Report of the Reclamation Service* (n. 26 above), pp. 99–102.

[85]Report on "Expense of Operation and Maintenance, Roosevelt Cement Plant, May 1905 to July 1910, Inclusive," NA RG 115, Entry 3, Box 519; also see F. H. Newell to A. P. Davis, March 4, 1904, and E. A. Hitchcock to Charles Walcott, March 4, 1904, both in Box 494.

[86]The quotation is taken from G. Y. Wisner and W. H. Sanders to F. H. Newell, March 3, 1905, NA RG 115, Entry 3, Box 490. Problems in erecting the cement plant are outlined in "Cement-Salt River Project" (no author), March 3, 1905, Box 519.

[87]See L. C. Hill to F. H. Newell, July 5, 1905; L. C. Hill to F. H. Newell, August 8, 1905; and A. P. Davis to James Kruttschnitt, July 17, 1907, all in NA RG 115, Entry 3, Box 489.

[88]Final costs for the cement production are given in "Expense of Operation and Maintenance, Roosevelt Cement Plant, May 1905 to July 1910, Inclusive." The Granite Reef Dam is located more than 40 miles downstream from the Roosevelt Dam. It is a concrete overflow structure that diverts water out of the Salt River and into distribution canals serving the valley. The dam's construction was not planned as part of the original Salt River Project but became necessary after the privately owned Arizona Dam failed in the heavy flooding of 1905. The 60,000 barrels of cement used in its construction helped reduce the unit costs of production at the Roosevelt cement mill by allowing fixed costs to be distributed over a greater number of barrels. Because of its close proximity to Phoenix, the costs of supplying cement to the Granite Reef Dam from

intentions, the service's effort to manufacture cement at Roosevelt proved to be, at best, a tepid success.

Costs for the Roosevelt Dam exceeded original estimates and the overall project proved much more expensive than anticipated in 1903. Originally estimated at about $1.9 million (including the power canal, power plant, cement mill, hydraulic gates, tunnels, roads, and "damage to private lands"), the dam ultimately cost more than $3 million dollars (a figure excluding more than $1 million for land purchases, placement and repair of hydraulic gates, tunnel excavation, and road construction, as well as more than $2.3 million for the entire electric power system).[89] Hopes for financing the Salt River Project with monies derived from public land sales in Arizona Territory quickly faded, and initial project costs eventually exceeded $10 million.[90] Some of this increase resulted from raising the height of Roosevelt Dam and from taking responsibility for the Granite Reef Dam and distribution canals in the valley. But much of it resulted from the Reclamation Service's failure to efficiently build and operate facilities such as the power canal and the cement mill and to develop plans feasible to implement given the financial and temporal framework adopted by the agency.

Even during the earliest phases of construction, Newell knew that expenditures were escalating out of control. By the end of 1905, almost $3.5 million had been contractually obligated for the Salt River Project, prompting Newell to express concern over "the enormous expenditures which have been made in Arizona" while bemoaning that "we have already allotted to this project a very large sum, more than can easily be defended, and the allotment has been so liberal that I did not suppose that we should come to an end so quickly. The end, however, has been reached."[91] This financial predicament did not abate after 1905, and, as O'Rourke fell further behind schedule, the notion of the Reclamation Service as an efficient engineering organization became difficult to sustain. But Newell could not directly confront the agency's failures because this would bring the Progressive attributes of the service into question. Residents of the Salt River Valley participated in the effort to reduce the repayment require-

---

private contractors would have been much less than offers made to supply cement to Roosevelt.

[89]Cost of the Salt River Project as of the end of 1910 is given in the *Ninth Annual Report of the Reclamation Service* (n. 76 above), pp. 69–70. Work on the Granite Reef Dam and on distribution canals in the valley came to a little more than $2 million.

[90]Determination of the Salt River Project's official initial cost of approximately $10.2 million is documented in Smith, *Magnificent Experiment* (n. 23 above), p. 140.

[91]F. H. Newell to L. C. Hill, December 26, 1905, NA RG 115, Entry 3, Box 480.

ments of the 1902 Reclamation Act, and, especially after the Roosevelt Dam began storing water, they did not hesitate to express dissatisfaction with the service.[92]

Newell insisted that beneficiaries of Reclamation Service projects pay their full costs (excluding interest charges), but the agency's moral standing in these matters had long since eroded. In his 1901 message to Congress, President Roosevelt had set a high standard for federal reclamation in assuring the American people that "no reservoir or canal should ever be built to satisfy selfish personal or local interests, but only in accordance with the advice of trained experts. . . . There should be no extravagance, and the believers in the need of irrigation will most benefit their cause by seeing to it that it is free from the least taint of excessive or reckless expenditure of the public moneys."[93] Unfortunately, cost overrun problems were endemic with service projects, and Newell, one of Roosevelt's "trained experts," could not easily justify what opponents might term "excessive or reckless expenditure of the public moneys."[94] While Newell staunchly demanded that all project costs be repaid as originally stipulated, the westerners who were to bear the brunt of increased construction costs sought ways to reduce their financial liability.[95] In large part, their political maneuvering was the result, not the underlying cause, of Newell's and the Reclamation Service's problems.

In retrospect, a conspicuous aspect of the early Reclamation Service concerns the numerous articles and promotional tracts published in

[92]Detailed discussion of how the Salt River Valley Water Users' Association sought to extend the project repayment period and reduce their financial responsibility is presented in Smith, *Magnificent Experiment,* pp. 92–124.

[93]Quoted in Newell, *Irrigation* (n. 25 above), p. 396.

[94]Another dramatic cost overrun afflicted the Strawberry Valley Project in Utah. In 1905, cost estimates were $1.25 million, but by the time of project completion almost a decade later costs had reached almost $3.5 million; See "Strawberry Valley Project," HAER, UT-26.

[95]A few months after the completion of Roosevelt Dam, Newell expressed dismay that Phoenix residents were attempting to evade payment: "It is not wise to let it be understood that the project is completed and that the Valley has no more to expect. This has been emphasized by the number of the citizens who have implied that now that there is nothing more to be had out of Uncle Sam they can concentrate their energies on securing deferred payments, or even the repudiation of a whole or a part of the debt. I have been astonished at the way this feeling has apparently spread" (F. H. Newell to L. C. Hill, November 25, 1911, NA RG 115, Entry 3, Box 490). In 1917 the federal government gave the locally controlled Salt River Valley Water Users' Association complete operational control over the Salt River Project system. However, the federal government retained legal ownership of the Roosevelt Dam and other major structures. Final payment for the association's initial $10.2 million project debt to the government was not tendered until the mid-1950s.

the popular press that extolled the virtues of federal reclamation. Frequently appearing under Newell's name, or one of his subordinates, this publicity highlighted the agency's noble ideals but gave little indication of the hard economic realities faced by the service.[96] To a cynic, Newell's promotional activity might appear as sophisticated hucksterism. To someone viewing Newell in a kinder, gentler fashion, such publicity could be interpreted as necessary to convince the American people of the Reclamation Service's great public import. Regardless, the emphasis that Newell placed on public relations is undeniable. Rather than let the agency simply be recognized for its attainments, he felt compelled to loose a barrage of publicity that frequently obscured the service's myriad difficulties.

Along with generating publicity to help maintain a positive image, Newell utilized other means to promote the agency, even if they might have hindered efforts by others to develop irrigation projects. For example, the Reclamation Service could prevent public land from being transferred into private hands, ostensibly to keep speculators from grabbing up territory that would later benefit from a federal irrigation project. In the Phoenix area, this power was seemingly irrelevant because the acreage within the Salt River Project was already privately owned. However, Newell prevented a substantial amount of public land outside of the Salt River Project area from being made available for development by privately financed irrigation companies. Newell's opposition to the Beardsley project on the Agua Fria River directly west of Phoenix helped delay for decades implementation of a large (35,000+ acres) irrigation system.[97] Similarly, hopes to develop Paradise Valley north of the city (and outside the area served by the Salt River Project) by using water from the Verde River were never realized, in part

[96]For example, see F. H. Newell, "The Reclamation Service," *Popular Science Monthly* 66 (December 1904): 106–16, "Reclamation, the Work of the Government in Providing Irrigation for Arid and Hitherto Useless Lands in the West," *Outlook* 83 (August 25, 1906): 933–41, "Work of the U.S.R.S.," *Journal of the Franklin Institute* 164 (July 1907): 29–42, and "Ten Years of Government Irrigation Work: A Review of the Activities and Experiences of the U.S. Reclamation Service since the Passage of the Newlands Act," *Engineering Record* 67 (February 1, 1913): 128–30; C. J. Blanchard, "Winning the West: An Account of the Marvelous Progress of Our Reclamation Service in Reclaiming the Desert," *National Geographic Magazine* 17 (February 1906): 82–99, and "National Reclamation of Arid Lands," *American Geographic Society* 21 (April 1906): 209–23. Blanchard officially served as "Statistician" for the service.

[97]See David Introcaso, "History of Waddell Dam," HAER, AZ-8.

because of Newell's lack of interest.[98] Newell promoted himself as a champion of public works, but his public service was tempered by an element of private interest. Irrigation endeavors that did not involve the Reclamation Service were, at best, irrelevant and, at worst, represented competition that could heighten awareness of his agency's limitations. Thus, reclamation projects that did not specifically relate to his program were often ignored and at times were bureaucratically suppressed.

* * *

Although the Reclamation Service was responsible for building several notable large-scale dams in the West, the agency's long-term significance apparently resides more in the financial than in the technical realm. Nongovernment engineers could prove equally adept in advancing the art of dam engineering, but the private sector could not compete with the government in devising financial arrangements that stretched out repayment schedules for decades and forgave interest charges.[99] As

[98]Newell's lack of interest in efforts by the Verde Water and Power Company (successor to the Rio Verde Canal Company) to irrigate Paradise Valley is noted in Donald C. Jackson, "History of Horseshoe Dam," HAER, AZ-24. Conflict between the Salt River Project and organizations such as the Verde River and Power District who wished to develop Paradise Valley continued into the 1920s and 1930s and was not ended until the Bureau of Reclamation supported construction of the Bartlett Dam (completed in 1939) that stored Verde River floodwaters for use by the Salt River Project.

[99]Many other American engineers demonstrated their skills in building massive gravity dams during the early 20th century; for example, Stone & Webster built several large dams of this type in the West including three for the Big Creek hydroelectric power system in California in 1912–13, financed by Henry Huntington's Pacific Light and Power Corporation. See David H. Redinger, *The Story of Big Creek* (Los Angeles, 1949). Although the Reclamation Service built arch designs for the Pathfinder and Shoshone dam sites in Wyoming, the most innovative work in this field was carried out by Lars Jorgensen, who designed the world's first constant-angle arch dam, built near Juneau, Alaska, in 1912–13. Jorgensen later formed the Constant Angle Arch Dam Company to promote construction of his designs. See Lars Jorgensen, "The Constant Angle Arch Dam," *Transactions of the American Society of Civil Engineers* 78 (1915): 685–733; and Nancy Farm Mannikko, "Seattle City Light's Constant-Angle Arch Dam at Diablo Canyon," *IA: The Journal of the Society for Industrial Archeology* 17, no. 1 (1991): 17–31. I do not wish to convey the idea that hydraulic engineers associated with the Reclamation Service and/or the Bureau of Reclamation were devoid of technical skill or failed to provide useful service to residents of the western states. For example, in the late 1920s the bureau became a leading proponent of the trial-load method of analyzing stresses in arch dams. However, I do believe that their accomplishments have heretofore received a disproportionate amount of attention.

much as Newell might have wished to deny it, in large part fiscal adaptability (in the sense of easily extended federal loans) represents the service's dominant legacy, not technical achievements that exceeded the abilities of other engineers.[100]

My emphasis on the Reclamation Service's technological limitations is intended as a partial corrective to the abundance of written material that places the agency at the center of western water resources development with little reference to any other engineers or organizations. In 1920 less than 7 percent of the West's irrigated land bore any relation to the Reclamation Service.[101] And a 1934 survey of western hydroelectric plants with capacities exceeding 20,000 horsepower included only one facility (at Roosevelt Dam) built by the Reclamation Service. In contrast, the survey documented more than 60 operating hydroelectric plants built by private utilities or municipalities with a combined capacity exceeding 3.6 million horsepower.[102] The agency's limited role in developing the West's water resources during the early 20th century is not reflected solely in statistical aggregations but also extends to issues of large-scale engineering design.

For example, in the 1910s and 1920s the American West became the focus of development for a new type of dam that dramatically reduced the amount of concrete required for construction and hence offered to reduce the cost of water storage significantly. In 1908–9, the world's first reinforced-concrete multiple-arch dam was built by John S. Eastwood for a lumber company at Hume Lake, California; during the next two decades over thirty structures of this type were built in the United States, and the technology received extensive notice in the technical press.[103] It also prompted publication of a

[100]See n. 34 above. Sherow, *Watering the Valley* (n. 4 above), p. 137, echoes this view by averring: "A common theme in western history [is]: Let the federal government pay the means to development but keep control of it in private hands in the West."

[101]The Fourteenth Census (1920) reported that federal resources were being used in the irrigation of 2 million acres while the West as a whole had 27 million irrigated acres; of these, 80 percent were watered by private interests while over 12 percent utilized state assistance. These statistics are reported in Joseph B. Lippincott, "A National Reclamation Policy: Explanatory Statement regarding Report of the Committee of the Irrigation Committee," *Transactions of the American Society of Civil Engineers* 95 (1931): 1307–20.

[102]"A Survey of Hydroelectric Developments," *Electrical Engineering* (July 1934): 1087–94. The survey covered the entire United States, but the above statistics only refer to power plants west of Denver.

[103]See John S. Eastwood, "Hume Lake Dam," *Journal of Electricity Power and Gas* 23 (October 30, 1909): 398–404, "The New Big Bear Valley Dam," *Western Engineering* 3 (December 1913): 458–70; Lars Jorgensen, "Multiple Arch Dams on Rush Creek, California," *Transactions of the American Society of Civil Engineers* 81 (1917): 850–906;

special chapter focusing on "Mathematical Discussion and Descriptions of Multiple Arch Dam" in the 1927 edition of Edward Wegmann's *The Design and Construction of Dams*.[104] Multiple-arch designs appeared to offer an excellent way to promote structural and economic efficiency, but, aside from one diversion dam in Oregon, the Reclamation Service avoided association with the technology.[105]

Multiple-arch dams conserve material by utilizing a series of buttresses to support numerous short-span arches that form the upstream face of the dam. Known generically as buttress dams (or sometimes "hollow dams"), multiple-arch structures are not solid monoliths; this design feature eliminates a problem associated with massive gravity dams in which water can seep through the foundations, push upward, and help cause the structure to slide downstream. In 1915, both the economic savings and structural stability of the technology figured into Salt Lake City's decision to select a multiple-arch design for the 150-foot-high storage dam at Mountain Dell, Utah (see fig. 7).[106] Bids on comparable concrete gravity and multiple-arch dams were solicited from several outside contractors, and, because the latter technology required over 66 percent less concrete (approximately 15,000 cubic yards vs. 45,000 cubic yards), the low bid on the multiple-arch design was over $85,000 less than the low bid on the gravity design (approximately $139,000 vs. $226,000).[107] Salt Lake

B. F. Jakobsen, "Stresses in Multiple Arch Dams," *Transactions of the American Society of Civil Engineers* 87 (1924): 276–341; and Fred A. Noetzli, "Improved Type of Multiple Arch Dam," *Transactions of the American Society of Civil Engineers* 87 (1924): 342–413. More references on multiple-arch technology are available in Jackson, "A History of Water" (n. 66 above), pp. 853–77.

[104]In the final edition of Wegmann (n. 67 above), pp. 439–536, Fred Noetzli authored the new chapter on multiple-arch dams.

[105]See H. D. Newell, "Multiple Arch Diversion Dam at Three Miles Falls, Oregon," *Engineering News* 73 (May 27, 1915): 1009–12. In the 1930s the Bureau of Reclamation also used a multiple-arch design for the Bartlett Dam in Arizona.

[106]City engineer Sylvester Q. Cannon characterized the Mountain Dell site as lying only a few miles "above a section of the city that is rather thickly settled . . . and a break in the dam would be very serious." He also indicated that the site's relatively poor quality of bedrock prompted construction of a buttress dam and that the multiple-arch design provided "practical elimination of upward pressure, the practical impossibility of overturning on its base and the ready facilities for the inspection of the dam at any time." See Sylvester Q. Cannon to John S. Eastwood, [July] 10, 1915, Mountain Dell Dam Folder (no. 16), Berkeley, Water Resources Center Archives, John S. Eastwood Papers; and Sylvester Q. Cannon, "The Mountain Dell Dam," *Monthly Journal Utah Society of Engineers* 3 (September 1917): 223–30.

[107]Data on bids submitted by eight different contracting firms are available in "Summary of Bids—Mountain Dell Dam," November 24, 1915, Eastwood Papers. Wegmann, p. 481, also provides data on the low bids for the competing designs.

FIG. 7.—A 1971 view of the downstream side of the Mountain Dell Dam showing buttresses used in the multiple-arch design. (Jack Boucher, Historic American Engineering Record Collection, Prints and Photographs Division, Library of Congress, Washington, D.C.)

City's experience with the Mountain Dell Dam is noted in this article to highlight how an early-20th-century public works bureaucracy could respond positively to a new form of water storage technology that provided real economic savings. Significantly, Newell's inclination to "adhere to the older, more conservative type of solid dam" because it would be "recognized by the public as in accordance with established practice" was expressed in a letter specifically responding to an inquiry concerning the construction of buttress dams.[108] Other engineers were willing to explore development of the multiple-arch dam, but the Reclamation Service largely opted to ignore this technology in order to meet a preconceived notion of what "the public" would think a dam should look like.[109] Beyond being involved in only a small

[108]Newell to Dimock, April 16, 1912 (n. 71 above).

[109]Multiple-arch dams were criticized by some engineers, most notably John R. Freeman, because they did not offer a visually massive appearance, and it appears that the Reclamation Service declined to explore the possibilities of the technology for similar, nontechnical reasons. Freeman became the most outspoken opponent of multiple-arch dams, and he attacked them not on technical grounds, but because the "psychology" of their appearance was ill-suited to a public that supposedly expected dams to look massive. Freeman's opposition is discussed in my article "Considering the

fraction of the West's irrigation projects and hydroelectric power systems during the first third of the 20th century, the agency also demonstrated a decided lack of interest in developing a new water-storage technology well suited to conditions imposed by the region's lack of financial resources.

This article has not sought to analyze Newell's relationship to America's larger economic order, a capitalist system that both Donald Worster and James Sherow perceive as dramatically influencing the course of western development because it transformed water, along with other aspects of the natural environment, into a commodity of business.[110] However, in terms of the way Newell's orientation to technology relates to the business world, a few comments are warranted. Given his manipulation of the family farm/public domain issue in promoting the Reclamation Service and the subsequent transformation of federal reclamation into a supporter of agribusiness and urban growth, one might be tempted to portray Newell simply as a servant of the corporate interests that came to control much of the region's hydraulic infrastructure. Although it is to some degree plausible, I believe such a portrayal would be misleading. True, he should have recognized the hyperbole inherent in his advocacy of a federal solution to the family farm/public domain issue that energized the 19th-century irrigation movement. But he became enthralled with the idea of a centralized, scientific bureaucracy solving the West's water problems, and this quickly took on a momentum of its own.

Although Newell would have liked to operate in a physical environment devoid of existing economic and political interests, such an idealized environment bore little resemblance to actual conditions in the West at the start of the 20th century. Rather than abandon his dreams of creating large-scale reservoirs and water-supply systems for yeoman farmers on the arid public domain, he accommodated them to political reality. The Roosevelt Dam offered an opportunity to develop one of the West's best reservoir sites, and, although it did nothing to open up the public domain to new farms, Newell actively pursued and promoted the project because of its superior technical attributes. In many ways, he let his enthusiasm for large-scale

---

Multiple Arch Dam: Theory, Practice and the Ethics of Safety in a Case of Innovative Hydraulic Technology," *Natural Resources Journal* 32 (Winter 1992): 77–100. Freeman was a close colleague of Arthur P. Davis and frequently corresponded with both Newell and Davis; see "Discussion Related to Dams," NA RG 115, Entry 3, Box 287.

[110]Donald Worster, *Dust Bowl: The Southern Plains in the 1930s* (New York, 1979), pp. 5–8; and Sherow (n. 4 above), pp. 1–7.

technological projects determine his actions because this was what gave him greatest satisfaction.

As reflected in his concern for monumentality in dam design, Newell was driven to oversee the erection of huge structures that would visually testify to the power of his vision. When the Reclamation Service began to flounder in a financial morass, he became deeply disturbed that the joy of creation was being squeezed out of his work. In 1910 he expressed this in a private letter: "The outlook is very dubious, and we do not know from day to day what will occur. I am keeping the work going as best I can under the circumstances, but, of course, there is not the feeling of satisfaction or of enthusiasm which formerly existed."[111] Newell's motivations were more complicated than simply being "pro-family farm" or "pro-business," and many of the problems associated with federal reclamation were rooted in his naive faith in the seemingly unconstrained bounty of technology. His zeal in espousing the rhetoric of scientific efficiency helped catalyze political support for the formation of the Reclamation Service, but, coupled with a relative lack of experience in the actual design and construction of dams and water-control structures, it also fostered expectations beyond what he and his agency could realistically provide to westerners, and to the nation as a whole.

In conclusion, historians should be wary of accepting uncritically the self-professed "progressive" and technologically efficient image that Newell and the Reclamation Service projected. Similarly, our understanding of Progressivism and its relation to western development is not enhanced by magnifying Newell's engineering capabilities while ascribing his shortcomings to the political maneuverings of others. And for historians of technology, a clearer understanding of why Newell left the Reclamation Service might prompt a different view of his accomplishments as founder/administrator of the American Association of Engineers. Frederick Haynes Newell warrants continued attention as a major figure in engineering history, but scholars should be encouraged to look deeper when assessing the significance of his work.

[111]F. H. Newell to J. B. Lippincott, February 11, 1910, Newell Papers.

13

A narrow window of opportunity: the rise and fall of the fixed steel dam

Terry S. Reynolds

Between 1890 and 1910 a few dam designers seriously considered steel as an alternative to such traditional dam-building materials as masonry, earth, rock, and concrete. Three fixed steel dams were constructed; two still survive. Using evidence from the two surviving structures, especially the steel dam at Redridge, Michigan, as well as written records, this article challenges previous explanations for the demise of the steel dam and suggests that the neglect of steel construction is better explained by perception and personal factors than by objective, scientific factors.

"Steel dams? Dams are made of earth, or rock, or masonry, or concrete—not steel!" Such is the typical reaction to the mention of steel dams. Yet two fixed steel dams stand in the United States: one in the northern Arizona desert, the other near Lake Superior on Michigan's sparsely populated Upper Peninsula. Moreover, even though they are now largely forgotten,[1] the fixed steel dam once attracted considerable attention as a viable alternative to conventional dam forms (see figure 1).

Using both written records and a surviving steel dam in Michigan, this article seeks to answer two questions: Why did steel attract attention as a dam-building material between 1890 and 1910? And why did interest in steel dams cease?

The Steel Dam, 1890-1910

At the turn of the century, a handful of civil engineers argued that steel had numerous advantages over more conventional dam-building materials. They believed that the joints between steel plates could be made, as in boilers, almost perfectly watertight—more watertight than masonry, earth, or concrete embankments. Dams made of steel, they argued, could be built for less than comparable structures, with a price advantage that increased with the dam's height. Calculations could be made more accurately for steel structures than for concrete, earth-fill, masonry, or rock-fill structures, and the uniformity and quality of materials were much more certain because steel was fabricated under shop conditions. Finally, steel dams offered easier accessibility for inspection, maintenance, and repairs, faster construction,

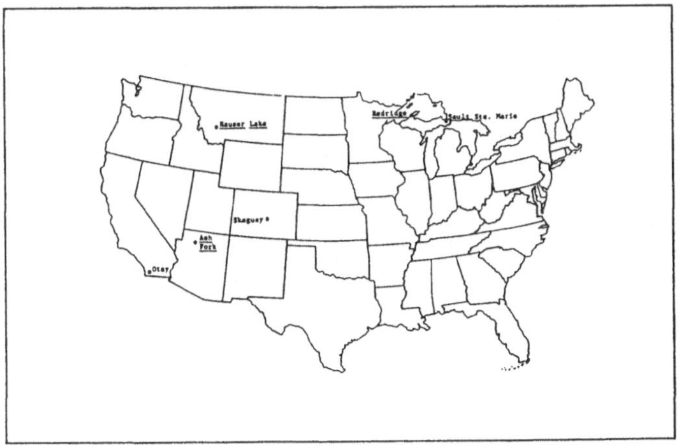

Figure 1. *Location of fixed steel dams (underlined) and other early U.S. dams with substantial steel component (1890-1910).*

Industrial Archeology

and (compared to concrete) greater ability to handle thermal expansion and contraction and uneven settling.[2]

These arguments were generally valid and made steel a realistic contender as a dam-building material. In 1893, for instance, the commission reviewing the Quaker Bridge dam in the New York water-supply system considered a steel dam, noting that it could be both economical and safe.[3] That same year F. H. Bainbridge, of the Chicago & North Western railroad, made estimates on constructing steel dams, concluding that they were economically feasible. In 1894 Bainbridge went further and prepared preliminary plans and estimates for a 210-foot-tall steel dam across the Santa Ana River in California.[4] The dam was never built, but Bainbridge patented his design in 1895.[5]

In 1897 Henry Goldmark, future designer of the ship canal locks for Panama and New Orleans, then at the beginning of a distinguished career as a consulting engineer,[6] studied the use of a steel superstructure for a dam near Ogden, Utah. Goldmark published several designs for structural steel dams, which "attracted considerable attention," even though they were not adopted.[7]

In 1897 the Santa Fe Railroad built the first American fixed steel dam near Ash Fork, in the arid northern Arizona desert. To secure a reliable water supply for its steam-powered locomotives, the Santa Fe began constructing a system of dams and storage reservoirs there in 1894. The first three dams were conventional masonry structures. In 1897, however, Bainbridge proposed steel. His proposal was accepted—probably because the site of the fourth Ash Fork dam was not easily accessible. Steel beams and plates could be moved there easier than building stone.[8]

Bainbridge designed the Ash Fork steel dam in collaboration with James Dun, the Santa Fe's chief engineer. They used conventional masonry construction for the shallow wings of the 300-foot-long overflow structure. But they installed steel plate for the deeper 184-foot-long central section. Twenty-four triangular bents, or frames, made from I-beams and spaced on eight-foot centers, supported ⅜-inch thick steel plates at a 45° angle. The plates were curved to permit expansion and contraction without affecting the joints between plates and bents and were riveted to the bents with their concave portions facing upstream. Because of very hard igneous rock at the site, Bainbridge and Dun anchored the steel bents and the toe of the dam directly to bedrock, although some concrete was used as a sealant.[9] In brief, the Ash Fork dam was a buttress dam with an inclined upstream face, similar in form to some dams built using wood frames earlier in the 19th century.[10] Santa Fe crews began work at Ash Fork in October 1897; they completed the dam in March 1898. (see figure 2).

Figure 2. *Downstream view of the Ash Fork, Arizona, steel dam during closing phases of construction, 1898. Note the steel bents used to support the curved steel plates facing upstream.* From *Engineering Record* 37 (April 9, 1898): 507.

The second fixed steel dam appeared shortly after in a completely different climatic zone—at Redridge on Lake Superior in the copper-mining region of Michigan's Upper Peninsula. In 1894 the Atlantic Mining Company built a stamp mill where the small Salmon Trout River flows into Lake Superior. The lakeside location gave the company an unlimited dump for its stamp sand, while the nearby river offered a gravity-fed water supply for stamping operations. To develop this supply Atlantic erected a timber crib dam across the Salmon Trout. In 1901 the Baltic Mining Company, another copper company and one whose board of directors interlocked with Atlantic's board, built a stamp mill on the opposite side of the river. Since the timber dam provided insufficient water for both mills and needed repair, the two companies agreed to jointly construct a new and larger dam (see figure 3).[11]

The Redridge site posed several problems. First, the immediate area had insufficient conventional dam-building materials. It had no good building stone for a masonry dam and too little topsoil for an earth-fill dam. Second, swift construction was essential. The rapid expansion of the electrical industry had driven up copper prices, so neither company wished to suspend stamping operations for long. Moreover, the severe climate of the region shortened the building season, making the slow pace of construction of conventional dams troublesome. These problems prepared the mining companies to consider the unconventional. In 1899 or 1900 J. F. Jackson, the Wisconsin Bridge and Iron Company's engineer in the region, proposed a steel dam as a solution to these problems. Wisconsin Bridge and Iron had supplied the steel for the Ash Fork steel dam, and Jackson had been responsible for its erection.[12] The companies accepted.

The Redridge steel dam is 1,006 feet long, including concrete-core, earthen embankments on both wings. The central steel section is 464 feet long and 74 feet high at its deepest point. In appearance it is similar to Ash Fork: bents of steel I-beams on eight-foot centers support an inclined layer of ⅜-inch curved steel plates. Redridge differs, however, in several respects. First, Redridge is not a structural dam. At Ash Fork inclined bents carry the lateral thrust of the impounded water directly to bedrock. The weight of the

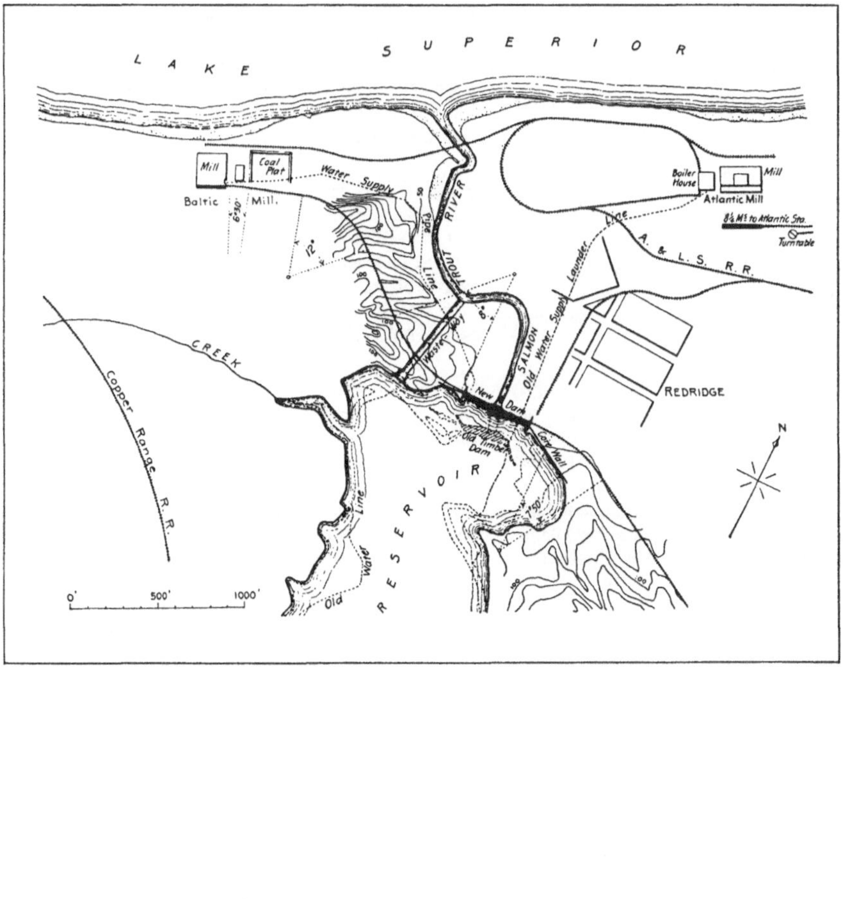

Figure 3. *Site plan of the steel dam at Redridge, Michigan, 1901.* From *Engineering News* 46 (August 15, 1901): plate following p. 101.

impounded water, not that of the dam, provides the resistance to uplift and lateral thrust. This approach could not be used at Redridge—the bedrock there was weak sandstone. Thus Jackson had to build a reliable foundation for his steel superstructure. He took advantage of sand from the nearby stamping operations and rock aggregate from mining to erect a large concrete base (8,000 cubic yards). This base, not the arrangement of steelwork and impounded water (as at Ash Fork), provides the force needed to resist uplift and sliding.

The use of the large concrete base sacrificed some of the vaunted economy of steel dams, but Jackson regained part of that cost. The water face on the Ash Fork dam was inclined 45°, so that the weight of the water on the plates could substantially contribute to the dam's resistance to sliding. The heavy concrete base performed this function at Redridge, so Jackson placed the steel plates closer to the vertical—around 55°. Less steel plate was required at that angle. In addition, the heavy concrete base gave the Baltic Mining Company a ready-made foundation for its trestle across the Salmon Trout River. The company placed the trestle's supports on the dam's base amid, but not connected to, the steelwork of the dam (see figure 4).

Another difference between the two dams lies in their water discharge systems. Bainbridge designed Ash Fork as a weir, or overflow, dam. Jackson and Foster Crowell, his consultant on hydraulic matters, designed the Redridge dam otherwise. Excess water at Redridge flowed out either through four large valve-controlled pipes in the concrete base of the dam or through a large waste-channel at the west end of the structure.[13]

Crews began construction at Redridge in May 1900 and completed it in November 1901.[14] The finished structure created a lake of 150 acres, submerging the old timber crib dam under 20 feet of water (see figure 5).

The completion of Ash Fork and Redridge stimulated further interest in the use of steel. In 1903 Charles Steiner, an engineer in the U.S. Reclamation Service, described a steel dam "under consideration" by the office of the Chief of the Hydrographic Bureau of the U.S. Geological Survey "in connection with the extensive construction of future storage reservoirs for irrigation." Steiner estimated that steel dams could be built for half the cost of masonry dams. His description, published in *Engineering News,* prompted a number of letters to that journal on steel dams.[15]

Shortly after, Jackson, the engineer who had erected the steelwork at Ash Fork and designed the steel gravity dam at Redridge, persuaded the Helena Power and Transmission Company to erect a steel dam on the Missouri River, 16 miles from Helena, Montana, to provide hydroelectric power to nearby copper mines, smelters, and urban areas.

The Hauser Lake dam was the third American fixed steel dam. It resembled the earlier ones in general design: steel bents carried an inclined face of curved steel plates. But there were important differences, especially in the foundations. At Ash Fork the steelwork rested directly on bedrock; at Redridge it rested on a large block of concrete, which, in turn, rested on bedrock. At Hauser Lake, however, the steelwork rested on longitudinal footings placed in a bed of water-bearing gravel, because bedrock, which lay 40 to 60 feet deeper, would have been difficult to reach. The leading, or lower, edge of the steel plates rested on a triangular concrete footing, or toe, which also extended the length of the dam. To prevent water from undermining the footings, Jackson used steel sheet piling. This piling, driven down 35 to 40 feet, extended the length of the dam and was anchored to the toe (see figure 6).

To provide increased stability for a structural dam not anchored in bedrock, Jackson gave the Hauser Lake dam a low angle of inclination—around 35° from horizontal, versus around 45° at Ash Fork and 55° at Redridge. This permitted him to use more of the weight of the water for stability against sliding.

Begun in 1905 and completed in March 1907, the Hauser Lake steel dam was around 670 feet long by 81 feet high. It was an overflow structure, for Jackson provided a spillway 500 feet long and 13 feet deep at the center of its 630-foot-long steel section. When placed in operation the Hauser Lake steel dam was both longer and higher than its sister dams at Ash Fork and Redridge (see table 1 and figure 7).[16]

The contemporary use of steel in association with other dam-building materials provides further evidence of widespread interest in steel as a dam-construction material between 1890 and 1910. For example, the Lower Otay dam in California, completed in 1898, used a vertical steel plate curtain in the center of a rock-fill embankment to prevent seepage.[17] The Skaguay dam in Colorado, built between 1901 and 1903, used steel plates to face a rock-fill dam.[18] And the combination hydroelectric dam and powerhouse erected between 1898 and 1902 at Sault Ste. Marie, Michigan, used curved steel plates anchored to reinforced concrete turbine chamber walls to restrain a 20-foot head of water.[19]

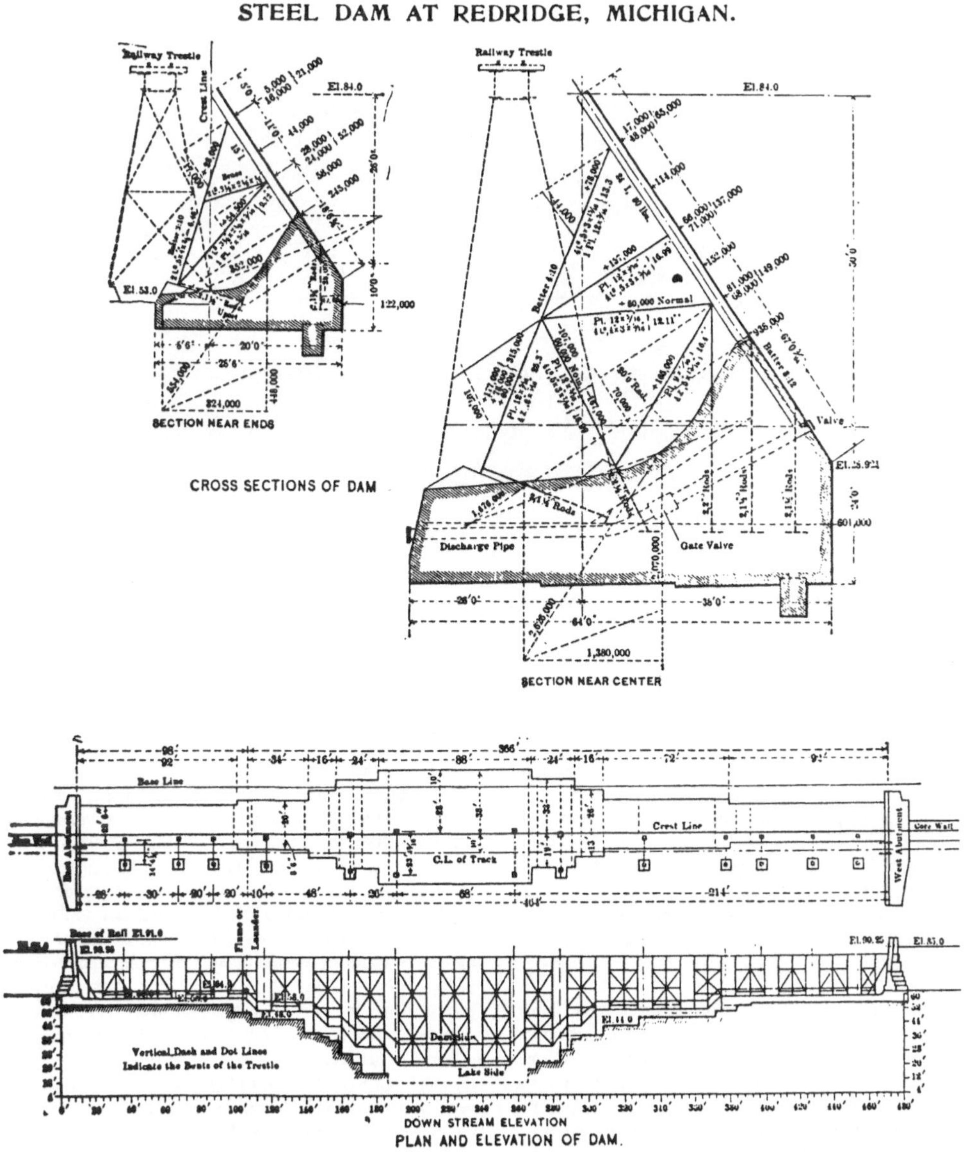

Figure 4. *Section, plan, and elevation of the Redridge, Michigan, steel dam. Note on upper right sectional view the outline of the railroad trestle which used the same foundation as the dam. Note also the size of the dam's concrete foundation.* From *Engineering News* 46 (August 15, 1901): plate following p. 101.

Figure 5. *View of upstream, or water face, of the Redridge steel dam as it neared completion in 1901. Figures in upper left background are standing on the railroad trestle which was placed on the same foundation as the dam but was not linked directly to the dam. The three pipes on the lower left were part of the dam's wastewater system.* From *Proceedings of the Lake Superior Mining Institute* 7 (1901): frontispiece.

Steel thus attracted widespread interest as a dam-construction material near the turn of the century. By 1910, however, this interest had vanished.

What Happened to the Fixed Steel Dam?

Why was this? The abrupt termination of interest in steel dams shortly before 1910 seems paradoxical in light of the numerous advantages that steel dams seemed to have over comparable concrete, masonry, rock-fill, and earth-fill structures. If steel dams enjoyed notable technical and economic advantages, and if three were actually constructed, demonstrating their feasibility, what happened? Why did steel dams abruptly enter a long period of near-total neglect around 1910?

In the limited published literature that discusses steel dams, five explanations, offered singly or in combination, commonly have been given for the demise of the fixed steel dam: higher maintenance expenses, leakage problems, the failure of the Hauser Lake dam, impermanence (especially due to corrosion), and "force of habit," or, in the jargon of historians of technology, the "technological momentum" enjoyed by more traditional forms of dam construction. I will review these explanations individually to determine whether artifactual and documentary evidence support them, and suggest other possible contributing factors.

Higher Maintenance

Creager's classic *Engineering for Dams* (1945) and Golze's more recent *Handbook of Dam Engineering* (1977) suggest that steel dams require greater and more constant maintenance than those made of other materials and that this was a factor in their disappearance.[20] But this explanation has problems. Reports published in 1916, in the 1930s, and in the 1950s on the Ash Fork dam indicate that maintenance costs were minimal. Ash Fork was painted only every seven to nine years and required little additional care.[21] Redridge was even less regularly maintained. When Jackson visited the site in 1930, he reported that it had been repainted only once—in 1913. Nonetheless, he noted: ". . .a hundred dollars or so would put everything, both steel and concrete, in perfect condition."[22] In 1935 A. L. Engels, Superintendent of the Copper Range Company (then owner of the dam), reported that Redridge had little rust even though it had last been painted in 1913. "To my knowledge," he added, "no repairs or replacements have been made to any of its steel work. . . .Its maintenance cost has been negligible."[23] A 1986 interview I conducted with Ray Franz, a former official with the Copper Range Company, revealed similar informa-

The Rise and Fall of the Fixed Steel Dam

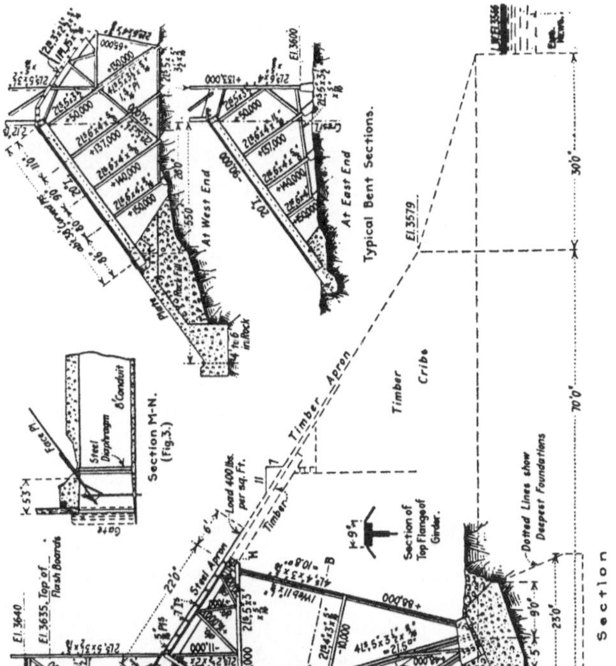

Figure 6. *Cross-section of the Hauser Lake, Montana, steel dam, 1907. The footings on which the steel bents rested were embedded in water-bearing gravel since bedrock was 40 to 60 feet deep at the dam site. The steel sheet piling driven in front of the dam (far left) was intended to prevent water from seeping under the dam and undermining the footings and hence the steelwork.* From *Engineering News* 58 (November 14, 1907): 508.

Industrial Archeology

Figure 7. *The Hauser Lake dam under construction, 1907. View from upstream, or water face.* From *Engineering News* 58 (November 14, 1907): 507.

Table 1
Comparison of the Three American Fixed Steel Dams

Dam	Total Length	Length (steel portion)	Maximum Height (steel portion)	Angle (from horizontal)	Storage Capacity (millions of gallons)
Ash Fork (1897-98)	300 ft.	184 ft.	46 ft.	45°	36
Redridge (1900-01)	1,006 ft.	464 ft.	74 ft.	55°	600
Hauser Lake (1905-07)	670(?) ft.	630 ft.	81 ft.	35°	?

tion—the steel portion of the dam had never posed serious maintenance problems.[24] An interview with Bill Brinkman, a longtime resident of the community that adjoins the dam, confirmed this: the dam had been painted, at best, only every 20 years or so.[25] Clearly, maintenance costs were not a serious factor in the demise of the steel dam.

Leakage Problems

As early as 1901 James Dix Schuyler, a prominent turn-of-the-century dam engineer, noted that the Ash Fork experiment was not likely to be repeated because: "It has been found difficult, and in fact impossible, to make a tight joint between the steel and masonry work."[26] In 1905 Lyman Cooley, a prominent Chicago consulting engineer and a prolific writer on waterways, cited the "extraordinary care" required in the junction of the steelwork of steel dams with the bottom and sides of the dam site as a key reason for opposing steel dams.[27] Schuyler and Cooley were supported by others in viewing leakage as the Achilles' heel of the steel dam. H. M. Hadley, a regional representative of the Portland Cement Association, asserted in 1933 that one of the principal problems with steel dams, and presumably one of the principal reasons for their demise, was the difficulty of linking, without leakage, the steel superstructure to the substructure (concrete or bedrock).[28]

The surviving artifacts, however, undermine this explanation as well. On the two surviving steel structures, leakage between the steelwork and the foundation materials—bedrock at Ash Fork, concrete at Redridge—does not appear to have been a serious or insoluble problem. Bainbridge, the engineer who designed the Ash Fork dam, admitted some early problems with leakage at the base, but noted that a coat of asphalt placed in 1900 on the concrete used to seal the footing had completely eliminated it.[29] His testimony has been subsequently supported by other observers.[30]

At Redridge, leakage between the concrete foundation and steel superstructure was *never* a problem. In 1903 Jackson was specifically asked about leakage at that junction and reported no problems. The only leakage in the entire dam was minor seepage through disintegrated sandstone beneath the concrete foundations, with no relationship to the questionable joints.[31] Later, in 1930, Jackson reported that the steel portion of the dam was "tight as a bottle."[32]

Personal inspection confirms Jackson's statements. If water had leaked between the steelwork and the concrete foundation at Redridge, evidence would still be visible in the form of rust stains on the concrete. Close personal inspection of the dam in the fall of 1987 and the spring of 1988 revealed no evidence of such stains. The only indication of leakage, past or present, was some slight seepage through the concrete base itself, not at the joint between steel and concrete. Interviews with a former Copper Range Company official and a long-term Redridge resident provided further confirmation. Neither was aware of any problem with leakage at the dam.[33] Leakage, like maintenance expenses, was not a key factor in the steel dam's disappearance.

The Hauser Lake Dam Failure

In 1908, after a year's service, the steel dam at Hauser Lake failed. Published accounts suggest that the key problem was the dam's unusual foundation, not its steelwork. As noted previously, the Hauser Lake dam, unlike those at Ash Fork and Redridge, was built on water-bearing gravel instead of bedrock. It used sheet piling to prevent water from undermining the concrete footings that supported the steelwork. Water, however, apparently seeped under the sheet piling or, according to others, seeped between the steel piling and the concrete footing at the toe of the dam, and washed the gravel from beneath the footings. This caused the central portion of the dam to collapse (see figure 8), with damages estimated at $250,000 to $300,000.[34] The dam was replaced between 1909 and 1911 with a more conventional concrete gravity structure.[35]

No fixed steel dam was built after the Hauser Lake disaster. This coincidence has provided several authors with a seemingly open-and-shut explanation for the disappearance of steel dams. For example, Jackson, the engineer involved in the construction of all three fixed steel dams, accepted this explanation. He noted in 1930 that after Hauser Lake "there seems to have been little if anything said or published in reference to this very special type of dam. . . ," implying a cause-and-effect relationship.[36] Similarly, C. E. Grunsky in 1910 and H. M. Hadley in 1933 cited Hauser Lake, and the problem of sealing between steel skin and concrete foundations, as critical reasons for abandoning steel dams.[37]

Yet the failure of Hauser Lake is not a sufficient, or even a good, explanation. In 1903 Nils F. Ambursen introduced a reinforced-concrete flat slab dam. Several of these dams failed, one in 1909 near Pittsfield, Massachusetts, for example.[38] This attracted as much attention as the Hauser Lake failure. But hundreds of Ambursen flat slab dams were built *after* the failure of the Pittsfield dam.[39] Similarly, a multiple-arch dam collapsed in northern Italy in December 1923.[40] While the failure had an impact, the construction of multiple-

Industrial Archeology

Figure 8. *The Hauser Lake dam after April 1908 failure.* From *Engineering News* 59 (April 30, 1908): 491.

arch dams did not come to an abrupt halt following it.[41] Finally, literally dozens upon dozens of masonry dams and earth- and rock-fill dams have failed, some doing far more damage than the Hauser Lake catastrophe. But these dam types continued to be built.[42] A single dam failure, in and of itself, cannot explain why steel dams have been neglected.

Impermanence

Probably the most frequently cited explanation for the demise of steel dams, and the one that likely comes to mind to *IA* readers, is impermanence. For example, *Engineering News* in 1916 noted that one "pronounced objection" to the steel dam was the great possibility of corrosion: "It seemed probable that the constant wetting to which the dam is subjected would hasten its rusting and early failure."[43] In 1932 C. M. Stanley, attempting to analyze why steel dams had not been more frequently used, noted that "the question of the permanence of steel dams has in all probability been one of the principal objections to the adoption of such construction."[44] And Hovey in a treatise on steel dams in 1937 cited "lack of confidence in its durability" as the "principal reason" for the death of the steel dam.[45]

Were steel dams impermanent? Did they seriously suffer from corrosion? The artifacts suggest that the answer is no. Consider Ash Fork: As early as 1916 a report on Ash Fork indicated that corrosion had *not* been a problem on the then-18-year-old structure.[46] Additional reports published in the 1930s and 1950s found that corrosion remained negligible. Moreover, the 1950s report pointed out that there was *less* deterioration in the steel portion of the dam than in its adjacent masonry abutments.[47] When T. Lindsay Baker inspected Ash Fork in the early 1970s he reported it in a "very good state of preservation." The upstream face had a coat of rust, but the dam's air face was "so well preserved that the date 1897 can still be clearly read on some of the curved steel sheets that form the face."[48]

The primary form of corrosion at Ash Fork was not general rust but pitting.[49] In 1916 Ash Fork's steelwork had 50 to 75 pits of 0.031-inch average depth and 0.062-inch average diameter per square foot. On steel plates 0.375 inches thick this was not much. Portions of the steel dam might look as if they had had smallpox, but there was little appreciable loss of section, and later reports on Ash Fork did not indicate that the pitting had become a serious problem.[50]

The Redridge dam also suggests that the "impermanency" explanation for the demise of the steel dam requires qualification. In 1930 Jackson visited the site after a long absence. He reported the plates in "first-class" condition, covered (and thus protected) by a light layer of slime on the water side and "entirely dry" and "in perfect condition" on the under, or air, side.[51] In 1935 A. L. Engels, Superintendent of the Copper Range Company, which by then owned the structure, reported that Redridge had "very little rust exposed," even though it had not been painted since 1913 (see figure 9).[52]

My personal inspection of the Redridge dam in the summer and fall of 1987 revealed that the structure remained in reasonably good condition despite almost total neglect for many decades and total exposure of the steelwork on the water face to air for the past eight years.

The Rise and Fall of the Fixed Steel Dam

Figure 9. *Aerial view of the Redridge steel dam around 1947. The small former milling town of Redridge is in the background.* Photo by Bill Brinkman, from Bill Brinkman Collection, Copper Country Archives, Michigan Technology University, Houghton, Michigan.

Micrometer measurements (see table 2) that I made on the steel plates at Redridge revealed several things. First, although all published accounts indicate that the steel plates were ⅜ (0.375) inches thick, the plates actually installed were thicker. Measurements made near the top of the structure, where practically no corrosion was evident, suggest that the plates were not precisely rolled and probably had originally thicknesses varying between 27/64 (0.422) and 7/16 (0.438) inches. Second, corrosion had not measurably diminished the thickness of the steel plate in many sections. And third, as at Ash Fork, pitting, not uniform rust, was the primary form of corrosion, especially in sections periodically exposed to both water and air.

This last point requires further elucidation. Pitting is evident on large areas at Redridge, but this pitting does not seem, despite many years of non-maintenance, to have seriously undermined the dam's integrity. In 1979 the Copper Range Company cut four large openings in the bottom of the steel section of the dam. These openings permitted me to take micrometer readings of the steel's thickness just above the dam's foundations at a point where pitting was evident. Table 2 provides the results of those readings, taken at six-inch intervals, one inch above two of the openings. Seventeen of the measurement points fell on portions of the steel plate which had not been pitted. These points averaged 0.416 inches thick, only 0.014 inches below the probable average original thickness of the two plates. Fifteen of the measurement points fell in points that were clearly pitted. The average thickness at these points was 0.329 inches, or 0.101 inches below probable average original thickness.

This diminution of cross-section was minor and did not seriously weaken the dam. In addition, I found a half-dozen pits near the very top of the dam that had completely penetrated the steel plate, but these pits were nothing more than pinpricks and could very easily have been spot-welded closed.

In brief, my inspection of corrosion and steel plate thickness at Redridge indicates that rust has *not* seriously undermined the Redridge dam, 87 years after construction and after many decades of near-total neglect.

In addition to corrosion, steel dams may have been considered impermanent due to their flimsy appearance when compared to the more massive concrete, earth, and rock gravity dams. But the Redridge dam's performance in flooding suggests that this form of the impermanence explanation also needs qualification. Unlike Ash Fork and Hauser Lake, which were overflow structures, the Redridge dam was *not* designed as an overflow dam. Due to inadequacies in the valves and waste way and the need of the stamp mills for a maximum head of water, the Redridge dam was subjected to the most severe test of a dam's stability—overtopping—on a number of occasions. The most severe test came at Easter in 1941 when a combination of snow melt (the area receives over 200 inches of snowfall annually), heavy rains, and the

Industrial Archeology

Table 2
Thickness of Steel Plate at Redridge Dam

Measurements taken in September 1987 from the center two of four openings cut in 1979 in the steel plates just above the concrete foundations. Measurements taken one inch from top of opening at six-inch intervals, east-to-west.

Specification Thickness: ⅜" = 0.375

Plate No. 1

Reading no.			
1	0.359 inch	9	0.417 inch
2	0.310	10	0.414
3	0.374	11	0.414
4	0.395	12	0.409
5	0.402	13	0.402
6	0.328	14	0.398
7	0.412	15	0.285
8	0.411	16	0.270

Average: 0.375
High: 0.417
Low: 0.270
Average of 9 readings where no clear pitting evident: 0.408
Average of 7 readings taken in clearly pitted areas: 0.332

Plate No. 2

Reading no.			
1	0.402	9	0.432
2	0.394	10	0.437
3	0.309	11	0.424
4	0.369	12	0.327
5	0.332	13	0.417
6	0.426	14	0.315
7	0.428	15	0.311
8	0.434	16	0.257

Average: 0.376
High: 0.437
Low: 0.257
Average of 8 readings where no clear pitting evident: 0.425
Average of 8 readings taken in clearly pitted areas: 0.326

collapse of beaver dams upstream led to a rush of water and debris that totally destroyed the dam's waste way after clogging it with floating timber, and that battered at and over-topped the dam for hours. The dam withstood the test with no measurable damage.[53]

The Ash Fork steel dam has survived more than 90 years; the Redridge dam nearly as long. In 1976, when the Redridge dam was approaching 75 years of age, a Copper Range Company official noted that the structure "could last another 75 years."[54] One-hundred-fifty years of life for a structure that has received practically no maintenance is commendable. In other words, the record of steel dams suggests that, especially if properly maintained and perhaps even if not, they have a life potentially as long as dams constructed of rival materials.

So, the impermanence explanation for the steel dam's demise has to be qualified. The two surviving dams indicate that steel structures were durable, and that, objectively, fixed steel dams should not have been neglected because of impermanency. But the frequency with which impermanency has been offered as an explanation suggests, I think, that it *was* a factor, albeit indirectly. Engineers may have rejected steel dams not because of solid, empirical evidence that they were impermanent, but because of a non-rational, intuitive belief that steel did not or could not offer security, or because of a fear that the general public would react to steel dams in that way. In other words, perception, or perception of others' perceptions, may have played a more important role than reality in the neglect of steel dams within the supposedly scientific and rational engineering community.[55]

The concern expressed by prominent engineers over the fragile appearance of the steel dam and the surprise expressed by others on discovering its durability support this contention. For example, the committee investigating the possibility of a steel dam at the Quaker Bridge site in 1893 rejected it because they *felt* that a steel dam did "not give *the idea of permanence* which should attach to the water supply of a great city like New York."[56] In 1903 H. M. Wilson of the U.S. Geological Survey referred to steel dams as "evidently ephemeral" and urged adoption of more permanent structures.[57] Lyman Cooley, prime mover behind the Chicago Ship and Sanitary Canal, and a leading figure in American hydraulic engineering, argued in 1905 that "frail or perishable structures" might be acceptable for temporary services in isolated regions, but added: "I. . . do not sympathize with the present tendency toward bric-a-brac structures. . . I regard steel as one of the most perishable materials for hydraulic work requiring. . . constant solicitude in maintenance. Timber is more satisfactory. . . ."[58] In a similar vein, in 1932 James B. Girand, a consulting engineer and an advocate of rock-fill dams, was "surprised at the state of preservation" when he visited the Ash Fork dam, probably because he did not expect it to be so well preserved.[59] Finally, a recent authority on dams, James Sherard, noted "the *fears* which have existed" about steel plate's corrosion.[60]

The large role that perceptions and intuitive fears can play on occasion in engineering decision-making may also partially explain why high maintenance expenses and the Hauser Lake failure have been seen as causes for the demise of the steel dam. As noted above, neither of these explanations can be supported on objective grounds. Maintenance expenses were very low, and single dam failures do not, in and of themselves, abruptly halt dam-building technologies. Yet the intuitive, unsupported *perception* that steel would have high maintenance expenses in hydraulic structures may have been more important than the contrary evidence quickly offered by Ash Fork and Redridge. And the *perception* that steel structures could not provide stability and safety may have led many in the engineering community to regard the Hauser Lake failure as a steel failure rather than what it was—a collapse due to faulty foundations that had nothing to do with the steel design.

The steel dam was not the only novel and light dam type to suffer from erroneous perceptions in the early 20th century—John Eastwood's multiple-arch concrete dams encountered similar reactions. His Big Meadows dam in California was abandoned before completion in large part because of "psychological" objections offered by prominent eastern civil engineers John R. Freeman and Alfred Noble. Freeman, for example, noted that "the psychology of these airy arches and the lace curtain effect of [Eastwood's] stiffening props is not well suited to inspire confidence."[61] Such a statement, modified to refer to thin steel sheets and airy steel frames, would likely describe Freeman's reaction to steel dams.

Force of Habit or Technological Momentum

Related to the "perception" of impermanency as an explanation for the demise of the steel dam is force of habit or the tendency to follow precedent. An *Engineering News-Record* editorial offered such an explanation in 1932.[62] What that editorial called "force of habit" is, in a sense, what historians of technology have termed "technological momentum": the power of an old, established technology to turn back challenges from technically superior solutions and survive, even after the factors that stimulated its emergence have disappeared.

Thomas Hughes introduced the concept of technological momentum in his account of work on hydrogenation in Germany between 1898 and 1933. He pointed out how World War I produced a body of engineers, chemists, and managers experienced in high-pressure hydrogenation processes and corporations that had invested heavily in such processes. After the war these groups successfully sought to continue development in the area, even after the need for and economic viability of hydrogenation processes had vanished.[63]

The analog to hydrogenation in dam building in the early 20th century was the massive gravity dam constructed with traditional materials–earth, rock, and masonry. That tradition negatively influenced the reception of steel dams in several ways. First, the tradition provided engineers and the general public with a paradigm of what a dam should look like–a massive, solid structure. Steel dams, by comparison, were bound to appear flimsy and unsound, no matter what calculations indicated; hence the comments, cited above, about steel dams being "evidently ephemeral" and "bric-a-brac structures." Second, the technological momentum of the gravity dam tradition meant that many prominent engineers had a ready-made interest in preserving its dominance. Massive gravity dams had emerged in an era when the lack of theoretical understanding made them the only safe structures to build. By 1900 there was a body of prominent engineers trained and experienced in their design and construction.[64] They were prepared to defend the traditional construction against other forms because it retained some advantages over its challengers, because intuitively it seemed safer and more permanent, and because they had experience with it. Economic reasons may also have intruded. Lighter, more economical dam designs could have eaten into incomes. Dam designers were traditionally paid a percentage of the total cost of construction. This created a built-in bias against economy and toward bulky, expensive, overly safe designs.

The steel dam was not the only dam form affected by the technological momentum of the traditional massive gravity dam. Donald C. Jackson, in a solid dissertation on John Eastwood and the multiple-arch dam, sees this as one of the factors behind the limited application of the economical multiple-arch structure. In considerable detail, he shows how eastern engineers in prominent positions who had long built in the massive gravity tradition undermined Eastwood, who sought to promote the more economical multiple-arch configuration.[65]

Comments made by engineers opposing the use of steel in dams lend support to the argument that attachment to the massive dam-building paradigm undermined the steel dam, just as it undermined Eastwood's multiple-arch design. Burr Bassell, resident engineer for the Kern River Company, commented in 1903 that he failed "to appreciate the necessity of resorting to purely metal construction" when earth or rock were nearly always available.[66] That same year, H. M. Wilson of the Geological Survey urged Reclamation Service

Industrial Archeology

engineers, who were considering the use of steel dams, to realize "the necessity for the most conservative design and most permanent construction in government works" and not to consider things as "untried" as steel.[67] Lyman Cooley, mentioned already as one suspecting the permanency of steel dams, provides an even better example of the attitude of the dam "establishment." In 1905 Cooley argued that dams were the most important and responsible class of structures engineers could be called on to design or construct and that therefore they should "have all the elements of permanence and safety which the engineering art will permit." To Cooley, permanence and safety implied massive gravity structures. He made it clear that steel structures, in particular, did not fit the bill.[68] Similarly, M. M. O'Shaughnessy, for two decades San Francisco's city engineer, responded to the *Engineering News-Record*'s 1932 editorial by asking: "Why experiment with steel when successful dams of the rock-fill type have been built and faced with concrete. . . ." He argued that engineers "should . . . adhere to successful materials and practice rather than experiment with novel types."[69] Finally, H. M. Hadley, a west coast engineer, commented in 1933 that he found "it very difficult to believe that any other type of dam was superior in respect to safety, security and service to the gravity dam."[70]

Thus the technological momentum possessed by the massive gravity dam-building paradigm, along with the associated perception of steel's impermanency, help explain the steel dam's demise. I believe, however, that two additional factors, not generally cited in the literature on steel dams, also contributed to the disappearance of the steel dam-building tradition: lack of a product champion and diminishing cost advantages.

Lack of a Product Champion

While there were multiple-arch dam and Ambursen slab dam failures, both traditions enjoyed a longer and fuller life than the steel dam, even if they did not displace the massive gravity tradition. While the fixed steel dam died an abrupt and complete death, these other dam types survived and continued for some decades to be built at least occasionally, partly because of their product champions.[71] Both the flat slab and multiple-arch dams had advocates who devoted their careers to proclaiming their advantages and encouraging their construction. The multiple-arch dam's great proponent was John Eastwood, who for two decades enthusiastically argued the merits of this design against significant opposition from prominent eastern engineers.[72] Ambursen played a similar role for the flat slab dam.

The steel dam had no such product champion. One of the candidates for such a role should have been F. H. Bainbridge, who patented the steel dam in 1895 and constructed the dam at Ash Fork. But after Ash Fork, Bainbridge apparantly withdrew from dam building. He wrote an article advocating structural steel dams in 1905,[73] but he played no observable role in either the Redridge or Hauser Lake dams. After 1905 little is heard from him.

A more serious candidate for product champion was J. F. Jackson. Jackson played a role in all three fixed steel dams. He was in charge of steel erection at Ash Fork and designed the Redridge and Hauser Lake dams. But after Hauser Lake, Jackson ceased active promotion of the steel dam, with the exception of a 1909 article proposing a new steel dam design and a short 1930 article discussing a long-dead dam-building tradition.[74] Perhaps the disappointment of the Hauser Lake collapse discouraged Jackson and his company, Wisconsin Bridge and Iron, from further promoting this use of steel. After Hauser Lake, Jackson continued to be very active as a structural engineer, designing or supervising construction of several large bridges and docks and designing ore-crushing plants and concentrating mills. He became a respected advocate of steel in mining and milling structures and a pioneer in the area.[75] But neither he nor his company was ever again intimately involved in dam design or construction.

Without an active product champion, without a technological enthusiast to push the new technology, steel dams had no one to demonstrate that the greater engineering community's perceptions about steel were false. Steel dams thus had little chance of derailing the technological momentum enjoyed by the massive gravity dam tradition.

Declining Cost Advantages

Economic factors also clearly played a role in the rise and fall of the fixed steel dam. They are particularly important, I believe, in explaining why interest in steel dams emerged when it did—in the mid-1890s.

The achievement of integrated, large-scale production in steel in the 1880s, the low demand and depressed market for steel following the panic of 1893, and the price wars between Carnegie and other steel manufacturers later in the decade brought steel prices to an all-time low in the mid- to late 1890s.[76] Steel billets, for example, averaged $31.03 a ton between 1884 and 1888. Following the economic collapse of 1893, they dropped to an average of $16.73 a ton between 1894 and 1898. The first serious proposals for steel dams appeared at this time, when steel prices were at an all-time low.[77]

After 1898 the price of steel rose, perhaps reflecting the consolidation of the American steel industry around U.S. Steel and increased demand. The average price for steel billets between 1904 and 1908, for instance, was $25.82 a ton, 54 percent higher than between 1894 and 1898, though still less than in the 1880s. Because of the technical advantages of steel construction in dams, the rising price of steel might not, by itself, have proven fatal to the steel dam or to the broader use of steel in hydraulic structures, if the prices of alternate materials for dam construction had also risen. But they did not. The prices of alternate materials, particularly concrete, steadily dropped, even after steel prices rose (see table 3).

Table 3
Comparison of Cost of Portland Cement and Steel Billets at 5-Year Intervals, 1889-1918

Average 1889-1893 price = 1.00 for index

	Portland Cement 376-lb. barrel		Steel Billets gross ton	
Interval	**Average Price**	**Index**	**Average Price**	**Index**
1889-93	$1.99	1.00	$25.73	1.00
1894-98	1.63	0.82	16.73	0.65
1899-03	1.20	0.60	27.46	1.07
1904-08	0.98	0.49	25.82	1.00
1909-13	0.87	0.44	23.83	0.93
1914-18	1.17	0.59	41.05	1.60

Sources:
Portland Cement: United States Bureau of the Census, *Historical Statistics of the United States* (Washington, D.C.: Government Printing Office, 1976), pp. 597-598.
Steel Billets: *Metal Statistics 1940* (New York: American Metal Market, 1940), p. 90, and Peter Temin, *Iron and Steel in Nineteenth-Century America: An Economic Inquiry* (Cambridge, Mass.: MIT Press, 1964), p. 284.

In contrast to steel, portland cement prices were relatively high in the early 1890s but were just at the beginning of a long decline due to the introduction of new production technologies like the rotary kiln.[78] Between 1889 and 1893 a 376-pound barrel of portland cement was sold f.o.b. at the mill for $1.99. Between 1904 and 1908 the same barrel cost only $0.98, or 51 percent less. Between 1909 and 1913 the average price dropped even lower, to $0.87, or 56 percent less than in the early 1890s. Thus, between the early 1890s and the middle of the first decade of the 20th century the price of steel rose 54 percent, while the price of cement fell by approximately the same percentage.[79]

What the comparative prices of cement and steel suggest is that steel dams enjoyed a brief "window of opportunity" from around 1893 to some point very early in the 20th century. In this period steel enjoyed cost advantages over concrete and had an opportunity to demonstrate its non-cost advantages, such as speed of construction and resistance to cracking from expansion or contraction. But this window closed in the early 20th century with declining cement prices.

Conclusion

Economic factors provide the best explanation for the sudden burst of interest in steel as a dam construction material in the mid-1890s and *help* explain its disappearance toward 1910. But they do not provide a sufficient explanation for the near-complete disappearance of steel from dam construction. Steel's cost advantage declined after 1900 but not for all situations. Moreover, steel continued to have economic advantages as a supplementary material, for example, to provide a watertight membrane over rock fill. Yet after 1910 steel ceased to be used, even in a supplementary role. So the causes of the demise of the fixed steel dam are also non-economic.

Most of the non-economic explanations offered, however, have focused on supposed technical shortcomings — leakage between steel and foundation, the need to protect steel from the elements with high maintenance expenses, and steel's impermanency. This is perhaps to be expected, since the ideology of engineering assumes that engineers always make technical decisions based on objective, measurable, scientific criteria. But the surviving artifacts and the literature on steel dams suggest that these technical shortcomings were not really evaluated by the engineering community in an objective manner, based on quantitative evidence. If they had been, the supposed leakage, maintenance, and impermanency problems and the failure of the Hauser Lake dam would *not* have led to the abandonment of steel dam construction.

What this study of the fixed steel dam suggests is that non-rational, non-measurable factors–particularly intuitive perceptions, some of which may be false–play a far greater role than is usually presumed in the supposedly objective and rational decision-making of the engineering community. The steel dam did not die because of measurable leakage, maintenance, or impermanency problems or because of the Hauser Lake failure. Steel dams died because of the *perception* that these should be problems with steel in hydraulic structures and the *perception* that steel dams were inherently unsafe. These perceptions, although false, were no doubt reinforced by the "force of habit," or technological momen-

Industrial Archeology

tum, enjoyed by the traditional massive gravity dam. Finally, the absence of a long-term product champion, an individual willing to stake reputation and career on promoting a new technological idea, enabled the false perceptions to prevail.

Engineering practice is often discerned by the engineering community and by the general public as a very rational, scientific endeavor, where technologies live or die for sound, objective, technical reasons. The case of the fixed steel dam implies that intuitive perceptions and personal factors can play a more important role than objective technical and economic considerations in engineering decisions.

Postscript

The steel dam at Ash Fork is likely to continue in use for some time. Isolated and still useful for stock watering, it seems in no danger of destruction.

It is less certain how much longer the Redridge steel dam will continue to stand. It is no longer a useful structure. The two stamp mills that it supplied with water are long gone; one closed in 1912 and the other in 1922. In the early 1950s the dam's owner, the Copper Range Company, considered dismantling it because it no longer had utility and had become a potential liability hazard.[80] Before pursuing demolition the company offered the structure to the Michigan Department of Conservation, hoping it would maintain the dam and reservoir for public use, but the department refused the offer.[81] Copper Range then tried to interest Houghton County, the Houghton County Road Commission, and Stanton Township in taking possession of the structure and its reservoir and using them as a public park. All three refused.[82] However, the Redridge dam escaped demolition, apparently because this would have compelled Copper Range to make extensive modifications to the roadway located below the dam and because estimates of the return from salvage were disappointing.[83]

In 1976 the Copper Range Company again considered dismantling the Redridge dam, fearing that spring flooding, the abandonment and overgrowth of the lateral discharge channel, the blockage of several of the discharge pipes, and general neglect of the structure might someday lead to dam collapse.[84] But the company found a more economical alternative. To relieve water pressure on the dam, in 1979 Copper Range cut four four-by-eight-foot openings in the steel plates just above the dam's concrete foundation.[85] These large openings drained most of the reservoir behind the dam, eliminating the water load on the steelwork. They have also completely exposed the steel on the dam's water face to the elements and have probably accelerated the dam's aging (see figures 10 and 11).

Presently the fixed steel dam is a dead technology, despite the two surviving examples. But the idea of using steel plate to supplement other dam-building materials, an idea that

Figure 10. *Water face of Redridge steel dam, 1988. View from the south taken from remnants of 1894 timber crib dam (foreground). Note sections of steel plate removed just above waterline (center of photo). These were cut out in 1979 to reduce water load on the dam. The railroad trestle which shares the concrete foundation of the dam is visible above the dam.* Photo by Author.

emerged with the fixed steel dam and apparently died with it, revived in the 1930s. Since then engineers have constructed a number of rock- and earth-fill dams faced with steel plates, both in the United States and abroad.[86]

Acknowledgments

This paper was presented in preliminary form at the 1987 Troy meeting of the Society for Industrial Archeology. Charles Hyde provided me with information on the Redridge dam early in the course of my research. Donald Jackson furnished me with several key ideas, with literature references, and with a host of detailed suggestions in a number of conversations in which we discussed the similarities between the fate of the fixed steel dam and the multiple-arch dam. Bruce Seely, my colleague at Michigan Tech, provided several valuable suggestions on organizing my argument. Finally, Laurence Gross, my session leader at Troy and one of my not-so-anonymous referees, forced me, through his suggestions, to rethink the organization and emphasis of the paper. All of them have made this a better article than it otherwise would have been.

Notes

1. Steel dams, for example, are not mentioned in Carl W. Condit, *American Building* (Chicago: University of Chicago Press, 1968), or Norman Smith, *A History of Dams* (Secaucus, N. J: The Citadel Press, 1972). Movable dams made of steel, though often not recognized as dams in the conventional sense, are more common than fixed steel dams. The steel gates on canal locks, for instance, can be considered movable steel dams. Other forms of movable steel

Figure 11. *Air face of Redridge steel dam, 1988. View from the north. Railroad trestle built on the same foundation as the dam projects above rim of dam. Openings cut in steelwork of the dam just above concrete foundation are visible in center.* Photo by Author.

Industrial Archeology

dams are reviewed in Leland R. Johnson, *The Davis Island Lock and Dam, 1870-1922* (Pittsburgh: U.S. Army Corps of Engineers, 1985), pp. 34-41.

2. The advantages of steel dams are discussed by numerous authors, including: Charles R. Steiner, "Proposed Steel Dam, Irrigation Reclamation Service, U.S. Geological Survey," *Engineering News* 49 (June 11, 1903):526-527, and "Steel Dams for Storage Reservoirs" (Letter to Editor), *Engineering News* 50 (August 6, 1903): 123; F. H. Bainbridge, "Structural Steel Dams," *Engineering News* 54 (September 28, 1905: 323-324; J. F. Jackson, "Four Steel Dams—Their Design and History," *Engineering News-Record* 104 (February 13, 1930): 281; C. Maxwell Stanley, "Why Not Steel Dams?" *Engineering News-Record* 109 (December 1, 1932):653-654; Otis E. Hovey, *Steel Dams* (New York: American Institute of Steel Construction, 1935), pp. 12-13, and "Principal Considerations in the Design of Steel Dams," *The Canadian Engineer* 73 (August 24, 1937):6; and William P. Creager, Joel D. Justin, and Julian Hinds, *Engineering for Dams*, vol. 3 (New York: John Wiley & Sons, 1945), p. 834.
3. Bainbridge, "Structural Steel Dams" (n. 2 above), p. 323. This article was printed in a slightly more extended form with additional illustrations and discussion in *Journal of the Western Society of Engineers* 10 (1905): 615-637. Since *Engineering News* is more widely available, all subsequent citations to Bainbridge's article will refer to the *Engineering News* article unless otherwise noted.
4. Ibid., p. 323.
5. U.S. Patent no. 537,520.
6. For a biographical sketch, see "Henry Goldmark," *Transactions of the American Society of Civil Engineers* 106 (1941):1588-1593.
7. Henry Goldmark, "The Power Plant, Pipe Line and Dam of the Pioneer Electric Power Company at Ogden, Utah," *Transactions of the American Society of Civil Engineers* 38 (December 1897):246-314. The discussion of possible steel construction appears on p. 293. See also Appendix B: "Designs for a Structural Steel Dam," pp. 302-305. The notice that Goldmark's consideration of steel "attracted considerable attention" comes from "Steel Weir, Ash Fork, Ariz.," *Engineering Record* 37 (April 9, 1898):404. Goldmark had some interest in steel dams as late as 1929, when he mentioned them briefly in a professional meeting: "Steel Fabricators Hold Record Meeting," *Engineering News-Record* 103 (November 21, 1929):811.
8. For the background of the Ash Fork dam, see: "Steel Weir, Ash Fork, Ariz." (n. 7 above); "Steel Dam at Ash Fork, Arizona; A., T. & S.F. Ry.," *Engineering News* 39 (May 12, 1898):299; James Dix Schuyler, *Reservoirs for Irrigation, Water-Power, and Domestic Water-Supply* (New York: John Wiley & Sons, 1901), pp. 214-215; and Hovey, *Steel Dams* (n. 2 above), p. 44.
9. For the technical details of the Ash Fork steel dam, see: Bainbridge, "Structural Steel Dams" (n. 2 above), pp. 323-324; Schuyler, *Reservoirs for Irrigation* (n. 8 above), pp. 222-224; "Steel Dam at Ash Fork" (n. 8 above), pp. 299-300; "Steel Weir" (n. 8 above), pp. 404-405; Edward Wegman, *The Design and Construction of Dams*, 7th ed. (New York: John Wiley & Sons, 1922), pp. 294-297; and Hovey, *Steel Dams* (n. 2 above), pp. 44-51.
10. See James Leffel, *Leffel's Construction of Mill Dams* (Springfield, Ohio: by the author, 1881), pp. 16-19, for similar timber buttress dams.
11. For the background of the Redridge steel dam, see Atlantic Mining Company, *Report of the Directors to the Stockholders*, for the years ending December 31st 1890 to 1900, esp. 1900, pp. 6-7. See also Baltic Mining Company, *Report of the Directors to the Stockholders for the Year Ending December 31st, 1900*, pp. 18-19. On the earlier timber crib dam, see "Michigan. Copper. Atlantic Mining Company," *Engineering and Mining Journal* 57 (May 16, 1894):494; "Atlantic Mining Company, Michigan," ibid. 59 (March 16, 1895):246; and Charles K. Hyde, *The Upper Peninsula of Michigan: An Inventory of Historic Engineering and Industrial Sites* (Washington, D.C.: Historic American Engineering Record, 1978), pp. 207-208.
12. Bainbridge, "Structural Steel Dams" (n. 2 above), p. 324.
13. For technical details on the Redridge dam, see: "The Redridge Dam," *Engineering News* 46 (August 15, 1901): 101-102; Schuyler, *Reservoirs for Irrigation* (n. 8 above), pp. 456-459; Horace J. Stevens, comp., *The Copper Handbook*, 1st ed. (Houghton, Mich.: Horace J. Stevens, 1900), pp. 254-256 (or 4th ed., 1904, pp. 209-210); Bainbridge, "Structural Steel Dams" (n. 2 above), p. 324; Wegman, *Design and Construction of Dams* (n. 9 above), p. 297; "The Redridge Dam," *Mining and Scientific Press* 83 (August 31, 1901):88-89; and Hovey, *Steel Dams* (n. 2 above), pp. 54-62.
14. Construction progress is outlined in: Baltic Mining Company, *Report*, 1900, pp. 15, 18-19; 1901, pp. 7, 16, 19; Atlantic Mining Company, *Report*, 1900, p. 14; 1901, pp. 6, 16; *Daily Mining Gazette* (Houghton, Michigan), August 24, 1900.
15. Steiner, "Proposed Steel Dam" (n. 2 above), pp. 526-527. For the letters, see *Engineering News* 50 (July 2, 1903): 13; (July 16, 1903): 59; and (August 6, 1903):123.
16. For technical details on the Hauser Lake dam, see: "The Hauser Lake Steel Dam in the Missouri River Near Helena, Mont.," *Engineering News* 58 (November 14, 1907):507-509; and Wegman, *Design and Construction of Dams* (n. 9 above), p. 298.
17. W. S. Russell, "A Rock-Fill Dam with a Steel Heartwall at Otay, Cal.," *Engineering News* 39 (March 10, 1898):157-158.
18. W. P. Hardesty, "The Water and Electric Power System of the Pike's Peak Power Co., Colorado," *Engineering News* 49 (January 1, 1903): 2-5. A later report on the excellent condition of this structure is J. R. Wemlinger, "Facing Dam with Steel Plates," *Engineering News-Record* 109 (August 25, 1932):233.
19. Terry S. Reynolds, "The Soo Hydro: A Case Study of the Influence of Managerial and Topographical Constraints on Engineering Design," *IA* 8,1 (1982): 37-56, or *Sault Ste. Marie: A Project Report* (Washington, D.C.: Government Printing Office, 1982).
20. Creager, *Engineering for Dams* (n. 2 above), p. 834; also Alfred R. Golze, ed., *Handbook of Dam Engineering* (New York: Van Nostrand Reinhold, 1977), p. 338.
21. George W. Lamb, "Steel Dam in Good Shape after 50 Years," *Steel Construction Digest* 7, 2 (April 1950): 14.
22. J. F. Jackson, "Four Steel Dams" (n. 2 above), p. 281.
23. Hovey, *Steel Dams* (n. 2 above), p. 60, citing a report of January 1935 from A. L. Engels, Superintendent, to W. H. Schact, President of the Copper Range Company.
24. Interview with Ray Franz, June 27, 1986. Franz is a former official, now retired, of the Copper Range Company, which owned the dam from 1905 until recently.
25. Interview with William Brinkman, May 11, 1987.
26. Schuyler, *Reservoirs for Irrigation* (n. 8 above), p. 223.
27. In discussion of Bainbridge, "Structural Steel Dams," *Journal of the Western Society of Engineers* (n. 3 above), p. 636.
28. H. M. Hadley, "In Favor of Gravity Dams" (Letter to the Editor), *Engineering News-Record* 110 (March 30, 1933): 415.
29. Bainbridge, "Structural Steel Dams" (n. 2 above), p. 324.
30. "Steel Dam Eighteen Years Old Still Well Preserved," *Engineering News* 75 (June 1, 1916): 1035; Hovey, *Steel Dams* (n. 2 above), pp. 47-53.
31. J. F. Jackson, "Copper Mining in Upper Michigan," *Journal of the Western Society of Engineers* 8 (February 1903): 20-21.

32. J. F. Jackson, "Four Steel Dams" (n. 2 above), p. 281.
33. Interview with Ray Franz, June 27, 1986, and with William Brinkman, May 11, 1987.
34. F. L. Sizer, "The Break in the Hauser Lake Dam, Montana," *Engineering News* 59 (April 30, 1908):491-492.
35. Barrett Smith, "The Construction of the New Hauserlake [*sic*] Dam," *Engineering News* 66 (August 10, 1911):162-164; George A. McKay, "Difficult Deep Foundation Work at the Hauserlake [*sic*] Dam, Montana," ibid. 65 (June 22, 1911):743-747.
36. J. F. Jackson, "Four Steel Dams" (n. 2 above), p. 281.
37. C. E. Grunsky in discussion on John S. Eastwood, "The Hume Lake Multiple-Arch Dam, *California Journal of Technology* 15, 3 (March 1910): 27 (which blames the Hauser Lake failure on difficulties of linking the steel skin with the foundations); Hadley, "In Favor of Gravity Dams" (n. 28 above), p. 415; J. F. Jackson, "Four Steel Dams" (n. 2 above), p. 281; Hovey, "Principal Considerations" (n. 2 above), p. 5; Stanley, "Why Not Steel Dams?" (n. 2 above), p. 652; and Donald Jackson, "A History of Water in the American West: John S. Eastwood and 'The Ultimate Dam'," Ph.D. diss. (University of Pennsylvania, 1986), p. 114.
38. "The Undermining of a Reinforced-Concrete Dam at Pittsfield, Mass.," *Engineering News* 61 (April 1, 1909):345-347. A second Ambursen dam failed in 1914: F. W. Scheidenhelm, "The Reconstruction of the Stony River Dam," *Transactions of the American Society of Civil Engineers* 81 (1917):907-1024.
39. D. Jackson, "Eastwood and 'The Ultimate Dam'" (n. 37 above), pp. 114, 115.
40. "Official Report on the Collapse of Gleno Dam," *Engineering News-Record* 93 (August 7, 1924): 213-215. D. Jackson, "Eastwood and 'The Ultimate Dam'" (n. 37 above), pp. 695-698, points out that the failure here was not due to the multiple-arch design, but to other circumstances such as foundations and poor-quality masonry. The same was the case, of course, with the Hauser Lake steel dam.
41. D. Jackson, "Eastwood and 'The Ultimate Dam'" (n. 37 above), pp. 747, 760.
42. American Society of Civil Engineers, *Lessons from Dam Incidents USA* (New York: American Society of Civil Engineers, 1975).
43. "Steel Dam Still Well Preserved" (n. 30 above), p. 1035. Among others who have cited corrosion as a major objection against steel dams or as a factor contributing to their failure to catch on are: D. Jackson, "Eastwood and 'The Ultimate Dam'" (n. 37 above), p. 114; John S. Fielding, "The Use of Steel in the Construction of Dams and Reservoirs," *The Canadian Architect and Builder* 10 (August 1897):148; Hovey, "Principal Considerations" (n. 2 above), p. 5, and *Steel Dams* (n. 2 above), p. 51 ("the principal objection"); Creager, *Engineering for Dams* (n. 2 above), p. 834; and Golze, *Handbook of Dam Engineering* (n. 20 above), p. 338.
44. Stanley, "Why Not Steel Dams?" (n. 2 above), p. 653.
45. Hovey, "Principal Considerations" (n. 2 above), p. 5.
46. "Steel Dam Still Well Preserved" (n. 30 above), p. 1035, and Bainbridge, "Structural Steel Dams," *Journal of the Western Society of Engineers* (n. 3 above), pp. 629-630. Bainbridge quotes from a letter written by A. F. Robinson, Bridge Engineer of the AT&SF system, noting that after the joint between steel and masonry at Ash Fork had been treated in the second season with asphalt at a cost of $150, "there has not been the slightest indication of leaks of any kind in or around the dam."
47. J. F. Jackson, "Four Steel Dams" (n. 2 above), p. 281; James B. Girand and W. B. Storey, "Steel-Faced and All-Steel Dams" (Letters to the Editor), *Engineering News-Record* 109 (October 27, 1932): 505-506; Lamb, "Steel Dam in Good Shape" (n. 21 above), pp. 13-14, and "Steel Dam in Good Shape after 50 Years," *Engineering News-Record* 144 (January 12, 1950): 36-37. For the comment on less deterioration in steel portion than masonry portion, see Lamb, "Steel Dam in Good Shape" (n. 21 above), p. 37.
48. T. Lindsay Baker, Steven R. Rae, Joseph E. Minor, and Seymour V. Connor, *Water for the Southwest: Historical Survey and Guide to Historic Sites* (New York: American Society of Civil Engineers, Historical Publication no. 3, 1973), p. 33.
49. The pitting is probably due either to ice action against the face of the steel or to air dissolved in the water at low temperatures. Generally, however, the light slime which covers the steel on a steel dam prevents entrained air from reaching the steel to cause rust.
50. Lamb, "Steel Dam in Good Shape after 50 Years," *Engineering News-Record* (n. 47 above), p. 37.
51. J. F. Jackson, "Four Steel Dams" (n. 2 above), p. 281.
52. In Hovey, *Steel Dams* (n. 2 above), p. 60.
53. An account of the 1941 flood with an illustration of the water overtopping the Redridge dam can be found in *Daily Mining Gazette* (Houghton, Michigan), March 22 and April 9, 1982. Also: interview with William Brinkman, May 11, 1987. Brinkman has lived in Redridge near the dam for most of his 81 years.
54. Raymond A. Kenney to Edward R. Bingham, October 22, 1976 (Copper Range Company records). Copper Range Company records relating to the Redridge dam apparently date back only to 1951. The surviving records were kindly loaned to me by John Lasio of Laurium, Michigan, who is currently engaged in co-authoring a history of the Copper Range Company. The Copper Range Company contacted Wisconsin Bridge and Iron Company about surviving records on the Redridge dam in 1951 (H. A. Kiesow to William Nicholls, July 10, 1951) but was informed that the contract files on the dam had either been lost or destroyed.
55. Fielding, "The Use of Steel in Dams" (n. 43 above), p. 148, pointed out that rationally there was not much in the argument that water in contact with steel would create a major corrosion problem. If this were so, he pointed out, ". . . why use iron water mains or steel bridge piers or even steel ships, steel lighthouses, steel piles, . . ."
56. Bainbridge, "Structural Steel Dams" (n. 2 above), p. 323, emphasis added. See also Cooley's comments on Bainbridge's paper in the version in the *Journal of the Western Society of Engineers* (n. 3 above), p. 635.
57. H. M. Wilson, "Steel Dams and Concrete-Steel Diaphragms for Earth or Rock-Fill Dams" (Letter to Editor), *Engineering News* 50 (July 16, 1903):59.
58. Comment by Lyman E. Cooley on Bainbridge, "Structural Steel Dams," *Journal of the Western Society of Engineers* (n. 3 above), p. 636.
59. James B. Girand to M. N. Baker, August 22, 1932, letter published in M. N. Baker, "Steel-Faced and All-Steel Dams," *Engineering News-Record* 109 (October 27, 1932): 505-506.
60. James L. Sherard et al., *Earth-Rock Dams: Engineering Problems of Design and Construction* (New York: John Wiley, 1963), p. 480, emphasis added. Sherard's work notes that experience with steel plate membranes indicates that steel plates have approximately the same life as reinforced concrete, that corrosion of the plates has never been a serious problem, and that maintenance costs on steel dams have been very low.
61. D. Jackson, "Eastwood and 'The Ultimate Dam'" (n. 37 above), pp. 368-379.
62. "Force of Habit," *Engineering News-Record* 109 (December 1, 1932): 658.
63. Thomas P. Hughes, "Technological Momentum in History: Hydroge-

nation in Germany 1898-1933," *Past and Present* 44 (August 1969): 106-132.

64. Smith, *History of Dams* (n. 1 above) is practically the only survey history of dams.
65. D. Jackson, "Eastwood and 'The Ultimate Dam'" (n. 37 above), esp. chap. 8, pp. 747-823. Eastwood's position was further undermined by two other developments that reinforced the technological momentum already possessed by the massive gravity dam tradition: (1) the cautious nature of the emerging state regulatory bureaucracies, often dominated by more traditional dam designers, which placed safety far above economy, and (2) the replacement of private industry, with its emphasis on economy, by the federal government as the major builder of dams.
66. Burr Bassell, "The Type of Steel Dam Proposed by the U.S. Irrigation Reclamation Service," *Engineering News* 50 (July 2, 1903): 13.
67. Wilson, "Steel Dams and Concrete-Steel Diaphragms" (n. 57 above).
68. Comment by Cooley on Bainbridge, "Structural Steel Dams," *Journal of the Western Society of Engineers* (n. 3 above), p. 636.
69. M. M. O'Shaughnessy, "Steel-Faced Dams," *Engineering News-Record* (July 28, 1932): 113.
70. Hadley, "In Favor of Gravity Dams" (n. 28 above).
71. A product champion is an individual or company willing to invest heavily in time, money, or reputation to support a particular product or concept. The concept was used, for example, in J. Langrish, M. Gibbons, W. G. Evans, and F. R. Jevons, *Wealth from Knowledge* (London: Macmillan, 1972).
72. D. Jackson, "Eastwood and 'The Ultimate Dam'" (n. 37 above), provides an excellent overview of Eastwood's career-long promotion of the multiple-arch dam.
73. Bainbridge, "Structural Steel Dams" (n. 2 above), pp. 323-324.
74. J. F. Jackson, "Some Observations on the Stability of Dams," *Engineering News* 62 (July 29, 1909): 120-121, and "Four Steel Dams" (n. 2 above).
75. Raleigh W. Gamble, "John Franklin Jackson," *Transactions of the American Society of Civil Engineers* 104 (1939): 2042-2043.
76. See Peter Temin, *Iron and Steel in Nineteenth-Century America: An Economic Inquiry* (Cambridge, Mass.: MIT Press, 1964), pp. 188, 208; and W. David Lewis, *Iron and Steel in America* (Greenville, Del.: Hagley Museum, 1976), pp. 50-51.
77. During the Great Depression of the 1930s, steel prices again dropped sharply. As a result, the steel industry began to seek new markets for its product and attempted to revive the fixed steel dam. This explains the American Institute of Steel Construction's sponsorship of Hovey's *Steel Dams* (n. 2 above) in 1935.
78. On the importance of the rotary kiln in the manufacture of concrete, see, for example, Jasper O. Draffin, *A Brief History of Lime, Cement, Concrete, and Reinforced Concrete* (Urbana, Ill.: University of Illinois, Bulletin 40, June 29, 1943), p. 12; A. W. Skempton, "Portland Cements, 1843-1887," *Transactions of the Newcomen Society* 35 (1962-1963): 149; and P. Gooding and P. E. Halstead, "The Early History of Cement in England," *Proceedings of the Third International Symposium on the Chemistry of Cement, 1952* (London: Cement and Concrete Association, 1954), p. 26.
79. Bainbridge, "Structural Steel Dams" (n. 2 above), p. 323, is the only author to have pointed to the changing prices of steel and concrete. He noted that structural steel was as low as one cent per pound in 1894, but in 1905 was 75 percent higher. On the other hand, he pointed out, portland cement's price was about ⅓ of what it had been in 1894. He concluded: "Had the relative values of steel and Portland cement in 1894 been maintained, the subject of steel dams would have received far greater attention than has been the case." Low prices definitely encouraged the use of steel. J. F. Jackson, "Copper Mining in Upper Michigan" (n. 31 above), p. 21, noted: "It was believed, at the time the work was under way [on the Redridge Dam], that a steel dam was considerably cheaper than a dam of masonry or entirely concrete."
80. William P. Nicholls, Manager, Lands and Forestry Division, Copper Range Company, to F. P. Strusaker, Secretary, Michigan Department of Conservation, August 7, 1951 (Copper Range Company records, n. 54 above).
81. Ibid., and Strusaker to Nicholls, September 28, 1951 (Copper Range Company records, n. 54 above).
82. P. F. Beaudin, Vice President, Copper Range Company, to W. E. Romig, General Manager, Copper Range Company, October 4, 1951; Memo, Copper Range Company, November 14, 1951 (Copper Range Company records, n. 54 above); *Daily Mining Gazette* (Houghton, Michigan), July 9, 1952.
83. William A. Waara, Superintendent, Copper Range Company, to William P. Nicholls, Manager, Lands and Forestry Division, June 30, 1952, and handwritten memo on expected salvage revenue, July 10, 1961 (Copper Range Company records, n. 54 above).
84. Memo, Coppper Range Company, February 9, 1976; Edward R. Bingham, Vice President, Copper Range Company, to Kathryn B. Eckert, Michigan History Division, March 17, 1976; and "Notes on Meeting: Redridge Dam and County Road," September 7, 1976 (Copper Range Company records, n. 54 above).
85. *Daily Mining Gazette* (Houghton, Michigan), April 23, 1979, and April 9, 1982.
86. For later use of steel as a facing on earth- and rock-fill dams, see, for example, H. I. Reid, "Steel Plates with Welded Joints Seal Rockfill Dam," *Engineering News-Record* 108 (May 26, 1932):761-763; J. D. Galloway, "The Design of Rockfill Dams," *Transactions of the American Society of Civil Engineers* 104 (1939): 21, 91; Charles P. Seger, "Steel Used Extensively in Building El Vado Dam," *Engineering News-Record* 115 (August 15, 1935):211-215; H. P. Bunger, "Constructing the First Large Steel Faced Dam for Irrigation Storage in New Mexico," *Western Construction News* 10 (February 1935): 36-41; H. P. Bunger, "Metal-Faced Gravel Dam Being Built in New Mexico," *Engineering News-Record* 111 (October 26, 1933): 504-505; "Metal-Faced Dams," ibid.: 510; Sherard, *Earth-Rock Dams* (n. 60 above), pp. 479-487. Sherard describes, in addition to the El Vado dam in New Mexico (1934), later dams such as the Alazar dam in Portugal (1948) and the Caracas Water Supply dam in Venezuela (1958).

14

Considering the multiple arch dam: theory, practice and the ethics of safety in a case of innovative hydraulic engineering

Donald C. Jackson

ABSTRACT

Various factors influence the technological decisionmaking process and lead engineers to devise solutions to technical problems that cannot be construed as "single-best" answers that respond to the desires of all interested parties with equal alacrity. In this article, the technology of dam building in the early 20th century is examined with specific attention given to the work of John S. Eastwood and his development of the reinforced concrete multiple arch dam. This type of dam offered a way of reducing the cost of dam construction by as much as 50 percent over more traditional technologies and was promoted by Eastwood as part of his search for a better, more efficient means of storing water in the arid western United States. Beyond introducing the reader to a distinctive type of water storage technology, the article elucidates some of the vagaries that intercede into the space separating the world of engineering from the world of science. In this context, the role of practice (or experience) is contrasted with more mathematically oriented approaches to design that are often characterized as "scientific." In conclusion, the article stresses that the scientific and rational nature of modern technology is not always as absolute as generally believed. Safety remains a relative value that only has meaning in terms of how people choose to evaluate a wide range of factors and influences. A key lesson illustrated by Eastwood's experiences in promoting the multiple arch dam is that society, and the legal system, cannot afford to relinquish its responsibility over the technological decisionmaking process to "experts" without appreciating the traditions and technical prejudices that may color their view of what is safe and/or acceptable.

When considering water resources development and the decisionmaking process that controls how hydraulic technologies are implemented, historians and public policy professionals often view engineering projects as representing a "single-best" solution to a problem. After all, modern engineers are supposedly trained to provide efficient, scientific answers which, although some parties may find them to be politically or socially undesirable, comprise rational reactions to a situation.

A difficulty with this conception of engineering is the presumption that all engineers ultimately will discern a similar "objective" design so long as they devote enough time and effort to solving a problem. Of course, this is not always the case. However, society as a whole is reluctant to abandon the notion that engineering is an exact science capable of developing ideal responses to technological quandaries.

Although we should not be enamored by a blind faith in technical proficiency, we often take comfort in thinking that adherence to a "scientific method" will foster value-free solutions that, at the very least, are optimal within an engineering context. As part of such thinking, a faith has developed that mathematically-based, theoretical approaches to technical problem solving are more rigorous, and hence more valid, than methodologies stressing the importance of qualitative considerations based upon practical experience. In a legal environment where the influence of expert opinion on judicial proceedings is often quite great, this tendency can have far reaching ramifications affecting the very heart of how our society governs itself.

This paper examines a type of hydraulic technology developed and promoted in the early twentieth century as part of a search for a better, more efficient means of storing water. The first reinforced concrete multiple arch dam was built in 1908–09 by John S. Eastwood in response to the problem of erecting large storage dams while using a very small amount of material.[1] Eastwood's work in developing the multiple arch dam reflected both an interest in designing inexpensive, economical structures and a desire to devise a technically superior method of water storage. In promotional literature, he unabashedly extolled the virtues of his designs by counselling potential clients with advice such as, "Why Build Any Other Type of Dam When You Can Get the Best For Less Money."[2] He further asserted that his self-proclaimed "Ultimate Dam" could be "fitted to any site long, short, high, low, straight, curved, angular, wide or narrow, with greater economy of cost for safety than any other type, because it is designed on TRUE SCIENTIFIC PRINCIPLES."[3]

1. *See* D. Jackson, A History of Water in the American West: John S. Eastwood and the 'Ultimate Dam' (1908–1924) (1986) (unpublished Ph.D dissertation, University of Pennsylvania; for general biographical information on the life of John S. Eastwood). This dissertation contains much of the material presented within this article and is the most comprehensive document available concerning Eastwood's career as a dam engineer. *See also* Whitney, *John S. Eastwood: Unsung Genius of the Drawing Board*, 19 Montana: The Magazine of Western History 38 (1969) (for a good condensed treatment of Eastwood's life).

2. Eastwood, *Eastwood Bulletin*, 5 W. Engineering (1915) (a four-page promotional leaflet was distributed with this technical journal).

3. *Id.* (emphasis in original).

Between 1908 and 1924, when Eastwood died, 17 dams were built using Eastwood's multiple arch designs. A few other engineers became multiple arch dam advocates in the late teens, 1920s and early 1930s, but by World War II the technology had passed out of the design lexicon of American engineering.[4] It became nothing more than an arcane structural type that received, at best, cursory notice in hydraulic engineering textbooks. No multiple arch dams have been built in the United States since World War II but the technology continues to play a role in the world's water resources development. Many of Eastwood's dams, as well as multiple arch dams designed by other engineers, remain in service as they enter their seventh decade or more of operation.

Beyond introducing the reader to a distinctive type of water storage technology, this article elucidates some of the vagaries that separate engineering from the world of science. After providing essential background on the basic principles of dam design, attention focuses on how theoretical methods of design based upon mathematical formulas were used (and later judiciously ignored) by Eastwood in his search for a "scientific" design. In this context, the role of practice, or experience, in the evolution of his design method will be explained and contrasted to more mathematically oriented methods of structural design employed by other early twentieth century dam engineers. In addition, the importance of visual appearance as a criterion by which the engineering community assessed the validity of dam designs will also be discussed.

The distinction between a mathematically rigorous, yet not necessarily imaginative, method of[5] engineering design and a more visually intuitive, yet still carefully reasoned, approach to design has most recently been brought to public attention by Professor David Billington of Princeton University. A structural engineer by training and a leader in the development of thin shell design theory, Billington has devoted much of his career to the study of technological history. In his 1983 book *The Tower and the Bridge: The New Art of Structural Engineering,* Billington examines a range of developments in bridge, roof and building design that have occurred since the early 19th century. His overall analysis includes discussion of how the proliferation of wrought iron, steel and reinforced

4. Aside from Eastwood, the most prominent engineers associated with the promotion of the multiple arch dam were Lars Jorgensen, Fred Noetzli, and B.F. Jakobsen. Interestingly, all three were Europeans who immigrated to California in the early 20th century after receiving advanced technical training at European universities. *See* Jorgensen, *Multiple Arch Dams on Rush Creek, California*, 81 Transactions Am. Soc'y of Civil Engineers 850 (1917); Noetzli, *Improved Type of Multiple Arch Dam*, 87 Transactions Am. Soc'y of Civil Engineers 342 (1924); Jakobsen, *Stresses in Multiple Arch Dams*, 87 Transactions Am. Soc'y of Civil Engineers 276 (1924). The disappearance of the multiple arch dam from American engineering practice is reflected by the lack of articles on the subject since the early 1930s in the Transactions of the American Society of Civil Engineers.

5. D. Billington, The Tower and the Bridge: The New Art of Structural Engineering (1983).

concrete during the industrial revolution opened up possibilities for engineers to develop new structural forms which differed dramatically from those suited for traditional stone masonry work. In addition, Billington develops a concept of "structural art" based upon the ideals of material conservancy, economic efficiency and visual elegance. Without restating at length the principles underlying Billington's idea of "structural art," his views on the evolution of engineering design in the late nineteenth and early twentieth century are worth noting:

> [Following World War I,] everything from factory management to the unconscious sex drive was thought to be amenable to scientific analysis. The truths of life were bound up in formulas which great minds, in academic settings, were carefully setting about to refine and solve . . . In structural engineering this faith in formula led to an emphasis on mathematical analysis which . . . seemed to require that designers justify their works by sophisticated calculation . . . [but] the leading structural artists resisted that trend and sought rather to base design on simplified calculations and on observations of physical behavior.[6]

In his book, Billington does not focus on dam design or any specific issues surrounding the development of the multiple arch dam. Nonetheless, in discussing the evolution of thin shell concrete design, he makes observations directly relevant to an analysis of American dam building:

> [A]s the 1920s wore on, science-based formulas gained prestige. The major results of these formulas were to create a scientific discipline of analysis isolated from design, to set up standard methods of analysis which would come to have quasi-legal status in codes of practice . . . this was not a question of intuition versus rigor, or of approximate and uncertain estimates versus precise and scientific predictions. The primary uncertainties in reinforced concrete behavior have never been removed by mathematical analysis however rigorously consistent.[7]

In other words, the structural engineering establishment of the early twentieth century vigorously embraced the idea that the use of mathematically sophisticated analysis necessarily would foster more accurate methods of design even if mathematical models did not always correlate with physical reality. Billington clearly rejects the position that more complicated mathematical techniques necessarily leads to a more accurate calculation of stresses within a structure. In fact, his notion of "structural art"

6. *Id.* at 213–14.
7. *Id.* at 214–15.

promotes a view of engineering in which practical experience supplements and guides the use of mathematical theory as a means of developing more efficient and socially useful designs. However, the more pervasive trend in 20th century engineering has been antithetical to the principles of "structural art" and has embraced mathematical theory as a primary determinant of structural design rather than as a significant, yet somewhat subordinate, tool in the overall design process.[8] As described below, Eastwood's experiences in promoting multiple arch dams adhere closely to Billington's more generalized observations on the development of structural engineering in the 20th century. But first, in order to relate Billington's views to the particular nature of Eastwood's approach to dam design, some historical background is in order.

THE MASSIVE AND STRUCTURAL TRADITIONS

Dam design can be divided into two basic traditions extending back hundreds of years.[9] The first, and most common, is the massive tradition oriented toward building up such massive quantities of material in a dam that the horizontal thrust of the water in the reservoir is insufficient to move or dislodge the structure from its foundations. The massive tradition is exemplified by earthen or masonry dams of huge proportions erected in accordance with the dictum "more is better." Beginning with dam builders in ancient Egypt and Mesopotamia, the massive tradition represented a simple and logical response to the problem of water storage technology: assemble a large quantity of material in order to insure safety. If a dam washed away, builders would learn that more material was necessary in order to insure structural integrity. Eventually, the massive tradition led to the development of several basic types of earthen, rock and masonry designs that proved their stability through actual construction and long term service. Because massive dams rely solely upon their weight for stability they are usually called gravity dams in recognition of how the force of gravity acting upon the structure's mass is what provides resistance against hydrostatic pressure.

By the mid-nineteenth century European engineers were using mathematical formulas to determine the proportions of masonry gravity dams in an effort to develop a more "rational" method of design for these

8. In discussing the work of the great Swiss bridge engineer Robert Maillart (arguably the greatest of all structural artists), Billington notes that the conflict between mathematical complexity and interest in choice of structural form "would engage Maillart over his entire career, and the power of his designs both scientifically and visually would grow as he went further and further from the mathematical tradition which was increasingly dominating the profession [in the 1920s and 1930s]." *Id.* at 215.

9. N. Smith, A History of Dams (1972); Schnitter, *Roman Dams*, 3 Water Supply & Mgmt. 29 (1979); Schnitter, *The Evolution of the Arch Dam*, 19 Water Power & Dam Construction 34 (1976) (for basic history of early dam construction).

types of structures.[10] Beginning with the Furens Dam in France in the 1860s and continuing with prominent structures such as the Gilleppe Dam in Belgium, the Vrnwry Dam in England, and the New Croton Dam outside of New York City, by the early twentieth century the masonry gravity dam became recognized as the "design-of-choice" for large municipalities and wealthy manufacturing interests.[11] The use of mathematical formulas helped rationalize the design of gravity dams, but structures of this type still required large quantities of material (as befit the massive tradition) and thus were quite expensive.

In addition to solid masonry structures and their antecedents formed out of concrete, the massive tradition includes "fill" dams composed of loose earth, rock or a combination of these two materials. Earthfill and rockfill dams invariably contain huge quantities of construction material and they assume the form of giant sloped mounds large enough to resist the horizontal thrust exerted by water stored in a reservoir. Obviously, earth and loose rock are susceptible to erosion and this limit s their desirability in dam construction projects. However, they often are extremely inexpensive materials that can be located so near a potential dam site they comprise the most economical material for a project. Modern earthen and rockfill dams are not irrationally designed, but the proportions employed for these types of structures primarily are determined by empirically-based experiences rather than a seemingly rigid application of mathematical theory.[12] In many ways, earthfill and rockfill dams exemplify why dam technologies that rely upon excessive size are best characterized as being part of a massive tradition.

In contrast to the massive tradition, the structural tradition of dam design relies upon the shape of a dam rather than mere size in order to insure stability. Although far less common than massive dams, structural dams date back to Roman times and have been built in many parts of the world.[13] In the simplest variant, structural dams rely upon a curved up-

10. The famous "profile of equal resistance" that presented a mathematically rational means for designing masonry gravity dams was first proposed by the French engineer M. De Sazilly in 1853; it was first implemented by M. Delocre for the Furens Dam in France in the 1860s. Smith, *supra* note 9, at 197–200. *See* E. Wegmann, The Design and Construction of Dams 1–45 (3d ed. 1911) (for subsequent development of this profile). The term "equal resistance" derives from the concept of equalizing the compressive stresses at the upstream face of the dam when the reservoir is empty with the compressive stresses at the downstream edge of the dam when the reservoir is filled. The profile (or cross-section through the dam) developed in response to this criteria takes on a characteristic triangular shape with a height-to-width ratio of approximately 3 to 2.

11. Wegmann, *supra* note 10, at 63–209 (for descriptions of these dams).

12. *E.g.* W. Creager, J. Justin & J. Hinds, 3 Engineering for Dams 222–23 (1945) ("precise and highly mathematical methods of design or analysis of earth dams are seldom justified because we know in advance that the assumptions which must be made are sufficiently wide so that no precision is justifiable").

13. Schnitter, *Roman Dams, supra* note 9; Schnitter, *Evolution of the Arch Dam, supra* note 9; Schnitter *The Evolution of the Buttress Dam,* 26 Water Power & Dam Construction (1984) (for a well synopsized history of structural dams).

stream shape, or arch, for stability. Gravity dams can be built along a curved alignment but such curved gravity dams do not necessarily rely upon their curvature in order to resist hydrostatic pressure. For example, the famous Hoover Dam across the Colorado River is curved in plan but its cross-sectional profile is so thick the dam could stand safely even if built without any upstream curve. Consequently, the 726-foot high Hoover Dam is a curved gravity dam and should not be considered part of the structural tradition.[14] In contrast, the 78-foot high Kebar Dam built in 13th century Persia (present day Iran) is built with a maximum cross-sectional profile capable of safely supporting a gravity design only about 50 feet in height.[15] Consequently, the Kebar Dam is a true arch dam exemplifying the structural tradition because its curvature is mandatory in order to achieve stability under full hydrostatic load.

Early arch dams such as Kebar and the seventeenth century Elche Dam in Spain were built empirically and without any theoretical analysis to help determine the appropriate thickness of the arch.[16] In the early nineteenth century the French mathematician Louis Navier developed a simple formula (often called the "ring formula" or the "cylinder formula") that could help engineers determine suitable arch thickness depending upon hydrostatic pressure (i.e. the depth of water in a reservoir) and the curvature or arc of an arch.[17] It appears likely that Navier's theory was used by the French engineer Francois Zola in designing the 130 foot-high Zola Dam near Aix-en-Provence in the 1840s; the theory was certainly utilized in designing the 64-foot- high Bear Valley Dam built in southern California in the 1880s.[18] As a result, by the end of the 19th century the use of rational, mathematically-based design methods had been brought into the realm of the structural tradition.[19]

The other primary variant within the structural tradition is the buttress dam. In basic form, buttress dams do not constitute designs with cross-sectional profiles that remain constant across the length of the dam. Instead, buttress dams use a series of discrete thick walls (or buttresses) that absorb the hydrostatic pressure and allow the overall quantities of material in the dam to be reduced.[20] Primitive buttress dams consist of simple masonry designs which have been strengthened by the erection of buttress walls along the downstream face. But to achieve substantial re-

14. Schnitter, *Arch Dams, supra* note 9 (describing the Hoover Dam as a "massive curved gravity structure").

15. *Id.* (for description and illustration of the Kebar Dam).

16. Smith, *supra* note 9.

17. *Id.*

18. Smith, *supra* note 9, at 181–83, 207–08; Schnitter, *The Evolution of the Arch Dam, supra* note 9, at 36–38.

19. Schnitter, *Evolution of the Arch Dam, supra* note 9.

20. Schnitter, *The Evolution of Buttress Dams, supra* note 13.

Figure 1: The downstream side of the curved gravity Hoover Dam (completed 1935). The conservative proportions of this 726-foot high structure exemplify the massive tradition of dam design. Note how its appearance contrasts with the downstream side of the Littlerock Dam. [Source: Prints and Photographs Division, Library of Congress]

ductions in the amount of material required, buttress dams usually use designs with an inclined upstream face. This upstream face forms a barrier that holds back the water in the reservoir; in turn, the upstream face is supported by buttresses carrying the hydrostatic forces down to the foundations. By inclining the upstream face into the reservoir, it is possible to reduce the amount of concrete or masonry necessary for overall structural stability. If the buttresses are used to support a flat slab that forms the upstream face, then the design is known as a "flat slab buttress dam." If, instead, the upstream face is formed by a series of arches that are supported on the buttresses, then the design is referred to as a "multiple arch dam."

The multiple arch dam represents a particularly efficient form within the structural tradition because it uses both the arch and the buttress principle in its design. The use of discrete buttresses allows the amount of material within the body of the dam to be reduced significantly, while the use of arches for the upstream face allows this part of the design to be built with a minimum amount of material. Small examples of multiple arch dams using stone masonry were built in Spain in the eighteenth century, but the blossoming of multiple arch dam technology needed the development of reinforced concrete construction techniques in the late nineteenth and early twentieth centuries.[21]

The massive and structural traditions of dam design have existed for centuries and attracted the interest of a wide range of builders. Although some engineers have prepared designs within both traditions, it is more typical for an individual (or engineering bureaucracy) to devote their professional energies to erecting structures within a single tradition. As a consequence, an engineer skilled and knowledgeable in the practice of one tradition may be quite unsuited to rendering a valid, unbiased assessment of a design developed out of another tradition. As we shall see, John Eastwood became an impassioned advocate of the structural tradition in dam design and this ultimately brought him into conflict with John R. Freeman, a prominent advocate of the massive tradition. The two engineers believed each other to be responsible for promulgating designs which were either unsafe, uneconomical, or both.[22] In truth, the roots of their antipathy for one another came from the fact that each approached issues of dam design from different perspectives, different traditions and from different self interests.

EASTWOOD AND THE MULTIPLE ARCH DAM

Born in Minnesota in 1857, John Eastwood attended the University of Minnesota as an engineering student before heading West to work on railroad construction in the early 1880s.[23] In 1883, he set up shop as a civil engineer/surveyor in Fresno, California and for the next 29 years worked in the heart of the state's Central Valley. During the 1880s and early 1890s, he undertook a variety of jobs ranging from the city engineer for Fresno to surveyor for lengthy lumber flumes in the Sierra Nevada mountains.[24] Eastwood's interest in water storage extended from his work as chief engineer for the San Joaquin Electric Company in the mid-1890s when, under his direction, the company built a sophisticated hydroelectric power system that transmitted 11,000 volt alternating current to Fresno over a transmission line 34 miles long. At the time operations began in April 1896, this

21. Smith, *supra* note 9, at 118–20 (for description of several small buttress dams built by Don Pedro Bernardo Villareal de Berriz); Garcia-Diego, *The Chapter on Weirs in the Codex of Juanelo Turriano: A Question of Authorship*, Technology & Culture 217 (1976) (for description of a sixteenth-century multiple arch dam design).

22. Jackson, *supra* note 1, at 330–424.

23. *Id.* at 151–70. The primary source of material on Eastwood's life is the John S. Eastwood Papers at the Water Resources Center Archives, University of California, Berkeley, CA. These papers are divided into a series of numbered folders that refer to either a specific project or a particular file. Hereafter, material from folders in the Eastwood Papers will referenced as JSE, WRCA.

24. *Id.* at 151–70.

constituted the longest commercial power system in the world.[25] Unfortunately, the company's plant depended upon a steady flow of water in the North Fork of the San Joaquin River to power its turbines. Because the company initially could not afford to build a large dam for storing spring flood waters, operation of the plant was hindered severely by a drought that hit central California in the late 1890s. This drought ultimately forced the San Joaquin Electric Company into bankruptcy in 1899, and Eastwood retained no financial interest in the reconstituted San Joaquin Light and Power Corporation that took over the insolvent system.[26] Eventually the new company became a very profitable component of the San Francisco-based Pacific Gas and Electric Company and proved the viability of Eastwood's pioneer electric power project.[27] Although blocked from participating in the long term economic success of the San Joaquin facility, Eastwood did not lose interest in either hydroelectric power technology or the subject of dam design.

During the first decade of the twentieth century, Eastwood planned a new hydroelectric power system for the main stem of the San Joaquin River. Today known as Big Creek, this development is operated as part of Southern California Edison Company's power grid and provides a generating capacity of almost one million horsepower.[28] Eastwood's development of the reinforced concrete multiple arch dam resulted directly from efforts to devise an inexpensive means of storing water for the Big Creek system.[29] Eastwood had no desire to repeat the experience of the drought-plagued San Joaquin Electric Company, and he set out to develop a dam type that would minimize construction costs. To accomplish this, he sought a design that would reduce the amount of material necessary for stability. Perforce, he became an advocate of the structural tradition of dam construction.

25. Low, *The Fresno Transmission Plant*, 2 J. Electricity 79 (1896).

26. Jackson, *supra* note 1, at 209–11.

27. F. Fowler, Hydroelectric Power Systems of California and Their Extensions into Oregon and Nevada 454–58 (1923) (for the financial history of the San Joaquin Electric Company); C. Coleman, P. G. and E. of California: The Centennial Story of the Pacific Gas and Electric Company (1952) (for description of the later success of the San Joaquin Light and Power Company).

28. D. Redinger, The Story of Big Creek (1949); W. Myers, Iron Men and Copper Wires: A Centennial History of the Southern California Edison Company (1983) (for more on the Big Creek power system).

29. Henry E. Huntington Collection in the Henry E. Huntington Library, San Marino, CA. Within the Huntington Collection there are a few specific documents in file No. 11/7/(2) providing evidence of Eastwood's development of a multiple arch buttress dam design for use at Big Creek. *E.g.* J. Eastwood, Report on the Comparative Cost of Plant Showing the Advantages of the Use of Buttressed Dams (Aug. 3, 1906). *See* Dams (circa 1907) (unpublished report, in JSE, WRCA, No. 21; for evidence of Eastwood's desire to build multiple arch dams at Big Creek).

Figure 2: View showing the multiple arches that form the upstream face of John S. Eastwood's Big Bear Valley Dam (completed 1911). The vignette on right shows Eastwood circa 1912 [Source: Eastwood Papers, Water Resources Center Archives, Berkeley, California]

Because of financial uncertainties caused by the Panic of 1907 and because of corporate machinations surrounding the management and control of the Pacific Light and Power Corporation, Eastwood was prevented from implementing his innovative multiple arch dam technology at Big Creek.[30] However, in 1908, he was given the opportunity to design and build the 64-foot-high Hume Creek Dam as part of the Hume-Bennett Lumber Company's new logging plant in the Sierra Nevadas about 50 miles east of Fresno.[31] Completed in 1909, this structure was the world's first reinforced concrete multiple arch dam. In 1910, he began work on the 92-foot high Big Bear Valley Dam in southern California to be used by irrigation farmers to increase crop production in the Redlands/San Bernadino region.[32] These first two dams were important structures for the

30. Whitney, Dollars and Genius Built Southern California (circa 1970) (unpublished manuscript; a copy of this document is on file at the Water Resources Center Archives, Berkeley, CA). *See* Jackson, *supra* note 1, at 217–69. Actual construction of the initial Big Creek system was undertaken by the Boston-based engineering firm of Stone & Webster (using Eastwood's plans) in 1912–1913. H. Johnston, The Railroad the Lighted Southern California (1966).

31. Eastwood, *Hume Lake Dam*, 23 J. Electricity Power & Gas 398 (1909). *See* JSE, WRCA, No. 12.

32. Eastwood, *The New Big Bear Valley Dam*, 3 W. Engineering 458 (1913). *See* JSE, WRCA, No. 1.

people and organizations that commissioned Eastwood to build them, but they were not the types of projects that attracted widespread attention. However, in late 1911, he began working on a project for the Great Western Power Company that brought his dam design work to the attention of the national civil engineering community.[33]

In the early 20th century the Great Western Power Company was a major competitor of the Pacific Gas and Electric Company (P.G.& E.) in northern California.[34] During the 1920s, the two companies eventually merged together, but in 1911, Great Western sought to expand its system on the North Fork of the Feather River (about 100 miles north of Sacramento) in order to compete better with P.G.& E. As part of the new Feather River system, the company planned to build a large storage dam at a reservoir site known as Big Meadows in order to regulate water flow through a series of hydroelectric power plants along the lower part of the river. Originally, the company planned to build a concrete gravity dam at this important reservoir site, or at least this had been the hope of engineering consultant John R. Freeman. Freeman was a graduate of the Massachusetts Institute of Technology and a prominent hydraulic engineer who had advised Boston, New York City and Los Angeles on the design and construction of their municipal water supply systems. In addition, Freeman had served as president of the American Society of Mechanical Engineers and as vice-president of the American Society of Civil Engineers.[35] Although well connected with the New York financial community, Freeman discovered that his influence upon the corporate leadership of the Great Western Power Company seriously waned when H.H. Sinclair, an experienced California-based electric power entrepreneur, became vice-president in charge of the firm's California operations in 1909.[36] Much to Freeman's chagrin, in early 1911 Sinclair gave Eastwood responsibility for developing a multiple arch design for the Big Meadows site.[37]

Within this article it is impossible to recount the full story of the controversy involving Eastwood, Freeman and the Big Meadows Dam.

33. Jackson, *supra* note 1, at 341.

34. Coleman, *supra* note 24, at 211–24, 267–76 (for history of the Great Western Power Company).

35. John Ripley Freeman, 98 Transactions Am. Soc'y of Civil Engineers 1471 (1933) (for Freeman's obituary with a description of his professional career).

36. Sinclair's experience with California's electric power industry dated back to the early 1890s when he served as vice president for the Redlands Electric Light and Power Company, a firm which later became a key component of the Southern California Edison Company. *See* Henry Harbinson Sinclair Papers, Henry E. Huntington Library, San Marino, CA. (for evidence of Sinclair's difficulties with Freeman). *See also* John Ripley Freeman Papers (MC51), Institute Archives and Special Collections, Massachusetts Institute of Technology, Cambridge, MA (for evidence of Freeman's disagreements with Sinclair).

37. *The Big Meadows Dam*, 27 J. Electricity Power & Gas 287 (1911). *See* JSE, WRCA, No. 18 (for evidence of Eastwood's extensive involvement in planning and building the Big Meadows Dam).

However, two aspects of the events surrounding Eastwood's attempts to build the Big Meadows Dam are relevant to issues addressed in this article: 1) How Eastwood responded to technical criticisms of his design that were made on the basis of mathematical analysis, and 2) Freeman's ability to use non-technical arguments (or what he termed "psychological" concerns) to force abandonment of Eastwood's half-built multiple arch design in favor of a massive earthfill structure. Each of these topics will be considered separately within the context of Eastwood's overall career and the development of twentieth century dam engineering.

THE ROLE OF MATHEMATICAL THEORY

The first issue concerns how Eastwood used mathematical theory in his design method and how practical experience and observation of structural behavior affected the form of his subsequent design. Eastwood's most obvious and immediate use of mathematical formula came in determining the thickness of the arches that enclosed the space between the buttresses. To calculate the thickness of the arches at various elevations within the dam, Eastwood used the formula developed by Navier in the 1820s known as the cylinder formula.[38] Using this formula, the thickness T of an arch under hydrostatic pressure is equal to the pressure P multiplied by the radius R of the arch divided by the allowable stress Q, or:

$$\mathbf{T = P \times R / Q}$$

With concrete having an ultimate strength of about 3,000 pounds per square inch (psi), Eastwood determined that concrete could easily withstand stress equaling ten percent of this ultimate strength, thus providing an allowable stress Q of 300 psi. With Q held at a constant, and with hydrostatic pressure P constant for any given depth of water, Eastwood was able to vary the arch thickness T in direct proportion to the radius R of the arch.[39] Taking the distance between the buttresses to be a constant, any variation in the radius R necessarily changes the arc encompassed by the arch. For example, with a distance of 20 feet between buttresses, a radius R = 10 feet will create an arch with an arc of 180 degrees (or a semicircle). As R becomes larger, the arch will necessarily become flatter and the arc will decrease.

In using the cylinder formula, engineers have long understood that it represents an idealized model that only approximates conditions

38. Schnitter, *The Evolution of the Arch Dam*, *supra* note 9, at 36.

39. Eastwood, *The Ultimate Dam*, 3 W. Engineering 175 (1913); Eastwood, *The Eastwood Multiple-Arched Dam*, 33 J. Electricity Power & Gas 49 (1914); Eastwood, *The Multiple-Arched vs. the Single-Arched Dam*, 38 J. Electricity 149 (1917). *See* Jackson, *supra* note 1, at 486–546 (providing a complete discussion of Eastwood's method of design).

within actual structures. For example, to be completely accurate the cylinder formula requires that the foundations supporting the arch be absolutely rigid and unyielding. Although this condition is unattainable in practice, reinforced concrete and hard rock foundations are usually sufficiently rigid that the formula retains much of its validity despite discrepancies between theoretical assumptions and actual conditions. Another limitation of the cylinder formula that has caused many engineers to question its value as a predictive mathematical model relates to the formula's inability to consider the effect of structural deformation (sometimes called "rib-shortening") and temperature changes on the distribution of stresses within the arch. To cope with such effects in analyzing arch stresses, by the early twentieth century engineers were using the elastic theory of arches to provide a mathematical means of calculating the stresses created by rib-shortening and temperature changes [note: sometimes these two type of stresses are jointly referred to as secondary stresses].[40] Use of the elastic theory indicates an increase in stresses for arches with very flat arcs. In other words, to reduce rib-shortening and temperature stresses to a minimum, the elastic theory directs engineers to use an arch with an arc of 180 degrees. The flatter the arch, the greater the effect of secondary stresses; consequently, arches with an arc of 120 degrees will (at least according to the elastic theory) always be more stressed than comparable arches with an arc of 140 degrees. This is significant because there can be good economic reasons, such as reduced costs for concrete and reduced costs for wooden formwork, that might lead an engineer to use a flat arch despite being informed by the elastic theory that flatter arches induce higher stresses.

While the Great Western Power Company was considering the advisability of using a multiple arch dam for Big Meadows, Eastwood evaluated the effect of secondary stresses upon his arch designs. Utilizing the elastic theory, he determined that the effect of rib-shortening and temperature changes could be minimized by the use of deep arches encompassing an arc of over 140 degrees.[41] Eastwood understood that his multiple arch dams employed relatively thin arches for the "water face" [i.e. the barrier that held back the reservoir] and he wanted to be certain that his arch designs were strong enough to resist any forces that might cause them to crack or collapse. Consequently, when the seemingly sophisticated elastic theory indicated that deep arches could reduce stresses that might precipitate cracking, Eastwood incorporated this design feature into his plans for the Big Meadows Dam.

40. I. Baker, A Treatise on Masonry Construction 670–703 (10th ed. 1909) (chapter 23 is devoted to the subject of analyzing arches which are considered to support their loads "by virtue of the internal stresses developed in the material"). *See* Cain, *The Circular Arch Under Water Loads*, 85 Transactions Am. Soc'y of Civil Engineers 233 (1922).

41. J. Eastwood, Discussion of Temperature Stresses in the Multiple Arch Dam (Oct. 1912) (unpublished report making specific reference to the tenth edition of Baker, Treatise on Masonry Construction, as a source for Eastwood's method of analysis; JSE, WRCA, No. 12).

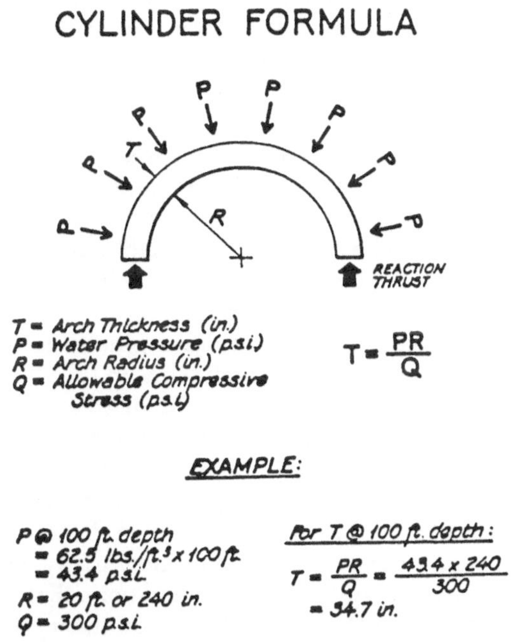

Figure 3: Illustration showing how the cylinder formula can be used in arch design. [Source: Courtesy Richard K. Anderson from D. C. Jackson Collection]

When Eastwood continued his dam design career after working on the Big Meadows project, he remained committed to using deep arches as a way to reduce the deleterious effect of secondary stresses. In fact, with his Los Verjels Dam (completed in late 1913), he employed arches with an arc of over 144 degrees, the deepest of any of his structures.[42] For his next commission (the Kennedy Dam erected in 1914) he adopted a new structural tactic to eliminate secondary stresses in the arches.[43] Rather than merely rely on a deep arch as a means of countering such stresses, Eastwood devised a "three-hinge" design that would allow the arches to adjust themselves (by minute internal rotation) in response to any rib-shortening or temperature stresses that might cause cracking. At the time, "three-hinge" designs were becoming popular among reinforced concrete bridge engineers as an innovative way to reduce uncertainty in stress analysis.[44] Eastwood picked up on this trend and in 1914 began adapting it for

42. Eastwood, *Los Verjels Dam: A Multiple Arch Structure*, 5 W. Engineering 7 (1914).

43. Eastwood, *The Kennedy Dam*, 5 W. Engineering 407 (1915).

44. *E.g. A Three-Hinged Concrete Arch Bridge Over the Danube at Eringen*, 47 Engineering News 35 (1902) (for one of the earliest references to three-hinged arch technology).

his dam designs.[45] The three-hinge arch appeared to offer an ideal way of counteracting the effect of secondary stresses and in 1915 Eastwood employed the feature as a key part of his 150-foot-high Mountain Dell Dam design for Salt Lake City.[46] However, he soon abandoned the technology and moved toward the use of rigid [i.e. non-hinged] arches with very shallow arcs. In fact, during the final years of his career he developed designs with arches as shallow as 100 degrees in apparent defiance of the elastic theory. To understand why Eastwood changed his approach to arch design, we must consider how he perceived the relation of mathematical theory to actual construction experience (or practice) in his effort to develop more sophisticated, and useful, dam designs.

First, it is significant that almost all engineers (aside from Eastwood) who advocated use of multiple arch dams in the 1920s relied upon the elastic theory and considered the use of 180-degree arcs as mandatory if arch stresses were to be minimized.[47] Billington's comment about the engineering profession's slavish devotion to mathematical formulas after World War I is extremely accurate in regard to how most engineers approached the process of multiple arch dam design. In retrospect, what is most disconcerting about this phenomenon is how many engineers could turn a blind eye to practical, in-the-field observations that were not compatible with conjectures indicated by mathematical theory. In this context, Eastwood distinguished himself as being responsive to how his designs actually functioned in the real world, regardless of what mathematical analysis might predict.

This duality between theory and practice can be most clearly appreciated when analyzing why Eastwood dropped his use of three-hinge arches, despite the advantages this technology seemed to offer in countering the effect of secondary stresses. Specifically, during final construction of the Mountain Dell Dam, a thin coating (or wash) of cement was applied to the upstream face of the structure's three-hinge arches. This cement coating provided a smooth surface for the upstream face but was not intended to serve any structural purpose. Presumably, when the three-hinge arches would rotate in response to temperature and other rib-shortening stresses, small vertical cracks would appear in the cement coating and provide evidence of the hinges' movement. However, such cracking did not

45. JSE, WRCA, No. 14; and JSE, WRCA, No. 35 (these files describe two unbuilt designs prepared by Eastwood in 1914–15 that used three-hinged arches).

46. *High Multiple-Arch Concrete Dam for Salt Lake City Water Supply,* 80 Engineering-News Record 455–57 (1918); Cannon, *The Mountain Dell Dam,* 3 Monthly J. Utah Soc'y of Engineers 223 (1917).

47. *See* Jakobsen, *supra* note 4. *See also* E. Wegmann, Design and Construction of Dams (8th ed. 1927) (this edition of Wegmann's book contains a special section on designing multiple arch dams written by Fred Noetzli).

occur. In fact, Fred Noetzli, a multiple arch dam designer and advocate of the elastic theory, noted in 1927:

> It was found [during a 1925 inspection] that in the seven years of operation of the [Mountain Dell] dam and reservoir none of the hinges had moved a sufficient amount to break the thin coating of cement wash, except for a distance of one to two feet from the crest of the dam down. The hinges were painted with asphalt and probably offered relatively little resistance to rotation due to deformation of the arches. It appears, therefore, that the deformation of the arches was too small to break the cement coating.[48]

Amazingly, Noetzli's observation did not lead him to question the validity of relying upon the elastic theory when designing arches for multiple arch dams. Instead, he declined to use the actual physical behavior of a functioning structure as a rationale for modifying his reliance upon mathematical theory in the dam design process. Despite undeniable indications that rib-shortening and temperature stresses were negligible in the arches of multiple arch dams (otherwise the cement wash on the face of the Mountain Dell Dam would have exhibited cracks), Noetzli continued to advocate the use of 180 degree arches as a means of countering the supposed secondary stresses revealed by the elastic theory.

In contrast to Noetzli's reaction, by late 1916 Eastwood expressed dissatisfaction with the elastic theory based on "the result of experience and many trials."[49] In response to adherents of the elastic theory who claimed that high arch stresses existed in many multiple arch dams, Eastwood defended his design method by stating: It is undoubtedly the proper procedure to fit theory to practice but the factors of the theory must be based on the actual physical conditions as they are in practice or else theory will lead us astray.[50]

After constructing three-hinge arches as part of the Kennedy and Mountain Dell Dams and then observing their actual performance, Eastwood came to appreciate several variances that existed between theoretical assumptions and physical realities in relation to the structural performance of his designs. These variances included such conditions as: a) the hydrostatic loadings on the arches are not uniform because the upstream face of most multiple arch dams are built on a slope; b) the supports for the arches (i.e. the buttresses) are not rigid and inelastic, but are similar in elasticity to the arches themselves; and c) with the reservoir filled, the

48. Wegmann, *supra* note 42, at 482.
49. Letter from J. Eastwood to P. Downing (Sept. 27, 1916) (JSE, WRCA, No. 4; discussing limitations of the elastic theory).
50. *Id.*

upstream side of the arches are saturated with water which causes the arches to expand in volume (just as a slice of bread will expand in size if soaked in water); this expansion will consequently help counteract any rib-shortening stresses caused by the hydrostatic loading.[51] The precise technical significance of the factors listed above need not be explicated in this article. However, what is important to realize is that the ability of the elastic theory to accurately determine arch stresses is very much affected by variance in these factors.

Rather than adhere to the dictates of seemingly advanced mathematical theory or attempt to assess quantitatively the inaccuracy of the elastic arch theory, Eastwood chose another approach to the problem of dam design. As such, he opted to use the relatively simple cylinder formula and design thin arches that provided relatively good compatibility with the theoretical assumptions underlying this formula. When the elastic theory led to conclusions incompatible with the actual structural performance of his dams, he did not hesitate to reject the universal validity of this theory in favor of the less complicated cylinder theory. The cylinder formula may not have been absolutely accurate in its ability to predict stresses within multiple arch dams, but Eastwood recognized that it could provide good estimates that were more than sufficient to insure structural safety.

Eastwood sought to forge a logical, but oftentimes misunderstood, middle ground between highly abstract, academic approaches to structural design and engineering based upon practical field experience. After being confronted with numerous studies that seemingly showed his designs to be unsafe or unstable, he drafted a memo entitled "Limitations to Formulae" in the early 1920s in which he denigrated the notion that mathematical theory could always provide exact answers to structural problems involving the physical environment.[52] Recognizing how mathematical formulas could only provide answers that were relatively accurate, Eastwood aptly compared this situation with the newly-popularized theories of Albert Einstein by observing: "Einstein is right! All lines are curves and all things mundane are relative."[53] Eastwood's pragmatism produced stable dams but ran counter to the notion that engineers could, and should, precisely calculate stresses in man-made constructs. Consequently, his design methods represented a challenge to the idea that mathematical formulas can establish beyond cavil the magnitude of all stresses

51. *Id.*; and Letter from J. Eastwood to W.S. Post (March 27, 1917) (Carroll Dam Folder, Box 8, Ed Fletcher Collection, Special Collections, University of California, San Diego) (describing in detail Eastwood's thoughts on why the elastic theory did not provide accurate results when used to calculate arch stresses in multiple arch dams).

52. J. Eastwood, Limitations to Formulae (circa 1922) (JSE, WRCA, No. 45; unpublished report).

53. *Id.*

within a structure. Rather than claim that the limitations of such formulas stripped them of any validity, he championed the idea that careful consideration of how theory related to physical reality would lead to the true ideal of scientific design.

Unfortunately for Eastwood, engineers generally have moved away from his practical approach to dam design in the decades since his death. The divergence was well demonstrated in the late 1960s, when the Bechtel Corporation, a major American engineering firm, analyzed the safety of Eastwood's 1924 Littlerock Dam in southern California.[54] This modern, "scientific" engineering firm utilized a special mathematical methodology of its own to analyze the stresses in the buttresses of multiple arch dams. On the basis of this simplified, yet seemingly precise, investigation the firm determined that dangerous stresses would exist at the upstream face of the buttresses at times when the reservoir was filled.

These stresses were calculated to be so great as to precipitate possible collapse if the buttresses were placed under the loading of a full reservoir. Based upon this analysis the firm proposed a complicated means of "strengthening" the buttresses with external steel cables, all the while conveniently forgetting that for more than 40 years the reservoir frequently had been filled with no damage resulting to the structure.[55] In this context, a comparison between Noetzli's and the Bechtel Corporation's approach the problem of analyzing stresses in multiple arch dams is illuminating.

Noetzli may not have fully appreciated the implications of the Mountain Dell Dam's non-moving hinges, but at least he refrained from attacking the safety of the design based upon his interest in promoting the value of mathematical analysis over practical insight. However, by the 1960s a major engineering firm chose to ignore the 40 year performance history of the Littlerock Dam in deference to mathematical analysis which indicated the existence of supposedly dangerous buttress stresses; this predilection to consider the results of mathematical analysis to be somehow sacrosanct ultimately led to a bitter court battle focused around the issue of whether or not the Littlerock Dam represented an imminent threat to public safety.[56]

54. Bechtel Corporation, Littlerock Dam Safety Study (Jan. 1968) (copy available in the records of the Littlerock Creek Irrigation District, Littlerock, CA). *See* Jackson, *Controversy in Concrete: The Littlerock Dam (1918–1977)*, Proc. Am. Soc'y of Engineering Education 471 (1982); and D. Jackson, Great American Bridges and Dams, at 74–78 (1988) (providing detailed discussion of engineering issues surrounding the operational history of the Littlerock Dam).

55. Bechtel, buttress strengthening page ref.

56. Based upon the 1968 Bechtel Corporation report, the California Division of Safety of Dams took legal action to force revocation of the state permit that allowed the Littlerock and Palmdale Irrigation Districts to store water behind the Littlerock Dam. The decision to prevent the state from revoking the storage permit without complying with the requirements of the California Environmental Quality Act (CEQA) is recorded in Citizen's Committee to Save the Littlerock Dam, Inc. v. Robie, c184 269 (Superior Court, County of Los Angeles, Apr. 20, 1977).

Figure 4: View showing the buttresses that form the downstream side of the Littlerock Dam (completed 1924). With a maximum height of 170 feet, it stands as Eastwood's tallest dam. [Source: D. C. Jackson Collection]

THE "PSYCHOLOGY" OF THE MULTIPLE ARCH DAM

As discussed previously, the Big Meadows Dam first prompted John Eastwood to confront the theoretical possibility of using the elastic theory of arch analysis in his search to discover better ways of designing multiple arch dams. In contrast, when John R. Freeman sought to convince the Great Western Power Company to abandon construction of Eastwood's half completed dam, he primarily drew upon non-technical, non-mathematical arguments to influence the corporation's decision making process. Whereas Eastwood focussed his creative energies on the technical and mathematical issues that surrounded the process of dam design, Freeman discerned a dramatically different approach to the subject involving arguments which were difficult, if not impossible, to counter with strictly technical counter arguments.

After H. H. Sinclair became vice-president of the Great Western Power Company, Freeman exercised little control over planning for the Big Meadows Dam.[57] However, this situation changed in February 1912 when Edwin Hawley, president of the company and the person responsi-

57. Executive Committee of the Great Western Power Company, Relating to Construction [of] Permanent Dam and Method of Future Expenditure (May 20, 1910) (unpublished report, Sinclair Papers, *supra* note 36; for evidence of Freeman's lack of influence within the Great Western Power Company).

ble for bringing Sinclair into the firm, passed away. In his place Mortimer Fleishhacker, a San Francisco banker who recently had become involved with Great Western because of his control over San Francisco's City Electric Company, assumed the presidency. Fleishhacker held no special relationship with either Sinclair or Eastwood, and within a few weeks after Hawley's death Freeman began a campaign to gain Fleishhacker's confidence and force abandonment of Eastwood's Big Meadows design.[58]

This campaign extended over the next several months with Freeman claiming that while he "fully believe[d] in the multiple arch principle for many sites . . . [he had] doubts about [Big Meadows] being the best place [for it]."[59] Although he expressed concern over technical matters such as "imperfect theories of stress" that related to the cylinder formula, Freeman's most devastating attack on the multiple arch concept did not rely on technical arguments *per se*.[60] After making a one-day visit to the construction site in August 1912 Freeman began castigating Eastwood's design because of its visual appearance. Referring to undefined "popular apprehensions and misapprehensions," Freeman hypothesized how the general public might react to the distinctive appearance of multiple arch dams with the assertion:

> [T]he psychology of [the multiple arch dam's] airy arches and the lace curtain effect of Eastwood's stiffening props [between the buttresses] is not well suited to inspire public confidence.[61]

Freeman's perspective in arguing against the multiple arch dam derived directly from his adherence to the massive tradition of dam design, a tradition which placed little emphasis on conserving the amount of construction material required for a structure. Freeman reported that he had "repeatedly informally urged [the Great Western company] . . . to build a big massive lump of a dam." because he considered dams to be a technology in which the appearance of bulk was a natural and desirable feature, regardless of whether or not this bulk served any useful purpose.[62] Because Eastwood's design seemed to present an "airy" appearance, Freeman went so far as to recommend formally to the company that they consider "lessen[ing] the apparent height of the slender buttresses... [by placing a] fill of loose fragments of rock and gravel" along the downstream side of the dam where it would serve no structural purpose.[63] Interestingly, he defended the value of such a costly action purely on the

58. Letter from J. Freeman to A.W. Bullard (Feb. 23, 1912) (Freeman Papers, *supra* note 36).
59. *Id.*
60. *Id.*
61. Letter from J. Freeman to A. Davis (Sept. 30, 1912) (Freeman Papers, *supra* note 36).
62. *Id.*
63. Letter from J. Freeman to H.P. Wilson, secretary of the Great Western Power Company (Sept. 30, 1912) (Freeman Papers, *supra* note 36).

grounds of "diplomatic or psychological reasons."[64] Thus, Freeman advocated an expensive alteration of the appearance of Eastwood's Big Meadows Dam simply so that it would look like a massive earthfill dam. As he put it in October 1912:

> [P]lainly, it is worthy of some considerable expenditure beyond that necessary to satisfy engineers... in order to satisfy the more or less ignorant public... [who would] regard [Eastwood's] dam not from a technical standpoint, but by comparison with the solid gravity type of masonry or earth."[65]

Eastwood made a valiant attempt to defend the integrity of his design, but Freeman's attack on the "psychology" of the multiple arch dam found an audience with Fleishhacker and other corporate leaders of Great Western.[66] Apparently these businessmen were more inclined to listen to the highly respected Freeman than to an engineer who had spent most of his career in the remote hinterland of Fresno. In March 1913, the Great Western Power Company officially abandoned Eastwood's multiple arch design, despite having already expended hundreds of thousands of dollars on its construction.[67] In its place they chose to build a completely new, and much more expensive, massive earthfill dam.[68] Officially, the company's decision to change plans was justified in terms of minor foundation problems that led them to opt away from a daring new design.[69] However, any foundation problems that might have affected Eastwood's design would have been equally (or more) troublesome for a gravity masonry or earthfill dam. At the root, Eastwood's design was abandoned because of Freeman's success in castigating its unorthodox appearance as being ill-suited to the proper image of the Great Western Power Company, an appeal that struck a resonant chord in the corporate hierarchy.

After the debacle at Big Meadows, Eastwood entered a period in which he had to scramble intensely in order to obtain even small-scale design commissions; in fact, it was not until the early 1920s that he began to work on projects that approached the scale of Big Meadows. Meanwhile,

64. *Id.*

65. Letter from J. Freeman and A. Noble to H.P. Wilson (Oct. 17, 1912) (Freeman Papers, *supra* note 36; this letter was officially signed by both Freeman and Alfred Noble, a prominent New York City-based consulting engineer, but it was written almost entirely by Freeman).

66. J. Eastwood, Description and General Specifications of the Big Meadows Dam (circa Oct. 1912) (unpublished report); J. Eastwood, Non-Technical Description of the Big Meadows Dam (circa Oct. 1912) (unpublished report); Letter from J. Eastwood to M. Fleishhacker (Feb. 13, 1913) (JSE, WRCA, No. 18).

67. Jackson, *supra* note 1, at 382–84.

68. *Raising the Big Meadows Dam, Plumas, Cal.*, 3 Modern Irrigation 42 (1928) (describing earthfill dam built at Big Meadows).

69. Jackson, *supra* note 1, at 330–424 (for a complete analysis of all aspects of the Big Meadows controversy).

Freeman continued to gain influence in the engineering establishment as he ascended to the presidency of the American Society of Civil Engineers in 1923–24.[70] Throughout the final 20 years of his career Freeman remained a staunch proponent of massive gravity dams and refused to consider multiple arch dams as worthy of serious attention.[71] Without question, Eastwood's long-term success in promoting his dams was seriously impeded by Freeman, and this opposition derived from the latter engineer's personal and professional interest in advocating the massive tradition as a superior method of dam design.[72] Most remarkably, in light of his scientific training at the Massachusetts Institute of Technology, Freeman did not hesitate to brandish the most non-technical, "unscientific" arguments in order to stop construction of a dam design that did not adhere to a tradition and style of his own choosing.

CONCLUSION

The purpose of this article is not merely to rescue the good name of John S. Eastwood and bring his career greater public attention. Nor is it oriented toward getting people to think that multiple arch dams comprise the only type of water storage technology that should ever be built. As a technology, multiple arch dams have their limitations, and in the latter 20th century the economics of building such labor-intensive structures has not been particularly compelling. However, in terms of assessing Eastwood's accomplishments it is important to understand that everyone of his 17 dams constituted the least expensive design for a project (there is no evidence that any of his designs were ever underbid) and that none of his designs ever failed in any manner which caused loss of life or property.[73]

Given both the inability of the engineering establishment to accommodate Eastwood's design method with a more mathematically oriented approach to structural design and the manner in which engineers such as Freeman castigated Eastwood's dams on the grounds of visual appearance, a few comments on the ethics of technological safety are in order. As a society we are eager to live in a world where our physical safety is never threatened and where people responsible for technological mis-

70. Eastwood, *supra* note 32.

71. J. Freeman, Summary of Recommendations Regarding Future Extensions of Water Supply for the City of San Diego, Cal. 81–82 (May 16 and 22, 1924) (a copy of this report, which provides evidence of Freeman's long-standing disinterest in multiple arch dams, is on file at the Water Resources Center Archives, Berkeley, CA).

72. Letter from A.G. Wishon, General Manager of the San Joaquin Light and Power Corp., to J. Eastwood (Mar. 29, 1922) (JSE, WRCA, No. 48; containing Wishon's observation that "it is a tremendous undertaking to sell the idea [of multiple arch and single arch dams] to the financier who slips behind your back and consults a man like John R. Freeman").

73. Jackson, *supra* note 1.

haps are held legally responsible for their miscalculations. This is all well and good. But how do we decide how safe is safe? And how do we judge the potential benefits of a new technology that may foster unforeseen problems as well as anticipated benefits. We usually perceive the "ethics of safety" in terms of unscrupulous individuals trying to foist an unsafe product off on an unsuspecting public. But what about the ethics of having someone declare a new technology to be unsafe when, in fact, it is just as safe as the technologies it is attempting to supplant.

Eastwood faced a difficult challenge in simultaneously defending the suitability of his designs from both the realm of excessively rigorous mathematics and from the realm of "psychological" intuition. In several instances, he proved capable of withstanding his opposition and his dams were able to provide successful, long-term service as water storage structures. But in other cases, most notably at Big Meadows, he succumbed to the forces of conservative hydraulic design and failed to implement his projects for the benefit of potential clients. The reasons for his failure to win design commissions never related to economic inefficiency. Rather, they concerned the perceived weakness of his designs as determined by seemingly precise (but in fact erroneous) mathematical analysis or by non-rational considerations related to visual appearance.

The larger lesson to be drawn from Eastwood's experiences is that the scientific and "rational" nature of modern technology is not always as absolute as we might like to think. Regardless of our desires, safety remains a relative value that only has meaning in terms of how people choose to evaluate a wide range of factors and influences. Engineering "experts" used to bolster legal arguments in technically-oriented court cases may bring huge amounts of knowledge and experience to a problem, but this background cannot insure the cogency or validity of their professional opinions. A key insight to be gleaned from Eastwood's work in promoting multiple arch dams is that society, and the legal system, cannot afford to relinquish responsibility over the technological decisionmaking process to experts without appreciating the traditions and technical prejudices that may color their view of what is safe and/or acceptable. Dams may appear to represent a simple, straightforward technology in which all engineers can agree on a few basic mathematical principles that govern the design process. But the history of the multiple arch dam, replete with concerns over visual appearance, "psychology" and the truthfulness of advanced mathematical theory, provides compelling testimony to the contrary.

15
Historical development of British embankment dams to 1960

A.W. Skempton

SYNOPSIS. The geotechnical history of British earth dams is traced up to 1960, with reference to specific case records. Particular attention is given to the evolution of puddle core dams during the 19th century, a period in which more than 200 large dams of this type were successfully completed, and to the application of soil mechanics in design and analysis from 1937.

18th CENTURY DAMS

1. Before about 1770 probably no earth dams in Britain exceeded 6 or 7 m in height. Where reasonably good impervious material existed nearby in adequate quantity the embankment simply consisted of 'homogeneous' fill, with suitable outlet arrangements, and usually some form of protection on the upstream slope. Lancelot (Capability) Brown built a dam of this type in the 1770s for a large ornamental lake at Sherborne Castle, and doubtless there are many earlier examples for mill ponds and even for medieval fish ponds of which the most notable is the 6 m high 'Great Weir' at Alresford in Sussex dating from about 1200 (ref. 1).

2. If the bank fill had to be gravel or other pervious material, clay would be brought, if necessary from a distance, and rammed in thin layers to form a central core or 'wall'. This, too, may be an ancient method of construction but the earliest case known to the writer is the Serpentine dam in Hyde Park, built 1730–31 with a height above stream bed H = 6.5 m and a 'wall' of rammed London Clay (ref. 2). John Grundy's well designed dam at Grimsthorpe (1746–47, H = 5.5 m), shown in Fig. 1, is the best known example (ref. 1).

3. Alternatively, clay could be used to form a lining on the upstream slope, suitably protected from wave action and, given a porous foundation, continued as a lining over part or all of the reservoir bottom. This seems to have been the scheme adopted by Capability Brown for Petworth dam (1755–56, H = 7 m, ref. 3) and at one or two other sites including Burghley House (c. 1775). Clay would also be rammed around the outlet pipe and valve well to prevent leakage.

4. With the advent of canal reservoirs somewhat larger dams were required. Smeaton's Townhead dam, completed in 1772 on the Forth & Clyde is 8 m high (ref. 4). Knypersley No. 1 dam on the Caldon Canal, acting as a feeder to the Trent & Mersey summit level, was built by Thomas Dadford about 1786 with a height of 9 m (ref. 5) and Clattercote dam on the Oxford Canal, originally dating from 1776, was

enlarged by James Barnes in 1787 to a height of 11 m (ref. 6).

Homogeneous clay fill canal dams

5. The 8 m high Bath Pool dam, built in the late 18th century for the Trent & Mersey Canal, is known to have had a 'homogeneous' clay fill (ref. 7). Clattercote dam is founded on and surrounded by Lias clay, and may well be of this type, while Aldenham dam built 1795–97 (with H = 7.5 m) for the Grand Junction Canal by Barnes under Jessop's direction, certainly consists entirely of brown (weathered) London Clay fill; a clay of high plasticity with typical index properties LL = 80 and PL = 30. Recent tests have shown that the fill has an average undrained shear strength of about 20 kPa and a water content w = 35 (ref. 8), discounting the somewhat drier and stronger material above the phreatic surface. Knipton dam on the Grantham Canal, built 1794–96 with a height of 10 m and 2:1 slopes using Middle Lias clay (ref. 9) is a good example of Jessop's homogeneous clay fill dams of which at least two others are known. One of these is the main dam of Hollingworth Lake (1796–1801, H = 10 m) for the Rochdale Canal with Boulder Clay fill, also designed with 2:1 slopes (ref. 10). Combs dam on the Peak Forest Canal is another instance of Boulder Clay used to form a 'homogeneous' section; built 1804–06, by Thomas Brown, with H = 11.5 m and 5:3 slopes (before being raised in the 1830s to 16 m). Index properties and strength parameters of the original Combs fill are: w = 19, LL = 35, PL = 16, γ_d = 17.3 kN/m^3, c' = 9 kPa and φ' = 25° (ref. 11).

6. John Farey in 1806 (ref. 12), having stated that 1.5:1 and 2:1 are the usual slopes for reservoir embankments, goes on to say 'but if the soil should prove to be a slippery clay, as at the Aldenham reservoir ... a flatter slope should be given'. He is referring to a slip in the 2:1 upstream slope which occurred after a drawdown of water level in 1802. The outer slope seems originally to have been at 2.5:1 but with a berm possibly added soon after or even during construction. Knipton dam also had a drawdown slip, in 1804.

7. Nevertheless the homogeneous clay fill dams generally gave little trouble and in particular tended to be free of leakage problems owing to the low hydraulic gradients involved in the bank and its foundations: typically less than 0.3, as shown by piezometric observations at Aldenham (ref. 13) and at Combs.

8. The bank fill in these homogeneous section dams would have been placed in layers probably not more than 60 cm thick, after dumping from barrows or small carts, and trodden in by men (ref. 12); though in some cases stiff clay was 'trampled, layer by layer, by heavy horses' (ref. 14).

9. It may be noted in passing that earth dams in France, throughout the 19th century, were invariably of homogeneous clay fill construction. Up to about 1830 compaction was achieved simply by the passage of horse and carts; hand punners were then introduced, to be followed after 1860 by horse-drawn rollers and from about 1890 by steam operated or petrol driven rollers (ref. 15). Several of these dams suffered drawdown slips where unduly steep slopes had been adopted but again there appears to have been little trouble with leakage as the embankments had a broad base or at least a wide compacted-fill taken down in trench through superficial pervious strata. By comparison with

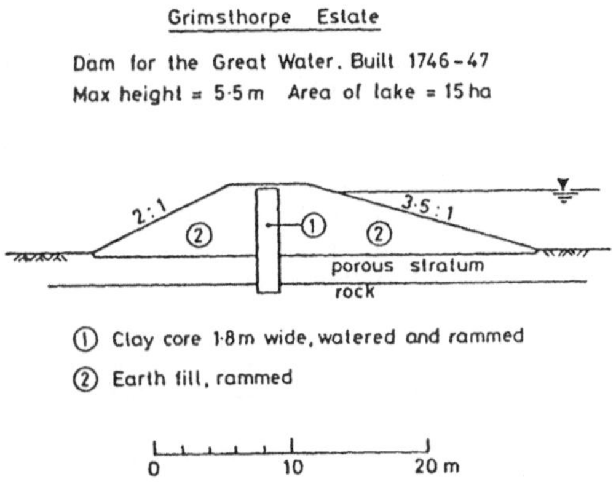
Grimsthorpe Estate
Dam for the Great Water. Built 1746-47
Max height = 5·5 m Area of lake = 15 ha
2:1
3·5:1
①
②
②
porous stratum
rock
① Clay core 1·8 m wide, watered and rammed
② Earth fill, rammed
0
10
20 m

Fig. 1

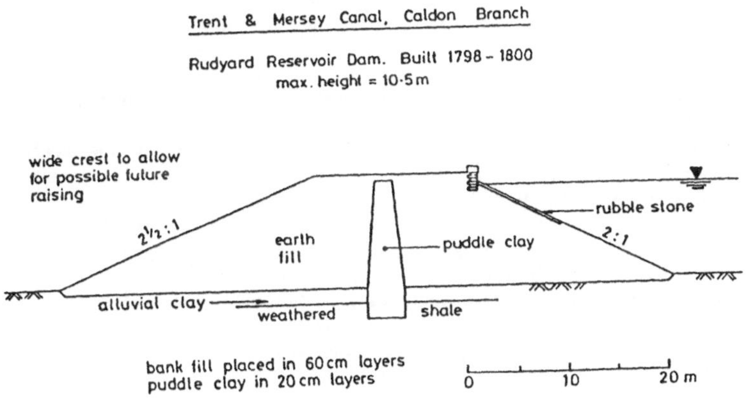
Trent & Mersey Canal, Caldon Branch
Rudyard Reservoir Dam. Built 1798 - 1800
max. height = 10·5 m
wide crest to allow for possible future raising
2½:1
earth fill
puddle clay
rubble stone
2:1
alluvial clay
weathered
shale
bank fill placed in 60 cm layers
puddle clay in 20 cm layers
0
10
20 m

Fig. 2

British practice, however, these dams were few in number and never more than 20 m in height; nor were they built in situations requiring a deep cut-off.

EARLY PUDDLE CORE DAMS

10. The systematic use of puddle clay for lining canals in porous ground is generally attributed to Brindley in the 1760s. Moreover puddle clay was also used in the side embankments of canals either as a sloping core, as on the Forth & Clyde canal in 1768 (ref. 4) or, probably more often, as a vertical central core (ref. 12). The idea of using puddle clay when an impervious core had to be incorporated in a dam would therefore have been obvious to the canal engineers.

11. Specifications for dams dating from the end of the 18th century and the first half of the 19th century all indicate that the clay (after being left to weather, if necessary) was brought to the embankment to be placed in layers 15 to 20 cm thick, watered and allowed to soak overnight, or for 24 hours, and then trodden, cross-cut and heeled in . In the second half of the 19th century more usually the clay was watered and worked up at the borrow pit (or 'clay field') before being brought on to the dam. Finally, in the 20th century, the clay was prepared in pugmills and placed generally in 10 to 15 cm layers. The second, and especially the third of these procedures resulted in more uniform material but in all three methods the average strength must have been much the same: suitable for the unaltered process of treading and heeling. Tests on several puddle cores soon after placing show an undrained shear strength around 10 kPa (ref. 16).

12. Butterley. Though very probably not the first, the earliest puddle core dam of which any details are known to the writer is at Butterley; 10 m high built 1793-95 to Jessop's design for the Cromford Canal (ref. 17). The bank fill consisted mostly of spoil from a deep approach cutting to the Cromford summit tunnel in shales and sandstones of the Coal Measures; the core, vertically-sided, was 2.7 m wide and extended in trench a few metres below ground level.

13. Rudyard. A improved form of puddle core, tapering symmetrically from 6 ft (1.8 m) at the top to a width of 12 ft (3.6 m) at ground level, was adopted by Rennie for the 10.5 m high Rudyard dam; see Fig. 2. Built 1798-1800 for a branch of the Trent & Mersey (ref. 18) the dam displays logical design in all its features and is maintained today in excellent condition after nearly two centuries of uninterrupted service. Rennie's specification includes a requirement that the bank fill should be placed in 60 cm layers with large stones excluded. It is probably made of weathered sandstone and shale of the Millstone Grit Series, obtained from borrow pits within the reservoir area. From this date symmetrically tapered cores became standard practice.

14. Glencorse. The first deep cut-off was constructed at Glencorse dam (1819-23, H = 23 m) for Edinburgh's water supply: design by James Jardine with Rennie and Telford as joint consultants. The puddle trench had to be taken to a depth of 38 m below crest level, in water-bearing coarse gravel, before reaching sound rock. To avoid any risk the core (and top of the puddle cut-off) was made 18 m wide at ground level at the deepest section, while the upstream and downstream slopes of the embankment were built at 3:1 (ref. 19). The trench, excavated in open cut, had a bottom width of at least 4.5 m.

15. Blackbrook. Other engineers less capable of resisting pressure to economise, or simply lacking in judgement, ran into considerable trouble. A particularly bad example is Blackbrook dam for the Leicester Navigation. Built 1795–97 to a height of 13 m it had a vertical–sided puddle core only 1.8 m wide, apparently made of poor quality clay, and banks of loose rubbly material (ref. 1). The dam collapsed in 1799, fortunately with no loss of life. Failure was attributed to overtopping and inadequate spillway capacity, but the overtopping almost certainly resulted from crest settlement caused by internal erosion of the core following hydraulic fracture (ref. 20).

16. Rotton Park. Built 1825–28 with a height of 14 m (ref. 1) this is one of the last canal dams in England, but set standards followed by the best works during the next thirty years. It formed part of Telford's notable improvements of the Birmingham Canal. There is little doubt that the dam could have been made of homogeneous section, as plenty of Boulder Clay appears to have been available. However, Telford chose to use a puddle clay core (with any large stones removed) and to rely on the bank fill only for stability; making the upstream and downstream slopes at 3:1 and 2:1 respectively. A cross section of the original design was published in 1838 along with other details (ref. 21) and from this time it can be said that British engineers, at home and abroad, preferred puddle core dams to any other type of earth dam for more than a century. An analysis of the contractor's account book shows that the bank fill at Rotton Park was placed by horse and cart, except for a strip on each side of the core wheeled by barrows on planks; puddle clay was also barrowed before being watered and heeled in.

17. The core has a top width of 1.8 m and batters at 1:10 to a bottom width B = 4.6 m at the deepest section, designed to be exactly one-third of the water head h = H − f (see Fig. 5) assuming, as will nearly be the case, that full reservoir pressure acts on the core, and tail water is at stream bed level on the centre line. Actually the proportion is slightly less since Telford raised the dam (and top water level) by 1.2 m immediately after completion.

18. A core of these proportions ensures that the maximum hydraulic gradient h/B to be sustained by the core does not exceed 3.0; and the ratio h/B can also be regarded as an index of its slenderness. For comparison, the index for Butterley dam is 3.3 and for Rudyard 2.6, in contrast to an apallingly high 6.7 for Blackbrook dam.

VICTORIAN DAMS (i) 1835–1865

19. The ICOLD World Register lists 60 earth dams built in Great Britain during this period, of which nearly half exceed 20 m in height. Almost all of them had a central puddle clay core, and the majority were perfectly successful. However, some of the dams experienced trouble due to leakage around culverts; for tunnel outlets, first introduced by Jessie Hartley at Entwistle dam in 1833 (ref. 22), were still fairly uncommon before 1865. Outlet pipes through the embankment and core also gave trouble, an interesting case being the fracture of the pipes in Bateman's Torside dam (H = 30 m) at the end of construction in 1854 due to base spreading (ref. 23). A few slips occurred during construction in the (steeper) downstream slopes; for example at Arnfield dam (1849–54, H = 20 m) in Boulder Clay fill

(ref. 23), at Calf Hey dam (H = 16 m) in 1856 (ref. 24) and at Piethorne dam (H = 25 m) in 1862 (ref. 25). These were repaired by adding a toe berm, a traditional remedial measure going back to the canal era and often used on the early railway embankments.

20. As the height of dams increased during this period, so the cores tended to be made narrower in proportion to height, and in extreme cases the bottom width, instead of being about one-third of the water head, as in early puddle core dams (para. 18), was only about one-sixth. Such narrow cores were susceptible to hydraulic fracture. Disastrous failures of this type occurred at Bilberry dam in 1852 and at Dale Dike dam in 1864 (both around 30 m high) while serious trouble arose at Rhodes Wood dam (H = 24 m) in 1858 and at Doe Park (H = 18 m) in 1863.

21. Bilberry. The history of Bilberry dam is given by G.M. Binnie (ref. 26) but very recently it has been discovered that a plan and sections of the dam were published by James Leslie, the Edinburgh water engineer, who visited the site shortly after the failure (ref. 27). Fig. 3 is redrawn from Leslie's paper. Briefly, the dam, designed by George Leather, was built 1839-41 in a steep-sided valley in sandstone of the upper Millstone Grit Series containing beds of mudstone a few metres thick. Its maximum height was 30 m, not 20 m as usually quoted; this latter figure being the height at outlet level, located well above river bed at an elevation giving an adequate head of water to the uppermost mill downstream. The puddle core had a top width of 2.4 m but, with side batters of 1:24, the bottom width at the deepest section was only 4.9 m; thus h/B = 5.6. On first filling the reservoir a muddy leak, indicating erosion of the core, developed around the culvert. After unsuccessful attempts to cure the leak, Leather proposed to line the upstream face of the dam with puddle keyed in to a trench at the toe. The owners, however, countermanded his orders and appear to have been content to keep the reservoir normally about two-thirds full, amazingly without lowering the rim of the waste pit. In 1846 a flood filled the reservoir, almost overtopping the dam where subsidence had occurred above the culvert, and a new leak developed in the central section. Bateman, consulted by the owners, also proposed an upstream lining but again nothing was done. Evidently the puddle core had now become fractured, for the new leak continued and led to subsidence with a swallow hole just upstream of the core at or near the deepest section. Finally in February 1852 another, larger, flood filled the reservoir up to the swallow hole; subsidence rapidly increased, the dam was overtopped and 81 people lost their lives as a result.

22. From what he could see in the remaining parts of the dam Leslie considered the puddle core to be of good quality and the banks, though rather stony, were not unsubstantial. However, Captain Richard Moody, R.E., appointed by the Government to investigate the failure, realised that the fundamental weakness lay in having a highly permeable fill immediately adjacent to the core.

> The heaviest portion of the materials (he says) should be placed on the outside, and the more binding materials on the inside. Close also to the puddle wall, on both sides, the material should be very binding in its quality and well rammed ... In the construction of Bilberry dam, however, this careful selection has not been made. The material is similar on both sides, and loose

in its nature. The inner portion is permeable throughout ... and a dry open backing appears to be carried up fromthe bottom, on both sides of the puddle, inviting the water, as it were, to act on the whole inner surface of the puddle, and to escape with greater ease at any leaks or fissures arising from settlement or bad execution of the work.

Introduction of zoned fill

23. The desirability of having transitional material between puddle clay and coarse bank fill had been recognised earlier in a few cases; for example, when writing specifications for Dingle dam (1848–50, H = 12 m) Joseph Jackson, water engineer of Bolton, stipulated that the 'strongest' (most cohesive) material should be reserved for fill adjoining the puddle core, stones and shale being used for the outer parts (ref. 22). But the Bilberry failure and Moody's emphatic criticism brought this aspect of earth dam design into sharp focus.

24. In this and other respects Leslie's specifications, dating from about 1854, were regarded as a model (ref. 28): 'The most clayey and earthy part of the stuff from the excavations to be used for the heart of the embankment (adjacent to the core) and the more stony, gravelly or sandy portion to be used in forming the slopes'. In addition he ordered the bank fill to be placed in 15 cm layers by horse and cart (wagons on rails being forbidden) and rolled or pounded if necessary to ensure solidity. He also introduced (possibly for the first time) the improved process of wetting and working the puddle clay in the clay field, and allowing time for 'moistening', before it was carried to the embankment to be placed in thin layers and trodden down to form the core. Leslie's dams for Edinburgh at Harperrig (1856–59, H = 12 m) and Crosswood (1864–68, H = 15 m) and for Dundee at Crombie Dean (1866–69, H = 18 m: specification in ref. 29) were built in this way.

25. Bateman's early dams such as Kentmere (1845–48, H = 18 m ref. 26) and Torside (H = 30 m) and Rhodes Wood (both in Longdendale and built 1849–55, ref. 23) had the more adhesive material, carted in 60 cm layers, in the upstream bank but more stony material, in 120 cm layers, in the downstream bank. Rhodes Wood dam with a maximum height H = 24 m and an exceptionally narrow core (h/B = 6.0) suffered turbid leakage and associated crest settlement in 1858. On investigation Bateman found a sand-filled crack across the core at a depth of 17 m, which he attributed to faulty construction but in retrospect can be identified as a case of hydraulic fracture.

26. However, by 1862 in specifications for Fishmoor dam he adopted a zoned section with select cohesive material on both sides of the core and coarse-grained fill in the outer parts (ref. 1). Bateman used the same form of construction for the 25 m high Dean Head dam for Halifax Waterworks (specification dated September 1864, quoted in ref. 29) and all his later dams.

27. Several other examples of zoned dams from the early 1860s can be quoted. After 1865 very few well designed large earth dams in this country were built in any other manner, except for those cases where the entire bank fill was clay.

Dale Dike

28. An epoch in the history of earth dams came to a close with

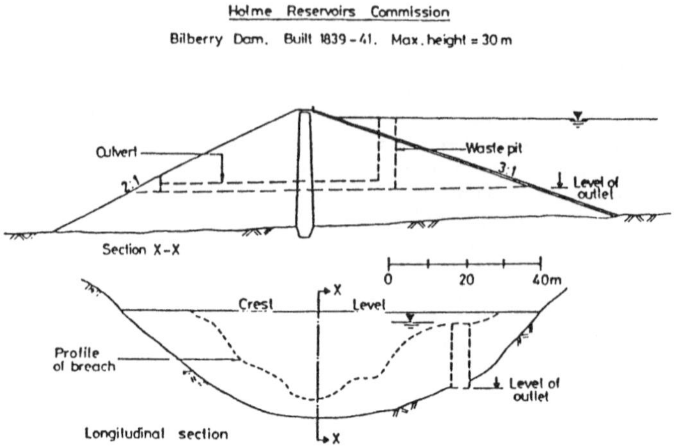
Holme Reservoirs Commission
Bilberry Dam. Built 1839 - 41. Max. height = 30 m
Culvert
Waste pit
2:1
3:1
Level of outlet
Section X-X
0
20
40m
X
Crest
Level
Profile of breach
Level of outlet
Longitudinal section
X

Fig. 3

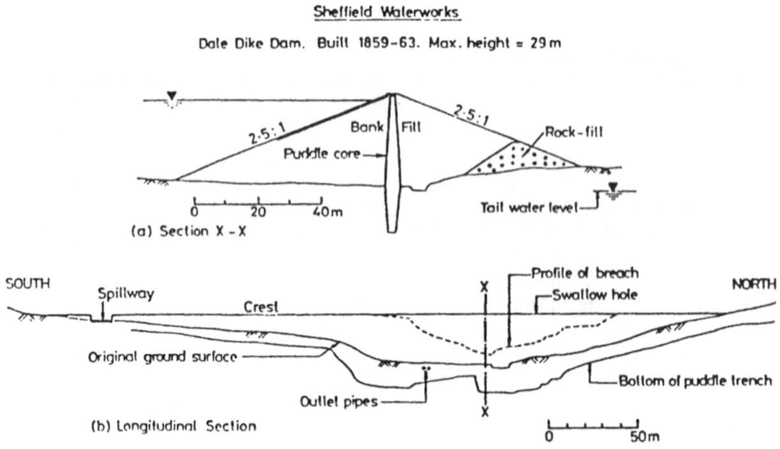
Sheffield Waterworks
Dale Dike Dam. Built 1859-63. Max. height = 29 m
2.5:1
Bank
Fill
2.5:1
Rock-fill
Puddle core
0
20
40m
Tail water level
(a) Section X - X
SOUTH
Spillway
Crest
X
Profile of breach
Swallow hole
NORTH
Original ground surface
Bottom of puddle trench
Outlet pipes
X
(b) Longitudinal Section
0
50m

Fig. 4

Britain's worst civil engineering disaster: the collapse of Dale Dike dam in March 1864, with the loss of 250 lives. The story of this terrible event is well known (refs. 30 and 31) and leading engineers, reporting on the failure, agreed in identifying several defects in design and construction; though they were not able to agree on a single unambiguous cause of the collapse. However, in a recent study of the case Binnie (ref. 32) attributes the collapse to hydraulic fracture of the core; a conclusion now supported by an analysis of the mechanics of this process (ref. 33). To Mr. Binnie's account the present writer would add only three remarks: (i) the h/B ratio of the core is 5.6, (ii) from written evidence by an eyewitness it appears that failure began near the point marked 'swallow hole' in Fig. 4, (iii) it is almost inevitable that hydraulic fracture also occurred in the deep puddle trench (ref. 33) and, as Binnie suggested in a later contribution (ref. 34), there may have been cracking due to differential settlement at an abrupt change in level of the bottom of the trench near section X–X in Fig. 5. However such occurrencies probably could not lead to such rapid propagation of internal erosion as evidently happened.

29. To summarise the case record, Dale Dike dam, 29 m high, was built 1859–63 for Sheffield Waterworks in a valley in sandstones and shales of the upper Millstone Grit Series. The engineer was J.T. Leather. He had previously been responsible for five other dams for the city's water supply, at Redmires and in Rivelin valley. Constructed between 1834 and 1854, typically with heights of 15 m, they have puddle cores 1.5 m wide at the top and side batters at 1:16, and the dams stand today in first class condition. Practically the same shape of core was adopted at Dale Dike but, with a far greater height, its bottom width at the deepest section was only 16 ft (4.9 m) or about one-sixth of the water head of 90 ft (27.5 m). Also, in contrast to clayey fill (weathered mudstone) at Redmires, a lot of coarse material was used in the banks. The intention seems to have been to place the finer part of the fill adjacent to the core, in layers not exceeding 90 cm, but in practice this was not done systematically, and the upper half of the bank is said to have been composed mostly of 'rubble material' (broken sandstone) tipped from earth wagons of 1 cu.yd. capacity in layers up to 150 cm thick. On the other hand, from what could be seen of the puddle clay in those parts of the dam remaining after the collapse it was 'sound and tight' and of good material.

30. The failure started about 10 p.m. on 11 March, during a storm, with reservoir level on first filling marginally below the spillway weir, when water began 'boiling through the bank'. Soon afterwards the crest at this point subsided and within half an hour the dam was breached. Earlier, in the afternoon, a longitudinal crack was noticed several metres below crest level on the downstream side. Though still not fully explained, the crack can be discounted as a direct cause of the failure.

31. As would be expected by 1864 several of the engineers called in to report on the disaster said the core should have been backed with clay. They also considered the core to be too narrow; a point seemingly not made explicit before this time, though the Dale Dike core was certainly narrow by comparison with those in most dams of the period. However, Robert (later Sir Robert) Rawlinson, reporting to the Government (ref. 35), went much further and spelt out the danger

of the process now known as hydraulic fracture.

> The puddle-wall (he says) is too thin for a work of this magnitude; the material placed on either side is of too porous a character, and much of it was tipped from railway waggons, which is the worst way of constructing a water-tight embankment. No puddle-wall should ever be placed betwixt masses of porous earth, as puddle, under such conditions, is liable to be fractured by pressure of water ... It requires to be backed up, on both sides, to at least its own thickness, with fine selected material. Six feet depth of water will puncture 12 inches of puddle if laid over rubble-stone and loose earth, like this bank; so that 16 feet of puddle would be liable to be punctured by a head of water of 90 feet, if the least flaw existed in any point, and the full pressure could act on it.

Criteria for hydraulic fracture

32. That a hydraulic gradient of 6.0 is critical for fracture of a puddle core with the 'least flaw' is borne out by data plotted in Fig. 5. An empirical rule, it nevertheless serves well to distinguish between cases of hydraulic fracture and non-failures. Two apparent exceptions with h/B > 6 can easily be explained: the points refer to narrow cores backed by clay fill.

33. Doe Park. If, as seems probable, the failure at Doe Park dam is an example of hydraulic fracture it represents a departure from the rule. Built 1858-61, with a height of 18 m, a serious leak developed through the core at a depth of 12 m after second filling of the reservoir in 1863 (ref. 36). The core had similar proportions to those at Dale Dike but, with a smaller height of dam, h/B was 4.4. The reason for the failure may be that while the possibility of hydraulic fracture in these narrow cores exists at depths of more than about 18 m there is only a small margin of safety above that depth (ref. 33) so a small fault in construction could lead to failure. For this reason the line in Fig. 5, corresponding to cores with a top width of 1.5 m and side batters at 1:16, as at Doe Park, is a safer limiting criterion; though of course internal erosion cannot propagate unless the core is backed by material failing to meet filter requirements.

34. According to recent field measurements the horizontal (minor) stress is puddle cores at a depth z below crest level probably drops below a value equal to reservoir pressure at that depth, and hydraulic fracture is therefore possible, when the ratio of z to core width b exceeds about 5 or 6 (refs. 37 and 38). For cores with a top width of 1.5 m and 1:16 batters a z/b ratio of 5 is reached at depths of about 19 m, with h/B = 4.5; figures which compare well with those for Doe Park dam and appear to confirm this to be a limiting case.

35. Roddlesworth Upper dam (1861-65, H = 21 m) is more worrying from both the theoretical and practical points of view. The first serious trouble seems to have developed so late as 1904 in the form of a swallow hole upstream of the core just below top water level. Further subsisdence occurred nearly two years later. In 1908 a new swallow hole appeared some 30 m from the first one, again in front of the core and extending, so far as could be discovered, to a depth of about 10 m (ref. 39). This would seem to be a case of hydraulic fracture, except for the delay of 40 years and the presence of a fairly

wide core with a top width of 1.8 m and 1:12 batters (ref. 40) giving h/B = 3.8. However, the maximum depth of puddle trench is 36 m, and perhaps internal erosion slowly developed from cracks initiated below ground level.

VICTORIAN DAMS (ii) 1865-1905

36. In an appendix to the Dale Dike report of May 1864 Rawlinson remarked that puddle clay backed by 'tight and sound' material will resist any ordinary depth of water, but if the supporting material is 'open, fissured or porous' only a moderate pressure can be carried. 'To secure a puddle-wall remaining sound (he says) the entire area of support, below the ground line in the puddle-trench, and above, in the made embankment, must be in fine close material or there must be such an excess of thickness in the puddle as to render it capable of resisting any possible searching, puncturing or wasting power of water'; and he goes on to recommend that the thickness should be not less than one-third of the water head (ref. 41).

37. The 'one-third' rule, implicit in the early (small) puddle core dams (paras. 17 and 18), had rarely been followed in high dams before 1865; James Leslie's Torduff dam (1849-51, H = 26 m) is perhaps the only example. By contrast, after that date, during a period of 40 years in which about 100 large dams ($H \geq 20$ m) were built in Britain, the rule was often quoted (refs. 29, 40 and 45 for example) and in practice the great majority of dams had a core width not less than 25 percent of the water head (h/B = 4). Moreover, as already mentioned, select fine-grained material was usually placed on both sides of the core, thus providing an extra safeguard. Consequently failures of the core were eliminated except in a few cases of thoroughly bad design.

38. However, several sources of trouble remained. Chief among these were (i) slips during construction, usually caused by weak clays in the foundation (examples listed in ref. 20), (ii) leakage through bedrock around the ends of the cut-off, cured by constructing wing trenches, (iii) leakage and internal erosion in puddle trenches beneath the dam due to hydraulic fracture or cracks resulting from differential settlement.

Deep puddle trenches

39. Grizedale dam (1862-66, H = 22 m) provides a well-documented case record of hydraulic fracture in a deep puddle trench (ref. 42). The engineer was Thomas Foster who had worked under Bateman on the reconstruction of Bilberry dam. The cut-off trench, width C = 3 m, had a maximum depth of 36 m; the bottom 6 m, in rock, being filled with concrete followed by puddle clay extending to a depth D = 30 m below crest level through what appears to be landslide debris overlying gravel (Fig. 6). In June 1867, the reservoir having been full for 7 months, a small leakage hitherto of clear water rapidly increased to about 45 ℓ/s and became charged with clay particles; the dreaded signal of internal erosion, though as yet with no visible settlement of the crest. Investigations showed the leak to be through the puddle trench at a depth of z = 24 m, opposite a layer in the ground described as 'loose rock and earth'. Water head with reservoir full is h = 19.5 m, so the hydraulic gradient across the trench h/C = 6.4, and D/C = 10. By analogy with the z/b ratio for cores, at the point of failure in the trench z/C = 8.

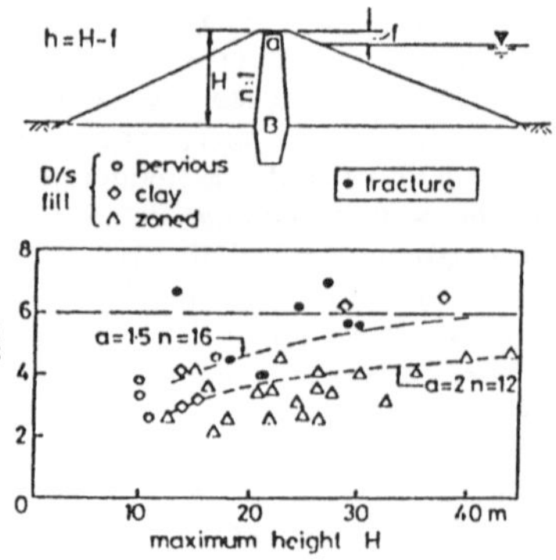

Fig. 5. Susceptibility to hydraulic fracture in puddle clay core

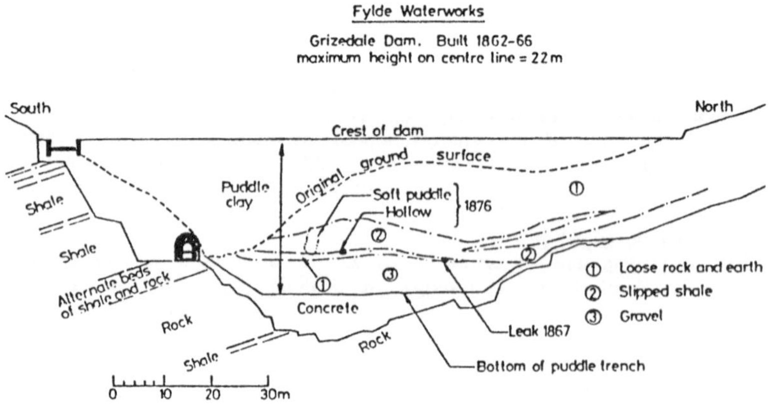

Fig. 6

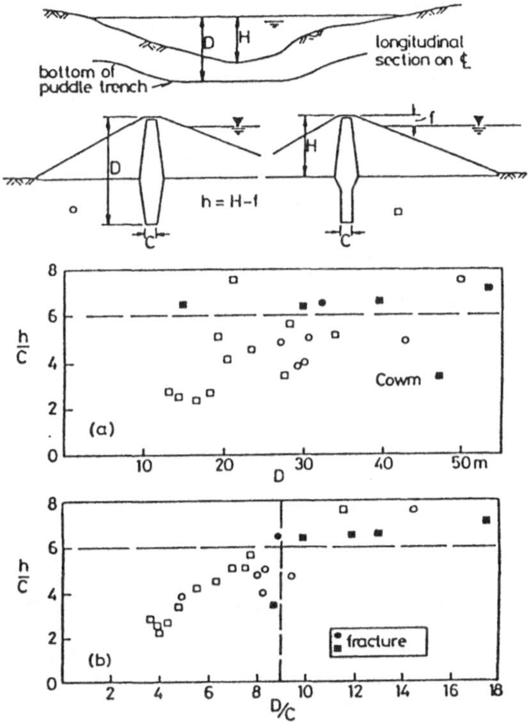

Fig. 7. Susceptibility to hydraulic fracture in puddle clay trench

40. The damaged part was rebuilt in a wider trench (C = 3.8 m) and proved satisfactory. But in December 1876, ten years after first filling, another turbid leak developed; this time with associated crest settlement. It was traced to a sand-filled cavity through the original puddle trench, again opposite the 'loose rock' layer, some 20 m from the first leak at much the same depth (Fig. 6).

41. Walshaw Dean Lower. Confirmation that hydraulic fracture can occur in puddle trenches of dimensions similar to those at Grizedale comes from Walshaw Dean Lower dam (1901–07, H = 22 m). The trench extends to a depth D = 40 m through fissured sandstone and is 3 m wide with h/C = 6.6. Impounding began in February 1907; the reservoir would have been filled by the end of the year and in February 1908 leakage increased, apparently with loss of material, leading to the need for extensive rock grouting immediately upstream of the puddle trench in 1911–15 (ref. 43). This was partially successful, but settlement continued and a thorough grouting programme both in the puddle and in rock downstream of the trench was carried out in 1982. Recent measurements show that, even after this treatment, horizontal stresses in the puddle at depths of 23 m (z/C = 7.7) and 33 m (z/C = 11.0) are respectively 94 and 90 percent of reservoir pressure (ref. 37).

42. Points representing the two cases mentioned above and three others where hydraulic fracture has clearly been identified are plotted in Fig. 7, together with numerous cases in which there is no evidence of fracture. With some important exceptions the graph shows that failure is likely if the hydraulic gradient h/C exceeds 6, thus justifying Rawlinson's perception (para. 36) that 'puncturing' of a thin puddle wall could occur in trenches, just as in cores, if the supporting material is 'open, fissured or porous'.

43. The observations at Walshaw Dean (para. 41) suggest that susceptibility to hydraulic fracture increases with increasing values of the ratio D/C. This can be taken an index of the tendency for arching to develop, and from Fig. 7 a safe limit appears to be about D/C = 9.

44. Fracture of a puddle trench can also be caused by differential settlement; an effect most pronounced at sudden changes in level. Rawlinson in 1872 (ref. 44) warned against the danger of having vertical 'steps' in the bottom of a puddle trench, and Alexander Binnie in 1877 (ref. 45) refers explicitly to cracking associated with 'unequal strains'. He goes on to say 'the practice now is to form the bottom of the trench into long inclines and so avoid abrupt changes in depth of the puddle'.

45. This was not always done, however, and a remarkable case is Walshaw Dean Middle dam (1901–07, H = 24 m) where the trench steps up 20 m from a depth of 53 m below crest level, in quite a short distance; grouting had to be carried out soon after filling the reservoir, and again in 1940–41 (ref. 46).

46. Evidently Binnie in 1877 knew of specific cases of cracking due to differential settlement, but the classic example is somewhat later in date; at Neuadd Lower dam (1880–84, H = 16 m) where leakage was eventually traced to a sand-filled shear crack rising up from the culvert structure, as shown in Fig. 8 (ref. 47).

47. The severe reduction in pressure in a puddle trench immediately adjacent to culverts is demonstrated by measurements at Gorpley dam

(1900–04, H = 26 m) where the horizontal stress in this situation was found to be only 65 percent of reservoir pressure (ref. 48).

48. The best way of avoiding problems associated with culverts was to eliminate them by providing an outlet tunnel through bedrock. Such tunnels had been sparingly used before 1865 (para. 19) but afterwards became almost standard practice for large dams, at least by the leading engineers. Examples include Vale House (1865), Yarrow (1868), Woodhead No. 2 (1871), Tunstall (1873), Leeshaw (1875), Barden Upper (1876) and many subsequently.

49. Cowm. Returning to the history of deep puddle trenches, two apparently exceptional cases must be considered. The first refers to serious trouble experienced by Thomas Hawksley at Cowm dam (ref. 1). This was to be 21 m high above stream bed and the cut-off had to reach a maximum depth D = 47 m below crest level in order to bottom in sound rock beneath open-jointed sandstone and shale of the Upper Millstone Grit Series (Fig. 9). The trench, excavated 1868–69, was given the generous width of 18 ft (5.5 m), thus limiting the hydraulic gradient to h/C = 3.5. The rock joints were of course 'packed up' with cement, and Hawksley took exceptional care in the worst section to replace the sandstone with a blanket beneath the upstream slope, composed of puddle clay over a concrete slab. Work reached completion in 1875; but not long after first filling, turbid leakage and more than usual subsidence of the crest occurred near section X–X (Fig. 9). Hawksley finally cured the trouble in 1879 by rock-grouting in front of the trench at this location; a process he introduced for the first time, earlier in the same year, at the wing trench of Tunstall dam (refs. 49 and 50). Further leakage at Cowm occurred in 1884 near the spillway; this too was successfully treated by grouting.

50. From an examination of the Cowm puddle trench failure it is clear that sealing all fissures in rock is a practical impossibility. A more positive 'stop' to the washing through of clay particles has to be provided, and in later works Hawksley used a thin concrete wall on the downstream face of deep puddle trenches in rock, as in the wing trench of Spring Mill dam c. 1880 (ref. 1). It is also evident that with a trench 18 ft wide no trouble could reasonably have been anticipated; the point representing Cowm in Fig. 7(a) is a conspicuous exception. Two possible explanations of the failure are (i) the trench is so deep that, despite its width, the D/C ratio is nearly 9, see Fig. 7(b), and arching action may have reduced the horizontal stress below reservoir pressure, (ii) differential settlements at the rather steep bottom slopes near section X–X and, to a lesser degree, under the spillway might have been sufficient to cause cracking.

51. Yarrow. The second case arises at Yarrow reservoir, for Liverpool's water supply. This is formed by two dams built 1868–75 under the direction of Joseph Jackson (ref. 22). The two embankments together contain nearly 1 million cu.m of fill and both have very deep puddle trenches (completed by 1872) which taper down from a width of about 7 m at stream bed level to C = 3.5 m above a 2 m thick concrete base (longitudinal and cross sections in ref. 40). For the northern embankment H = 27 m and D = 50 m, giving h/C = 7.5 and D/C = 14.5 (Fig. 7). Comparison with vertical-sided trenches is not exact, but even allowing for a somewhat larger effective width the

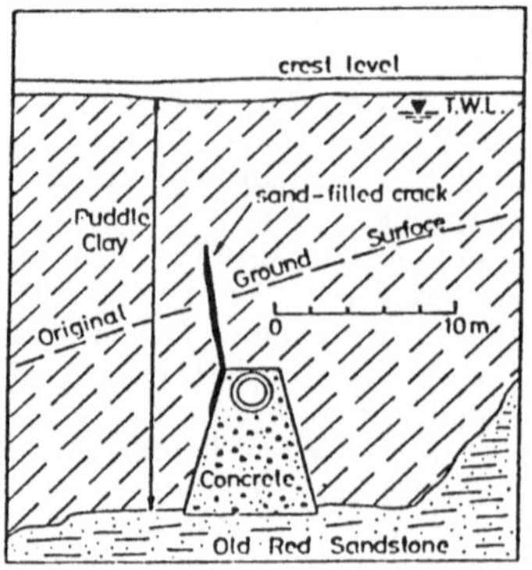

Fig. 8. Longitudinal section through puddle trench and core at lower Neuadd dam (redrawn from Deacon, 1902)

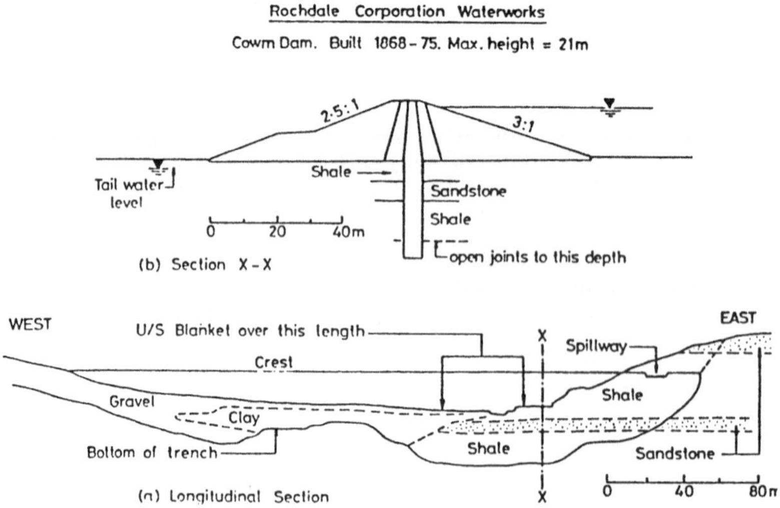

Fig. 9

ratios are alarmingly high; yet there seem to have been no appreciable leakage or settlement problems. The difference from Cowm is probably due to the trenches at Yarrow being mostly through sand and gravel which could act as a filter preventing internal erosion even if, as seems probable, hydraulic fracture had occurred.

Concrete cut-off trenches

52. Woodhead No. 2. Meanwhile, with the second Woodhead dam, in Longdendale, Bateman faced the construction of a cut-off which had to reach 50 m below crest level in difficult strata: open-jointed rock, a large fault zone, and a deep gravel-filled channel buried beneath an ancient massive landslide (section in ref. 23). Preliminary work started in 1870 and by mid-1871 a 3 ft gauge railway was in operation, with the first steam locomotive used in reservoir construction, bringing men and materials from Tintwistle and the nearby clay-field 6 km from the site (ref. 51). Originally the intention may have been to use puddle clay in the trench but if so it was not long before Bateman decided to rely entirely on concrete, judging that no practical thickness of puddle clay 'could withstand the wasting or penetrating power' of reservoir pressure at such great depths. Accordingly the trench was made 2 m wide, broadening out to 6 m at rock head to form a seat for the puddle core, and filled with concrete. This first complete concrete cut-off was finished in 1874 and the dam in 1876.

53. Probably the next example is the 38 m high Barden Upper dam (described in para. 56) where work started in 1876 and the cut-off reaches a depth of 47 m below crest level. This was followed by several dams for Oldham Corporation, completed between 1880 and 1883 and designed by G.H. Hill who had been Bateman's principal existant on the Longdendale works. William Watts, the resident engineer, gives a geological section of one of the Oldham dams, at New Years Bridge (H = 20 m), where the trench was taken through open-jointed sandstone to a depth of 36 m below crest (ref. 52). Another example of the same date is Ramsden dam (1879-83, H = 26 m) with a concrete cut-off reaching a maximum depth of 48 m (ref. 53).

Late Victorian dams

54. By about 1880 design principles had evolved leading to safe and resaonably economic large earth dams. So far as the geotechnical aspects are concerned the main principles may be summarised as follows: (i) The puddle core ideally to have a thickness not less than one-third of the water head and in no case to have dimensions less than a top width of 2 m and 1:12 side batters. (ii) Select fine-grained fill to be placed on both sides of the core, preferably with a zone of clay 2 to 3 m wide immediately adjacent to the core. (iii) Coarse fill to be placed in the outer parts of the embankment. (iv) The upstream slope to be not steeper than 3:1 and the downstream slope not steeper than 2:1 with berms in high dams. (v) A drainage mattress to be provided under the downstream bank. (vi) Any 'slippery or compressible' material in the ground to be removed before construction of the embankment. (vii) For dams requiring a cut-off not more than about 30 m below crest a puddle trench of normal width (say up to 4 m) is satisfactory, all obvious fissures and open joints in rock having

been cemented. (viii) Sharp changes in depth of the puddle trench are not permissible and, in rock, the bottom of the trench should be concreted. (ix) For deeper cut-offs the trench should be filled with concrete, a width of 2 or 3 m being sufficient but broadening at the top to make a seat for the puddle core. (x) For large dams there should be a tunnel outlet, not a culvert, with the valve tower at the upstream toe of the embankment.

55. The above paragraph is essentially a precis of the relevant parts of lectures on water supply by Alexander Binnie, published in 1878 (ref. 45). The sole addition is the limiting depth for puddle trenches. Binnie simply advises a concrete cut-off where 'trenches of great depth' have to be constructed, though there is an implication elsewhere that this means about 30 m. Of the methods of 'working up' puddle clay (see para. 11) he prefers to carry out this operation in the clay field, rather than on the embankment, to avoid a tendency to produce wet patches or layers from occasional excessive watering of the clay. On the placing of fill he remarks that some engineers specify that no earth wagons should be allowed on the bank, all work being done by wheelbarrows and carts; but whereas high tips, as used on railway embankments, are wholly objectionable (leading to segregation and poor compaction) he thinks 'that if the layers be kept down to a maximum of 75 cm in thickness, the continual passage of the laden wagons, followed as it should be by heavy rollers, will be found to have a beneficial effect rather than otherwise'.

56. Barden Upper. Binnie illustrates the lectures by various cases, but chiefly refers to his Barden Upper dam, then under construction for Bradford Corporation: the highest (H = 38 m) of 19th century earth dams, with a concrete cut-off reaching 47 m below crest level. The specification dated July 1876 (ref. 54) incorporates all the points summarised above including a requirement that the fill should be placed, in layers not exceeding 75 cm in thickness, by (horse-drawn) 1 cy.yd side-tip waggons and compacted with a steam-hauled 8t roller. The core has a top width of 3 m and 1:12 batters; select fill on each side of the core is battered at 1:1; the slope inclinations decrease from the top downwards, averaging about 4:1 upstream and 3:1 downstream, with berms; and there is a drainage mattress beneath the downstream bank fill. Clay for the puddle core is to be dug and exposed to the weather in ridges at the clay field, then worked up by cutting, watering and treading before it is carried to the embankment, to be placed and trodden in 15 cm layers. Work began soon after July 1876 and finished in 1881 by the contractor Easton Gibb. In the schedule of prices unit cost of puddle clay in the core is given as £0.175 and rolled bank fill £0.067 per cu.yd.

57. Barden Upper dam may be taken as inaugurating the late Victorian period (1880-1905) during which some 30 dams exceeding 25 m in height were successfully completed: an achievement unparalleled elsewhere in the world at that time. Not all the dams of this period were built to so careful a specification, the use of heavy rollers in particular being a feature more characteristic of the early 20th century: and mistakes were made, usually in a misguided attempt to save cost or through not determining ground conditions with sufficient clarity before letting a contract. But relative to the amount of construction, failures were few and far between and none of them ranked as disasters.

58. Langsett. An example of these well designed late Victorian works is Langsett dam: see Figs. 10 and 11 (refs. 55 and 56). Built 1897–1904 under the direction of William Watts (previously of Oldham, para. 53) for Sheffield Corporation, the dam is 33 m high above stream bed. The concrete cut-off trench, excavated between September 1897 and July 1899, reaches a maximum depth of 57 m below crest in order to bottom in sound shale beneath strata contorted in a 'valley bulge', and almost the same depth below ground at the south and to get beneath open-jointed shale. Groundwater ran into the trench at a rate of about 65 ℓ/s. Puddle clay, worked up at the clay field, and bank fill, came in earth wagons (1 cu.yd. capacity, of which 60 were employed) on a 3 ft gauge railway from borrow pits close to the end of the reservoir area (ref. 51). Clay would have been tipped from the wagons on timber staging and barrowed on to the core for treading and heeling. Select fill adjacent to the core was probably handled in a similar manner. The rest of the select fill and all the bank fill was tipped from the wagons in layers not exceeding 90 cm in thickness and compacted by frequent slewing of the rails as construction proceeded. Apart from a relatively small amount of rock from the trench and tunnel excavations, placed in the outermost parts of the banks, the 680,000 cu.m embankment was built in rather less than 4 years at an average rate of about 170,000 cu.m per year. The dam, with its 60 m spillway and valve tower, reached completion in October 1904. Three years later, crest settlements were one-quarter of the post-construction allowance of 1.5 percent of the embankment height.

EARLY 20th CENTURY

59. Three periods can be recognized in the history of earth dams in Britain after the Victorian era: (i) Up to the late 1930s puddle core dams continued to be built to the same design rules as before but generally with better compaction of bank fill, increased use of mechanical plant including pugmills (rare before 1900) for preparing puddle clay, and the incorporation of rock grouting as an integral part of design. (ii) The next twenty years are characterised by the application of soil mechanics and the introduction of 'modern' earth-moving plant. (iii) Around 1960 rolled core dams made a belated appearance following developments overseas, particularly in USA, and the long tradition of puddle core construction comes rather abruptly to a close.

60. Grassholme. The opening of the 20th century is typified by Grassholme dam built 1901–11 (engineer James Mansergh) with a height of 30 m and a concrete cut-off extending 62 m below crest level (ref. 57). It resembles Langsett dam in many respects but clay for the puddle core, after weathering in the clay field for a month, was worked up in a steam-driven pugmill before being delivered to the embankment for treading in 12 cm layers, and both the select fill and bank fill were compacted in 30 cm layers by 10 t steam rollers. Wagons were not allowed within 2.5 m of the edge of the puddle, select fill in this width being barrowed and rammed.

61. Burnhope. Full details of the construction of this dam in Weardale are given by the resident engineer (ref. 58). It has a height of 41 m above stream bed, contains 1.2 million cu.m of fill, and was built 1930–35 by direct labour to the design of W.S. Nicholson (of T.

KEYNOTE ADDRESS

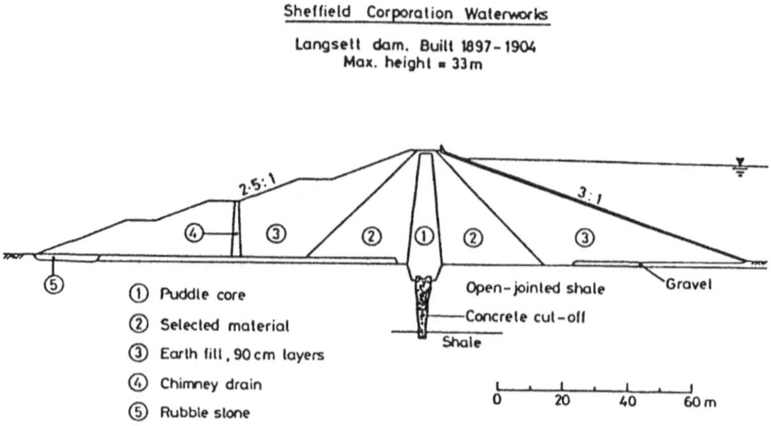

Fig. 10

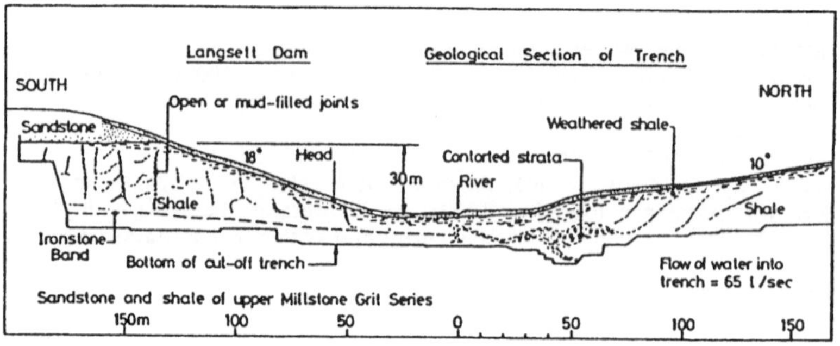

Fig. 11

& C. Hawksley) with W.J.E. Binnie as joint consultant. The cut-off trench, reaching 65 m below crest level, was excavated between October 1931 and May 1932, concreted by February 1933, and the underlying rock (Carboniferous shale and limestone) was grouted. Meanwhile two roll mills and a pugmill had been installed, and rock from the trench and tunnel excavations was used to form the embankment toes. Construction of the main bank and puddle core began in January 1933 and proceeded with little interruption until completed in June 1935, at an average rate of 490,000 cu.m per year.

62. The borrow pits were excavated by four diesel face shovels (1.5 cu.yd buckets); twenty 8 t locomotives hauled cu.yd wagons to the mills and on to the embankment on 2 ft gauge rails; Boulder clay for the puddle core, with the bigger stones removed by hand, was passed through the roll mills to crush any stones larger than 15 mm before being fed into the pugmill and carried to the dam, where it was trodden and heeled in 10 cm layers. Clay adjacent to the core was compacted by benzol rammers. Fill materials, tipped from the wagons and spread in 45 cm layers, were compacted by repeated slewing of the rails followed by a 17 t wide-wheeled steam roller on the select fill (Boulder Clay) and by two 10 t and 13 t rollers on the bank fill (hard mudstone). Control tests on the puddle clay gave consistent results, averaging w = 23, sand content = 30%, silt and clay = 70%, and permeability $k = 2 \times 10^{-9}$ m/s. Accurate total costs amounted to £0.375 per cu.yd for the puddle core and £0.115 for fill.

63. Settlement records and an investigation in 1959 (ref. 59) show the mudstone fill to be a blocky almost free-draining material with w = 11 after 25 years, compared to w = 5 in the borrow pits, and index properties of the Boulder Clay LL = 46, PL = 18. Post-construction settlements of the crest were 1.2 percent after 4 years and after 22 years 1.8 percent of the height (39 m of puddle clay at the point of measurement) compared to an allowance of 2 percent. Settlements of the upper berm, at the same dates, were 0.4 and 0.5 percent of the 26 m height composed of 10 m of mudstone fill above 16 m of select clay fill.

64. Cheddar. Two dams of slightly later date illustrate the transition to modern methods of placing and compacting the bank fill. For Cheddar dam (ref. 60) the 690,000 cu.m embankment was built 1933-35 of weathered Keuper Marl, excavated by dragline and conveyed in side-tip wagons on standard (4'8" gauge) rails, then spread by small face shovels fitted with scoop buckets, and compacted in 45 cm layers by a tractor-drawn roller. In the later stages, in 1935, a bulldozer was used (perhaps for the first time on a British dam) to spread the material before rolling in 23 cm layers. The rate of placing averaged about 340,000 cu.m per year: a figure which is about double the rate at Langsett, an embankment of almost exactly the same volume.

65. Hollowell. Work on the embankment of Hollowell dam (ref. 61) began in November 1936. Weathered Jurasic sandstone, won by dragline, was carried on to the bank by locos hauling standard gauge wagons, as at Cheddar, and diesel bulldozers spread and compacted the fill. For the puddle core and trench Upper Lias clay was taken in narrow gauge 'jubilee' wagons from the clay field to two pugmills and thence in trucks to the dam where it was cast by hand and spread in 15 cm layers for treading and heeling. After a three month shut-down

work restarted in March 1937 and proceeded until 31 October when, the bank having reached a height of 12 m, a slip occurred in alluvial clay beneath the downstream bank. Remedial works comprised a toe berm and steel sheet-piling driven through the bank and clay into firm Lias clay.

66. Ladybower. In 1936 construction also began at Ladybower, in north Derbyshire, to designs by G.H. Hill & Sons. The highest of all puddle core dams (H = 44 m), it has a concrete cut-off reaching 96 m below crest in order to extend beneath sandstones and mudstones crumpled in a valley bulge of exceptional depth (ref. 62). The embankment was nearly complete in 1940, but wartime conditions imposed delays and filling of the reservoir did not start until March 1943.

APPLICATION OF SOIL MECHANICS AND TRACTOR EQUIPMENT

67. A new era was inaugurated in 1937/38 with the first application of soil mechanics in Britain to the investigation and re-design of an earth dam, and the introduction of 'modern' earth-moving plant. Two distinct stages can be recognised before 1960.

First Stage 1937-1945

68. Chingford No. 2 (William Girling reservoir). Built by the contractor John Mowlem Ltd for the Metropolitan Water Board, the embankment of this reservoir is of great length (5.7 km) but only between 11 m and 13 m in height. The puddle trench passes through about 1 m of alluvial clay and 5 m of river gravel into stiff blue London Clay (Fig. 12). The trench was excavated and filled with puddled London Clay from May 1936 to February 1938, working in lengths of 210 m at a time. The clay, excavated by a 3.5 cu.yd cableway-dragline (or tower-scraper) from a long rectangular pit, was stockpiled to weather for at least two weeks, then transported in jubilee wagons to the pugmills and carried thence in standard gauge side-tip MSC wagons (4.5 cu.yd) drawn by steam locos to the trench. All this departed little from the traditional practice, but when bank building started in 1937 the fill (mixed alluvial clay and gravel) was dug and brought on to the embankment by tractor-scrapers (probably 8 cu.yd capacity) and spread by bulldozers, while clay for the puddle core came from the pugmills in tractor-drawn Athey wagons (ref. 63).

69. With this tractor equipment, imported from America and used at Chingford for the first time in British dam construction, work proceeded rapidly until 30 July 1937 when an 85 m length of the embankment failed on a slip surface passing through the core and a layer of soft clay 90 cm thick beneath the outer bank (Fig. 12). Within a few days the top of the bank sank 0.6 m and the toe moved out horizontally by 4 m. At this section, near the N.W. corner, the embankment had been built to a height of 8 m in 11 weeks. The slip provided a striking contrast to the stable slope of King George V reservoir (Chingford No. 1) just across the road, which was completed in 1912 and stood 9.5 m high on similar foundation strata.

70. The soil mechanics group of the Building Research Station, under L.F. Cooling, immediately began an investigation (ref. 64). Samples were tested to measure the undrained strength or 'cohesion' of the alluvial clay (13 kPa at the toe and 20 kPa under 8 m of bank), the

SKEMPTON

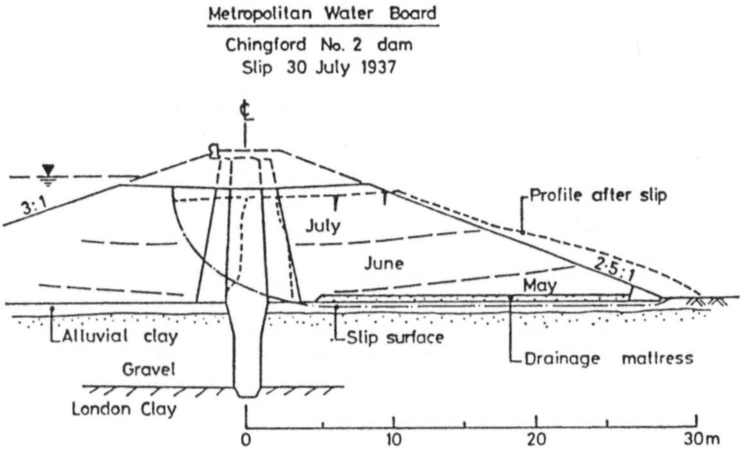

Fig. 12

puddle core (10 kPa), select fill (15 kPa) and bank fill (about 24 kPa). Some preliminary tests were also made to gauge the rate of gain in strength of the alluvial clay due to consolidation under load. In September H.Q.Golder developed a two-circle stability analysis (ref. 65) from which, using measured strengths in the embankment materials, he found the cohesion of the alluvial clay for a condition of limiting equilibrium to lie between 14 and 17 kPa depending upon assumptions in the analysis. Though uncertainty still existed regarding the effect of partial consolidation in the clay during construction, there could be no doubt that the measured and calculated strengths were broadly in agreement. Then, in October-December, the writer made a series of 'equilibrium' shear box tests (approximating to drained tests) and, knowing the rate of construction, was able by Terzaghi's theory of consolidation to calculate the effective pressure in the centre of the clay layer, and hence the strength, at various positions under the bank at various times. With a coefficient of consolidation c_v = 0.4 m^2/yr and strength parameters c' = 10 kPa and φ' = 15° the average shear strength at the date of failure came out to be 18 kPa. Substituting this value in Golder's analysis the factor of safety is around 1.15; an acceptable result, now known to be a remarkably good approximation to the correct value when allowance is made for progressive failure (ref. 33).

71. Meanwhile on 8 Dec 1937 a delayed slip occurred in the inner slope at a section 150 m from the first one, where the embankment was 7 m high. Here the bank fill contained less gravel and (probably softened by winter rain) had a cohesion of only 16 kPa. With no drainage mattress beneath the inner bank there would have been a smaller degree of consolidation in the clay layer, the strength of which at the toe was found to be almost identical with that at the first slip. A two-circle analysis gave a factor of safety F = 1.15, or slightly higher if a small φ value was allowed in the bank fill, and a simple check using Taylor's curves for homogeneous material, taking an average cohesion of 14 kPa throughout, gave F = 1.1.

72. Cooling summarised all this work in a report dated January 1938, with some reservations arising from lack of knowledge of the true nature of cohesion and friction in clays (ref. 66). Soon afterwards the Water Board supplied information from their records on King George V reservoir embankment. At the south end, opposite the July 1937 slip, the bank had been built in 8 or 9 months at little more than one-third the rate of Chingford No. 2; so there was no great difficulty in accounting for its stability.

73. By this time tests had shown the alluvial clay at many other points around the perimeter to be no stronger and in places considerably thicker than at the slips. Moreover, large (30 x 60 cm) shear box tests by Mowlems on bank fill at the S.W. corner, reported to BRS on 3 March 1938, gave cohesion values (15-18 kPa) for clayey material similar to those previously found at the second slip and confirmed that even with a fairly high gravel content the cohesion was not more than about 25 kP/a. Evidently problems existed in producing a design for the full height of embankment, and three revised sections prepared between January and March proved to be less than adequately safe when subjected to analysis.

74. Whatever the final solution might be, the embankment at the first

slip had to be demolished to repair or rebuild the core. During this operation, in April 1938, the distorted shape of the core was recorded (Fig. 12) and the slip surface examined; it passed almost as a horizontal plane through the clay from the toe, where it had already been seen in trench immediately after the slip, to about 5 m from the centre line. Samples were also tested, showing (8 months after construction) a notable gain in strength and, in accordance with consolidation theory, a minimum strength at the centre of the layer.

75. No acceptable solutions having yet been found in April, Mowlems asked Professor Terzaghi for his advice. He came to the site at the end of that month, examined the test results and produced a radical re-design involving wide berms, on both slopes, with keys of gravel fill taken down through existing bank material and alluvial clay to the underlying gravel stratum; to obtain a minimum factor of safety of 1.5 (ref. 67). Work was actively resumed later that year following Terzaghi's recommendations, but the outbreak of war delayed completion until the late 1940s.

76. Staines No. 2 (King George VI reservoir). The first of London's large reservoirs had been built 1897-1901 at Staines, with a maximum height of 10 m, incidentally using a pugmill for preparing the London Clay puddle (ref. 68). Mowlems started operations on the second Staines reservoir, on an adjacent site, in August 1937. Here the embankment, 5.25 km in length, was to be between 12 and 16 m in height, founded on about 1 m of sandy clay (Brickhearth) overlying 4 to 5 m of terrace gravel over London Clay. The section was to be similar to Chingford (Fig. 12) but with Brickearth used for the select fill and gravel for the bank fill.

77. Following the Chingford slip the Metropolitan Water Board naturally wanted a stability check on Staines. Samples of Brickearth were taken in October 1937. Tests at BRS (ref. 69) showed a relatively high coefficient of consolidation $c_v = 8$ m^2/yr compatible with the rather sandy nature of the clay (LL typically around 45) and the layer, lying between gravel fill and a gravel deposit, would consolidate fully during construction. The 'equilibrium' shear strength parameters ($c' = 5$ kPa, $\varphi' = 25^o$) could therefore be used in stability calculations, along with an assumed $\varphi' = 33^o$ in the bank fill and $c = 10$ kPa ($\varphi = 0$) for the puddle core. Golder again carried out the analyses, at the beginning of April 1938, and found the factor of safety to be not less than 1.7 by slip circles or 1.6 assuming full active pressure from the puddle core as given by Bell's equation.

78. Construction of the embankment began shortly afterwards, and its total volume of 3.4 million cu.m was due to be completed in 2.25 years at an average rate of 1.5 million cu.m per year. Up to September 1939 this rate was maintained, using about thirty 10 cu.yd Athey side-tip wagons hauled in tandem pairs by a D8 (15 t) tractor, loaded by elevating graders supplemented by one or two large face shovels. Bulldozers spread the fill in 30 cm layers, compaction being achieved by the heavy construction traffic. The puddle trench was made exactly as at Chingford, except that clay from the pugmills came in jubilee wagons, and puddle clay for the core was delivered in tractor-drawn Athey wagons (ref. 70). During this period, working double shifts, the average number of men employed had been 700. Afterwards, under war conditions, a greatly reduced work force

completed the embankment in 1945.

79. Muirhead. Construction of the embankment of Muirhead dam began in April 1940; the engineers being Babtie Shaw & Morton. Boulder Clay fill (LL = 43, PL = 18) was excavated by face shovels, transported to the bank in 5 t end-tip lorries, and spread by two D8 bulldozers. At the end of the first season the fill had reached a height of 9 m. In April 1941 the lorries were supplemented by 20 narrow gauge diesel locos hauling 170 wagons (1 cu.yd compacity), and four bulldozers were used for spreading and compacting; while in September the railway wagons were replaced by three 13 cu.yd Euclid trucks and two 12 cu.yd Athey wagons. During the 1941 season an almost constant rate of placing at 12,000 cu.m per week had been maintained up to 22 September when the embankment reached a height of 21 m and very noticeable deformations developed in both the upstream and downstream slopes, with tension cracks some 6 m from the centre-line on each side. Surveys a few days later showed maximum horizontal displacements of the order 0.5 – 1.2 m at a height of several metres above ground level, but no movement of the toes (ref. 71).

80. Investigations by BRS during October and November revealed that the stiff Boulder Clay (w = 16.5 and c = 130 kPa in the borrow pits) had been placed in the 1940 fill at a dry density not more than 95% of Proctor maximum (18.2 kN/m^3) with about 6 percent air voids into which rain water could percolate; and the fill, after about a year, had softened to a water content w = 20.5 (3.3 percentage points above Proctor optimum) and a shear strength c = 35 kPa. These are average figures from tests showing a rather wide scatter of results in the upper half of the fill; softening in the lower part was less pronounced. By contrast the better compacted and recently placed 1941 fill had a dry density 97% of Proctor, with 2 percent air voids, a water content w = 18.3 and a shear strength c = 70 kPa; and if fully softened c would not drop below about 47 kPa.

81. These tests (ref. 72) provided the first quantitative data on the softening of clay fills with only moderate compaction; a process leading to slips in clay embankments a year or two after construction, as often encountered in early railway works. The measured strengths were consistent with limiting equilibrium on a slip surface passing from the tension crack through select fill, the puddle core and the upper part of the 1940 bank fill. Proof that the embankment existed in a state of limiting equilibrium was given by adding 50 cm of fill, when immediate outward movements of 15 cm occurred in both slopes, increasing to 30 cm before coming to rest after several days.

82. The re-design was based on a shear strength determined by back analysis of the existing embankment, knowing the factor of safety to be 1.0, and this strength could then be used with confidence in analysing a proposed new section. In order to get a factor of safety of 1.25 (judged to be ample in the circumstances) the bank had to be kept at its present height, with a berm added on the upstream side and material removed from the upper part of the downstream side. The work was successfully completed by November 1942.

83. Knockendon. When the Muirhead slip occurred the embankment of Knockendon dam, in the same locality and built of similar Boulder Clay (LL = 37, PL = 16), had reached a height of 6 m. As the dam was planned to be 27 m high, its stability obviously needed checking

and Messrs Babtie Shaw & Morton asked BRS to help. Investigations early in 1942 showed that the fill already placed had a shear strength of about 40 kPa with a water content 4 to 5 percentage points above Proctor optimum (w = 15) and a dry density of 17 kN/m^3 or 93 percent of the Proctor maximum (ref 73). With this strength there was little hope of being able to build the dam to its full height. The engineers therefore modified the design to incorporate a key of granular fill in the downstream bank and a large berm on the upstream slope, giving a factor of safety not less than 1.33 (ref. 74). Analysis of the re-design was completed by July and work proceeded.

84. At the end of the 1943 season, when the bank had reached a height of 15 m, a check was made on the strength of the 1942 fill. Soil Mechanics Ltd put down several borings, and tests proved the 40 kPa strength had been maintained: but the foreman noted that water level in two of the boreholes rose above bank level when the tubes were left projecting overnight. This clearly indicated the existence of high pore pressures in the bank fill. To measure the pressures four standpipe piezometers were installed early in 1944. In each of the following seasons the pore pressure increased with increasing height of fill, with little dissipation during winter shut-down, and at the end of construction in 1946 the ratio of pore pressure to total vertical pressure at the piezometer tip had a maximum value of r_u = 0.55 (ref. 74). These are the first observations of construction pore pressures in Britain.

85. King George V reservoir. For some years before 1938 water in this reservoir had not been allowed to rise within 0.6 m of normal top level, this being the typical amount by which the crest had sunk below its design level; ranging from 0.4 m at the north to 0.7 m at the south (highest) end (ref. 75). As the bank would have been built with a settlement allowance of about 0.3 m, the total settlement at the south end was therefore around 1 m in the 26 years since completion in 1912; or roughly 6 percent of the total depth of puddle clay in the core and trench. Work was then put in hand to raise the crest and puddle core to their design levels.

86. The remedial works were duly carried out, but as a precaution against enemy action the reservoir was kept 1.5 m below normal top water level (and 3 m below crest) from September 1939. However, on restoring the full depth of water in February 1945 leakage occurred at several points. Investigation proved the cause to be shrinkage cracking of the puddle core to a depth of 2.5 m below crest, resulting chiefly from plant transpiration (ref. 76). The core was made of alluvial clay of very high plasticity (LL = 130, PL = 45) the water content of which had been reduced in the drying zone from about 75 to 60. The high plasticity also accounts for the exceptionally large settlements. A puddle core of London Clay (w = 43, LL = 76, PL = 26) in another Metropolitan Water Board reservoir embankment, subject to the same water level restriction during the war years, suffered no appreciable shrinkage.

Second Stage 1950-1960

87. Apart from completing the several dams on which construction had been interrupted by hostilities, little work was undertaken until about 1950. Important advances, however, were being made in soil

mechanics; and during the post-war stage of dam building this new knowledge became available in the form of improved testing procedures and stability analysis, systematic piezometer observations, and more advanced applications of the principle of effective stress.

88. Usk. The next large puddle core dam, after Knockenden, was built 1950-54 on the river Usk, in south Wales, with a maximum height of 34 m (ref. 77). A typical section is given in Fig. 13. The first problem was a lens of silt in the glacial drift foundation. Tests at BRS in October 1950 showed a moderately high permeability $k = 6 \times 10^{-7}$ m/s, but no significant consolidation could occur under the embankment load as the silt was sealed by Boulder Clay above and below. The engineers, Messrs Binnie, Deacon & Gourley, decided, as an economic alternative to removing the silt, to introduce a row of sand drains connected via a gravel-filled trench with a drainage mattress. This was done between June and August 1951, with six piezometers installed by BRS to monitor pore pressures (ref. 78). The sand drains proved so efficient that scarcely any excess pressures developed in the silt during construction of the embankment.

89. The bank fill is composed of Boulder Clay (LL = 25, PL = 17). To guard against the possibility of a bank failure as at Muirhead, and mindful of the Knockendon observations, the engineers asked BRS to install piezometers in the first season's fill (ref. 78). At the end of that season, in October 1952, the pore pressure ratio r_u at mid-height of the 12 m of fill which had then been placed was 0.75, and five months later, no more fill having been added during the winter shut-down, r_u had fallen only to 0.6 at a rate corresponding to a coefficient of consolidation c_v = 12 m²/yr, and field tests gave a permeability $k = 2 \times 10^{-9}$ m/s (ref. 79). The writer and Dr. A.W. Bishop were consulted in March 1953. Analyses based on pore pressure predictions and measured strength parameters of the fill (c' = 24 kPa, $\varphi' = 30^o$) showed that building could not safely continue without a modification in design, and that the most expeditous method of achieving adequate stability was to introduce drainage blankets before adding the second, and then the third, season's fill (see Fig 13). In this way the future build-up of pore pressures would be restricted, leading to a factor of safety at full height of not less than 1.45. Work accordingly began again in May 1953 and reached completion on schedule in September 1954. The pore pressures during construction are recorded in refs. 77, 80 and 81.

90. With an embankment volume of 660,000 cu.m the contractor Richard Costain Ltd experienced no difficulty in placing the fill in three seasons. Initially they used 15 cy.yd scrapers and 11 cu.yd Euclid end-tip trucks, the latter being loaded from draglines and face shovels and weighing 27 t when fully laden. But as the scrapers left a wide area of clay exposed in the borrow pit to the rather high summer rainfall at this site, thus tending to bring wet material on to the bank, they were not employed in the second and third seasons (ref. 80). Throughout construction the fill was spread in 30 cm layers by bulldozers and smoothed with a 4 t roller to throw off rain. A 15 t wobbly wheel roller was available, but generally the construction traffic alone gave sufficient compaction: 96 percent of the Proctor maximum dry density (20 kN/m³) with fill placed in 1952 on average at a water content 2.5 percentage points above optimum (w = 9.8) and 1.5 above

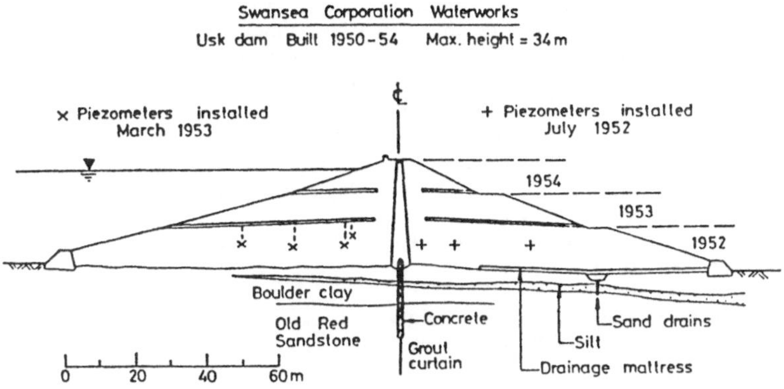

Fig. 13

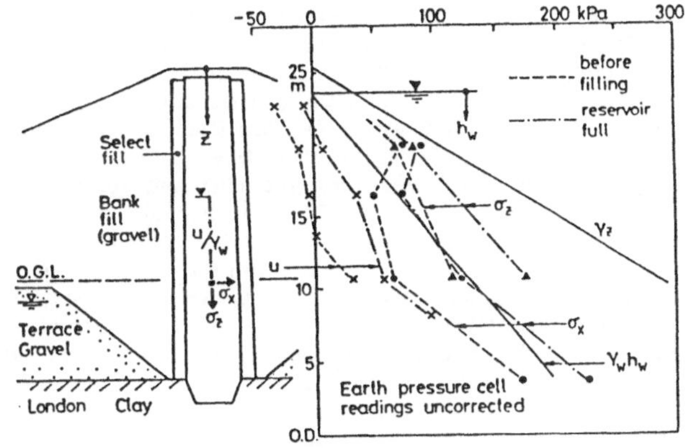

Fig. 14. Queen Elizabeth II reservoir. Built 1957-61
Stresses and pore pressures in rolled clay core

optimum in 1953–54.

91. Daer. When no clay suitable for puddle exists within a reasonable distance from the site a reinforced concrete core wall could be used, as at Daer dam in southern Scotland (ref. 82). With a height of 41 m above stream bed the large embankment, containing 1.8 million cu.m, was built in four seasons 1951–54 by direct labour under the direction of the resident engineer, to the design of Binnie, Deacon & Gourley. Geologically the fill is Boulder Clay, but with a higher proportion of sand and gravel than the material at Usk and a very small clay content. It was excavated by face shovels, carried to the dam in 13 cu.yd Euclid trucks, bottom-tipped in windrows, spread in 60 cm layers by D8 and D7 (11 t) bulldozers and constantly tracked over by another D7 bulldozer; and at the end of each day's work, or earlier if rain was imminent, sealed by flat-track tractors. In this way the winter shut-down period could be reduced to three months and the required average rate of placing was not more than 11,000 cu.m per working week.

92. The resulting dry density of the fill was 98 percent of the Proctor maximum (19.7 kN/m^3) at a water content 2 percentage points below the optimum of 9.6 (ref. 83). Piezometers installed by BRS in May 1951, and in May 1952 in the second season, showed only small pore pressures at any time during construction. The extreme difference in this respect between the fills at Usk and Daer was studied by a series of laboratory tests, details of which are given in ref. 81.

93. Chew Stoke. This puddle core dam, built 1951–54 to designs by T. & C. Hawksley, for Bristol Waterworks, has a height of 13 m above a wide floodplain, the foundation strata being 2.5 m of sandy clay over 2 m of soft grey clay. Tests early in 1951 showed the grey clay (w = 31, LL = 40, PL = 20) to have an undrained shear strength of 20 kPa, a low permeability $k = 1.4 \times 10^{-10}$ m/s and coefficient of consolidation $c_v = 0.8\ m^2/yr$ (ref. 84). As the clay is underlain by Keuper Marl its consolidation under the bank would be restricted to the condition of single drainage impeded by the sandy clay above, which itself had c_v of only 5 m^2/yr. Rather than excavating the clay and replacing by bank fill, an operation adding about 30 percent to the volume of embankment, it was therefore decided to install a system of sand drains calculated to reduce the build-up of pore pressures in the grey clay to a value giving a minimum factor of safety of 1.5 when the bank reached its full height; taking measured strength parameters of the clay, $c' = 0$ and $\varphi' = 32^o$. The required degree of consolidation, about 60 percent, could be achieved with sand drains at 3 m centres connected to drainage mattresses under the upstream and downstream bank fill; but not under the select fill. Some 500 drains were installed, with six piezometres (placed in July 1951) to monitor pore pressures in the clay during construction of the embankment.

94. Bank fill of hard marl was tipped from scrapers and lorries, spread by bulldozers and compacted in 30 cm layers by sheeps-foot rollers. To avoid a tendency for the puddle clay to dry out and crack when placed, about 30 percent of sand was added before passing material through the pugmill (ref. 85). The embankment was successfully completed in 1954, the pore pressure in the grey clay being rather less than predicted; even so the downstream toe moved outwards by 60 mm during construction.

95. Stability calculations were based on the resistance to sliding offered by the soft clay layer to horizontal thrust exerted by the puddle core with full allowance for (undrained) side shear or arching action (ref. 86). This is justified in a limit equilibrium analysis applied to the conditions at Chew Stoke, but generally the stresses within a core, and more especially in a puddle trench, can be very different from those predicted by from simple plasticity theory (ref. 87).

96. Glen Shira. The embankment of Glen Shira Lower dam (H = 17 m) has a concrete core wall. There is an upstream rock-fill zone, and the main bank fill is glacial moraine: a silty gravel compacted in 23 cm layers with 8 t smooth rollers (ref. 88). The fill was placed at the natural water content of the borrow pit material, 4 or 5 percentage points above Proctor optimum, but the rollers nevertheless produced a dry density close to Proctor maximum (19 kN/m^3). Piezometers installed during construction in 1954 showed no significant pore pressures; a result compatible with the rather high permeability k = 1 x 10^{-7} m/s. Valuable observations were made in 1957 of the pore pressures in the upstream bank fill during a controlled rapid drawdown of reservoir level (ref. 89).

97. Selset. Built 1955-59 by Balfour Beatty to the design of Sandeman, Kennard & Partners, Selset dam has a height of 39 m above stream bed (ref. 90 and 91). The puddle core is very narrow (h/B = 6.5), with a permeability of k = 2 x 10^{-10} m/s, but the adjacent fill on both sides is made of Boulder Clay (LL = 30, PL = 16) placed about 1 percentage point above Proctor optimum (w = 10) at 99 percent of Proctor maximum dry density (19.5 kN/m^3). As the fill had a coefficient of consolidation c_v = 3 m^2/yr and a permeability k = 6 x 10^{-10} m/s drainage blankets were necessary, as at Usk (para. 89), but with a closer spacing of 4.5 m. An extensive system of sand drains was also installed to accelerate consolidation in the upper zone of rather soft Boulder Clay foundation.

98. The fill material was excavated by face shovels (1.5 to 3.5 cu.yd capacity) and conveyed to the embankment in rear-dump trucks of 8 to 18 cu.yd capacity. It was spread by a D8 bulldozer usually making three passes, then smoothed by a grader and given two or three passes of a 15 t smooth roller. Work generally stopped in wet weather to avoid churning up the clay and was restricted to the six months from April to September. In the 1958 and 1959 seasons on average 29,000 cu.m were placed per week or about 700,000 cu.m per year. The embankment has a total volume of 1.86 million cu.m. The concrete cut-off extends through Boulder Clay and gritstone into shale at a maximum depth of 87 m below crest level.

99. At Selset, as at several other dams going back to Chingford in 1939, attempts were made to mechanise the placing of puddle clay. All were unsuccessful, however, and the era of puddle cores ended with the labour-intensive process little changed (except for the pugmill) from early times.

100. Greenbooth. One of the last puddle core dams, Greenbooth (1958-61, H = 35 m) was designed by G.H. Hill & Sons (contractor Sir Alfred McAlpine) in a tradition which can be traced back in a direct line of development from the Pennine dams of J.F. Bateman. A concrete cut-off extends through glacial drift, gritstone and shale to a depth of 60 m below crest; clay for the core (h/B = 4.0) was carried

from the clay field in jubilee wagons to the rolls and pugmill, and heeled in 10 cm layers; on each side of the core are wide zones of select shale fill placed in 60 cm layers, and the outer bank fill is composed on stony material, these being excavated by face shovels and brought on to the embankment in 15 cu.yd trucks, then spread and compacted by bulldozers (ref. 92).

101. When the embankment reached a height of 27 m pore pressures in the glacial drift beneath the upstream slope were still quite low, but two of the piezometers beneath the downstream bank showed rather high pore pressures (r_u = 0.7), and small displacements beyond the downstream toe indicated an incipient slip. A large berm was therefore added on this side, during April and May 1961, and the embankment reached completion satisfactorily in September of that year. Dr. P.W. Rowe acted as geotechnical consultant.

102. Walton (Queen Elizabeth II reservoir). By the mid 1950s, when experience had proved that compacted clay fill could be a good 'impervious' material, the change to rolled core construction became inevitable on economic grounds, even if the core had to be somewhat wider, as puddle clay cost about three times as much as rolled fill (paras. 56 and 62). The embankment for Queen Elizabeth II reservoir, built 1957–61 at Walton for the Metropolitan Water Board, is the first example in England; the contract having been won by W. & C. French with a lower tender than for a puddle core design (ref. 93). The bank, 4.4 km long, is typically 15 m high with a volume of 3.1 million cu.m. The core wall, 3.6 m wide tapering above ground level to a top width of 3 m, is keyed into blue London Clay and brought up in a back-filled open-sided trench excavated by dragline through 6 m of sandy terrace gravel (Fig. 14). On each side of the core there is select fill of mixed clay and gravel 0.9 m wide. This latter feature and the rather narrow width of the core for rolled clay (h/B = 3.5) point to a transitional design.

103. For the core, London Clay (LL = 65, PL = 24) was dug from three deep borrow pits and, after spraying to raise its water content to about 26, spread by bulldozer and compacted in 10 cm layers with eight passes of a 20 t rubber-tyred roller to give a fill having 2 percent air voids and a shear strength around 80 kPa (ref. 94). The bank and trench backfill is gravel, dug and carried on to the embankment by scrapers, which were push-loaded, with no compaction additional to that given by construction traffic.

104. Subsequent tests on this well compacted clay fill showed a small gain in strength in the lower part of the core wall, due to consolidation, and little if any softening in the upper part.

105. Earth pressure cells (of the vibrating wire type) and piezometers were installed at various depths on the centre line during construction. When the embankment at the instrumented section had been completed the pore pressure ratio at a depth of 15 m below crest was about r_u = 0.3 and the pressure cells indicated strong arching within the core. About a year later, just before reservoir filling began in .April 1962, appreciable dissipation of pore pressure had occurred. The piezometer and pressure cell readings at that time are plotted in Fig. 14 from which it will be seen that the horizontal (and also the mean) stresses at depths between about 8 and 18 m are much less than the future reservoir pressures.

106. During the reservoir filling all gauge readings increased and continued to increase for about eight months after top water level was reached in September 1962. The final steady 'reservoir full' readings are also plotted in Fig. 14 and show horizontal stresses within this depth range closely equal to reservoir pressure. Near the upstream face of the core the stresses will be slightly larger, however, and in addition a correction of perhaps 10 percent has to be made to the cell readings for under-registration (ref. 95). Thus there is a small margin of safety and this, coupled with the cohesive and homogeneous nature of the clay fill, is evidently sufficient to prevent hydraulic fracture as pore pressures on the centre line are about one-half of reservoir pressure; an indication of perfectly sound conditions within the core. Moreover, only minor changes were recorded during the next eight years.

DEVELOPMENTS SINCE 1960

107. An account of research and case records of the past thirty years would amount to a state-of-the-art review of modern practice, far exceeding the limits of this historical survey. However, as a postscript, some of the developments in this period may be summarised very briefly as follows.

108. About two dozen large rolled core dams have been built in Britain since 1960. Important early examples include: Balderhead (1961-64, H = 48 m), Backwater (1964-69, H = 43 m) and Scammonden (1966-69, H = 73 m). Typically these dams have a grout curtain cut-off and a wide or moderately wide core ($h/B < 3$) backed by filters.

109. Cracking of rolled clay fill due to differential settlement became recognised to an increasing extent in America during the 1950s, but hydraulic fracture was discovered (or, more accurately, re-discovered) only following failures at Hyttejuvet, in Norway, in August 1966 and Balderhead in April 1967.

110. Experimental studies of filters for clay fills subject to possible cracking began around 1964 and received increased attention after the incident at Balderhead, where a filter of conventional design had not prevented propagation of internal erosion in the rolled clay core.

111. Meanwhile it was generally assumed that clay fills placed well on the 'wet side' of Proctor optimum, and the much softer puddle cores, were immune from cracking; lessons from the 19th century having been forgotten until 1978 when G.M. Binnie identified the Dale Dike failure as an example of hydraulic fracture. His classic study of Victorian dams, published in 1981, revealed further instances of cracking and fracture in puddle cores and trenches, together with the logical measures taken to prevent such failures. Since that date studies of old dams have intensified; both from historical records and field investigations.

112. Also in the past decade finite element analysis has led to an increased understanding of the mechanics of embankment dams, with reference to arching action, hydraulic fracture and progressive failure; the latter process having been thrown into prominence by the massive slip at Carsington dam during construction in June 1984.

REFERENCES

1. BINNIE, G.M. Early dam builders in Britain. Thomas Telford Ltd, London, 1987.

2. RUTTON, W.H. The making of the Serpentine. Home Counties Mag. 5 (1903) 81-91.
3. File on Historical Engineering Work (HEW) 1691: Petworth earth dam. Inst. Civ. Eng. library.
4. SKEMPTON, A.W. (ed.). John Smeaton, F.R.S. Thomas Telford, London, 1981.
5. LINDSAY, J. The Trent & Mersey Canal. David & Charles, Newton Abbot, 1979.
6. COMPTON, H.J. The Oxford Canal. David & Charles, Newton Abbot, 1976.
7. KENNARD, M.F. Examples of the internal condition of some old earth dams. J. Inst. Water Eng. 26 (1972) 135-147.
8. Data courtesy of British Waterways Board and Professor P.R. Vaughan. See also FAULKNER, A.M. The Grand Junction Canal. David & Charles, Newton Abbot, 1972.
9. HEW 1616: Knipton dam.
10. Information courtesy of Dr. Andrew Hughes.
11. FERGUSON, P.A., LEMESURIER, M. & STEAD, A. Combs reservoir: mesures taken following an embankment slip. 13th Cong. Large Dams 2 (1979) 233-245.
12. FAREY, J. Canals. Article in vol 6 of The Cyclopaedia (ed. Abraham Rees). Longman, London, 1806.
13. WALBANCKE, H.J. Pore pressures in clay embankments and cuttings. PhD Thesis, University of London, 1975.
14. JESSOP, W. Third report to the Committee of the Rochdale Canal, April 1797. Jessop archive, Inst. Civ. Eng. library.
15. FRONTARD, J. Notice sur l'accident de la digue de Charmes. Ann. Ponts et Chaussées, 9th series, 22 (1914) 173-292.
16. BISHOP, A.W. Embankment dams: principles of design and stability analysis. In Hydro-Electric Engineering Practice (ed. J.G. Brown) 1, 349-406. Blackie, London, 1958.
17. HEW 1615: Butterley dam.
18. Ref. 1 and HEW 1472: Rudyard dam. With additional information from Mr. S.J. Goode.
19. LESLIE, A. The Edinburgh waterworks. Min. Proc. Inst. Civ. Eng. 74 (1883) 91-127.
20. CHARLES, J.A. & BODEN, J.B. The failure of embankment dams in the United Kingdom. Symposium on Failures in earth-works. Thomas Telford, London, (1985) 181-202.
21. Life of Thomas Telford, civil engineer, written by himself ... Edited by John Rickman. Hansard, London, 1838. See also the manuscript account books of the contractor Thomas Townshend. Inst. Civ. Eng. archives.
22. HOYLE, M. Reservoirs from Rivington to Rossendale. North West Water, 1987.
23. BATEMAN, J.F. History and description of the Manchester Waterworks. London, Spon, 1884.
24. Section kindly provided by Mr. Hoyle. See also ref. 22.
25. RAWLINSON, R. Report on the state of certain waterworks and canal reservoirs situated above Rochdale, Jan 1865. Parliamentary Paper: Rochdale and other waterworks reservoirs, pp. 1-6. House of Commons, London, 1865.
26. BINNIE, G.M. Early Victorian water engineers. Thomas Telford,

London, 1981.
27. LESLIE, J. Account of the bursting of Bilberry reservoir. Roy. Scottish Soc. Arts 4 (1852) 130-137.
28. CONYBEARE, H. Quotes Leslie's specification, Min. Proc. Inst. Civ. Eng. 18 (1858) 388-390.
29. HUMBER, W. A comprehensive treatise on the water supply of cities and towns. Crosby Lockwood, London, 1876.
30. HARRISON, S. A complete history of the great flood at Sheffield. Sheffield Times 1864.
31. AMEY, G. The collapse of Dale Dyke dam, 1864. Cassell, London, 1974.
32. BINNIE, G.M. The collapse of Dale Dyke dam in retrospect. Q.J. Eng. Geol. 11 (1978) 305-318.
33. VAUGHAN, P.R., DOUNIAS, G.T. & POTTS, D.M. Advances in analytical technique and the influence of core geometry. This volume.
34. BINNIE, G.M. Postscript to ref. 32. Q.J. Eng. Geol. 16 (1983) 357-358.
35. RAWLINSON, R. & BEARDMORE, N. Report on the failure and bursting of a reservoir embankment belonging to the Sheffield Waterworks Company. 20 May 1864. Parliamentary Paper: Dale Dike or Bradfield reservoir, pp. 4-7. House of Commons, London, 1864.
36. GOTT, C. Report on Doe Park reservoir. 5 Jan 1866. Parliamentary Paper: Bradford reservoirs, pp. 4-7. House of Comons, London 1866.
37. CHARLES, J.A. & WATTS, K.S. The measurement and significance of horizontal earth pressures in the puddle clay cores of old earth dams. Proc. Inst. Civ. Eng. Part 1 82 (1987) 123-152.
38. CHARLES, J.A. Deterioration of clay barriers: case histories. This volume.
39. ADAMSON, R.A. A history of the Rivington works. Internal report, dated March 1935, to Liverpool Corporation Waterworks. Copy in author's files courtesy of Dr. A. Hughes.
40. BELOE, C.H. On the construction of catch-water reservoirs in mountain districts. Spon, London, 1872.
41. RAWLINSON, R. Remarks on reservoir making. Parliamentary Paper: Dale Dike or Bradfield reservoir, pp. 11-13. House of Commons, London, 1864.
42. ARTHUR, C. History of the Fylde waterworks. Flyde Water Board, Blackpool, 1911. Copies of the original Grizedale dam drawings kindly made available by Dr. Hughes.
43. BOWTELL, H.D. Reservoir railways in the Yorkshire Pennines. Oakwood, Blandford, 1979.
44. RAWLINSON, R. Discussion in Min. Proc. Inst. Civ. Eng. 34 (1872) 103-105.
45. BINNIE, A.R. Lectures on water supply. Delivered February-March 1877. School of Military Engineering, Chatham, 1878. Much of the material in these lectures was repeated in the same author's Rainfall, reservoirs and water supply. Constable, London, 1913.
46. WOOD, J.N. Discussion in Trans. Inst. Water Eng. 51 (1946) 118-121.
47. DEACON, G.F. Water supply. Article in Encyclopaedia

Britannica, 10th edn., 33 (1902) 769-792.
48. TEDD, P., CHARLES, J.A. & BODEN, J.B. Internal seepage erosion in old embankment dams. Proc. 9th European Conf. Soil Mech. (Dublin) 1 (1987) 507-510.
49. ASKWITH, R. Description of the Weardale and Consett water-works. Trans. Assoc. Water Eng. 14 (1909) 59-64.
50. GLOSSOP, R. The invention and development of injection processes. Geotechnique 11 (1961) 255-279.
51. BOWTELL, H.D. Reservoir railways of Manchester and the Peak. Oakwood Press, Blandford, 1977.
52. WATTS, W. Concrete and puddle for reservoir embankments. Trans. British Assoc. Waterworks Eng. 9 (1904) 57-70.
53. TEDD, P., CLAYDON, J.R. & CHARLES, J.A. Detection and investigation of problems at Gorpley and Ramsden dams. Proc. 5th BNCOLD Conf. Manchester, 1988. Paper 5.1.
54. BINNIE, A.R. Upper Barden reservoir and conduits: specification and tender for work. Bradford Corporation Waterworks, Bradford, 1876. Longitudinal and cross sections are given in ref. 45.
55. WATTS, W. Geogical notes on sinking Langsett and Underbank concrete-trenches in the Little Don valley. Trans. Inst. Mining Eng. 31 (1906) 668-678.
56. WATTS, W. The design and construction of impounding reservoirs. Trans. Assoc. Water Eng. 13 (1908) 119-128.
57. Information by courtesy of Mr. John Brown and Mr. Julius Kennard.
58. ALLDERIDGE, S.S. The construction of Burnhope reservoir, Wearhead. Trans. Inst. Water Eng. 40 (1935) 63-107.
59. Unpublished report by Edward Sandeman, Kennard & Partners, December 1959; quoted by kind permission of Mr. M.F. Kennard.
60. HALL, R.W. The construction of Cheddar reservoir. Trans. Inst. Water Eng. 41 (1936) 27-47.
61. McLELLAN, A.G. The Hollowell reservoir scheme for Northampton. Trans. Inst. Water Eng. 49 (1944) 172-197. See also: SKEMPTON, A.W. Discussion on back-analysis of the slip and assessment of remedial measures. Ibid 204-208.
62. HILL, H.P. The Ladybower reservoir. Journ. Inst. Water Eng. 3 (1949) 414-425.
63. ANON. Chingford reservoir: new storage for Metropolitan Water Board. Civil Engineering 32 (1937) 164-167. And information kindly provided by Mr. R. Glossop.
64. Details of the Chingford investigation are taken from original files at the Building Research Station.
65. COOLING, L.F. & GOLDER H.Q. The analysis of the failure of an earth dam during construction. Journ. Inst. Civ. Eng. 19 (1942) 38-55.
66. COOLING, L.F. Summary of the present position in the investigation of soil samples from Chingford reservoir. BRS report to Metropolitan Water Board, January 1938.
67. TERZAGHI, K. Report on the stability of the embankment for reservoir No. 2 at Chingford. Report to John Mowlem & Co. Ltd., 21 May 1938. For a section of the revised design see Proc. Inst. Civ. Eng. Part 1, 1 (1952) 664.
68. MIDDLETON, R.E. The Staines reservoir for the water supply of

London. Trans. Assoc. Water Eng. 11 (1908) 81-105.
69. Details from BRS working files, Loc. 64. L.F. Cooling reported the results briefly to the Metropolitan Water Board in April 1938.
70. ANON. London's water storage: a new reservoir under construction at Staines. Civil Engineering 33 (1938) 305-310. See also: Journ. Inst. Water Eng. 2 (1948) 50-53.
71. BANKS, J.A. Construction of Muirhead reservoir, Scotland. Proc. 2nd Int. Conf. Soil Mechanics (Rotterdam) 2 (1948) 24-31.
72. SKEMPTON, A.W. Report on investigation on the stability of the earth dam at Muirhead reservoir, Dalry, Ayrshire. BRS report to Babtie Shaw & Morton, July 1942.
73. COOLING, L.F. Knockenden reservoir. BRS letter reports to Babtie Shaw & Morton, 21 April 1942 and 31 July 1942.
74. BANKS, J.A. Problems in the design and construction of Knockendon dam. Proc. Inst. Civ. Eng. Part 1 1 (1952) 423-443.
75. Notes of the month, November 1938. Civil Engineering 33 (1938) 397.
76. BISHOP, A.W. The leakage of a clay core-wall. Trans. Inst. Water Eng. 51 (1946) 97-116.
77. SHEPPARD, G.A.R. & AYLEM, L.B. The Usk scheme for the water supply of Swansea. Proc. Inst. Civil Eng. 7 (1957) 240-265.
78. PENMAN, A.D.M. A field piezometer apparatus. Geotechnique 6 (1956) 57-65.
79. SKEMPTON, A.W. Discussion on ref. 77. Proc. Inst. Civ. Eng. 7 (1957) 267-269.
80. LITTLE, A.L. Compaction and pore water pressure measurements in some recent earth dams. Proc. 6th Inst. Cong. Large Dams (New York) 3 (1968) 205-226.
81. PENMAN, A.D.M. Construction pore pressures in two earth dams. Proc. Conf. on Clay Fills, 177-187. Inst. Civ. Eng., London, 1978.
82. KERR, H. & LOCKETT, E.B. Daer water supply scheme. Proc. Inst. Civ. Eng. 7 (1957) 46-74.
83. LITTLE, A.L. Discussion on ref. 82. Ibid, 79-81.
84. SKEMPTON, A.W. & BISHOP, A.W. The gain in stability due to pore pressure dissipation in a soft clay foundation. Proc. 5th Cong. Large Dams (Paris) 1 (1955) 613-638.
85. PICKEN, J.A. The Chew Stoke reservoir scheme. Journ. Inst. Water Eng. 11 (1957) 333-369.
86. BISHOP, A.W. The stability of earth dams. Ph.D. Thesis. University of London, 1952.
87. DOUNIAS, G.T., POTTS, D.M. & VAUGHAN, P.R. Numerical stress analysis of progressive failure and cracking in embankment dams. Report to Dept. of Environment, June 1989.
88. PATON, J. The Glen Shira hydro-electric project. Proc. Inst. Civ. Eng. Part 1 5 (1956) 593-618.
89. PATON, J. & SEMPLE, N.G. Investigation of the stability of an earth dam subject to rapid drawdown. Conf. on Pore Pressure and Suction in Soils. Butterworths, London, 1961, 85-90.
90. KENNARD, J. & KENNARD, M.F. Selset reservoir: design and construction. Proc. Inst. Civ. Eng. 21 (1962) 277-304.
91. BISHOP, A.W. & VAUGHAN, P.R. Selset reservoir: design and performance of the embankment. Ibid, 305-346.
92. ORMEROD, D. Greenbooth reservoir of the Heywood and

Middleton Water Board. Journ. Inst. Water Eng. 16 (1962) 223-235.
93. ANON. Queen Elizabeth II reservoir. Civil Engineering 57 (1962) 753-755.
94. COOLEY, P. Queen Elizabeth reservoir: conditions of clay core wall. Proc. 11th Cong. Large Dams (Madrid) 3 (1973) 565-574.
95. KENNARD, M.F., PENMAN, A.D.M., VAUGHAN, P.R. Stress and strain measurements in the clay core at Balderhead dam. Proc. 9th Cong. Large Dams (Istanbul) 3 (1967) 129-151.

16
Machine-age iconography in the American west: the design of Hoover Dam

Richard Guy Wilson

In the 1930s and 1940s the massive concrete, multipurpose dam emerged as one of the most distinctive modern American structures. Scholars have tended to view great dams like Hoover in the Southwest, Shasta in northern California, Grand Coulee and Bonneville in the Northwest, and the Tennessee Valley Authority dams in the Southeast from political, economic, and natural-resource development perspectives. Constructed in the West by the Bureau of Reclamation and the Army Corps of Engineers, and in the East by the TVA, these dams were products of endless political negotiations and bureaucratic maneuvering. The great dams and many smaller ones transformed vast areas of the continent, tamed rivers and prevented floods, provided new water transportation routes, stored water for irrigation and drinking, created new agricultural land, generated electricity for industry and home, and substantially changed the living standards of millions of Americans. The political, economic, and so-

Some of the research for this article was done under a National Endowment for the Humanities grant to the Brooklyn Museum for an exhibition titled "Machine Age America" that will be seen in 1986–1987 in Brooklyn, Chicago, Los Angeles, and Pittsburgh. For assistance in research I wish to thank Virginia Fenton of Boulder City, Nevada; Julian Rhinehart, Public Affairs Officer, and his staff of the Lower Colorado Regional Office, Bureau of Reclamation, Boulder City, Nevada; and the staff of the Engineering and Research Center, Bureau of Reclamation, Denver, Colorado.

Figure 1 Hoover Dam, downstream face

Photograph by Ben Glaha *Courtesy of U.S. Bureau of Reclamation*

cial backgrounds and consequences of these dams has been thoroughly investigated. But as works of architecture and as symbols of American modernism they have not yet received substantial scholarly attention.[1]

At the time of construction, especially in the 1930s, large multipurpose concrete dams and earth-fill dams were widely regarded as prime symbols of American ingenuity, prowess, and modernity. A photograph by Margaret Bourke-White of spillway gate construction at the Fort Peck Dam in Montana graced the first cover *Life Magazine* in 1936. Dams caught the American imagination because they represented work for a depression-affected economy and the benevolent aspects of governmental planning. The dams suggested that man could alter and control his environment; they also signified the coming of age of a new source of power—hydroelectricity. Yet the dams were more than functional structures. They were symbols that observers struggled to understand. In 1935, as construction neared completion, journalist Theodore White stood on the rim of the Black Canyon and pondered Hoover Dam: "It is a beautiful tantalizing thing. It is complex. It has a meaning, not to be grasped in weeks, or perhaps years. It is subtle, sometimes cruelly obvious."[2]

[1]The extensive political and reclamation history will be cited in footnotes below. On scholarly attention to the design of dams, see Carl Condit, *American Building Art: The Twentieth Century* (New York, 1961), 219–273; and Marian Moffett and Lawrence Wodehouse, *Built for the People of the United States: Fifty Years of TVA Architecture* (Knoxville, 1983). The latter contains essays by William Jordy and Walter Creese, both of whom are preparing longer studies on the architecture and design of TVA. See also my earlier article on Hoover Dam, "Massive Deco Monument," *Architecture*, LXXII (Dec. 1983), 45–47, which contains a preliminary assessment of the visual character of the dam.

[2]Selected commentary includes: "Power and Industry," *Progressive Architecture*, XXXII (Nov. 1951), 56–62; "The Dam," *Fortune*, VIII (Sept. 1933), 74–88; "Remaking the World," *Collier's*, XCV (March 16, 1935), 60; "Boulder Dam: Engineering Triumph," *Literary Digest*, CXXII (Oct. 17, 1935), 19–20; Mildred Adams, "Taming the Untamable at Boulder Dam," *New York Times Magazine* (Feb. 24, 1935), 5, 19; editorial, *New York Times*, Dec. 7, 1935, p. 16; "Portfolio of Public Works," *Architectural Record*, LXXVII (June 1935), 340–341; "Undertakings without Precedent," *Engineering News-Record*, CXIII (Nov. 29, 1934), 675–677; Walter K. M. Slavif, "Monuments to the Living," *Reclamation Era*, XXX (Feb. 1940), 42–43. On the TVA, see, for example, "Tennessee Valley Authority," *Architectural Forum*, LXXI (Aug. 1939), 73–114; Talbot Faulkner Hamlin, "Architecture of the TVA," *Pencil Points*, XX (Nov. 1939), 720–731; *Life*, I (Nov. 23, 1936), front cover; Theodore White, "Building the Big Dam," *Harper's*, CLXXI (June 1935), 120.

White's attempt to understand Hoover Dam echoed Henry Adams's effort three decades earlier to understand what Adams considered to be the prime symbol of the twentieth century, the dynamo, "a symbol of infinity." Adams tried to understand the dynamo, but ultimately concluded that it was "silent and infinite." White reached a similar conclusion about Hoover Dam as he watched the sun go down over its wall.

> I stay on, fascinated. I stare at the thing trying to comprehend it, to fix it forever in mind's eye. I have been inspired and provoked in the weeks I have tried to know it. And now, the minds that invented it, the bodies that are building it, the complexity and spirit of it, the love of it which men feel—all, all of it bewilders me.[3]

Of the several large and impressive concrete dams of the 1930s, the first and the most famous is Hoover Dam constructed between 1931 and 1936 on the Colorado River between Arizona and Nevada. (Begun as Hoover Dam, the name was changed to Boulder Dam in 1933 when Roosevelt entered the White House; the name was changed back to Hoover Dam in 1947). Still one of the largest dams, the impact of Hoover Dam is overpowering. A massive wall of concrete—a large, seemingly plastic concave mass wedged between two nearly sheer canyon walls—holds back Lake Mead and twelve trillion gallons or 28,537,000 acre-feet of water that stretches for 115 miles up the Colorado River. A great functional structure resulting from the precise calculations of engineers, the dam is also a superbly designed and conscious work of architecture that takes advantage of every possibility to make a statement. It is one of the great designs of the 1930s, a true architectural landmark, that also spurred many of the public works programs of the Depression. The full story of Hoover Dam involves many thousands of individuals and many years of evolving ideas and political decisions. Hoover Dam—like all dams—can be considered from the standpoint of its function, its construction and engineering, and its architectural design. It may also be understood as a symbol. Hoover Dam is a modern work of architecture that contains several key themes of modernism: efficiency, the utilization of advanced technology, the use of abstract

[3]Henry Adams, *The Education of Henry Adams* (1918; New York, 1931), 380; White, "Building the Big Dam," 121.

and ahistorical images, and the emphasis on beneficial results. As a self-consciously modern structure the dam illuminates some of the themes of and myths about American modernism.[4]

Hoover Dam is located in what still must be considered one of the most inhospitable parts of the globe. Sited in the Black Canyon area of the Colorado River between Nevada and Arizona, its surroundings have not changed much. Desert sand and volcanic rock predominate. The little vegetation consists of greasewood. Rattlesnakes coexist with ground animals. In the late 1920s when the decision was made to locate the dam at Black Canyon, the

[4]Determining the size and relative ranking of dams is tricky, since they can be classified in many ways. At the time of its completion in 1936 Hoover Dam was the tallest dam in the world, had the greatest power capacity, and the largest reservoir. Today it has been surpassed in all categories. For the general history of dams, see Norman Smith, *A History of Dams* (London, 1971); Allen H. Cullen, *Rivers in Harness: The Story of Dams* (Phildelphia, 1962); Julian Hinds, "Continuous Development of Dams Since 1850," Paper No. 2605, and K. B. Keener, "Dams Then and Now," Paper No. 2606, in *Transactions of the American Society of Civil Engineers*, CI (1953), 489–535. See also T. W. Mermel, *Register of Dams in the United States* (New York, 1958).

The full story of Hoover Dam has not been written. Between 1938 and 1950 the Bureau of Reclamation issued a series of bulletins recording some of the dam's history, design, and technical features which have provided information for this study. See U.S. Dept. of the Interior, Bureau of Reclamation, *Boulder Canyon Project, Final Reports* (7 parts, Boulder City, Nev.; Washington, D.C.; and Denver, Colo., 1938–1950). The bureau's magazine, *Reclamation Era*, provides information for the 1930s. See also T. H. Watkins, "Conquest of the Colorado," *Idaho Yesterdays*, VI (April 1969), 4–9, 48, 60. Participants' viewpoints and local history can be found in Dennis McBride, *In the Beginning . . .* (Boulder City, 1981); Marion V. Allen, *Hoover Dam and Boulder City* (Redding, Calif., 1983); Angela Brooker and Dennis McBride, *Boulder City: Passages in Time* (Boulder City, 1981); and William J. Williams, "The Town . . . That Built . . . the Dam," *Reclamation Era*, LXVI (Spring–Summer 1981), 1–10. Because of charges that there was excessive overtime, the Six Companies commissioned George A. Pettitt to write *So Boulder Dam was Built* (Berkeley, 1935) which contains valuable information on construction. A short popular history is U.S. Dept. of the Interior, Bureau of Reclamation, *The Story of Hoover Dam* (Washington, D.C., 1976). The Bechtel Company has produced a video movie, *The Taming of Black Canyon* (San Francisco, [c. 1979]). Another film is the Bureau of Reclamation's *The Story of Hoover Dam* (Washington D.C., 1960). On the political context, see Alfred R. Golze, *Reclamation in the United States* (Caldwell, Idaho, 1961); Paul L. Kleinsorge, *The Boulder Canyon Project: Historical and Economic Aspects* (Palo Alto, 1941); Norris Hundley, jr., *Water and the West: The Colorado River Compact and the Politics of Water in the American West* (Berkeley, 1975); Hundley, *Dividing the Waters: A Century of Controversy between the United States and Mexico* (Berkeley, 1966); Linda J. Lear, "The Boulder Canyon Project: A Re-examination of Federal Resource Management," *Materials and Society*, VII (1983), 329–337; Lear, "Boulder Dam: A Crossroad in National Resource Policy," *Journal of the West* (forthcoming); and Remi Nadeau, *The Water Seekers* (New York, 1950). From a popular point of view, Hoover Dam has been considered by David O. Woodbury, *The Colorado Conquest* (New York, 1941); and Frank Waters, *The Colorado* (New York, 1946). Among other places, archival materials are located at the Bureau of Reclamation offices in Denver and Boulder City, the Boulder City Public Library, and the University of Nevada, Las Vegas.

nearest town was Las Vegas, thirty miles away. Then a ramshackle community, Las Vegas served primarily as a junction for the Union Pacific Railroad. The nearest electric power source was in San Bernardino, California, 222 miles away across the desert. Housing for the 5,000-person work force needed to construct the dam was nonexistent. The result was the construction of Boulder City, a model town for 6,000 laid out by the government and located about eight miles from the damsite.[5]

The reason for locating the dam at such a site was the necessity of controlling the Colorado River which drains rivers of seven western states. Every year, despite levees and canals, the Colorado River would overflow its banks, sometimes causing severe damage as with the overflow of 1905–1907 that created the Salton Sea and the 1916 inundation of the Yuma Valley. Flooding would occur in late spring or early summer; yet by late summer, fall, and winter, when the Imperial, Coachella, and Yuma valleys needed water, the Colorado River was dry or nearly so. A controlled river could also serve as a transportation link for these areas of the Southwest. Finally, the expanding population of southern California—Los Angeles in particular—badly needed water to survive.[6]

Plans for harnessing the Colorado River received major impetus with the establishment in 1902 of the Reclamation Service (later renamed the Bureau of Reclamation) under President Theodore Roosevelt. Charged with developing agricultural land in the West, the Reclamation Service focused on the Colorado River and by the early 1920s its geologists and engineers had concluded that the Black Canyon area was the best site for the first major dam on the Colorado. Other smaller dams—the Parker, Davis, and Imperial dams—would follow in a few years. Political squabbles among the neighboring western states over water rights continued through the 1920s, but by late 1928 Congress

[5] S. R. De Boer, a city planner, laid out the town of Boulder City in 1930–1931. See S. R. De Boer, "The Plan of Boulder City," *Architectural Record*, LXXIII (March 1933), 154–158. See also "Government Plans Model Town at Boulder City, Nevada," *New Reclamation Era*, XXII (Feb. 1931), 28–30, 41; and "Boulder City—Government's Model Town to Rise on the Nevada Desert," *American City*, VII (March 1931), 16–19. In execution, De Boer's plans were significantly modified. See Imre Sutton, "Geographical Aspects of Construction Planning: Hoover Dam Revisited," *Journal of the West*, VII (1968), 301–344.

[6] U.S. Bureau of Reclamation, *Final Reports, Part I—Introductory, Bulletin I, General History and Description of Project* (Boulder City, 1948), 22–42.

passed a bill authorizing the dam and President Calvin Coolidge signed it into law on December 21. Construction of the dam might have languished for several years except that Herbert Hoover, a professional engineer who had been Secretary of Commerce under Coolidge and had strongly supported the idea of a dam, now occupied the White House. By 1930 President Hoover could foresee the general outlines of depression and unemployment. Consequently, in 1930, a large appropriation bill was pushed through Congress for the dam's construction and preliminary site work by the government. Designs and specifications were hurriedly drawn up by December 1930; in March 1931 bids were opened for construction.[7]

The intention was to construct a large concrete dam at Black Canyon that would be capable of containing in reservoir about 30,000,000 acre-feet of water, the equivalent of two years of normal Colorado River flow. The engineering design of the dam had begun back in the 1920s with site investigations and a series of test models. Gradually a consensus emerged among the Bureau of Reclamation's engineers that of the various possible dam types—including earth fill, rock fill, and straight gravity dams—a concrete, arch-gravity dam would be best. Since concrete works better under compression than in tension, a wedge-shaped form had to be devised that was thicker at the bottom than at the top and which was convex on the reservoir side so that the structure would carry the stress or water weight by arch action to the abutting canyon walls and downward to the canyon floor. Between 1928 and 1931 extensive tests of models were carried out at the Denver laboratories of the Reclamation Bureau. These tests revealed that increasing the thickness of the dam at the top would also increase the horizontal tension, an effect to be avoided. Tests were also conducted on other parts of the dam's design to obtain the best profile for maximum velocity and minimum turbulence. This technical engineering design (figure 2), which opened to bids in December 1930 and then was modified and finally approved by the Colorado River Board on November 19, 1932,

[7]William E. Warne, *The Bureau of Reclamation* (New York, 1973) provides a general history of the reclamation program; U.S. Bureau of Reclamation, *Final Reports, Part I—Introductory*, 52, 54–56.

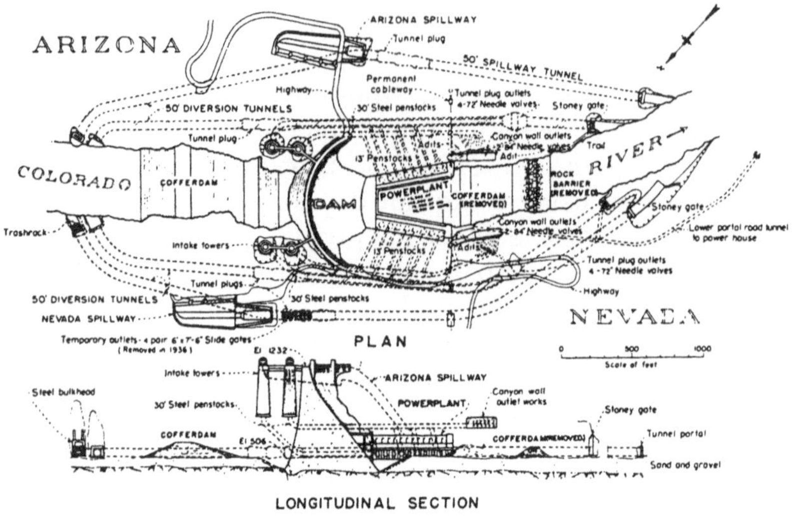

Figure 2 Hoover Dam, plan and section
Courtesy of U.S. Bureau of Reclamation

called for a massive concrete structure rising 726.4 feet from bedrock with a base thickness of 660 feet and a crest thickness of 45 feet, wide enough for four lanes of traffic and pedestrians. The crest length of 1,244 feet is nearly a quarter mile long, while the arch on axis has a radius of 500 feet.

Essentially the staggering construction plan involved diverting the Colorado River through tunnels drilled in the solid rock walls of the canyon, building temporary cofferdams to block the river, excavating the site, and then constructing the dam and power plant. This process translated into construction of four diversion tunnels fifty-six feet in diameter and three miles in length, excavation of more than 5.5 million cubic yards of rock and gravel, and the placement of about 5 million barrels of concrete. Two concrete batch plants were brought in along with eight steam locomotives. The concrete presented special difficulties. If it was simply poured and left to harden naturally, it would have taken about a century to cool and the shrinkage would have cracked the dam and left it unusable. The solution was to circulate refrigerated water through tubing placed in the wet concrete which cooled each pour in seventy-two hours. Grouting was then pressure injected into the space resulting from the contraction. The concrete was poured in blocks measuring five feet thick and between twenty-five and sixty feet square. Each block of concrete was notched or keyed to the surrounding blocks which became concrete columns as new pours were added. A monolithic concrete mass resulted. A "zipper" or eight-foot slot for the refrigeration pipes remained at the center of the dam as construction climbed upward. This was the last column to be poured. The concrete, composed of cement, water, and aggregate, was rigidly controlled. The often told tale of workers buried in the concrete during construction is apocryphal; the tolerances of the concrete could never stand such a messy water-filled object.[8]

Fundamental to any work of architecture is the process of construction, the putting together of various materials to build the design. While in most considerations of architecture this process is ignored, some reference must be made to the actual construction of Hoover Dam. Initially the government estimated the dam would take seven years to construct; it was finished in five years. The dam's builders were a consortium of contractors known as Six Companies Inc., consisting of the Utah Construction Company, the Pacific Bridge Company, the MacDonald and Kahn Company, the Morrison-Knudsen Company, the J. F. Shea Company, and the combined W. A. Bechtel and Henry J. Kaiser companies. The general superintendent for the Six Companies was a former Bureau of Reclamation and a Morrison-Knudsen construction engineer, Frank T. Crowe, who had supervised several earlier large dam construction jobs in the West. A legendary figure in his own right, Crowe was known as a man of action who wasted little time on paperwork and knew intimately the job of every one of the 5,000 men under his direction.[9]

---

[8]This tale has become folklore in the West. I remember hearing it as a child in Los Angeles and guides today at the dam repeat the tale. While many individuals did lose their lives, none were buried in the concrete.

[9]A popular history of construction companies is Neill C. Wilson and Frank J. Taylor, *The Earth Changers* (New York, 1957). Some history can also be found in Paul D. Nations, ed.,

Figure 3 Highliners at work on the canyon wall
Photograph by Ben Glaha *Courtesy of U.S. Bureau of Reclamation*

The size and scale of the undertaking along with the human effort involved in the project quickly assumed heroic dimensions. It was a major organizational undertaking to supervise at the height of construction several thousand men crammed into a canyon of less than 4,500 square feet. Men and equipment working on top of each other in the narrow gorge made the construction of the dam one of the great attractions of the day. Initially the insurance companies for the job estimated that over 200 men would lose their lives in the first year of construction; the actual loss for the entire project was ninety-six. Labor problems plagued the project in its early stages, especially concerns over working conditions and wages. A strike during August 1931 resulted in improved working and living conditions. Certainly the circumstances of working on the dam were daunting: summer temperatures at the canyon rim could hit 126 degrees and over 140 degrees in the bottom of the gorge and in the diversion tunnels. Strong winter winds could come at any time and cause temperatures to plummet below zero. The Colorado River, always dangerous, was a constant worry as it tried to break through the temporary cofferdams and flood the entire project. The heroic qualities of the "construction stiff," as the press termed itinerant workers on the project, were played up by writers. One journalist for the *New York Times Magazine*, reflecting on the use of machines, regarded construction at the dam as "Pioneering of the Machine Age."[10]

The actual building of the dam reenforced the heroic dimensions of the project. Black and white photographs record some of

---

River Tamers (Boise, Idaho, 1947); "50 Years of Construction Progress," *Em-Kayan*, XXI (1962), 1–36; "'The Earth Movers': The Epic of the Six Companies of the West," *Fortune*, XXVIII (Aug. 1943), 99–107, 210–214; and "Memoir No. (1818) Francis Trenholm Crowe, Hon. M. ASCE," *Transactions of the American Society of Civil Engineers*, CXI (June, 1946), 1–7.

[10]Guy Louis Rocha, "The IWW and the Boulder Canyon Project: The Final Death Throes of American Syndicalism," *Nevada Historical Society Quarterly*, XXI (1978), 2–24; Edmund Wilson, "Hoover Dam," *New Republic*, LXVIII (Sept. 2, 1931), 66–69; "Open Shop at Boulder Dam," *ibid.*, (June 24, 1931), 147–148; Elwood Mead, "Boulder Dam," *ibid.*, (Aug. 26, 1931), 48; "Engineers-Contractors Committee Finds Hoover Dam Conditions Satisfactory," *Reclamation Era*, XXIII (Feb. 1932), 32; and Pettitt, *So Boulder Dam was Built*, 38–39. See also Roosevelt Fitzgerald, "Blacks and the Boulder Dam Project," *Nevada Historical Society Quarterly*, XXIV (1981), 255–260; "The Dam," *Fortune*, 75; Duncan Aikman, "New Pioneers in the Old West's Deserts," *New York Times Magazine* (Oct. 26, 1930), 7.

the story, though the most memorable aspects of construction were captured through the motion-picture footage shot by Bureau of Reclamation photographer Ben Glaha. Construction went on twenty-four hours a day, every day of the year. At night the site was brilliantly lighted and struck some observers as a movie set. There were large machines, some of them new to construction sites: 250-horsepower Mack trucks with sixteen-yard dump bodies; dinosaur-like cranes, shovels, and draglines; Caterpillar tractors; "Jumbo" drilling rigs; and the cableways carrying men and equipment back and forth and up and down the canyon. While it seemed as if a new race of mechanical giants were inhabiting the earth, a choreographed dance step appeared to control the actions of many of the workers as they swung down the canyon walls with jack hammers and crow bars to scale away loose rock. Hand signals would direct the descent of the eight-yard "Crowe" concrete buckets until, when finally in place, bottoms of the buckets would open and concrete would rush out to be tamped and vibrated into place. The result, caught in photographs, were concrete cliffs, entirely abstract in the removal from both a human scale and a functional dimension. They were pieces of a larger whole, possessing an eternal beauty and finality all their own. These unfinished forms were one level of the aesthetic of large dams.[11]

The technical design of the dam was largely the work of the Denver office of the Bureau of Reclamation. The individual responsible for the conception of the project in 1902 and directing the early investigations was Arthur Powell Davis, a nephew of the first explorer of the Colorado River, John Wesley Powell. Davis was an engineer and a director of the Reclamation Service in the 1910s and 1920s. Although he had left the agency in 1923, his

[11]William H. Gater, comp., *Hoover Dam Including the Story of the Turbulent Colorado River* (Los Angeles, 1932), 14. There were numerous photographers associated with the project, but Ben Glaha (1899–1971) stands out as the best. See Ben Glaha, "Boulder Dam: The Photography of Engineering Works," *U.S. Camera*, I (Jan.–Feb. 1939), 18–23, 78–79; "Glaha Complimented," *Reclamation Era*, XXV (July 1935), 152; "Photographs Exhibited," *ibid.*, (Aug. 1935), 168, and (Sept. 1935), 185; and Willard Van Dyke, "The Work of Ben Glaha," *Camera Craft*, XLII (April 1935), 166–172. Glaha's work was exhibited nationwide in 1935 and praised by several individuals, including Ansel Adams. In 1938 Cliff Segerblom (1917-) became the official bureau photographer for the dam and is responsible for the excellent photographs of the test of the spillways and valve outlet houses. Interview with Cliff Segerblom, May 15, 1984.

Figure 4 Placing concrete in column forms
Photograph by Ben Glaha *Courtesy of U.S. Bureau of Reclamation*

Figure 5 Construction of dam and intake towers
Photograph by Ben Glaha *Courtesy of U.S. Bureau of Reclamation*

ideas formed the basis of the design. Elwood Mead, the commissioner of reclamation during most of the construction period and for whom Lake Mead—the reservoir—was named, approved the plans. To assign actual responsibility for the technical design to an individual is pointless, however. The chief engineer was Raymond F. Walter, while chief designing engineer was John. L. Savage. Walter R. Young served as the Boulder Canyon project office construction engineer and worked closely with Frank Crowe. These individuals, in turn, supervised a staff of over a hundred engineers who also had a hand in the technical design.[12]

Technical design for an engineer meant essentially working out plans so a structure could be built. Visual considerations did enter into engineering design, but the primary purpose was not to express beauty or abstract aesthetics. Rather, it was to create a functional structure. Some modern architectural theorists of the 1920s and 1930s, such as Le Corbusier, found great beauty in the designs of engineers, and certainly much of the power of the Hoover Dam form comes from engineering calculations and design. But in addition to the great concrete wedge such features as the crest, the towers, the bridges, the spillway gates, the power house, and the ornament and color of the dam resulted from self-conscious aesthetic decisions.[13]

The design of these features which added considerably to the dam's appearance and symbolic power was largely the work of the project's architectural consultant, Gordon B. Kaufmann. Why Kaufmann was selected for his post remains unknown. He had no prior experience as a dam architect and was generally known as an accomplished residential designer in southern California where he worked especially in the Spanish Colonial idiom. There existed, however, in Kaufmann's background certain experiences that might have influenced his selection. In 1931, for instance, he had received the commission for the Los Angeles Times Building from newspaper publisher Harry Chandler who was also a strong supporter of the dam. Whether this 1931 commission had any influence on Kaufmann's selection to work on

[12]"The Dam," *Fortune*, 75.

[13]Le Corbusier, *Towards a New Architecture* (1927; New York, 1960), 33.

Figure 6 Los Angeles Times Building, 1935, Gordon B. Kaufmann, architect
Courtesy of Los Angeles Times, Inc.

the dam is unknown, but it is significant that the Bureau of Reclamation looked to Los Angeles for an architect rather than to their Denver home office or elsewhere.[14]

Gordon Kaufmann (1888–1949) was born in London of a Scottish mother and German father. He attended the London Polytechnic and Royal College of Art before being apprenticed to a London architect. Apparently, he went to Germany to work for a short time in an architect's office before immigrating to Canada and working there for three years. In 1913 as a result of his wife's poor health, he went to southern California. By 1916 he was associated with Reginald Johnson and Roland E. Coate,

[14]Information on Kaufmann is scanty. See Alson Clark, "The 'Californian' Architecture of Gordon B. Kaufmann," *Society of Architectural Historians, Southern California Chapter Review,* I (Summer 1982), 1–8; Joseph Giovannini, "Gordon Kaufmann" in Baxter Art Gallery, California Institute of Technology, *Caltech 1910–1950: An Urban Architecture for Southern California* (Pasadena, 1983), "Gordon B. Kaufmann: Obituary," *Southwest Builder and Contractor,* March 4, 1949, p. 7; *Los Angeles Times,* March 2, 1949, part 2, p. 1; and Harris C. Allen, "It Can Happen Here: A Classical Scholar Learns a Modern Language," *Architect and Engineer,* CXXIX (May 1937), 13–34. Additional biographical information on Kaufmann can be found in his application for a fellowship from the American Institute of Architects in the AIA Archives, Washington, D.C.

suave designers in the Spanish Colonial Revival. Kaufmann struck out on his own in 1924 and designed a series of hotels and houses that, while reflecting the "period" styles, were extremely simplified. By the early 1930s his work had become more overtly "Art Deco" as in the Los Angeles Times Building (1931–1935) with its plastic set-back profile, strip pier buttresses, and low-relief, neoclassical semicubistic sculpture. Other work by Kaufmann through this period and later in the 1930s showed an increasing simplification of design and adoption of modernistic stylistic images, while still retaining vestiges of ornament and plastic modeling.[15]

The choice of Kaufmann as the architectural consultant was in retrospect excellent. He was willing to conform to the dictates of the technical design. As he wrote, the dam was the "result of pure engineering" and his "architecture followed the structural design, and it was considered as a complementary treatment rather than a dominant phase of the whole design." He was accustomed to molding broad plastic surfaces. In a sense this was his approach to the various appurtenances of the dam, integrating them into the whole and emphasizing the plastic qualities of concrete. As Kaufmann explained in an article:

> There was never any desire or attempt to create an architectural effect or style, but rather to take each problem and integrate it to the whole in order to secure a system of plain surfaces relieved by shadows here and there where the plan or requirements suggested.

Kaufmann, in other words, was not a radical modernist subscribing to the dictates of the International Style (as it was christened in February 1932), but he obviously knew enough about modernism to clean up the engineers' *retardataire* design.[16]

Kaufmann drew on several modernist sources for inspiration. Los Angeles had some of the country's most avant-garde architects led by Rudolph Schindler and Richard Neutra. Neutra had in fact worked for Kaufmann briefly in 1925. By the early 1930s several of the leading American architectural journals were show-

---

[15]In particular there is the San Pedro High School, 1935–1937; the Federal Building, Long Beach, 1932; and the Royal Laundry, Pasadena, 1927, and 1935 (all in California) and the Salisbury Building, Salt Lake City, Utah, 1936.

[16]Gordon B. Kaufmann, "The Architecture of Boulder Dam," *Architectural Concrete*, II, No. 3 (1936), 3. See also Wesley R. Nelson, "Ornamental Features of Boulder Dam," *Compressed Air Magazine*, XLIII (June 1938), 5615–5620.

ing with great regularity modern architecture, both European and American. Kaufmann's work at Hoover Dam reflected other modernist sources as well. Dams had appeared in Tony Garnier's *Une Cité Industrielle* (1901–1917) and in the 1912–1916 drawings of Italian Futurist Antonio Sant' Elia. Garnier's dam appeared in a distant panoramic view as a large cliff-like concrete structure with predominant vertical buttresses molded into the dam face and with a power house standing at the bottom. Since no details were shown, the overall impression was of a large monolithic concrete mass. Sant' Elia had projected several central power stations that bore similarities to dams. But Sant' Elia's drawings generally lacked a context and sense of how the buildings were supposed to function. His drawings glorified the mechanical parts as abstract objects; their stepped-back and molded characters perhaps influenced modernist massing of conventions of the 1920s and 1930s.[17] While it is uncertain whether Kaufmann knew of the dams of either Garnier or Sant' Elia, he certainly knew the work of the architectural illustrator, Hugh Ferris. Many of Ferris's drawings in his popular book, *The Metropolis of Tomorrow* (1929), projected immense building forms, plastically molded, with predominant vertical and horizontal setbacks. His charcoal drawings emphasized light and shadow and represented a distinctly modern approach. Finally, it can hardly be coincidental that Hoover Dam resembled a science fiction movie set. The emphasis Kaufmann placed upon drawing visitors' attention to such machinery as the

[17]Henry-Russell Hitchcock and Philip Johnson, *The International Style: Architecture Since 1922* (New York, 1932); see also Richard Guy Wilson, "International Style: The MoMA Exhibition," *Progressive Architecture*, LXII (Feb. 1982), 92–105; Thomas S. Hines, *Richard Neutra and the Search for Modern Architecture* (New York, 1982), 59. Two principal journals were *Architectural Record* and *Architectural Forum*, both of which converted from rather staid conservative magazines in 1928 and 1929 and began actively supporting modernism, though it should be noted there was no general agreement on what was modern. *American Architect*—the oldest American architectural magazine, founded in 1876—was slower to change, but was showing some modern work by 1930. *Pencil Points* remained very conservative. An agressive display and argument about modernism—both American and European—can be found in the *AIA Journal* from the mid 1920s.

Both Garnier and Sant' Elia had made minor appearances in books by such American authors as Henry-Russell Hitchcock, *Modern Architecture* (London, 1929), 146, 171, 199. Garnier's work was published in *Une Cité Industrielle* (Paris, 1917). See also Dora Wiebenson, *Tony Garnier and the Cité Industrielle* (New York, 1969). Futurist paintings were displayed at the San Francisco Panama-Pacific International Exposition in September 1915. However, Sant' Elia was evidently not shown. See Anne d'Haroncourt, *Futurism and the International Avant-Garde* (Philadelphia, 1981). For a record of Sant' Elia's exhibitions and bibliography, see Luciano Caramel and Albert Longatti, *Antonio Sant' Elia* (Como, 1962).

Figure 7 Artist's conception of Hoover Dam, power plant and Arizona outlet works, 1930

Courtesy of U.S. Bureau of Reclamation

turbines and on lighting the dam at night recalled Fritz Lang's expressionist movie, *Metropolis* (1927). As with Garnier and Sant' Elia, Fritz Lang's connection and possible influence on Kaufmann is circumstantial. But Kaufmann did have ties to Hollywood where he designed some buildings for various show business personalities. Since he also had spent time in Germany, he was probably aware of the Deutsche Werkbund's attempt to reconcile art and industry. Kaufmann's work at Hoover Dam, and the dam itself, surpasses any of the various Futurist-Expressionist-Art Deco influences that might be cited. Yet those sources existed in the background.[18]

[18]Hugh Ferris, *The Metropolis of Tomorrow* (New York, 1929). See also Jean Ferris Leich, *Architectural Visions: The Drawings of Hugh Ferris* (New York, 1980). Ferris published a drawing of Hoover Dam in Hugh Ferris, *Power in Buildings* (New York, 1953), pl. 49. *Metropolis* was filmed in Germany in 1925–1926 and released in 1927. It was shown in the United States and widely reviewed in 1927. Lang came to Hollywood in 1934. See Frederick W. Ott, *The Films of Fritz Lang* (Secacus, N.J., 1979), 123–141 and Lotte Eisner, *Fritz Lang* (New York, 1977).

Figure 8 Artist's conception of Hoover Dam and power plant as approved Nov. 19, 1932
From Reclamation Era *XXIV (Jan. 1933), cover* *Courtesy of U.S. Bureau of Reclamation*

Exactly why the Bureau of Reclamation hired an architect is apparent from the drawings the bureau submitted to contractors for bids in December 1930. On the crest of the dam, the balustrade and access towers appeared as attachments to the dam face with little attempt at integration. Two towers were topped by enormous eagles that also appeared in the pediments of the power house below. Aside from these pediments and the flagpole, the power house resembled a boring windowed factory. A variation on the factory fenestration and pediment even car-

Figure 9 Powerhouse and downstream face as completed
Photograph by E. E. Hertzog *Courtesy of U.S. Bureau of Reclamation*

ried over to the awkwardly shaped outlet houses perched on the canyon walls.[19]

At some point between December 1930 and November 1932, Kaufmann made the recommendations that resulted in a preliminary sketch published in January 1933. While some adjustments would continue to be made in the design of the dam, especially in the powerhouse, the direction of Kaufmann's work was apparent. He simplified and modernized the various parts of the dam. On the crest, the overhanging balcony and four unequal towers gave way to a series of observation niches and towers that rise from the wall and continue upward unimpeded. The emphasis, according to Kaufmann, was on "an orderly series of small vertical shadows

[19]U.S. Dept. of the Interior, Bureau of Reclamation, *Hoover Dam, Power Plant and Appurtenant Works, Specifications, Schedule and Drawings: Specifications No. 519* (Washington, D.C., 1930).

Figure 10 Nevada elevator tower with sculpture by Oskar J. W. Hansen
Photograph by Ben Glaha *Courtesy of U.S. Bureau of Reclamation*

Figure 11 Intake towers at night
Photograph by Cliff Segerblom *Courtesy of U.S. Bureau of Reclamation*

punctuated by the larger shadows of the elevator and utility towers." He treated these extrusions as continuations of the dam face, not as separate moldings. The four large towers have cutback corners and tops reminiscent of the set-back Los Angeles Times Building, but here were treated much more simply. The two outer towers were for utilities and public restrooms, while the two inner towers acted as public entrances to the dam since from them the elevator descended to the powerhouse and internal galleries. Consequently, these inner entrance towers contained the only ornament on the dam—two large cast concrete panels by sculptor Oskar J.W. Hansen. These panels depicted such subjects as flood control, irrigation, power, and the history of the area. Their low-relief, semiclassical, cubist style typfied Art Deco sculpture.[20]

The four large intake towers that provide water to the turbines were purposely placed in a symmetrical position even though the

[20] "Report of the Colorado River Board," *Reclamation Era*, XXIV (Jan. 1933), 4. The revised design is shown on the cover. Kaufmann, "Architecture of Boulder Dam," 3.

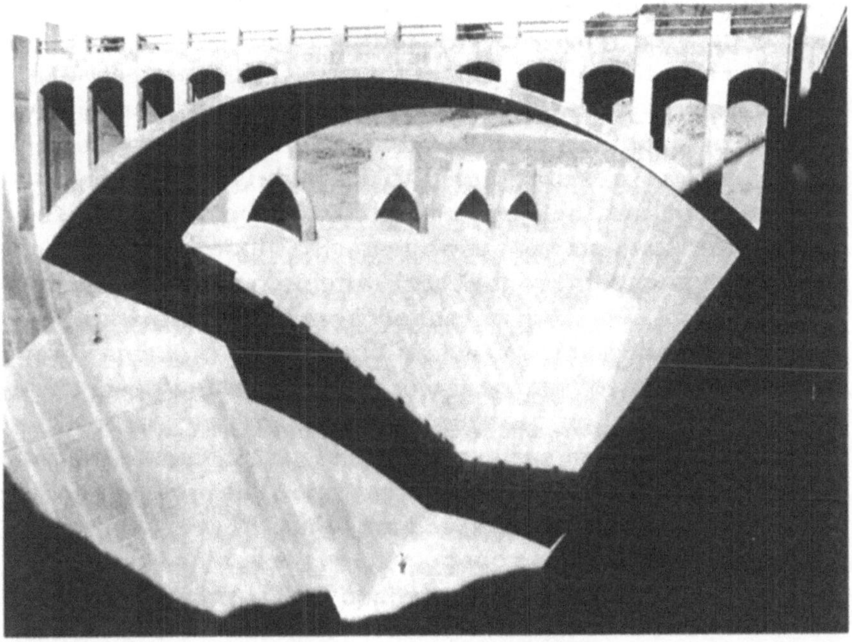

Figure 12 Arizona spillway and bridge
Photograph by G. Lawrence Ullom *Courtesy of U.S. Bureau of Reclamation*

site dictated asymmetrical placement. Anchored to bedrock about 250 feet above the original river bed, the towers stand 395 feet tall—the equivalent of a thirty-three-story building. The engineers' original design for the control house on top of each tower resembled crenelation, with the piers or fins creating a crown of thorns. As actually built following Kaufmann's suggestions, the control house appears as a cap on the tower with vertical buttresses terminating in a green-tinted concrete roof. Kaufmann, moreover, placed a light globe atop each tower so that at night the illumination of the interior of the house and the lighted globe conveyed an almost supernatural effect as if some hidden dynamo was pulsating energy.[21]

The spillways located on either side of the dam were intended for controlling the reservoir level, especially during the flood seasons. Kaufmann again had a hand in their redesign. Originally the bureau's engineers had projected a large overhead gate

[21] U.S. Bureau of Reclamation, *Final Reports, Part I—Introductory*, 78.

for each side. When they realized that this might be inadequate, they continued extensive model studies. The final design was for a series of four rolling drum gates for each side separated by substantial concrete piers which were carefully studied in profile to provide the smoothest flow of water. Their streamlined shape—consisting of a series of rounded forms built upon each other—conveyed a nautical feeling. The design of the spillways, especially on the Arizona side, was abstract and modern with the Art Deco stylisms receding in the face of the powerful masses of the concrete forms. The design resulted from both engineering and especially architectural considerations. As the bureau's report circumspectly noted, "the design of the bridge was governed to a considerable extent by architectural considerations, in order to provide a structure in harmony with the general massiveness of the dam and spillways. Consequently, some of its members are of greater size than actually required for utility."[22]

The powerhouse at the base of the dam used for housing the turbines was also substantially modified by Kaufmann. He noted that while "engineering principles" determined the overall size, shape, and plan of the powerhouse, his role was to mold the facades into a more harmonious statement. Taking the predetermined column spacing of the structure as a given, Kaufmann created recessed vertical fenestration panels in between the columns. Horizontally placed wood formwork (two-by-six-foot lagging) was used on the exterior walls to align them with the horizontal layering of the concrete work of the dam face. Along the front of the power house on either side, Kaufmann arranged smaller box-like switch houses with horizontal aluminum fins across their tops that contrasted with the predominant verticalism of the powerhouse. By placing transformers on the exterior in full view, Kaufmann obviously glorified the technological nature of the project.[23]

The outlet or valve houses located on the canyon walls and intended for overflow were also substantially modified from their earlier pedestrian mien. The bureau's report noted that "archi-

[22]U.S. Bureau of Reclamation, *Final Reports, Part VI—Hydraulic Investigation, Bulletin I, Model Studies of Spillways* (Denver, 1938); *Final Reports, Part IV—Design and Construction, Bulletin 3, Diversion, Outlet, and Spillway Structures* (Denver, 1947), 318–321.

[23]Kaufmann, "Architecture of Boulder Dam," 3.

tectural considerations suggested an appearance of stability and security, and a color in harmony with the surroundings." The result was a denser structure with fewer openings, especially openings containing references to traditional fenestration. Five vertical windows and six circular ports for water discharge were the basic features, while the form itself had horizontal grooves at the corners and a battered base. Kaufmann explained that the horizontal grooves gave "a feeling of stability—as if they served to hold the structure against the sheer wall of the canyon."[24]

Kaufmann also had some impact on the final color of the dam. During the curing and cooling of the concrete when all surfaces had to be kept moist—generally by continuous spraying—it was noticed that the concrete became colored. Reddish and brownish stains developed that came from the iron oxide in both the water and the pipes used for the curing. As a consequence, Kaufmann attempted to graduate the depth of color on both the dam's face, the powerhouses, and the valve houses. The valve houses were cured to a warm dark red at the base and a lighter red at the top. On the dam Kaufmann tried to graduate the color from the bottom to the top, though as the report stated, "the results . . . were not entirely satisfactory." Subsequent aging and weathering obliterated most of the color graduation and the dam now—fifty years later—is a light tan in color.[25]

Kaufmann's imprint is also obvious on the dam's interior. From the beginning it was assumed the structure would be a great publicity bonus for the Bureau of Reclamation and would attract visitors. Kaufmann recommended that the bureau engage Denver artist Allen True to act as color consultant for portions of the interior. True found the task conducive to furthering modernist aesthetics and declared: "it no longer suffices that a hydroelectric plant which serves millions of people be encased in a mere building—its housing must express through architectural planning the new understanding of functional form and satisfy in every respect

---

[24]U.S. Bureau of Reclamation, *Final Reports, Part IV—Design and Construction, Bulletin 3,* 202; Kaufmann, "Architecture of Boulder Dam," 4.

[25]U.S. Bureau of Reclamation, *Final Report, Part IV—Design and Construction, Bulletin 4, Concrete Manufacture, Handling and Control* (Denver, 1947), 171. See also Kaufmann, "Architecture of Boulder Dam," 4; and U.S. Bureau of Reclamation, *Final Reports, Part IV—Design and Construction, Bulletin 3,* 202.

Figure 13 Interior of the Nevada wing of the power plant
Photograph by E. E. Hertzog *Courtesy of U.S. Bureau of Reclamation*

the pride and esthetic joy that man finds in his accomplishments." Largely due to True's efforts, the four towers on the crest included small, yet elaborate, lobbies with floors of highly polished dark green and black terrazzo encased in walls of dark green and black marble with white stucco and aluminum ceilings and trim. Since the elevator towers served as public entrances, their bronze doors opened onto terrazzo decorative panels set in the floor. True based his design of these panels on decorative motifs of Southwest Indians and extended these patterns in the floors and around drains of the tunnels down in the dam through which visitors moved to view the turbine galleries and power house. True's patterns came from Pima baskets, Ácoma bowls, and sand paintings. While acknowledging their apparent primitiveness, True claimed that many of the patterns were also strikingly modern and appropriate for the dam. The swirling central motif, he claimed, "will reveal its striking similarity to what might be termed an engineer's basic diagram of a generator or turbine, with valves, gates, and suggestion of centrifugal motion."[26]

The visitor coming to the dam would enter the elevator lobby, drop some 528 feet in the elevator within the dam, walk into a gleaming white interior tunnel lined with glazed brick, stand on a floor patterned with True's Indian designs, and finally arrive at the platform for viewing the turbines. In the color scheme for the powerhouse interior, True selected ten colors that had "Indian characteristics" but were "made as brilliant as possible." The result was a striking interior set against repetitive concrete piers. Huge turbine galleries gleamed with the generators in their deep red casings and aluminum trim and piping. Overhead cranes painted a jade green along with vermilion, blue, and yellow valves, gray control panels with switches and dials in full view offered a far more dramatic setting than even Fritz Lang could have conceived.[27]

The physical fact of the dam coupled with the act of construction would seem testimony enough of the project's significance, but Americans have often wanted their achievements, whether small or large, to be seen in a cosmological light. Hoover Dam was no different. Under the guidance of Kaufmann, a competition was held for a monument to accompany the dam. The result was the commissioning of Norwegian-born Oskar J. W. Hansen. Hansen, who contributed the low relief plaques over the elevator towers, had also designed a memorial, placed on the rock canyon wall on the Arizona side, to workers who had lost their lives during construction. For the principal monument, located on the Nevada side, Hansen designed a base of polished black diorite for a 142-foot flagpole and two thirty-foot-tall seated winged figures. These surrealistic apparitions underscored the unreality of a dam and lake in the middle of a hostile desert. Made of bronze, the figures' wings and arms were molded into a single unit similar to the unity Kaufmann achieved between the crest towers and the dam face. Hansen claimed that the figures symbolized the universal aspirations of mankind and the achieve-

[26]Allen True, "Color and Decoration at the Boulder Power Plant," *Reclamation Era,* XXVI (Jan. 1936), 12–13.

[27]Allen True, "The Planned Use of Color at the Boulder Dam Power Plant," *ibid.,* (Feb. 1936), 48.

Figure 14 Winged figures of the Republic by Oskar J. W. Hansen,
Hoover Dam dedication monument
Photograph by R. C. Middleton *Courtesy of U.S. Bureau of Reclamation*

ment of Americans. More specifically, he intended their stoic visages "with the look of eagles" to represent those who settled the West. To further underscore the universality of the dam, Hansen embedded in the floor of the monument the Great Seal of the United States, signs of the zodiac, and a large star map indicating such important historical dates as the construction of the pyramids and the birth of Christ along with the precise astronomical time of the dam's dedication on September 30, 1935, at eight hours, fifty-six minutes, and 2.25 seconds in the evening. With constellations, stars, and planets placed in an intersecting pattern of circles, triangles, and straight lines similar to that used by the Russian Constructivists of the 1920s, Hansen, in effect, froze the entire universe at this one point in time. Looking at his creation, Hansen forecast that, "in remote ages to come, intelligent people may view the star map out of which rises the monument at Boulder Dam and from it learn . . . the astronomical time of the dam's dedication."[28]

On May 29, 1935, the last bucket of concrete was placed in the dam and on September 30 President Franklin Delano Roosevelt dedicated the dam. In February, water had already begun to be impounded in Lake Mead and in October the first electric power surged over 266 miles of transmission lines to Los Angeles, where a major celebration was held. The impact of the dam on the Southwest would be tremendous, for in a very real sense it allowed southern California and Los Angeles to continue to develop. Furthermore, the dam led to the growth of southern Nevada and Las Vegas as recreation spots, and permitted the continued agricultural development of the great food-producing valleys of California and Arizona. So great was the power potential of Hoover Dam that not until 1961, twenty-six years later, was the last hydroelectric generator placed in service. Not until then had the energy demand risen enough to consume all of the dam's output. If, when the dam was completed in 1935, all the generating units had been placed in service, their combined horsepower

[28]Oskar Hansen, "The Sculptures at Boulder Dam—Part I: With the Look of Eagles," *ibid.*, XXXII (Feb. 1942), 32; Hansen, "The Sculptures at Boulder Dam—Part II: Split Second Petrified in the Face of the Universal Clock," *ibid.* (March 1942), 57. These two articles by Hansen are reprinted in a pamphlet, *Sculptures at Hoover Dam* (Washington, D.C., 1968).

would have equalled one-eighth of all the horsepower available in the United States for generating electricity.[29]

Statistics continue to overwhelm and Hansen's monument still impresses visitors, but finally it is the dam as a physical entity that captures the imagination. One approach to the edifice is across the harsh desert, then down a winding canyon until around a corner there appears the massive, smooth cliff, a concave wedge of concrete—the machine hand of man—pushing apart the rough walls of Black Canyon and holding back trillions of gallons of water. Behind the dam a glistening mirage-like lake of deep blue stretches out into the far distance. Alternatively, the dam can be approached at night with the vast wall brilliantly lit. From the inky depths of Lake Mead the intake towers emerge, glowing with messages of energy and light. Hoover Dam is one of the few structures built in scale with the vast landscapes of the West. It exerts control over that landscape, both through size and design. As an architectural design, Hoover Dam belies one of the myths of modern architecture, as proposed by Le Corbusier and others, that the great functional structures were simply the work of engineers. Kaufmann—an architect—converted the dam's design from a banal, warmed-over classicism style to the realm of modernism. His modernist vocabulary not only respected the engineers' technological demands, it also enhanced them. The crest and intake towers have the setbacks of Art Deco skyscrapers. The spillways are streamlined with their rounded forms and thin incised "speed whisker" lines near the top of each pier. Far down at the dam's base the powerhouse—with its vertical fenestration, horizontal layering of form work, and gleaming aluminum and stainless steel transformers—captures the spirit of high technology and the inherent beauty of machines. In each case the different style fits the particular activity as well as the overall form.[30]

Equally impressive is the interior: the public lobbies of black, green, bronze, and aluminum; the hygenically white tunnels; Allen True's angular and circular Indian patterns set into the floors; and, finally, the great turbine chambers on either side of the powerhouse. The reiterated piers of the galleries and the

[29]"Boulder Dam: Engineering Triumph," *Literary Digest*, CXXII (Oct. 17, 1935), 19.

[30]Le Corbusier, *Towards a New Architecture*, 33, 42; Erich Mendelsohn, *Amerika: Bilderbuck Eines Architekten* (Berlin, 1926).

humming turbines encased in their smooth and glistening jackets recall Henry Adams's confrontation with the dynamo: "he began to feel the forty-foot dynamo as a moral force, much as the early Christians felt the Cross. The planet itself seemed less impressive, in its old-fashioned deliberate, annual or daily revolution, than this huge wheel, revolving within arm's-length at some vertiginous speed.... Before the end, one began to pray to it...." Artists like Charles Sheeler, Hugh Ferris, Ben Glaha, and Cliff Segerblom tried to capture the dam's force in paint, charcoal, and photographs. Writers from J. B. Priestley to Theodore White and Joan Didion have speculated on its moving power. For them, Hoover Dam represented a new modern world of special beauty that was simultaeously "strange and uncomfortable." In 1937, Priestley saw in the dam

> that world we catch a glimpse of in one of the later sequences of Wells's film, 'Things to Come,' a world of giant machines and titanic communal enterprises. Here in this Western American wilderness, the new man, the man of the future has done something, and what he has done takes your breath away. When you look down at that vast smooth wall, at its towers of concrete, its power stations, at the new lakes and cataracts it has created, and you see the men who have made it all moving far below like ants or swinging perilously in midair as if they were little spiders, and you note the majestic order and rhythm of the work, you are visited by emotions that are hard to describe, if only because some of them are as new as the great Dam itself.[31]

Ultimately Hoover Dam, like all great works of art and all important symbols, transcends description and interpretation. It is more than what can be written or expressed. As a work of art it tugs at the base of emotion, offers new experiences, and opens a new reality. Hoover Dam is one of the great architectural landmarks of American modernism in the 1930s.

[31] Adams, *The Education,* 380; "Power: A Portfolio by Charles Sheeler," *Fortune,* XXII (Dec. 1940), 73–83. On Ferris, see note 18; on Glaha and Segerblom, see note 11; see also Stanley Wood, "Boulder Dam, Portfolio of Watercolors," *Fortune,* IX (May 1934), 92–100; and William Woollett's lithograph of the dam under construction in *California Arts and Architecture* XLVII (Jan. 1935), front cover. Woollett did a number of lithographs on the dam (Woollett to author, July 15, 1984); White, "Building the Big Dam"; and Joan Didion, "A Piece of Work for Now and Doomsday," *Life* LXIII (March 13, 1970), 25, reprinted with changes in *The White Album* (New York, 1979), 198–201; J. B. Priestly, "Arizona Desert," *Harper's Monthly Magazine,* CLXXIV (March 1937), 365.

17
Hoover Dam: a study in domination

Donald Worster

The highway takes one northwest out of Kingman, Arizona, past the turnoff to Chloride, and runs along the Black Mountains, heading, it seems, farther into desolation.* Then it bursts through an opening in a rock wall, and there it sits—Hoover Dam—gleaming white in the hot desert sun, hurting one's eyes with its stark brilliance. Taking the highway across the dam, the traveler pulls up in a narrow parking lot and scrambles out to see this engineering wonder of the modern world, this answer to the pyramids and the ancient Colossus of Rhodes. When it was completed in 1935, Hoover Dam was the most massive structure of its kind in the world. It introduced a new era of building high-rise dams in country after country, remaking the face of the earth and altering the distribution of social and economic power on it. Among the leading symbols of twentieth-century technology Hoover Dam stands tall—near if not at the very top of the list of revealing "expressions of the age." Even now, more than fifty years after it was dedicated, in a time when many symbols and ideals of modernity have become discredited, Hoover still has the capacity to stir admiration and renew people's faith in the conquest of nature.

On a busy day thousands of tourists en route to enjoy the neon thrills of Las Vegas or the crowded, dusty campgrounds of the Grand Canyon pause to marvel at the dam. It has become one of the most popular travel attractions in the West—a ritualistic experience of "doing the West" and celebrating what its history has been. And what precisely do these tourists see when they stand on the dam's crest, looking down on Lake Mead, the man-made body of water created by the dam, flooding the Black Canyon of the Colorado River and over a hundred miles of upstream watershed? Is it the success of the westward movement or the defeat of wild nature they see and marvel at? Does there run through their minds a worry about the precariousness of human control, about its costs and dangers, or a rush of pride in the triumph of solid concrete over rambunctious river?

An elevator carries sightseers down into the bowels of the dam where immense turbines whir away, unceasingly generating electricity for cities like Los Angeles and Las Vegas, making the river pay for its own imprisonment. What do the visitors hear in that humming? The soft sound of money piling up, or the din of an anarchic, mindless acquisitive force let loose in American society? If the Bureau of Reclamation, the government agency conducting the tours and operating the dam, has its way, what people will see and hear will be completely reassuring. The official Hoover Dam story is supposed to convey a message of serene, benevolent order to travelers passing from one part of this unpredictable, irrational life to another. At least in this corner of the southwestern desert, nature is supposed to be under our firm control, working hard to make us all rich; and we can go on our way confident that responsible, competent people are in charge, assured that nature is no longer much of a threat to our welfare and that we are the very lords of life.

One long-dead traveler who will never see Hoover Dam but who might have had something less reassuring to say about its meaning was Henry Adams. When he was alive, the Black Canyon of the Colorado was still a remote, hidden country, all but inaccessible to civilization. Adams, therefore, had to go the Chicago and the great World's Fair of 1893, and from there to the Paris Exposition of 1900, to find his symbols of modernity. What he found in both places was a display of dynamos, or electric turbines, humming away as they still do inside the power plant of the dam. His reflections on those industrial displays, published in his autobiography, *The Education of Henry Adams,* might still be a provocative accompaniment to any Hoover Dam visit today. "The dynamo," he wrote, "became a symbol of infinity." It stood for a vast new power unleashed in human life, quite unlike anything before it, a mechanical power that acknowledged no limits and would seek to remake the whole world in its own image. "Before the end, one began to pray to it; inherited instinct taught the natural expression of man before silent and infinite force." The modern dynamo had come to be a religious-like symbol, driving out all the previous symbols of divine power that had ruled over human life for so long: nature, spirit, God, the Virgin Mary. What Adams also saw was that this cultural power of the dynamo represented a new king of rationality, one that did not deal with ultimate values and ends as the old religions had, but was focused instead on means, on instrumental effectiveness. The result of that shift was that the new, divine-like power, impressive and irresistible though it was, lacked any deep, carefully considered purpose or sense of direction. To those who sat down by such a machine and thought long and hard about its significance, as Adams did, the message was simply one of "more"—more profit, more comfort, more power for the sake of power.

In addition to the moral vacuum at the center of the coming technological society, Adams discerned another meaning: the power that was generated came to rest in fewer and fewer hands as centralized, bureaucratized structures of wealth and authority took command. "Once admitted that the machine must be efficient," Adams wrote, "society might dispute in what social interest it should be run, but in any case it must work concentration."[1]

The dynamo was, by itself, perhaps too technical and abstract a symbol of twentieth-century modernity to convey to ordinary minds all that Adams found in it. But put that machine inside the concrete wall of Hoover Dam rising 726 feet from bedrock, a dam three city blocks thick at its base, containing over three million cubic yards of cement, and the message becomes more visible. Put that concrete plug smack in the path of one of the most spectacular rivers on the planet, the Rio Colorado, which has carved the awesome Grand Canyon on its way to the sea, and the message gets harder yet to overlook. Surround that dam and river with an immense arid landscape, use the water and energy from them to conquer that landscape and create a urban civilization, an agribusiness empire, and you have the makings for an unrivaled study in the modern domination of nature and its consequences.

The Colorado River, though known to the Indians for tens of thousands of years, was not discovered by European civilization until 1540, when the Spanish naval commander Hernando de Alarcon sailed up the Gulf of California to the river's delta. Four hundred twenty-nine years later John Wesley Powell and a party of nine men put in at the headwaters of the river in Wyoming and raced off in the current, fetching up several months later, minus three of the men, at Grand Wash Cliffs, not far from the site of the future dam. Fifty years after that courageous exploring expedition, in the year 1922, several western American states signed a compact dividing the river among themselves, opening it for development. Then in 1928 the Boulder Canyon Act was passed by Congress, authorizing the construction of a dam that would be the first major step in controlling the river.[2]

Those five centuries of learning about the Colorado (actually most of the learning was concentrated in about five decades) seem almost never to have raised in people's minds a simple question: What changes in society would be required to master the Colorado's flow? Only Powell gave the matter much consideration; he recommended that both big government and big business stay away from the river, that ordinary settlers be encouraged to organize themselves into "cooperative commonwealths" and set about in their own way to make use of the watershed. Had his suggestion been followed, development would have been confined to the smaller side streams and upcountry valleys, for the main river was too strong for any local group of settlers, with limited capital and expertise, to harness.[3]

The Colorado runs for almost two thousand miles and drains an area of 244,000 square miles. Its volume is not great—far less than, say, the Mississippi or Congo or Amazon; even the Nile, another desert river to which the Colorado has often been compared, has a bigger average flow, discharging as it does 100,000 cubic feet per second at its mouth, whereas the Colorado, in its natural state before it was dammed and diverted, discharged one-fourth that amount. But the grandeur of the Colorado, its ability to defy human intervention, lay in the inexhaustible force it exhibited in its thrust to the sea. Over its course it drops from an elevation of 14,000 feet, generating in the process millions of horsepower of energy. From a time well before there were humans living on this planet, the river used that energy to grind its way seaward through the Colorado Plateau, a rising massif of sandstone and shale that makes the Hoover Dam impediment look like the merest pebble. Over the last thirteen million years the river has washed away more than a quadrillion tons of rock, sand, and gravel from that plateau, leaving behind it the most spectacular canyons found anywhere. It has wound its way deeper and deeper through that resistance, cutting meanders as much as a mile down, for hundreds upon hundreds of miles. All rivers are indefatigable workers, but this one has had a fanatic's determination in its industry. It would accept no compromises, make no detours, in the pursuit of its end but would push on night and day forever, until the plateau was gone, until there was no more work to be done, until it died of its own labors.[4]

Debouching from its multihued canyons, the Colorado carried one of the heaviest silt loads of all the world's rivers. The suspended sediment floating past the modern-day site of Yuma, Arizona, amounted to 160 million tons a year, more than all the railroad cars ever made could carry. For months at a time the river would run low, a slack, sluggish ooze to the gulf; but with the spring thawing of the mountain snowpack or with a torrential summer storm the sediment would become a mighty avalanche; great boulders would tumble along in the churning water. Overloaded by rock, dirt, and water, the current would slop over its banks onto the surrounding floodplain, where it would deposit some of its burden. The deeper the river sawed into that plateau upstream, the higher its bed rose above the lowlands downstream. Eventually the lower Colorado ran along a self-made ridge, well above the desert floor. In its struggle for equilibrium, tearing down to build up, the river was setting the terms for its conquest. All of it would have to be mastered, or none of it could be. And that comprehensiveness of control would necessitate a new social organization, a new set of instrumentalities, that would prove Henry Adams absolutely right.

A demand to bring the Colorado under human domination led in the late 1920s, as noted, to the scheme of building a gargantuan storage dam and reservoir. There were at least three distinguishable voices making that demand. The first of them was the collective clamoring of a westward-moving population, seeking new homes on the warm, dry California coast. That pressure, it should be emphasized, did not come from within the river basin itself. In the twenties the only sizable cluster of people living near the river and the Hoover Dam site was located up in Las Vegas, some thirty miles away, and they amounted to a minuscule 7000 inhabitants. Far more important was the pressure coming from outside the basin. In the last quarter of the nineteenth century Los Angeles grew from an adobe outpost of 10,000 people to a city of 200,000, with ambitions to bring in many, many more. After tapping the Owens Valley lying off to its northeast, draining it nearly dry, the city leaders began looking to the Colorado to slake their thirst. Yet theirs was mainly a contrived pressure. It was not so much a spontaneous demand from a mounting indigenous population as it was the deliberate strategy of a group of urban entrepreneurs to *induce* a migration boom. Their salesmen fanned out to midwestern and eastern states, promoting among Iowa farmers and New Jersey factory workers, often in the most extravagant style, the promise of a new life in the sun. Those who bought the pitch would most certainly need water where they were going, but the primary, decisive "need" was one hatched in the imagination of West Coast urban promoters. As William Mulholland, the chief water planner for Los Angeles, declared in 1907, "If we don't get the water, we won't need it. We have got to have water or quit growing."[5]

A second set of demands came from agricultural capitalists down in California's Imperial Valley, just north of the Mexican border. In 1901 they had begun to move into that area, which was one of the hottest, most sterile in the West. Soon they had farms, some of them over 3000 acres large; they had the beginnings of an army of Chicano laborers to plant and harvest their crops, and they had railway links to urban markets for shipping whatever they wanted to raise. What they did not have was a secure grip on the Colorado River, their only source of irrigation water. In 1905–7 the river left its high course and raged down into their valley, threatening to wipe out their entire investment. Then came dry years when there was not enough water to fill their ditches and canals. What they wanted, what they thought they "needed" to make money, was a steady flow and a bigger one—and that would require a great deal of skilled hydraulic engineering.[6] In the popular novel of 1911, *The Winning of Barbara Worth,* which is set in this valley, the characters are completely forthright about their motives toward the desert and the river. One of the community's leaders, a banker, says to an agriculturist: "We all want to make what we can in this new country; that's what we came in for."[7] In the pursuit of that economic ambition, they supported the construction of Hoover Dam, for it would stop the dangerous floods and enhance the flow from year to year.

A third group of voices came from within the federal bureaucracy, mainly from government engineers who wanted to exercise their technical expertise on the Colorado River. They were part of the Bureau of Reclamation, which had been established in 1902 with the assignment of developing irrigation projects in the American West. After a couple of decades of smallish, marginally successful undertakings, the Bureau was itching to get its hands on something big, something monumental. From 1914 to 1923 the director of that agency was Arthur Powell Davis, a nephew of the explorer John Wesley Powell; and more than any other single individual, it was Davis who conceived and pushed through the Hoover Dam idea. "I . . . considered problems in all of the Western States," he wrote, but there was "none which . . . excited my interest and imagination so much as the development of the Colorado River basin."[8] Early on he had envisioned placing the river under a single centralized control (that of his own agency, of course) in order to develop thousands of new irrigated farms for landless city people. Another federal engineer who shared Davis's passion for technological conquest was Francis Crowe, who joined the agency in 1904 and rose to become general superintendent of construction. He helped make the first rough surveys of the Hoover site, later recalling, "I was wild to build this dam." For Crowe the project in the Black Canyon meant a "wonderful climax—the biggest dam ever built by anyone anywhere."[9] In 1925 he resigned his government position to join the Utah Construction Company, which subsequently became one of the Six Companies, the private consortium that undertook to build the dam—and it was Crowe who directed the work.[10]

Those were the principal voices urging that the Colorado be taken in hand and taken apart. Disparate though their motives may have been, they had one thing in common: all wanted, with a desire that knew no bounds, to dominate nature. Only a river as wild and full of life as the Colorado could satisfy their desire. All of the voices spoke in terms of cool, calculating reason—of scientific planning, of market strategies, of hydraulic principles—but always under the rational surface there was an unfathomable layer of irrationality, a vague, unspecified longing, a will to power. If the river had a fanatical zeal about it, the men who came to control it were fanatics too, fanatics in George Santayana's definition of the term: men who redouble their efforts as they lose sight of their aim. They would not be satisfied until the river was under their total domination, from its headwater to its mouth—or, in other words, until the river was dead.

Hoover Dam was followed in 1938 by Parker Dam, located 150 miles downstream and furnishing water for the California Aqueduct, which supplied the city of Los Angeles. Then came the building of Imperial Dam and the All-American Canal to supply the agribusiness interests farther south on the Mexican border. Then Davis Dam began to go up in 1946, and soon after, Morelos Dam in Mexico. Also, there were Laguna Dam, Palo Verde Diversion Dam, and Headgate Rock Diversion Dam. And on the upper river there was Navajo Dam, finished in 1962, an earth structure containing 27 million cubic yards of material, followed by Flaming George Dam in Wyoming—and then Seedskadee, Savery-Pot Hook, Meeks Cabin, Vernal, Bonneville, Rifle Gap, Joes Valley, Paonia, Blue Mesa, Morrow Point, Crystal, Soap Park, Crawford, Silver Jack, Lemon, Frying Pan-Arkansas. The names proliferated endlessly, all of them echoing a lost, colorful frontier past, all of them in fact denoting a technological sameness and emerging industrial era. Last and most impressive in the Colorado control system was the immense monolith at Glen Canyon, a full 710 feet high, only a shade less than Hoover. More schemes remained on the drawing boards, but these were enough to make the dream of domination a reality at last. Blocked, trapped, stilled, and siphoned off, the Colorado finally ceased in normal seasons to reach the sea.

During the very period of Colorado river construction the German social philosopher Max Horkheimer argued that the conquest of the natural world has become the ruling passion of modern society. In terms strongly reminiscent of Henry Adams, he pointed out that such a passion has led us into a moral chaos. We know more and more about *how* to do things, less and less about *what it is that is worth doing.* Horkheimer called that state of affairs "the eclipse of reason." The work of reason ceases to be the discovery of the ultimate aims of life; it is reduced to the job of reducing whatever it encounters to the status of a mere tool. Nature becomes, at the hands of that instrumentalized reason, a mere miscellany of raw materials waiting to be worked up into something "useful." It ceases to have its own logic, order, or intrinsic value. Although Horkheimer was living as a refugee from fascism on the outskirts of Los Angeles when he wrote those thoughts, he never applied them directly to the case of Hoover Dam and the Colorado. And he might well have, for nowhere is there a clearer expression of what he was talking about than here.[11]

The social consequences of an instrumentalized river are the same as those of an instrumentalized nature everywhere. According to Horkheimer, "The human being, in the process of his emancipation [technology appears to offer a means of human freedom], shares the fate of the rest of his world. Domination of nature involves domination of man."[12] In other words, when the tourist looks into the flat water backing up behind a dam like Hoover, he is in fact seeing his own life reflected. What has been done to the Colorado has been done to him as well; he too has, in a sense, been conquered and manipulated, made to run here and there, made to serve as an instrument of production.

That human outcome becomes unmistakably clear when one asks, who gained power in the construction of the Colorado control complex and toward what purposes have they put that power? To ask those questions is not precisely to ask who has received some material benefits from the hydraulic control apparatus. Unquestionably, a lot of people have in some measure. Of course, most of the benefits have tended to collect in a few hands. But even if that were not true, even if the benefits had been distributed more equally than they have, that fact would not, as Henry Adams noted, alter the separate, equally important fact that power and authority have become increasingly concentrated in the pursuit of greater instrumental efficiency. The tourist may loll in the water as long as he or she likes, may bathe in it, drink it, ride a boat through it; may eat the fruit and vegetables it produces, may consume the electricity it generates—and do all those things without sharing in any real way the power that makes those benefits possible.

Among those who have gained considerable economic and social power through controlling the Colorado River have been the agribusinessmen of Imperial Valley. Even before Hoover and the other structures began to go up, they were a force to be reckoned with, thanks to building their own, early water works. The very name "Imperial," chosen deliberately to attract investors, spoke of their intentions: they were unabashedly imperialists, seeking power. Each increment they made in river control was a step closer to achieving that ambition of power. Today, there are fewer than 700 farm owners in the valley, and a mere 72 of them own more than half of the land, amounting to more than 300,000 acres. Those owners employ 12,000 people in their fields, the vast majority of them Mexican nationals, many of whom cross the border to work each day, or they are largely Americans of Mexican ancestry. By the standards of a modern urban, industrial society they are not well paid, which explains why this area, though one of the four leading agricultural producers among the nation's counties, has a low average standard of living. The farm owners, in contrast, have all risen to a state of rural grandeur, thanks to the federal guarantee of river water, and in some cases they have risen to the status of international food marketers. "Bud Antle," for instance, is no longer merely a single farmer but a large corporation, harvesting 20,000 acres of lettuce and celery in California and Arizona as well as in Africa. Without the managed Colorado River, Bud Antle, Inc., would soon be covered with desert dust.[13]

Another of the emergent powers in the Colorado River basin is the federal Bureau of Reclamation. In the twenties it was still a feeble organization, with a mere trickle of funds coming to it from the rural settlers on its projects. It was in some danger of being wiped out. The building of Hoover Dam saved it from that fate and helped make it the most famous and accomplished desert conqueror in world history. Eventually its name would be blazoned not only up and down the full 2000-mile length of the Colorado, the Bureau's own property as it were, but also along the San Joaquin, Sacramento, Rio Grande, Missouri, and Columbia, along virtually every stream in the American West. Congressmen and senators would come to seek its advice. Its engineers would be wooed and fêted in a thousand little towns and big cities. It would have immense quantities of electrical energy to sell, and in that role as engineer-supplier-manager would have much to say about the urbanization and industrialization of the West.

When it began to run out of construction sites to carry on its mission, when it was threatened with declining into a dull caretaker's job, the Bureau began to extend its reach in new directions—toward planning "attractive, healthful, and inspirational environments," as one of its publicists put it, toward building nuclear power plants, toward coal mining, toward pollution control, toward a coast-to-coast national role if possible.[14] The Bureau also eventually went abroad in search of work in Asia, Africa, and South America, encouraged by the fact that, as one of its commissioners, Michael Straus, pointed out, "the American concept of comprehensive river basin development . . . has seized the world imagination."[15] In all those ways the Bureau tried hard to be the indispensable service bureaucracy, and as it succeeded it was rewarded with a large share of power. After all, power in a society dedicated to the absolute conquest of nature does not belong to those who try to think rationally and systematically about the ends of domination or who challenge the unexamined, unspoken ends. Nor is it the exclusive possession of those who profit most by environmental domination, whether they be agribusinessmen, industrialists, or urban merchants. Power also comes to those who offer the means of conquest; it comes to all the principal agents and spokesmen of instrumentalism.

The chief political lesson of Hoover Dam then is that a new concentration of economic, social, and political power is the outcome of the domination of nature. And how could it be otherwise? Is it conceivable that a loose gathering of simple farmers could carry out such an engineering triumph? Or that a hundred little villages could achieve as much as the single, focused energies of a Los Angeles elite? Yet, strange to say, that obvious political lesson of the dam at Black Canyon and its companion structures has escaped observer after observer. Here, for example, is J. Kip Finch, late dean emeritus of Columbia University's engineering college, hailing the western hydraulic feats as evidence of "the increasing mastery of man over nature which has made possible our continuing progress toward a better life." "Nature," he goes on, "has been harnessed to meet man's needs on a scale that not so long ago would have been regarded as completely visionary and impossible."[16] There is not a word in that exuberant outburst about the fact that it is not "man" who has achieved mastery over western American rivers, but *some men,* while the rest of us have looked on in passive wonder. Or about the fact that the "progress toward a better life" of which the dean spoke so confidently has turned out to have been a gradual slide into what Horkheimer and his associate Theodor Adorno called "the world of the administered life," in which the average citizen has most of his or her life planned and administered by powerful external forces.[17]

The principal historian of Hoover Dam, Joseph Stevens, has fallen into the same mental trap as the dean of engineering. In the final pages of his work, he resorts to an almost hagiographical language that drives out all analysis or critical awareness, and leaves the most interesting and significant historical questions about the dam unasked. "In the shadow of Hoover Dam," he writes, "one feels that the future is limitless, that no obstacle is insurmountable, that we have in our grasp the power to achieve anything if we can but summon the will." Nowhere does Stevens try to define who that "we" is or ask what that "anything" ought to be. The feelings he expresses are in a sense religious, and the religion he calls us to follow is one of the domination of nature by instrumental reason. "Some try to suppress this feeling," he admits, "thinking that it is naive to be enthralled by an engineering marvel, dangerous to be seduced by twentieth-century technology, which cynics say is untrustworthy, exploitative, and destructive of the environment and the human spirit." But go stand in the presence of Hoover Dam, he admonishes, and all that doubting will disappear; you too will become a believer. A believer in what? In the destiny of the next generation, he vaguely replies; in "the romance of the engineer" and "the courage of the construction worker."[18] To doubt any of that romance, or to ask questions about the uses of courage, or even to wonder what kind of society Hoover Dam has helped us get in the West, is apparently not permitted; only "cynics" would ask such questions. Stevens is hardly unique in writing the history of water control in this reverential spirit, the spirit of a true believer espousing a cause that has no name but only a few empty slogans. But such history writing has always been dangerous. It tends to subordinate a full, critical, free exercise of human reasons, a reason looking for worthy ends, and lead readers into the worshipful, authoritarian zeal of fanaticism. The American West has a well-established tradition of such attitudes.

For a long while it looked as though the concentration of power along the Colorado was invulnerable to any challenge. Most historians were certainly not going to raise any hard questions. The workers had been paid off, their strikes busted, their jobs ended, and they were not around any more to challenge the project. And there was no one else in sight to mount that challenge. So warned Henry Adams and Max Horkheimer earlier, as they cast about for some escape from an unquestioning drive to "modernity" or for some "revolutionary" group that could seize control and disperse power. Every possibility of protest that came to their minds had one of two flaws: either it would be too ineffectual to mount a genuine challenge, or it would be too ready to set itself up as a substitute authority, rather than search for a radical alternative. But in the 1970s and '80s a new possibility for change appeared on the western river scene, though it was not necessarily a benign or happy one. Nature, it now appeared, might be pushed and pushed to the point where it refused to function any longer—where it might, that is, collapse ecologically and carry down with it the whole panoply of human instrumental control. Rather unexpectedly, after so long a period of human confidence in technology or of fatalistic acceptance before its advance, that prospect began to seem more and more plausible on the Colorado River as elsewhere.

The primary environmental threat on the river today, and it is now a very serious one, is from salinity buildup. In its undeveloped state the Colorado always carried lots of dissolved salts to the sea, mainly washed from exposed saline shale formations or drained from springs and seeps in the upper basin. Those natural sources of "salt loading" are estimated to have once produced over six million tons of salt per year.[19] From the beginning then the muddy red water was rather alkaline, enough so to corrode teakettles and damage more sensitive crops, but not enough so to make the water poisonous to all forms of life. Now, however, the technological domination of the river has increased significantly that alkalinity to the point that the water is becoming unfit to use, in cities and on farms alike.

The problem originates in part in the agricultural fields in the states of Colorado, Wyoming, and Utah. As farmers there withdraw water for irrigation, they leach salts out of their soils and put them into the hydraulic cycle. A typical irrigated acre in the Grand Valley of Colorado for example, adds eight tons of salt to the river each year. The resulting polluted water will be used again and again farther downstream, each use adding to the mineral load. Large storage reservoirs make the condition worse, for as their contents evaporate in the intense sun—the average rate of evaporation is 10 percent of the water per year—the salt becomes increasingly concentrated. Two thousand miles of river use, storage, more use, more storage, still more use and storage, and you have a sure recipe for the utter degradation of water quality.

The salinity problem first began to reach ominous levels in the 1950s. Then in late 1961 Mexico made a formal protest to the United States government that a water treaty between the two countries was being violated by the Americans; the Colorado River, which furnishes Mexican farmers with irrigation supplies too, had become so salty it could not be put on their fields. The immediate cause of the decline in quality was a Bureau of Reclamation project on the tributary Gila River in Arizona, the Wellton-Mohawk Irrigation and Drainage District, which was dumping its heavily polluted water across the border. As a consequence Mexico's salinity average jumped from 850 to nearly 1500 parts per million (ppm). The outcome of their protest was that the United States agreed to spend several hundred million dollars erecting and operating a desalinization plant for Mexican farmers. That was, however, only the beginning of the search for a solution to a problem that has steadily worsened year by year.[20]

At Lee's Ferry, the traditional dividing point between upper and lower basins, the average natural level of salinity would be 250 ppm, or roughly 5.1 million tons of dissolved salts in an annual flow of fifteen million acre feet of water. By 1969 the level had gone up to 655 ppm. By the end of this century, according to one estimate, the level will have reached 780 ppm, or three times the natural concentration. Downstream the problem gets, of course, much more severe. By the late 1960s the salt level at Imperial Dam was 850 ppm, and by the year 2000 it is projected to reach 1,210 ppm.[21] What that has meant, and will mean, for the agribusinessmen of Imperial Valley is heavy, unprofitable expenses. From 1929 to 1972 the Imperial Irrigation District was forced to spend $67 million on tile drains and on linings for their ditches and canals to relieve the salinity nemesis. Every few years and at considerable cost those drains must be cleaned. Each increase in lower Colorado salinity requires still more remedial measures from the district. By century's end, therefore, the increased annual expense may be as high as $40 million. Since this is a cost that farmers in areas with abundant rainfall do not have to pay, the Imperial growers must eventually find themselves at a serious disadvantage in the marketplace.[22]

The most popular panacea along the Colorado is to find a way of passing the clean-up cost on to American taxpayers across the nation. During the governorship of Ronald Reagan, the Colorado River Board of California (made up of his appointees) proposed that the Bureau of Reclamation be given the job and expense of controlling upper basin salt sources, that it undertake weather modification to increase streamflow to dilute the salt, and that it be authorized to bring fresh water to the watershed from California's northern coast and even from the Columbia River.[23] The first part of that advice was acted on by Congress when it passed, in 1974, the Colorado River Basin Salinity Control Act. How much farther American taxpayers will be willing to go to alleviate the salt problem will be a study in not only political power but public gullibility. At some point in their generosity toward the western water empire taxpayers may begin to realize that salinity is not a mere "side-effect" of a wonderful achievement but is an inevitable accompaniment of any large-scale river control scheme in an arid environment; that it is a systemic problem which can never be truly solved, one that has plagued every major irrigation system in history and destroyed many.[24] At some point too Americans may realize that there are cheaper ways to get food than this and that there are more sensible places to put cities than in the desert. At that point they may shake loose from their passive, administered life, and the Colorado pyramid of power may begin to crumble.

Nature has a way of undermining the most entrenched elites in history, especially when they have overreached themselves and, in their unbounded eagerness for more, made themselves ecologically vulnerable. The salt buildup in the Colorado plumbing works has the potential of doing just that. Otherwise, it is hard to see what else would be likely, within the next few decades, to have any revolutionary effect.

What began with Hoover Dam in the 1930s is now virtually complete: a work of environmental mastery that, for clarity, thoroughness and drama, is hardly equaled anywhere on earth. We are now in a better position to see the consequences, both social and ecological, of that work, to ask what the costs have been and whether they were worth paying, and to consider whether we would pay them again someplace else were we given the chance. Unfortunately, however, the casual tourist passing by the dam, eager to get on to the brightly lit casinos and an air-conditioned motel room, seldom takes advantage of that chance for enlightenment. He or she may, for a quick moment, lean out nervously over the dam's crest, peering down the steep, curving face to the water below, and in that moment gain a brief glimpse of human mortality; but he or she is unlikely, in a typical tourist visit, to wonder about the deeper meaning of the structure itself. That is, unless he or she happens to stop and contemplate at length the dark brown andesite breccia walls of the canyon on either side of the dam, asking what its history has been.

Running water, soft and yielding though it is in the human hand and so easily stopped by a concrete wall, is a power we do not have enough experience to appreciate. Before our species was born that water cut down through that brown rock for almost a thousand feet. Think on that fact for a while, and all the rhetoric of the Bureau of Reclamation, all the technology worship of many engineers and historians, begins to seem a little hollow. What an utter foolishness it is to believe that any lasting mastery over nature has really been won. Even at this very moment the Colorado River is busy preparing to saw through the Hoover Dam, laying down its silt behind the wall, gathering force to remove this new obstruction just as it has removed everything else ever put in its way by the forces of geology. Human domination over nature is quite simply an illusion, a passing dream by a naive species. It is an illusion that has cost us much, ensnared us in our own designs, given us a few boasts to make about our courage and genius, but all the same it is an illusion. Do what we will, the Colorado will one day find an unimpeded way to the sea.

5. Hoover Dam: A Study in Domination

1. *The Education of Henry Adams* (Boston: Houghton Mifflin, Sentry edition, 1918), 344, 380.

2. Wallace Stegner, *Beyond the Hundredth Meridian: John Wesley Powell and the Second Opening of the West* (Boston: Houghton Mifflin, 1954), part 1; and Norris Hundley, *Water and the West: The Colorado River Compact and the Politics of Water in the American West* (Berkeley: Univ. of California Press, 1975), chap. 7.

3. John Wesley Powell, "Institutions for the Arid Lands," *Century Magazine* 40 (May 1890): 16.

4. For the geological history of the river, see: E. Blackwelder, "Origin of the Colorado River," *Bulletin of the Geological Society of America* 36 (1934): 551–66; Henry James, "The Salient Geographical Factors of the Colorado River and Basin," *Annals* 135 (Jan. 1928): 97–101; Clarence Dutton, *Tertiary History of the Grand Canyon District*, U.S. Geological Survey Monograph No. 2 (Washington: Government Printing Office, 1882).

5. William Kahrl, *Water and Power: The Conflict over Los Angeles' Water Supply in the Owens Valley* (Berkeley: Univ. of California Press, 1982), 156–57; Remi Nadeu, *The Water Seekers* (Garden City, N.Y.: Doubleday, 1950).

6. Robert Schonfield, "The Early Development of California's Imperial Valley," *Southern California Quarterly* 50 (Sept. 1968): 297-307; (Dec.): 395-426; Helen Hosmer, "Triumph and Failure in the Imperial Valley," *The Grand Colorado,* T. H. Watkins, ed. (Palo Alto: American West Publishing, 1969), 205-21.

7. Harold Bell Wright, *The Winning of Barbara Worth* (1911; New York: Grosset & Dunlap, 1966), 145.

8. Arthur Powell Davis, quoted in Norris Hundley, "The Politics of Reclamation: California, the Federal Government and the Origins of the Boulder Canyon Act—A Second Look," *California Historical Quarterly* 52 (Winter 1973): 297.

9. Francis Crowe, quoted in T. H. Watkins, "Making an Empire to Order," *The Grand Colorado,* 172-73. See also Paul Kleinsorge, *The Boulder Canyon Project* (Stanford: Stanford Univ. Press, 1941), 185-230; Frank Waters, *The Colorado* (New York: Holt, Rinehart and Winston, 1946), 337-51; Imre Sutton, "Geographical Aspects of Construction Planning: Hoover Dam Revisited," *Journal of the West* 7 (July 1968): 301-44.

10. The story of building the project is well told in Joseph E. Stevens, *Hoover Dam: An American Adventure* (Norman: Univ. of Oklahoma Press, 1988). The Six Companies were comprised of the Utah Construction Company, Morrison-Knudsen (Crowe's employer, headquartered in Boise, Idaho), the J. F. Shea Company and Pacific Bridge (both of Portland, Oregon), MacDonald & Kahn of San Francisco, and a group known as Bechtel-Kaiser-Warren Brothers. Despite many difficulties with the workers they hired, the Six Companies completed the project in a little over two years, almost one month ahead of schedule, and made gross earnings of over $50 million. (Stevens, p. 252.)

11. For a general discussion of Horkheimer's critical philosophy, see Martin Jay, *The Dialectical Imagination* (Boston: Little, Brown, 1973).

12. Max Horkheimer, *The Eclipse of Reason* (New York: Plenum, 1974), 93.

13. Paul Barnett, *Imperial Valley: The Land of Sun and Subsidies* (Davis, Calif.: California Institute of Rural Studies, 1978), 30. See also Ernest Leonard, "The Imperial Irrigation District: Agency Behavior in a Political Environment," Ph.D. thesis, Claremont Graduate School, 1972.

14. William Warne, *The Bureau of Reclamation* (New York: Praeger, 1973), 207, 229-33.

15. Michael Straus, *Why Not Survive?* (New York: Simon and Schuster, 1955), 78.

16. J. Kip Finch, "Some Modern Wonders Named," *Civil Engineering* 25 (Nov. 1955): 33, 40.

17. Max Horkheimer and Theodor W. Adorno, *Dialectic of Enlightenment,* John Cumming, trans. (New York: Herder and Herder, 1972), ix.

18. Stevens, *Hoover Dam,* 266-67.

19. Myron Holburt and Vernon Valentine, "Present and Future Salinity of Colorado River," *Journal of Hydraulic Division, Proceedings of the American Society of Civil Engineers* 98 (March 1972): 505-7; Wesley Steiner, statement in *Salinity Control Measures on the Colorado River,* Hearings before Subcommittee on Water and Power Resources, Committee on Interior and Insular Affairs, U.S. Senate, 93 Cong., 2d sess. (Washington: Government Printing Office, 1974). Also see the international symposium on Colorado River salinity in *Natural Resources Journal* 15 (Jan. 1975); and Ralph Johnson, "Our Salty Rivers: Legal and Institutional Approaches to Salinity Management," *Land and Water Law Review* 13, No. 2 (1978): 441-64.

20. A good discussion of this issue may be found in Philip Fradkin, *A River No More: The Colorado River and the West* (New York: Alfred A. Knopf, 1981), 291-316. See also Norris Hundley, *Dividing the Waters* (Berkeley: Univ. of California Press, 1966); and Herbert Brownell and Samuel Eaton, "The Colorado River Salinity Problem with Mexico," *American Journal of International Law* 69 (April 1975); 255-71.

21. Holburt and Valentine, "Present and Future Salinity," 515.

22. Vernon Valentine, "Impacts of Colorado River Salinity," *Journal of Irrigation and*

Drainage Division, Proceedings of American Society of Civil Engineers 100 (Dec. 1974): 500-502.

23. Colorado River Board of California, *Need for Controlling Salinity of the Colorado River* (Sacramento: California Office of State Printing, 1970), 68-69, 78.

24. This conclusion is strongly suggested by Thorkild Jacobsen and Robert Adams, "Salt and Silt in Ancient Mesopotamian Agriculture," *Science* 128 (21 Nov. 1958): 1254-58.

Name and Place Index

Please note: Page numbers which appear in italics are references to tables or illustrations.

Subject Index